Processes in Microbial Ecology

Processes in Microbial Ecology

Second edition

David L. Kirchman

School of Marine Science and Policy
University of Delaware, USA

OXFORD
UNIVERSITY PRESS

OXFORD
UNIVERSITY PRESS

Great Clarendon Street, Oxford, OX2 6DP,
United Kingdom

Oxford University Press is a department of the University of Oxford.
It furthers the University's objective of excellence in research, scholarship,
and education by publishing worldwide. Oxford is a registered trade mark of
Oxford University Press in the UK and in certain other countries

First Edition published in 2012
Second Edition published in 2018

Impression: 5

Published in the United States of America by Oxford University Press
198 Madison Avenue, New York, NY 10016, United States of America

British Library Cataloguing in Publication Data
Data available

Library of Congress Control Number: 2017963785

ISBN 978-0-19-878940-6 (Hbk.)
ISBN 978-0-19-878941-3 (Pbk.)
DOI 10.1093/oso/9780198789406.001.0001

Printed in Great Britain by
Bell & Bain Ltd., Glasgow

Preface

The importance of microbial ecology has grown exponentially since the publication of the first edition of this book. The field continues to address some of the most pressing questions in science and the most serious environmental problems facing society today. As the field has forged ahead over the past decade, this book needed to be revised to avoid being left behind. It will be easy to see the differences between the first edition and the present one, starting with the order of the chapters. Very few of the paragraphs, tables, and illustrations have not been tightened, drastically altered, or just replaced with something better and up to date.

As with the first edition, this book targets students interested in learning some of the basics about microbes in natural environments and the processes they carry out. The emphasis is on biogeochemical and ecological processes, because the things that microbes do are what makes microbial ecology such an important field today. The book takes examples from all habitats, both terrestrial and aquatic, ranging from the deep biosphere to the open ocean. All microbes and viruses are covered. Readers will gain an appreciation for the diversity of the microbial world and of the questions being explored by microbial ecologists working on different microbes and processes in different habitats. In spite of this diversity, I hope readers will see the principles that microbes in aquatic and terrestrial environments have in common. Even if the focus is on one microbe or virus in one habitat, our understanding is immeasurably enriched by considering other microbes and viruses in several habitats.

Additional material for the book can be found at www.oup.co.uk/companion/kirchman. The material includes all of the figures, color versions of some of the figures (indicated in the figure caption), and things potentially useful if the book is adapted for a course. I welcome suggestions, comments, and feedback (kirchman@udel.edu).

Many people have helped and contributed in many ways to the book. Colleagues were very generous in providing photomicrographs or data, as acknowledged in the figure captions. I thank colleagues at the University of Delaware, especially Clara Chan and Tom Hanson, who patiently answered my many questions, helped with figures, or reviewed a chapter or two. Jennifer B. Glass at the Georgia Institute of Technology read the entire book and provided many insightful comments about all of the chapters. While translating the first edition into Japanese, Toshi Nagata brought to my attention several things that helped to improve this second edition. I thank the following people who reviewed a chapter, or more: Federico Baltar, Per Bengston, Ron Benner, Jennifer Brum, Dave Caron, Colleen Cavanaugh, Dreux Chappell, Matt Church, Kristen DeAngelis, Isabel Ferrera, Frank Ferrer-Gonzalez, Noah Fierer, Pep Gasol, Richard Geider, Stefan Geisen, Clara Gibson, Sara Hallin, Sebastian Horn, Wakako Ikeda-Ohtsubo, Gary King, Venu Kuo, Jay Lennon, Ramon Massana, Susanne Menden-Deuer, Aram Mikaelyan, Jim Mitchell, Mary Ann Moran, Brent Nowinski, Johannes Roush, Dave Royer, Clara Ruiz, Frank Stewart, Nadia Szeinbaum, Peer

Timmers, Mario Uchimiya, Lot van der Graaf, David Walsh, Bess Ward, Kurt Williamson, and Nathan Wisnoski. I have especially appreciated the comments from students and teachers. Tiffany Straza edited the entire book and made several excellent suggestions. Any mistakes remaining in the book are because I didn't follow the reviewer's advice, for inexplicable reasons.

A large part of this book was written while I was on sabbatical at the Institut de Ciències del Mar, Barcelona. Of the several colleagues there who helped with this book and my sabbatical stay, I need especially to thank two here. Ramon Massana provided much useful information about protist biology and threw an immemorial calçotada. Pep Gasol is the most gracious host and an unfailing source of insight into science, politics, and Barça football. I acknowledge financial assistance from the National Science Foundation and the University of Delaware.

Table of Contents

CHAPTER 1

Introduction

Microbes make up an unseen world, unseen at least by the naked eye. In the pages that follow, we will explore this world, the creatures that inhabit it, and the things they do. We will discover that processes carried out by microbes in the unseen world affect our visible world. These processes include virtually every chemical reaction in the biosphere that make up the elemental cycles of carbon, nitrogen, and all of the other elements necessary for life. The processes also involve interactions between organisms, both among microbes and between microbes and large organisms.

This chapter will introduce the types of microbes found in nature and some basic terms used throughout the book. It will also discuss why we should care about microbes in nature. The answers will give some flavor of what microbial ecology and this book are all about.

What is a microbe?

The microbial world is populated by a diverse collection of organisms, many of which having nothing in common except their small size. Microbes include by definition all organisms that can be observed only with a microscope and are smaller than about 100 μm. Microbes and its synonym "microorganisms" include bacteria, archaea, fungi, and many other types of eukaryotes (Figure 1.1). The microbial world also houses viruses, but arguably they are not alive and are not microbes, although some viral ecologists would argue otherwise. Whether dead or alive, viruses certainly are a big part of microbial ecology (Chapter 10). Bacteria and archaea are the simplest and the smallest microbes (excluding viruses) in nature, usually <10 μm in size, while microbial eukaryotes are structurally more complex and bigger, ranging from 1 to 100 μm.

Microbes are found in all three domains of life: *Bacteria*, *Archaea*, and *Eukarya* (eukaryotes). Bacteria and archaea look quite similar under the microscope, and in fact archaea were once thought to be a type of bacteria. An old name for archaea is "archaebacteria," while bacteria were once referred to as "eubacteria," the "true" bacteria. Now we know that bacteria and archaea occupy separate domains, accounting for two of the three domains of life found on Earth. The third domain, *Eukarya* (sometimes spelled *Eucarya*), includes microbes with a nucleus and other organelles. Examples include fungi and algae (but not blue-green algae, more appropriately called cyanobacteria), as well as higher plants and animals. Prominent members of the third domain are protists, which are single-cell eukaryotes. Protists do not include the fungi even though some forms of fungi, the yeasts, are single cells. All of these microbes and viruses are discussed in this book.

The diversity in the types of microbes found in nature is more than matched by the diversity of processes they carry out. Microbes carry out some processes that are similar to the functions of plants and animals in the visible world. Some microbes are primary producers and carry out photosynthesis similarly to plants ("photoautotrophy"), some are herbivores that graze on microbial primary producers, and still others are carnivores that prey on herbivores; both herbivores and carnivores are "heterotrophic." But microbes do many more things that have no counterparts among large organisms. These things, these processes, are essential for life on this planet.

Processes in Microbial Ecology. Second Edition. David L. Kirchman. Oxford University Press (2018). © David L. Kirchman 2018.
DOI 10.1093/oso/9780198789406.001.0001

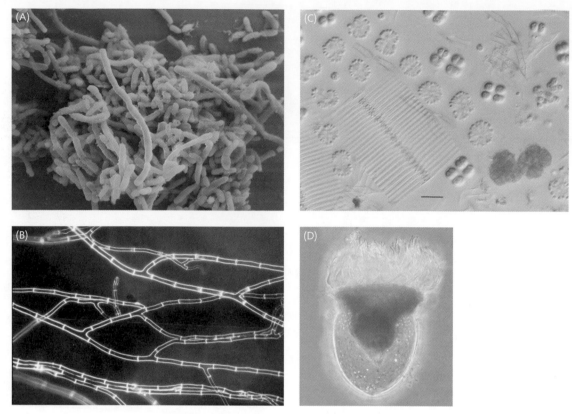

Figure 1.1 Examples of some microbes. Panel A: Soil bacteria belonging to the *Gemmatimonadetes* phylum, each about 1 μm wide. Image courtesy of Mark Radosevich. Panel B: Hyphae of the fungus *Sordaria*, commonly found in the feces of terrestrial herbivores. The width of each cell is about 10 μm. Image courtesy of George Barron, used with permission (http://hdl.handle.net/10214/7080). Panel C: Eukaryotic algae from a freshwater lake. The scale bar is 25 μm. Image courtesy of John Wehr. Panel D: A marine ciliate, *Cyttarocylis encercryphalus*, about 100 μm. Image courtesy of John Dolan. Color versions of panels B–D are available at http://www.oup.co.uk/companion/kirchman.

Why study microbial ecology?

The main reason has already been implied: microbes mediate many processes essential to the operation of the biosphere. But there are other reasons for studying microbial ecology. The following list of seven reasons starts with those that a nonscientist might give, if asked why we should learn about microbes. The list ends with the reasons that arguably are the most important, certainly in shaping the contents of this book.

Microbes cause diseases of macroscopic organisms, including humans

Most people probably think "germs" when asked how microbes affect their lives. Of course, some microbes do cause diseases of humans and other macroscopic organisms. The role of infectious diseases in controlling population levels of macroscopic plants and animals in nature is recognized to be important (Ostfeld et al., 2008), but its impact probably is still underestimated. Sick animals in nature are likely to be killed off by predators, or simply die while hidden to avoid predation before being counted as being ill. We know less about the impact of diseases on smaller organisms, such as the zooplankton in aquatic systems (Figure 1.2) or invertebrates in soils. These small organisms are crucial for maintaining the health of natural ecosystems. There is some evidence that diseases in the ocean are becoming more common (Burge et al., 2014), and amphibians on land are now declining worldwide due to infections by chytrid fungi, perhaps linked to global warming (Raffel et al., 2015).

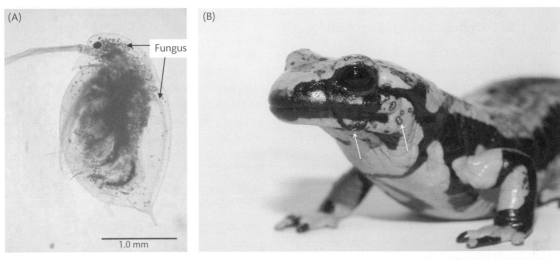

Figure 1.2 Two examples of animals infected by fungi. In panel A, the fungi are the small numerous dots visible in the transparent body cavity of a common freshwater zooplankton, *Daphnia pulicaria* (the common name is "water flea"). Taken from Johnson et al. (2006). Used with permission from the Ecological Society of America. In panel B, the arrows indicate the lesions affecting a salamander, a disease decimating these amphibians in Europe (Stegen et al., 2017). Picture provided by Frank Pasmans, Ghent University. A color version of panel B is available at http://www.oup.co.uk/companion/kirchman.

But pathogenic microbes are exceptions rather than the rule. The microbiologist John Ingraham pointed out that there are more murders among humans than pathogens among microbes (Ingraham, 2010). The vast majority of microbes in nature are not pathogenic, and that includes those living on and in us. The human body is host to abundant and diverse microbial communities. In fact, many of the cells in the human body are not human, but rather are bacteria. An average adult has about thirty trillion human cells but also thirty eight trillion bacteria (Sender et al., 2016), as well as numerous archaea and protists, not to mention viruses. Microbes inhabiting our skin and mucous membranes help to prevent invasion by pathogens, and the bacteria in the gastrointestinal tract do the same. Disruption of the microbial community in the colon, for example, allows the pathogen *Clostridium difficile* to flourish, causing severe diarrhea. One novel cure is "fecal transplantation" or "bacteriotherapy," in which a normal microbial community is "transplanted" into the colon of a diarrhea-suffering patient.

Bacteria also aid digestion in humans and in many other mammals, while these microbes and others help invertebrates live on otherwise indigestible food such as wood (Chapter 14). There is even a connection between the microflora of our digestive tract and our brain (Collins et al., 2012). A huge project is now examining the genomes of human-associated microbes, the "human microbiome" (http://www.hmpdacc.org/), using metagenomic approaches (Chapter 5) first designed for soils and oceans. The term "microbiome" is used by some investigators for any microbial community in a natural environment (Moran, 2015; Fierer, 2017), but here the term will be reserved for communities associated with larger organisms.

Microbes help to make our food and other useful products

Microbes produce several things that we eat and drink every day, including yogurt, cheese, kimchi, and wine. Some of these microbially produced comestibles were examined by early scientists who could be called microbial ecologists. Louis Pasteur (1822–95; see Box 1.1) was hired by the wine industry to figure out why some wines turned sour and became undrinkable. The problem, as Pasteur found out, was a classic one of competition between two types of microbes, one that produced alcohol (good wine) and the other, organic acids (sour, undrinkable wine). To this day, food microbiologists try to understand the complex microbial interactions and

---

Box 1.1 Two founders of microbiology

One of the founders of microbiology, Louis Pasteur, made many contributions to chemistry and biology, such as showing that life does not arise from spontaneous generation, a theory held in the mid-nineteenth century to explain organic matter degradation. Decomposition of organic material is still an important topic today in microbial ecology (Chapter 7). Pasteur also explored the role of bacteria in causing diseases, but it was a contemporary of Pasteur, Robert Koch (1843–1910), another founder of microbiology, who defined the criteria, now known as Koch's postulates, for showing that a particular microbe causes a specific disease. Koch developed the agar plate method for isolating bacteria, a method still used today in microbiology.

---

processes that affect our favorite things to eat and drink, an important job in applied microbial ecology. Microbes are also involved in meat and dairy products. Cows, goats, and sheep—ruminants—depend on complex microbial consortia to digest the polysaccharides in the grasses they eat (Chapter 14).

Microbes are also important in supporting life in lakes and the oceans, and eventually make possible the fish we eat. Microbes take over the role of macroscopic plants in aquatic environments and are the main primary producers, meaning they use light energy to convert carbon dioxide (CO_2) to organic material (Chapter 6). These microbes, the "phytoplankton," include cyanobacteria and eukaryotic algae. Phytoplankton are rarely eaten by fish, not even by their small, young stages, the fish larvae. Rather, the main herbivores in lakes and oceans are mostly microscopic animals (zooplankton) and protists. Zooplankton and protists in turn are eaten by still larger zooplankton and fish larvae as part of a food chain leading eventually to adult fish (Figure 1.3). There can be more

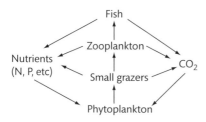

Figure 1.3 A simple food chain, from phytoplankton to fish, found in aquatic habitats. Microbes (phytoplankton) are at the base of this food chain, and other microbes (small grazers) make up the first few transfers. Still other microbes (mainly bacteria), not shown, contribute to the release of nutrients used by phytoplankton.

direct connections between microbes and fish. In some aquaculture farms, shrimp feed on "bioflocs" formed from bacteria growing on added wheat flour. The simple, linear food chain shown in Figure 1.3 is accurate in only some aquatic systems. But even in waters with more complex food webs, microbes are the base upon which the fisheries of the world depend. Consequently, there is a general relationship between microbial production and fishery yields.

Another important connection with our food is the role of microbes in producing the inorganic nutrients that are essential for growth and biomass production by higher plants in terrestrial environments (Figure 1.4) and by phytoplankton in aquatic environments. Essential inorganic nutrients, such as ammonium and phosphate, come from microbes as they degrade organic material (Chapter 7). Other microbes "fix" nitrogen gas, which cannot be used by plants as a nitrogen source, into ammonium, which can (Chapter 12). The fertility of soils depends on microbes in other ways. Organic material from higher plants, partially degraded by microbes, and other organic compounds directly from microbes, make up soil organic material. This organic component of soil contains essential plant nutrients and affects water flow, fluxes of oxygen and other gases, pH, and many other physical–chemical properties of soils that directly contribute to the growth of crop plants. So, our food indirectly and directly depends on microbes and what they do.

In addition to things we can eat or drink, microbes produce a vast array of other compounds that enrich our lives. These include antibiotics, chemicals such as butanol and citric acid, and enzymes used in the food,

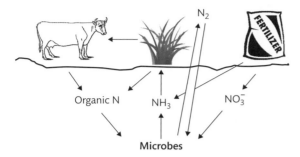

Figure 1.4 Some of the roles of microbes in supplying key nutrients for crops and plants in terrestrial habitats. The three N-cycle reactions depicted here include mineralization of organic nitrogen to ammonia (NH_3), the fixation of N_2 gas to ammonia, and the reduction of nitrate (NO_3^-) to N_2 gas or ammonia. The complete N cycle is discussed in Chapter 12.

pharmaceutical, and textile industries. Other microbially produced enzymes, such as the one used for the polymerase chain reaction (PCR; see Chapter 4), are essential in biotechnology. Most of "industrial microbiology" is far from the environmental sciences, but some aspects of the field touch on the basics of microbial ecology covered in this book.

Microbes degrade and detoxify pollutants

The modern environmental movement is often said to have started with the publication in 1962 of *Silent Spring* by Rachel Carson (1907-64). The book chronicled the damage to wildlife and ecosystems caused by the pesticide dichloro-diphenyl-trichloroethane, better known as DDT. Fortunately, the concentrations of DDT have decreased over time, in part due to regulations banning it in most developed countries, following the publication of *Silent Spring*. In addition, microbes, mostly bacteria, degrade DDT and other organic pollutants to innocuous compounds and eventually to CO_2, despite the pollutants being recalcitrant and difficult to degrade. With very few exceptions, bacteria and fungi are quite adept at degrading organic compounds, even those quite toxic to macroscopic organisms.

Inorganic pollutants cannot be removed by microbial activity, but by changing the chemical state of these pollutants, microbes affect their toxicity or movement through the environment or both. Microbes produce a

toxic form of mercury, methylmercury, that is more easily transferred to larger organisms than inorganic mercury. Another example is uranium (Williams et al., 2013). The most oxidized form of uranium, U(VI), moves easily through subsurface environments. When U(VI) is reduced by *Geobacter* and probably other bacteria, the resulting U(IV) is less mobile and less likely to cause environmental problems.

The study of pollutant–microbe interactions is a part of "environmental microbiology," but that term is also used more generally to describe the study of microbes in the environment regardless of whether a pollutant is involved or if the microbes are mediating a process of practical interest (Madsen, 2016). "Environmental microbiology" and "microbial ecology" are essentially synonymous.

Microbes are models for exploring principles in ecology and evolution

Microbes have served as models for exploring many questions in biochemistry, physiology, and molecular biology. They are good models because they grow rapidly and can be manipulated easily in laboratory experiments. For similar reasons, microbes are also used as models to explore general questions in ecology, population genetics, and evolution (Barberán et al., 2014). Virus–bacteria interactions, for example, have served as models of predator–prey interactions, and experiments with protozoa and bacteria were crucial for establishing Gause's "competitive exclusion principle" (Figure 1.5). The principle is that only one species can occupy a particular niche at a time because it excludes other species due to competition for a limiting resource. Experiments with microbes have demonstrated basic principles about natural selection and adaptation in varying environments. Richard Lenski and colleagues have explored the evolution of the bacterium *Escherichia coli* over 50,000 generations by growing it in new, fresh media every day since 1988, including weekends and holidays (Lenski et al., 2015). Genome sequencing (Chapter 5) has revealed how this bacterium has changed over time, providing insights into evolution not possible with large organisms.

Just as we can learn about large organisms from microbes, the flow of ideas can go the other way. General

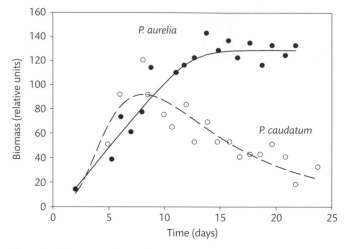

Figure 1.5 Experimental evidence for the competitive exclusion principle, which states that no two species can occupy the same niche at the same time. Here, two species of the protozoan *Paramecium* are forced to compete for the same food source, a bacterial prey (*Bacillus pyocyaneus*). Only one *Paramecium* species wins. Data from Gause (1964).

theories developed for exploring the ecology of plants and animals are often useful for exploring questions in microbial ecology. For example, microbial ecologists have used island biogeography theory, which was first conceived for large animals by R. H. MacArthur and E. O. Wilson in 1967, to examine the dispersal of microbes and relationships between microbial diversity and habitat size. Likewise, models about stability and diversity developed for large organism communities are now being applied to microbial communities and processes. Microbial ecologists look at microbial diversity for patterns that have been seen for plant and animal diversity, such as how diversity varies with latitude (Chapter 4). Not all large organism-based theories are applicable to thinking about microbes, but many are.

Microbes living today are models for early life on Earth and perhaps life on other planets

Life first appeared on Earth about 3.7 billion years ago and perhaps even earlier than that, not long after the planet was formed 4.5 billion years ago (Fig 1.6). The first cell on Earth was a microbe, perhaps a hydrogen-gas-powered acetogen or methanogen, types of bacteria and archaea that produce acetate or methane, respectively (Chapter 11), which may have evolved at hydrothermal vents (Martin and Sousa, 2016). Viruses have probably been around since cellular life began; some investigators

even hypothesize that life started in viral form. Cyanobacteria first appeared probably about 2.8 billion years ago, leading eventually to high oxygen concentrations in the Earth's atmosphere (Chapter 11).

The date when eukaryotes evolved is controversial, and estimates depend on the data being considered. Fossil structures perhaps from the first eukaryote have been found in 3.2 billion-year-old estuarine sediments, while steranes, derived from steroids or sterols produced only by eukaryotes, have turned up in shales estimated to be about 2.7 billion years old (Arndt and Nisbet, 2012). But the most convincing fossils, backed up by DNA sequence data, indicate that eukaryotes appeared on Earth about two billion years ago. Regardless of the exact dates, microbes were the only life forms on the planet for the first three billion years of Earth's history. Multicellular animals and plants did not appear until about a billion years ago, two to three billion years after microbes had invented most of the various metabolisms now known to operate on Earth. As Harvard University paleontologist Andrew Knoll once said, "Animals might be evolution's icing, but bacteria are really the cake."

We can gain additional insights into the evolution of early life by looking at microbes and environments around today that may mimic those on early Earth. These include bacteria and archaea living today with gene sequences that put them near the root of the Tree of Life and are perhaps closely related to the first cell that

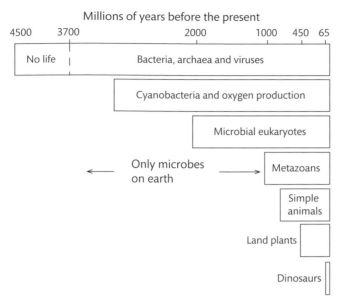

Figure 1.6 A few key dates in the history of life. For much of this history, only microbes were present, as multicellular eukaryotes appeared only one billion years ago, or 3.5 billion years after Earth had formed. Based on Humphreys et al. (2010), Payne et al. (2009), and Arndt and Nisbet (2012).

appeared on the planet. These microbes may have metabolisms similar to those of the first cell or two. Some of the first fossils of microbes have been found in stromatolites, which are large, laminated mats of different microbes and minerals abundant in the Precambrian when life was just starting on Earth (Chapter 13). These complex microbial mats can also be found today in some saline lakes and marine lagoons. Our planet still has many "warm little ponds" where Darwin speculated that life first evolved. Studying the Earth's first organisms is part of a branch of microbial ecology: geomicrobiology (Chapter 13).

In addition to helping us look at life millions of years ago, today's microbes may potentially provide insights into life on planets or moons of other planets millions of kilometers from Earth. Studying microbes in extraterrestrial-like environments on Earth is the main focus of "astrobiology." These environments are extreme, where only microbes, and often only bacteria and archaea, the "extremophiles," survive. Microbes live in hot springs and deserts, polar ice, permafrost of the tundra, and within rocks—unworldly habitats where it is hard to imagine life existing. Perhaps these earthly extremophiles are similar to life on other planets, and perhaps insights gained from astrobiological studies on this planet will help in the

search for life elsewhere in the universe. But the work would be worthwhile even if there were no extraterrestrial microbes. Extreme environments and extremophiles are often bizarre and always fascinating.

Microbes mediate biogeochemical processes that affect global climate

This reason shapes many topics appearing in this book. The role of microbes in degrading pollutants has already been mentioned, but microbes are involved in an even more serious "pollution" problem.

Humans have been polluting the Earth's atmosphere with various gases that affect our climate. These "greenhouse gases" trap long-wave radiation, better known as heat, from the sun. Most of these gases also have natural sources, and the Earth always has had greenhouse gases; fortunately. Because of these gases, the average global temperature is 16 °C, much warmer than the chilly −18 °C Earth would be without them (Schlesinger and Bernhardt, 2013). Mars does not have any greenhouse gases and is even colder (−53 °C). High concentrations of greenhouse gases, mainly carbon dioxide, in Venus's atmosphere, along with its proximity to the sun, explain why that planet has an average surface temperature of

474 °C. However, although some greenhouse gases are beneficial for the Earth's climate, their excessive levels are leading to global warming and many other related global problems.

Greenhouse gases have been increasing in Earth's atmosphere since the start of the industrial revolution in the early 1800s (Figure 1.7). Although water vapor is the dominant greenhouse gas, human society has a bigger, more direct impact on other gases, most notably carbon dioxide. Other important greenhouse gases include methane (CH_4) and nitrous oxide (N_2O). The concentrations of these gases in the atmosphere are much lower, but they trap more heat per molecule than does CO_2. Because of higher levels of greenhouse gases, the average temperature for the planet is about 1 °C warmer now

than it was in the nineteenth century (Figure 1.7), and is projected to increase even more in the near future. A degree or two warmer may not seem like much, but it has already caused many changes in the biosphere, to say the least. Another problem caused by raising atmospheric CO_2 is the acidification of the oceans, which threatens corals and other calcifying organisms, including some microbes (Chapters 13 and 14).

Microbial ecology has an essential role in understanding the impact of greenhouse gases on our climate and the response of ecosystems to climate change, one reason being that nearly all of these gases are either used or produced, or both, by microbes (Table 1.1). Carbon dioxide, for example, is used by higher plants on land and by phytoplankton in aquatic ecosystems. This gas is

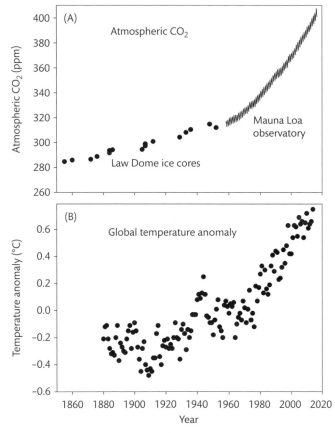

Figure 1.7 Atmospheric CO_2 concentrations (A) and global temperature anomaly (B) since the nineteenth century. Concentrations were estimated from ice cores (Etheridge et al., 1996) or measured directly at the Mauna Loa Observatory, Hawaii, provided by Pieter Tans (NOAA Earth System Research Laboratory and Ralph Keeling (Scripps Institution of Oceanography), used with permission (http://www.esrl.noaa.gov/gmd/ccgg/trends/#mlo_data). The global temperature is expressed as the difference between the average temperature for a year minus the average over 1951–80 (Hansen et al., 2010). Updated temperature data from http://data.giss.nasa.gov/gistemp/, accessed on November 20, 2016.

Table 1.1 Some greenhouse gases and how they are affected by microbes. Concentrations are expressed as parts per million (ppm), per billion (ppb), or per trillion (ppt). Data from the Mauna Loa Observatory (http://www.esrl.noaa.gov/gmd/) and Myhre et al. (2014).

Gas	Concentrations in 2005	Concentrations in 2016	Greenhouse effect*	Microbial process
Carbon dioxide (CO_2)	379 ppm	401 ppm	1	Primary production and organic carbon mineralization
Methane (CH_4)	1774 ppb	1834 ppm	21	Methanogenesis and methanotrophy
Nitrous oxide (N_2O)	319 ppb	329 ppb	270	Denitrification and nitrification
CFC-11 (CCl_3F)**	250 ppt	230 ppt	4660	Degradation by heterotrophs?

* Relative to CO_2.
** One of many hydrochlorofluorocarbons. Some of these other gases have even greater greenhouse effects than that of CFC-11.

released by heterotrophic microbes in all ecosystems. The impact of biological activity can be seen in the yearly oscillations of carbon dioxide in Figure 1.7; it goes down in the summer when plant growth is high and it increases in winter when carbon dioxide produced by respiration exceeds carbon dioxide use by plant growth. Methane, another gas that has been increasing in the atmosphere, is also produced and consumed by microbes (Chapter 11). It and another powerful greenhouse gas, nitrous oxide, are produced in anoxic environments which have expanded over the years, mainly due to the growth in agriculture.

What complicates our understanding of these greenhouse gases is that nearly all are produced or consumed by natural processes mediated by microbes, in addition to the anthropogenic inputs. Although the input of CO_2 to the atmosphere by human activity is substantial, it is still small compared to the input and uptake by natural processes. In contrast, production of the important plant nutrient, ammonium, directly by humans (fertilizer synthesis) or aided by humans (microbial production in agriculture) rivals the natural production of ammonium by microbes (Chapter 12). To complicate things further, greenhouse gases vary with the seasons, as already seen for CO_2, and have varied greatly over geological time, independent of human intervention. So, the challenge is to separate the natural changes from those effected by humans and to understand the consequences of these changes.

Microbial ecologists cannot solve the greenhouse problem. But many of the topics discussed in this book can help us understand the problem and predict future changes. One job of microbial ecologists and other scientists studying the Earth system is to figure out the impact of increasing greenhouse gases and other global changes on the biosphere. How will an increase in global temperature affect the balance between photosynthesis and respiration? How much CO_2 and CH_4 will be released if the permafrost of the tundra in Alaska and Siberia melts? Will a decrease in diversity caused by climate change affect the C cycle? Answering these and other questions depends on the work of microbial ecologists.

Microbes are everywhere, doing nearly everything

Most of the reasons discussed so far for studying microbial ecology have focused on practical problems facing human society. But microbial ecology would be an exciting field even if all of these problems were solved tomorrow (Box 1.2). One goal of this book is to show the importance of microbial ecology in explaining basic processes in the biosphere, even if they may appear to be far from any problem facing us today. Regardless of what microbes do, we should want to know about them

---

Box 1.2 Praise of microbial ecology from an ant ecologist

E. O. Wilson did most of his scientific work on ants and went on to write several award-winning books, including *Sociobiology* and *On Human Nature*. Near the end of his autobiography, *Naturalist*, published in 1994 (Island Press), Wilson mentioned that if he had to do it all over again, he would be a microbial ecologist. He ends with a paean to the microbial world: "The jaguars, ants, and orchids would still occupy distant forests in all their splendor, but now they would be joined by an even stranger and vastly more complex living world virtually without end."

because they are the most numerous and diverse organisms on the planet.

As a general rule, the smaller the organism, the more numerous it is (Figure 1.8). Viruses are the smallest and also the most abundant biological entity in both aquatic habitats and soils, whereas large organisms, such as zooplankton and earthworms, are rare, being 10^{10} less abundant than viruses. A typical milliliter of water from the surface of a lake or the oceans contains about 10^7 viruses, 10^6 bacteria, 10^4 protists, and 10^3 or fewer phytoplankton cells, depending on the environment. A typical gram of soil or sediment contains about 10^{10} viruses, 10^9 bacteria, and so on for larger organisms. Even deep environments, kilometers below the Earth's surface, have thousands of microbes. Even seemingly impenetrable rocks can harbor dense microbial communities. Overall, the biomass of bacteria and archaea rivals that of all plants in the biosphere, and is certainly greater than the biomass of animals.

Microbes are found even where macroscopic organisms are not, in environments with extremes in temperatures, pH, or pressure: "extreme" for humans, but quite

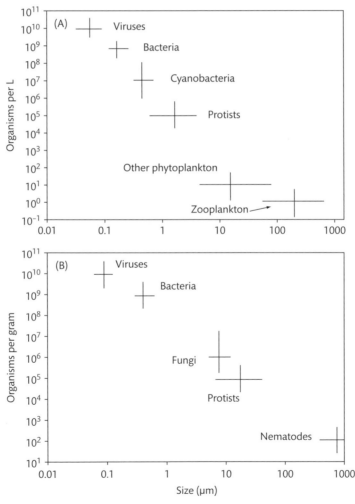

Figure 1.8 Size distribution and abundance of some microbes and other organisms in a typical aquatic habitat (A) and in soils (B). "Bacteria" here refers only to heterotrophic ones, while "other phytoplankton" are phytoplankton other than cyanobacteria or photoautotrophic protists. The size given for fungi is the diameter of the hyphae. Some fungi can be several meters long.

normal for many microbes. Some hyperthermophilic bacteria and archaea thrive in near boiling water (>80 °C), which kills all other organisms, including eukaryotic microbes. The hot springs of Yellowstone are famous for harboring dense and exotic microbes that not only thrive at high temperatures but also at low pH; these microbes live in boiling acid baths. At the other extreme, microbes live in the brine channels of sea ice where water is still liquid but very salty (20% versus 3.5% for seawater) and cold (−20 °C). Viable microbes have been recovered encased in salt millions of years old (Jaakkola et al., 2016). The deep ocean may be extreme to us with its high hydrostatic pressure, one hundred-fold higher at 1000 m than at sea level, and perennially cold temperatures (about 3 °C). Many microbes thrive and grow, albeit slowly, in these deep waters, one of the largest ecosystems on the planet. Over 70% of the globe is covered by the oceans of which 75% (by volume) is deeper than 1000 m.

In addition to being numerous, there are many different types of microbes, some with strange and weird (from our biased perspective) metabolisms on which the biosphere depends. In addition to plant- or animal-like metabolism, some microbes can live without oxygen and "breathe" with nitrate (NO_3^-) or sulfate (SO_4^{2-}) (Chapter 11). Compounds like hydrogen sulfide (H_2S) are deadly to macroscopic organisms, but are essential comestibles for some microbes. Several metabolic reactions, such as methane production and the synthesis of ammonium from nitrogen gas, are carried out only by archaea and bacteria. Other microbes are capable of producing chemicals, like acetone and butane, which would kill off eukaryotic cells and seem incompatible with life. Microbes are truly capable of doing nearly everything.

How do we study microbes in nature?

The facts about microbes in nature discussed so far have come from many studies using many approaches and methods. It is a great intellectual puzzle to figure out the actions and creatures in the unseen world and how they affect our visible world. This book will introduce some of the methods used in microbial ecology so that you can gain deeper insights and appreciation of the boundaries between the known and unknown. By learning a little about the methodology, you will also understand better

why some seemingly simple questions are difficult to answer.

Here we start with one of the most basic questions: how many bacteria are in an environment? We will focus on just two ways to answer this question, in order to introduce a fundamental divide between approaches for all microbes: "cultivation-dependent" versus "cultivation-independent." The number of bacteria in an environment was first estimated by the "plate count method," which consists of growing or cultivating organisms on solid agar media (Figure 1.9; Box 1.3). The assumption behind the plate count method is that each microbe in

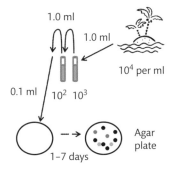

Figure 1.9 The plate count method. A sample from the environment is usually diluted first by adding 1.0 ml or 0.1 g of soil to 9.0 ml of an appropriate buffer, then sometimes diluted again by adding 1.0 ml of the first dilution to a new tube containing 9.0 ml of the buffer. Several more dilutions may be necessary. After a few days, if ten colonies grew up after two dilutions, for example, we could deduce that there were 10^4 culturable bacteria per ml or gram of sample.

Box 1.3 Able assistance with agar plates

Agar plates are made by pouring molten agar amended with various compounds into Petri dishes. Once cool, the agar solidifies and becomes a porous support on which microbes grow to form colonies—many, many cells piled together into a mound big enough to see. Although the approach is usually attributed to Robert Koch, two assistants of Koch came up with the key parts. Petri dishes were thought up by Julius Richard Petri, while Koch's wife, Fannie Hesse, suggested agar, which was used at the time to make jam.

the original sample will grow on the solid media and form a macroscopic clump of cells, or colony, that can be counted by eye or with a low-power microscope. Once isolated on the agar plate, the bacterium can now be identified by examining its response to a battery of biochemical tests. These tests provide some of the first clues about the bacterium's physiology and thus its ecological and biogeochemical roles in nature. The physiology and genetics of isolated bacteria in "pure culture" (cultures with only a single microbe) can then be examined in great detail. This method is at the heart of cultivation-dependent approaches.

The inadequacy of the plate count method became apparent when the abundance of bacteria it determined was found to be orders of magnitude lower than the abundance determined by one of the first cultivation-independent approaches, direct microscopic observation of microbes in nature. In seawater, for example, the plate count method indicates that there are about 10^3

bacteria per milliliter, a thousand-fold less than the number determined by direct microscopic counts (Jannasch and Jones, 1959). One of the first studies in soils did not find as dramatic a difference, but many more bacteria could be detected microscopically than by the plate count method (Skinner et al., 1952). This difference has been called the "Great Plate Count Anomaly" (Staley and Konopka, 1985). One explanation for the anomaly is that the uncultured microbes are dead, because a microbe must be viable and capable of growing enough to form a macroscopic colony if it is to be counted by the plate count method. For this reason, the plate count method is sometimes called the "viable count method." In contrast, a particle need only have DNA to be included in the direct count method, because the fluorescent dye used in the method binds to DNA (Figure 1.10). Dead bacteria could still have DNA and be counted. So, the discrepancy between direct and plate counts was first thought to be due to large numbers of dead or at least

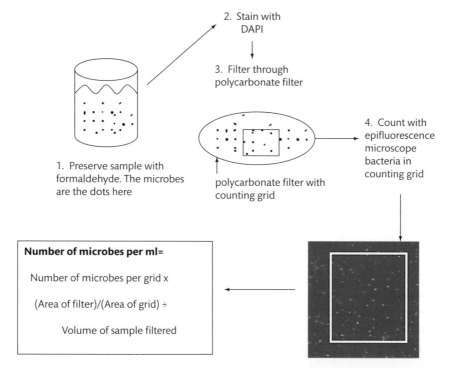

2. Stain with DAPI

3. Filter through polycarbonate filter

1. Preserve sample with formaldehyde. The microbes are the dots here

polycarbonate filter with counting grid

4. Count with epifluorescence microscope bacteria in counting grid

Number of microbes per ml=

Number of microbes per grid x

(Area of filter)/(Area of grid) ÷

Volume of sample filtered

Figure 1.10 Enumerating microbes by epifluorescence microscopy, the direct count method. "DAPI" is 4',6-diamidino-2-phenylindole, a stain specific for double-stranded nucleic acids. While under the microscope, the sample is exposed (the DAPI is "excited") to UV light (in the case of DAPI staining) and cells stained by DAPI fluoresce—they give off light, resulting in bright spots of light on a black background. The "epi" part of epifluorescence comes from the fact that the excitation light is above rather than below the sample.

inactive bacteria that were included in direct count methods, but not by the plate count method. There are similar problems with counting microbes other than bacteria.

In fact, to explain the Great Plate Count Anomaly, nearly all bacteria would have to be dead. If most bacteria were dead or dormant, it would have huge consequences for understanding the role of microbes in nature.

We now know that the discrepancy between plate and direct counts is not due mainly to dead or dormant bacteria (Chapter 8). Microbial ecologists still argue about the numbers of viable, dormant, and dead bacteria, but problems with the plate count method explain most of the discrepancy. The basic problem is that an agar plate is a very foreign habitat for most bacteria and other microbes, and many microbes are not adapted to growing in aggregates and forming macroscopic colonies, necessary for a microbe to be counted by the traditional plate count method. There are some problems also with the direct count method, such as confusing inert particles with real microbes due to non-specific staining. But overall, many of the particles observed by epifluorescence microscopy are active bacteria or archaea.

Regardless of what explains the Great Plate Count Anomaly, the difficulties in isolating microbes from nature and growing them in the laboratory have many consequences for microbial ecology. For starters, it means that most microbes cannot be identified by traditional methods. These methods depend on isolating and growing the microbe in the laboratory and then studying its phenotype: the capacity to grow with or without oxygen, for example, or to use specific organic chemicals. Even if they can be identified by other methods (Chapter 4), the physiology of uncultivated microbes cannot be studied by traditional laboratory approaches. Much about the physiology of some microbes has been learned from "enrichment cultures," in which growth conditions are set to select for or to "enrich" for the growth of one type of microbe. Although the target microbe dominates, others are still present. This approach has been fruitful for a few microbes, but it does not work for the vast majority of uncultivated microbes. This lack of information about physiology hinders our understanding of the ecological and biogeochemical roles of specific microbial taxa in nature. Fortunately, much can be learned about microbes as a whole by using approaches that examine processes and bulk properties of microbes in nature. For example, methods are available to examine the contribution of bacteria and fungi versus larger organisms in degrading organic material (Chapter 7).

One solution to the problem of identifying microbes without cultivation is to use sequences of certain genes obtained by cultivation-independent approaches. These gene sequence data are used to explore the "structure" of microbial communities: the taxonomic names and phylogenetic relationships of the microbes and their abundance. For reasons discussed in Chapter 4, the favorite phylogenetic marker gene of microbial ecologists and microbiologists is that coding for a type of ribosomal RNA (rRNA) found in the small subunit of ribosomes (SSU rRNA). There is no gene analogous to the rRNA gene for identifying viruses, but sequences of other genes and genomic sequencing are powerful tools for exploring viral diversity (Chapter 10).

Separating microbes by phylogeny: the three domains of life

Sequences of rRNA genes, first used by Carl Woese and colleagues in the 1970s (Woese and Fox, 1977), were instrumental in separating bacteria and archaea into two domains. Even before Woese's work, microbiologists knew that archaea had strange metabolisms that seemed to be advantageous for life on early Earth. For this reason, Woese called these microbes "archaebacteria," derived from the geological term Archaean, which is an early stage in the Precambrian, some 2–4 billion years ago. The "bacteria" part of the name was later dropped to emphasize the great differences between archaea and bacteria. We now know that archaea are not any more ancient than bacteria, but the name stuck anyway. Together, the two domains make up the "prokaryotes," which are organisms without a nucleus; bacteria and archaea are not eukaryotes. "Prokaryotes" has no taxonomic meaning, but it is useful shorthand for referring to the two domains in cases where they have a metabolism or physical feature in common or when a cell or process is clearly not eukaryotic and potentially could be either bacterial or archaeal or both (Figure 1.11).

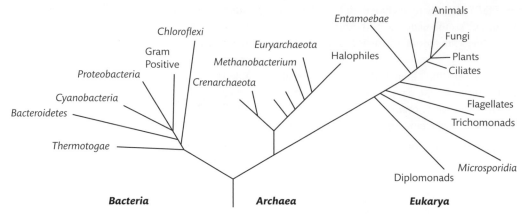

Figure 1.11 A phylogenetic tree showing the three domains of life: *Archaea*, *Bacteria*, and *Eukarya* (eukaryotes), as depicted by Olsen and Woese (1993) with updated phylum names. "Gram positive" now includes two phyla: *Actinobacteria* and *Firmicutes*. Much more complicated trees with many more phyla are now available (Hug et al., 2016).

Box 1.4 Molecular clock

The difference between sequences of phylogenetic marker genes can be translated to the amount of time that has passed since the two organisms diverged from a common ancestor. This molecular clock is very powerful for exploring the pace of evolution over geological time. The use of sequence data to date the appearance of eukaryotes has been mentioned earlier. For most applications, however, microbial ecologists use just the sequence data alone and do not convert dissimilarities to time.

Although bacteria and archaea cannot be distinguished by microscopy, the difference between the two prokaryotic domains and eukaryotes is easily seen microscopically. A prokaryotic cell appears to be empty when viewed by light microscopy, and in some sense it is, because it lacks a nucleus and all other membrane-bound organelles. The genome of prokaryotes is usually in a single circular piece of DNA in the cytoplasm (Chapter 5). The genome of a eukaryote, in contrast, is contained within a nucleus ("karyote" in Greek) and is organized into chromosomes. In addition to nuclei, eukaryotes have compartmentalized some metabolic functions into organelles, such as mitochondria and chloroplasts, which are absent in prokaryotes. These organelles are often visible under standard light micros-

copy and fill up the eukaryotic cell. The nucleus of a eukaryote is easily seen with epifluorescence microscopy after staining for DNA. In contrast, in the same epifluorescence photomicrograph, prokaryotes appear as solid dots with no internal structure.

One important characteristic distinguishing prokaryotes and eukaryotes is size (Figure 1.12). Because of the space needed for the nucleus and other organelles, eukaryotic cells, even microbial eukaryotes, are generally bigger than prokaryotes. There are some exceptionally large prokaryotic cells, visible even to the naked eye, but these microbes have a vacuole that pushes the cytoplasm to the outer perimeter of the cell, making the effective volume of these giants more like a typical bacterium. Size is not a useful taxonomic trait for distinguishing among organisms because there is much overlap among prokaryotes and eukaryotes. Even so, cell size has a huge impact in ecological interactions, such as in predator–prey interactions (Chapter 9) and in the competition between microbes for dissolved nutrients (Chapter 6).

Most bacteria and archaea are small, on the order of a micron, whereas most microbial eukaryotes are >3 μm, although there are exceptions, such as the <1 μm marine alga *Ostreococcus* (López-García et al., 2001). Bacteria grown on rich media in the laboratory are often bigger; the common laboratory bacterium *Escherichia coli*, for example, is about 1 × 3 μm and has a rod or bacillus shape. Other bacteria are spheres or coccus shaped (the plural is cocci), and the vibrios have a comma-like

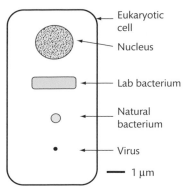

Figure 1.12 Size of eukaryotes, bacteria, and viruses. These organisms and viruses vary greatly in size and shape. A "lab bacterium" is one grown in nutrient-rich media in the laboratory.

appearance. In contrast, bacteria and archaea in most natural environments are much smaller, about 0.5 μm, and usually appear as simple cocci. There are reports of even smaller bacteria, called nanobacteria or ultramicrobacteria, with cells on the order of 0.1 μm. However, it is hard to fit all cellular components necessary for a free-living organism into a 0.1 μm cell. Just one important component, a ribosome, alone is typically about 0.025 μm in diameter.

The final general characteristic distinguishing prokaryotes and eukaryotes is metabolic diversity. Eukaryotes have two basic types of metabolism, one found in plants (autotrophy) and the other in animals (aerobic heterotrophy). Prokaryotes have many variations of autotrophy and heterotrophy and many unusual pathways with no analogs in eukaryotes. These pathways include the reduction of nitrogen gas to ammonium (nitrogen fixation is discussed in Chapter 12) and the synthesis of methane (see Chapter 11 for more on methanogenesis). The metabolic diversity of prokaryotes is vast and important in driving a great variety of biogeochemical processes in the biosphere.

This book uses the three-domain framework (bacteria, archaea, and microbial eukaryotes), depicted in the Tree of Life (Figure 1.11), to discuss differences among microbes, ecosystems, and biogeochemical processes. However, there are other hypotheses and versions of the Tree of Life. One version has only two domains, with *Eukarya* branching within the *Archaea* (Hug et al., 2016). These different versions have different implications for the evolution of eukaryotes. Regardless, this discussion illustrates again the key importance of microbes, here in exploring basic questions about the evolution of early life on the planet.

Separating microbes into functional groups

An entirely different way to divide up the microbial world is to sort microbes into various groups based on their metabolic capacity and physiology (Figure 1.13). The metabolism of a particular group will then help define its role in the ecosystem—its function. Before discussing some basic functional groups, it is useful to step back and remember what organisms need to survive and to reproduce. In the most basic terms, organisms need the raw materials that make up a cell, the most abundant being carbon. A microbe also needs a source of energy from which ATP can be synthesized. ATP, the universal currency of energy, is used to drive biosynthetic reactions that turn raw starting compounds into cellular components and ultimately more cells. Finally, microbes also need chemicals for various oxidation–reduction ("redox") reactions that transfer electrons from one compound to another. Biosynthesis sometimes requires elements in the starting material to be reduced. In this case, electrons from an "electron donor" are transferred to the starting material in a redox reaction. The most important example is the reduction of CO_2 to organic carbon, a reaction that all autotrophs carry out by definition. Other microbes need compounds to accept electrons ("terminal electron acceptor"), an example being oxygen, which receives the electrons produced by the oxidation of organic material. Functional groups can be defined by other aspects of microbial metabolism, but how a microbe satisfies these three needs—carbon, energy, and redox compounds—is an important first step in defining its role in the ecosystem.

A few functional groups are now discussed in order to define more terms used throughout the book.

Autotroph versus heterotroph

These terms refer to the source of carbon and are equivalent to "plant" and "animal" in the macroscopic world. However, "plant" and "animal" are not used to describe even eukaryotic microbes because they are misleading

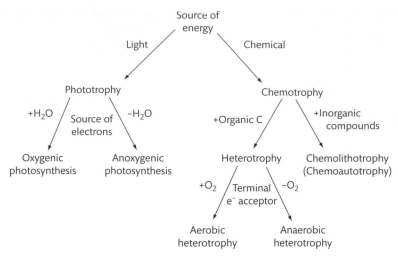

Figure 1.13 Microbial metabolisms that define many of the functional groups found in natural environments. "–H_2O" indicates sources of electrons other than H_2O, even though water is still present. "–O_2" indicates the use of terminal electron (e^-) acceptors other than O_2. Oxygenic photosynthetic microbes, anoxygenic photosynthetic bacteria, and chemolithotrophs use carbon dioxide as a carbon source, making them autotrophs, whereas heterotrophs use organic carbon. Adapted from Fenchel and Blackburn (1979).

and inaccurate. The terms do not reflect evolution (microbes gave rise to higher plants and animals), and many microbes have both plant-like and animal-like metabolisms. Rather, microbial ecologists use autotroph and heterotroph. The etymology of these words helps in remembering their definitions. "Auto" and "hetero" are from the Greek meaning "self" and "different," respectively, while "troph" refers to food. So, an autotroph uses CO_2 to make its own organic carbon, whereas a heterotroph uses organic carbon made by other organisms. When fueled by light or in some cases chemical energy, the autotrophs are primary producers upon which all heterotrophs directly or indirectly depend.

Phototroph versus chemotroph

The other major characteristic distinguishing microbes is the source of energy used for biomass synthesis. Phototrophs have devised means to capture light energy and convert it to chemical energy. Many phototrophs are also autotrophs, whereas some microbes are mainly heterotrophic and only supplement their energy supply with light energy; these microbes are called photoheterotrophs or mixotrophs. Microbes using light energy to fix CO_2 are photoautotrophs. Photoautotrophs can be

further subdivided depending on the source of the electrons used for reducing CO_2 to organic material. Higher plants, eukaryotic algae, and cyanobacteria use water and produce oxygen, making them oxygenic photoautotrophs. Other bacteria do not use water and do not evolve oxygen; these are called anoxygenic photoautotrophs. The source of electrons used by an important example of anoxygenic photoautotrophy is hydrogen sulfide.

The other major source of energy is from the oxidation of reduced compounds. Organisms living off these oxidation reactions are chemotrophs. Microbes using organic compounds are chemoorganotrophs, although heterotroph is the term used more frequently. *Homo sapiens* and other animals are chemoorganotrophs, specifically aerobic heterotrophs. Less common are those organisms capable of gaining energy from the oxidation of reduced inorganic compounds. This type of metabolism, chemolithotrophy, is restricted to the prokaryotes, and no eukaryote is known to use it for ATP synthesis, although some eukaryotes take advantage of chemolithotrophic symbionts (Chapter 14). Because the middle syllable, "litho," is derived from the Greek for "rock," chemolithotrophs could be called "rock eaters." Rather than rocks, however, chemolithotrophs use compounds like ammonium and hydrogen sulfide (Chapter 11).

Combining structure and function

Much of this book will be about the processes carried out by the functional groups just discussed and others. These processes can be discussed without detailed information about the types and abundance of microbes carrying out the process—their community structure. But answering many other questions in microbial ecology depends on knowing microbial community structure, and that work will be discussed here too. Today in microbial ecology, we want it all. We want to know about the processes, the functions, and the names and abundance of microbes carrying out those processes: the structure. Exploring processes and the microbes driving them is the heart of this book.

Summary

1. There are many reasons for studying microbial ecology, ranging from human health to the degradation of organic pollutants. The work of microbial ecologists is also important in examining the impact of greenhouse gases and other climate change issues.

2. Another rationale for studying microbial ecology is that microbes, especially bacteria, are the most numerous organisms on the planet and mediate many essential biogeochemical processes. Although not counted as organisms, viruses are even more abundant than bacteria.

3. Traditional cultivation-dependent approaches are often not successful for examining microbes in nature because many microbes cannot be isolated and maintained as pure cultures in the laboratory.

4. Sequences of SSU rRNA genes (16S rRNA in prokaryotes, 18S rRNA in eukaryotes) are used to define phylogenetic groups of *Bacteria*, *Archaea*, and *Eukarya*. These organisms also differ in several key aspects, including their cellular composition and metabolism.

5. Microbes can also be divided into functional groups, based on the mechanisms for acquiring carbon and energy and on the compounds used in redox reactions. Photoautotrophs use light energy and CO_2, whereas chemoorganotrophs, or more simply, heterotrophs, obtain both energy and carbon from organic compounds.

6. Linking community structure (the numbers and types of microbes in a community) with functions, such as food web interactions and biogeochemical processes, is a major theme in microbial ecology.

CHAPTER 2

Elements, biochemicals, and structures of microbes

This chapter will continue to introduce microbes by discussing the composition of bacteria, fungi, and protists. We will see examples of how composition can inform our understanding of microbial processes in natural environments. Here, "composition" includes virtually everything found in microbes, ranging from elements to complex macromolecular structures. This information will be used in later chapters to understand the ecology of microbes and to explain the role of particular microbes in biogeochemical cycles. Chapter 10 has a few words about the composition of viruses.

Microbiologists usually discuss the composition of microbes in terms of macromolecules, such as protein, RNA, and DNA. Microbial ecologists and biogeochemists often think of composition in terms of elements, most prominently carbon (C), nitrogen (N), and phosphorus (P), and ratios of elements, such as the C:N ratio. Both approaches are used here for a complete picture of what makes up a microbial cell. Some of the information presented here is from basic microbiology and experiments with laboratory-grown microbes. Although we can learn much from these experiments, we will see that the composition of microbes grown in the laboratory differs from that of microbes in natural environments. These differences give clues about how microbes survive and often thrive in nature.

Composition also helps to explain the imprint of microbes on the environment. The contents of microbial cells are released by various processes and left behind in soils and aquatic habitats. These remains can contribute to large geological formations, such as the White Cliffs of Dover, made of the calcium carbonate from an alga. On a smaller physical scale is the impact of microbial cellular components on the chemical make-up of aquatic habitats and soils. Although not as obvious as a cliff, this contribution has a large impact on the environment and on biogeochemical cycles. An understanding of these effects starts with knowing what makes up a microbe.

The elements discussed here and others found in microbial cells are just a few of the many elements affected by microbial activity. In addition to the major elements making up microbes (C, hydrogen [H], N, oxygen [O], P, and sulfur [S]), several other elements are required by microbes, albeit usually in very small amounts as "micro-nutrients", such as nickel (Ni) and boron (B). Still others are not incorporated into cellular structures but their chemical state can be modified by microbes. The example of uranium (U) was discussed in Chapter 1. In total, microbes have some use for 63 of 98 elements found in nature (Stolz, 2017).

Elemental composition of microbes

Figure 2.1 illustrates how the relative abundance of elements found in cells differs from the abundance of these elements in the Earth's crust, the ultimate source of the building blocks for life. The Earth's crust has high amounts of silicon (Si), but only some algae (diatoms— see Chapter 6), and a few other protists use Si. Although sodium (Na) and magnesium (Mg) are abundant in the Earth's crust and as major cations in natural environments, they are not large components of biochemical structures in microbes.

Processes in Microbial Ecology. Second Edition. David L. Kirchman. Oxford University Press (2018). © David L. Kirchman 2018.
DOI 10.1093/oso/9780198789406.001.0001

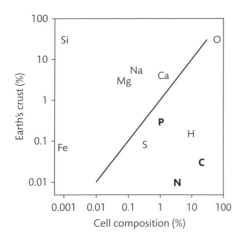

Figure 2.1 Elemental composition of the Earth's crust and of a typical cell. The line indicates equal percentages in both cells and the crust. The three elements in bold (C, N, and P) are those that commonly limit microbial growth in nature. Note that some elements are highly enriched in cells compared to the inert world (those below the line), whereas others are present in only low amounts in cells compared to the Earth's crust (those above the line). These differences are true even though the values vary with the organism.

Microbes do require these cations for growth to maintain osmotic balance and for some enzyme and membrane functions, but the cations make up a very small fraction of the average microbial cell. In total, inorganic ions make up only about 1% of the dry weight of a microbial cell. Use of these cations by microbes is also insignificant compared to the large concentrations usually found in natural environments. Another abundant cation, calcium (Ca^{2+}), is used only by select algae (coccolithophorids—see Chapter 6), but not by bacteria, archaea, and fungi, except as cationic bridges among polymers. The major biogenic elements are C, N, P, and S (Table 2.1).

Several other elements occur only in trace amounts in the Earth's crust and in cells. Although only vanishingly small amounts are needed, these trace elements or micronutrients are essential for microbial growth. Metals like zinc (Zn) and cobalt (Co) are important co-factors in some enzymes, such as the enzyme mediating urea degradation (urease needs Zn) and the vitamin B_{12}-requiring enzymes (they contain Co). A few microbes use more rare and strange metals such as tungsten (W) and nickel (Ni).

Table 2.1 Major and trace biogenic elements used by microbes in nature. The order roughly reflects the abundance in microbes. Trace elements are used in more components and enzymes than listed here. See Zhang and Gladyshev (2010) for more detail about some of the trace elements.

Element	Chemical form in nature[a]	Location or use in cell
Major biogenic elements		
C	HCO_3^-	All organic compounds
N	N_2, NO_3^-	Proteins, nucleic acids
P	PO_4^{3-}	Nucleic acids, phospholipids, ATP, NADPH
S	SO_4^{2-}	Cysteine and methionine, Fe–S clusters
Si	$Si(OH)_4$	Diatom frustules
Trace biogenic elements		
Fe	Fe^{3+} organic	Electron transfer system
Mn	Mn^{2+}, MnO_2, MnOOH	Superoxide dismutase, photosystem II
Mg	Mg^{2+}	Chlorophyll
Ni	Ni^{2+} organic	Urease, hydrogenase
Zn	Zn^{2+} organic	Carbonic anhydrase, protease, alkaline phosphatase, methyl coenzyme M reductase
Cu	Cu^{2+} organic	Electron transfer system, superoxide dismutase
Co	Co^{2+} organic	Vitamin B_{12}
Se	SeO_4^{2-}	Formate dehydrogenase
Mo	MoO_4^{2-}	Nitrogenases, nitrate reductase
Cd	Cd^{2+} organic	Carbonic anhydrase
I	IO_3^-	Electron acceptor
W	WO_4^{2-}	Substitutes for Mo in some oxidoreductases
V	$H_2VO_4^-$	Nitrogenases

[a] Those metals listed with "organic" occur mainly in organic complexes.

The most important micronutrient is iron (Fe). Nearly all microbes require iron but only relatively low amounts of it; the C:Fe ratio is on the order of 10,000 for most microbes. Iron is used mainly for electron transfer reactions, such as in respiration and photosynthesis. Because iron is abundant in the Earth's crust, there is enough of the element to support microbial growth in most environments. Iron is very abundant in soils, for example, varying from 0.05% in coarse-textured soils to >10% in highly weathered soils (oxisols) in the tropics. However, there are two important examples where iron concentrations are low and may be insufficient for microbial growth. Being far from terrestrial sources, the open

Box 2.1 Types of limitations

A general question in microbial ecology is about the factors setting or limiting growth rates and biomass levels for a microbial population in nature. The regulation of rate processes is due to Blackman-type limitation, while Liebig-type limitation refers to the control of biomass levels. The men who lent their names to these terms were not microbiologists. F. F. Blackman was a British plant physiologist who proposed his eponymous law of limiting factors in 1905. Justus von Liebig, a nineteenth-century German chemist, worked on crop yields and correctly hypothesized that plants get nutrients from inorganic rather than organic compounds, but thought, incorrectly, that decomposition was an abiotic process (Smil, 2001).

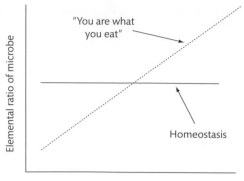

Figure 2.2 Potential relationships between the elemental ratio (e.g. C:N) of the resource used by a microbial consumer and the ratio of the consumer itself. With homeostasis (the horizontal line), a microbe is able to maintain a constant elemental ratio even though the resource varies. The other extreme is the lack of any control, such that the elemental ratio of the microbe varies proportionally with the ratio of the resource. In this case, "you are what you eat." Modified from Sterner and Elser (2002).

oceans do not receive much iron because its dominant form, Fe(III) oxide, is insoluble in oxygenated water at near neutral pH (Chapter 3). Uptake by microbes reduces iron concentrations to very low levels (10^{-12} M, or pM). Iron is also low in vertebrate serum because of the same chemistry keeping concentrations low in the open ocean. Concentrations of free iron are further reduced by binding of iron by a protein, transferrin, produced by the animal host. These low concentrations can limit the growth of pathogenic bacteria (Skaar, 2010). The bacterium causing Lyme disease, *Borrelia burgdorferi*, is successful in part because it does not appear to require iron.

Of the six most abundant elements in bacteria, two (O and H) are readily obtained from water, and a third (S) from a major anion in natural environments (sulfate; SO_4^{2-}). The remaining three elements, C, N, and P, are those thought to limit microbial growth most frequently in natural environments.

Elemental ratios in biogeochemical cycles and studies

Ecologists and biogeochemists often use elemental ratios to explore various questions in food web dynamics and in elemental cycles in the biosphere. We gain insights into what may control or "limit" the biomass and activity

of microbes by comparing their make-up in terms of C, N, and other elements with the composition in the resources (dissolved compounds, particulate detritus, or prey) available to them. Change in the elemental make-up of the microbe is also informative. Microbes have some capacity, termed "homeostasis," to maintain elemental ratios the same even in the face of changing growth conditions (Figure 2.2). In other cases, however, the elemental ratio of a microbe changes along with the change in ratio of the resource.

Elemental ratios can be used to examine a variety of biogeochemical and microbial processes in both terrestrial and aquatic ecosystems. For example, high C:P ratios could imply microbial growth is limited by P availability, and similarly for high C:N ratios, N limitation. Another example discussed in Chapter 7 is the use of C:N ratios to deduce whether heterotrophic microbes are net producers or consumers of ammonium. Latitudinal variation in soil and litter ratios has been used to explain changes in the make-up and functioning of soil microbial communities (Zechmeister-Boltenstern et al., 2015), and ratios of nitrate to phosphate concentrations have been used to identify regions of the oceans where denitrification (loss of nitrate to N gases) and N fixation are common (Chapter 12).

The use of elemental ratios to examine biogeochemical processes started with Alfred Redfield (1890–1983), who compared the composition of free-floating organisms (the plankton) in seawater and of major nutrients (Redfield, 1958). Redfield found that the ratio of C:N:P was 106:16:1 (in atoms) in the plankton, and that the N:P ratio was very similar to the ratio of nitrate to phosphate concentrations in the deep ocean. This observation was crucial in establishing the role of microbes in influencing oceanic chemistry. It is a great example of microbes molding their environment. Since then, the "Redfield ratios" of C:N = 6.6:1 and C:P = 106:1, have been used extensively in oceanography, limnology, and other aquatic sciences. The Redfield concept has also been used in soil sciences (Cleveland and Liptzin, 2007). The ratios for isolated soil organisms are remarkably similar to their aquatic relatives, with an important exception (higher plants), which will be discussed. The ratios vary greatly depending on the organism, soil type, vegetation, and climate regime (latitude) of the soils. These soil ratios are statistically different from the Redfield ratio and the elemental ratios found in aquatic systems (Zechmeister-

Boltenstern et al., 2015), reflecting the vast differences in environmental conditions between the two ecosystems. What is more remarkable, however, are the similarities.

Perhaps most elegantly, the Redfield ratio can be used to explore how primary production and respiration affect the concentrations of the major nutrients in oxic ecosystems. The following equation is built on the Redfield ratio:

$$106CO_2 + 16HNO_3 + H_3PO_4 + 122H_2O \leftrightarrow (CH_2O)_{106}(NH_3)_{16}H_3PO_4 + 138O_2 \tag{2.1}$$

The reaction proceeding left to right ($CO_2 \rightarrow CH_2O$) is primary production, while the opposite reaction ($CH_2O \rightarrow CO_2$) is aerobic mineralization of organic material back to its inorganic constituents. (Here and throughout the book, "CH_2O" symbolizes a generic organic compound.) An example exploring this mineralization is given in Figure 2.3, which was used to deduce the amount of phosphate released during mineralization. However, Equation 2.1 is not entirely accurate, and it hides some critical reactions. For example, the equation implies that nitrate (NO_3^-) is released during respiration and

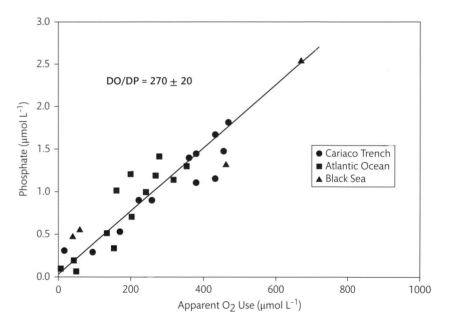

Figure 2.3 Illustration of the stoichiometric relations between elements set by biological processes, here the aerobic mineralization of organic material (apparent oxygen use) and an end product of mineralization, phosphate. Apparent oxygen use is the difference between the observed oxygen concentration at a depth and the concentration when the water was in equilibrium with the atmosphere. Data taken from Redfield (1958).

degradation of organic material, but in fact ammonium (NH_4^+) is the main nitrogenous compound produced during mineralization of organic material. A couple of additional steps, which are part of nitrification (Chapter 12), are needed to oxidize NH_4^+ to NO_3^-. Regardless, the equation and elemental ratios are very powerful and succinct descriptions of the interactions between microbes and the geochemistry of natural environments.

C:N and C:P ratios for microbes

The ratios C:N and C:P are used most frequently in microbial ecology and biogeochemistry. These ratios vary substantially among species of a particular microbial group (Chrzanowski and Foster, 2014) and within a species because of nutrient status. As previously mentioned, nitrogen starvation can lead to high C:N ratios, and phosphorus starvation has the same effect on C:P ratios. Before discussing this variation in more detail, it is useful to look at the grand averages and see if there are any fundamental differences among organisms.

Bacteria, whether from soils or the oceans, tend to be more N-rich and have lower C:N ratios than other microbes, such as fungi, algae, and other protists (Table 2.2). Heterotrophic bacteria tend to be more N-rich than even cyanobacteria. The C:N of bacteria is often <5, lower (more N) than that of fungi and eukaryotic algae, which have C:N ratios of about 9 and 7.7, the latter being not significantly different from the Redfield ratio (Table 2.2);

the precise values depend on the organisms and growth conditions used for calculating the average. Larger organisms, such as zooplankton (mainly copepods, which are a type of aquatic crustacean, roughly a millimeter in size) and *Collembola*, commonly called springtails (millimeter-sized insect-like invertebrates in soils), have higher C:N ratios and less N than bacteria and other microbes (Hunt et al., 1987). It is remarkable that insects and their planktonic version in lakes and the oceans, the zooplankton, have such similar C:N and C:P ratios (Elser et al., 2000).

The C:P ratios also differ substantially among microbes, and between microbes and larger organisms. Heterotrophic bacteria from soils or water also have more P than other organisms, including fungi and algae. The C:P ratio for heterotrophic bacteria (<100) is lower than that for fungi and eukaryotic algae (>100). The C:P ratios for two coccoid cyanobacterial genera common in the oceans, *Synechococcus* and *Prochlorococcus*, are much higher (less P) than the ratios for heterotrophic bacteria. Total P amounts are low in these cyanobacteria because their membranes have less P than the membranes of other microbes. Rather than the standard phospholipids, *Synechococcus* and *Prochlorococcus* have mostly sulfolipids (Van Mooy et al., 2009). An abundant heterotrophic marine bacterium, *Pelagibacter*, also can synthesize lipids without phosphate when starved for P (Carini et al., 2015). Low-P lipids are one reason why cyanobacteria and *Pelagibacter* are so abundant in oligotrophic oceans where phosphate concentrations are very low (<10 nM).

Table 2.2 Elemental ratios, expressed in moles, of some microbes and other organisms.

Organism	C:N	C:P	Comments	Reference
Autotrophs				
Marine coccoid cyanobacteria	5.4 ± 0.4	135 ± 22	Three species	Bertilsson et al. (2003)
Eukaryotic algae	7.7 ± 1.2	124 ± 19	15 species	Ho et al. (2003)
Higher plants	36 ± 23	970 ± 700	492 species	Elser et al. (2000)
Heterotrophs				
Soil bacteria	4.6 ± 1.5	75 ± 29	41 leaf litter isolates	Mouginot et al. (2014)
Marine heterotrophic bacteria	4.9 ± 1.0	87 ± 33	13 species	Zimmerman et al. (2014)
Soil fungi*	9.2 ± 4.8	120 ± 68	45 leaf litter isolates	Mouginot et al. (2014)
Protists	6.8 ± 0.9	102 ± 56	*Paraphysomonas imperforata*	Goldman et al. (1987b)
Zooplankton	6.3 ± 1.3	120 ± 47	43 freshwater species	Elser et al. (2000)
Insects	6.5 ± 1.9	116 ± 72	130 species	Elser et al. (2000)

* Zhang and Elser (2017) calculated a median C:N:P ratio of 250:16:1 for over 125 fungal species.

What stands out about the C:N and C:P ratios in Table 2.2 is the very high values for vascular plants on land. Large terrestrial plants have much more C per N or per P than seen for the aquatic primary producers, including cyanobacteria and eukaryotic algae. This difference is due to the large amount of cellulose, lignin, and other structural polysaccharides needed by higher plants to remain erect in air and to ward off herbivores. In contrast, cyanobacteria and algae do not need these structural polysaccharides for support (the surrounding water supports them), and they rely on other anti-herbivory defenses. This fundamental requirement for structural polysaccharides means that the detritus from terrestrial plants is much less rich in N and P than the detritus from microbial primary producers.

Biochemical composition of microbes

The elemental composition and ratios of organisms are mainly determined by the composition of the organic compounds making up the macromolecules of cells. The main macromolecules—proteins, nucleic acids, polysaccharides, and lipids—contain >97% of the elements found in microbes (Table 2.3). Concentrations of small compounds, like glucose, amino acids, and salts, are low (<3%), although fluxes through the pools of these compounds can be quite high.

Protein is the largest component of most microbial cells, most of the time, making up about 30–55% of the dry weight of algae, fungi (including yeasts), and bacteria (Table 2.3). Protein, the main metabolic machinery of all cells, occurs mostly as enzymes that catalyze reactions both within a cell and in the micron-scale environment surrounding the cell. In addition to enzymes, other proteins mediate transport of compounds across cell membranes (active transport proteins), while others make up flagella and cilia, the propellers used by motile microbes for moving through aqueous environments.

The relative amounts of the two classes of nucleic acids, RNA and DNA, vary greatly among microbes. Bacteria have relatively more RNA and DNA than do the other microbes, at least for the growth conditions used to generate the data in Table 2.3. The bacteria data used in Table 2.3 were taken from *E. coli* growing rapidly in the laboratory, more rapidly than the other organisms in the table. Bacteria and many other organisms generally have more ribosomes and thus more ribosomal RNA (rRNA) when growing rapidly in order to support fast rates of protein synthesis. Ribosomal RNA makes up about 80% of the RNA in all organisms, while 15% is taken up by transfer RNA (tRNA), which ferries amino acids to the ribosomes, and only 5% by messenger RNA (mRNA), which is used to direct the amino acids into the proper order to synthesize the protein of interest. Bacteria may seem to have more DNA than other organisms, but that is only true in a relative sense. The absolute amount of DNA—the genome size—in bacteria is less than that of eukaryotic microbes. For example, the genome of the

Table 2.3 Biochemical composition of various microbes. Data for a bacterium (*E. coli*) and two fungi (*Rhizoctoni solani* and *Sclerotium bataticola*) are from Ingraham et al. (1983) and Gottlieb and Van Etten (1966), respectively. The data for the eight yeast species and 117 algal species are from Finkel et al. (2016). NA: not available.

Biochemical component	% of dry weight			
	Bacteria	Yeasts	Fungi	Algae
Protein	55.0	32	10.7*	32.2
Lipids	15.3	5.4	34.2	17.3
RNA	13.7	4.8	5.4	5.6
DNA	10.0	0.1	0.4	1.0
Carbohydrates	4.1	24	42.8	15.0
Monomers	2.1	NA	2.1	NA
Inorganic components†	0	8.6	0	17.3
Total	100.2	74.9	95.6	88.4

* Other studies have found that the protein content of fungi is about 30% (Christias et al., 1975).

† Algae like diatoms and coccolithophorids have large amounts of silicate and calcium carbonate in their cell walls. The algae also had another 1% of dry weight attributable to chlorophyll.

archetypical bacterium, *E. coli*, with its 4.6×10^6 base pairs (4.6 Mb) is more than seven-fold smaller than the 34 Mb genome of the eukaryotic alga *Thalassiosira pseudonana*. Genome size is discussed in Chapter 5.

The biochemical data help to explain why different organisms have different C:N and C:P ratios. The low C:N ratios of bacteria are explained by their high amounts of N-rich protein, whereas the low C:P ratios of these microbes is consistent with their levels of DNA and RNA, which have high amounts of P as well as N. Bacteria have almost three-fold more nucleic acids, depending on growth rates, than large algae and fungi when expressed as a percentage of total cell mass (Table 2.3), while these microbial groups have nearly equal relative amounts of protein. Because nucleic acids are P-rich, whereas protein has no P, the difference results in lower C:P ratios for bacteria. Two other connections between elemental ratios and biochemical composition have already been mentioned: the high C:N and the polysaccharide content in higher plants, and the high C:P ratios and the replacement by sulfolipids of phospholipids in cyanobacteria.

It is worthwhile looking in more detail at the biochemical composition of one type of microbe, the heterotrophic bacteria, starting with one growing quickly in the laboratory (Table 2.4). The composition of this bacterium includes a number of different macromolecules, some present in multiple "copies," the term used to describe the presence of several molecules with the same molecular formula and structure. A rapidly growing *E. coli* cell has about a thousand different proteins, each ranging in abundance from one molecule or copy to multiple copies per cell. Each ribosome alone has about 55 proteins but only one 16S rRNA molecule. The *E. coli* cell growing at its maximum rate (doubling time of 24 min) has about 72,000 ribosomes. The number of unique mRNA molecules in this rapidly growing bacterium is only about 400. However, because some are present in several copies, the total number of mRNA molecules in a rapidly growing bacterial cell is over a thousand.

The numbers for a typical heterotrophic bacterium in a natural environment with low organic carbon concentrations are much different (Figure 2.4). First, this natural bacterium has fewer genes than bacteria like *E. coli* that can be grown easily in the laboratory, reflecting the genome sizes of these organisms. The typical natural bacterium also has fewer proteins, over half the number

Table 2.4 Biochemical composition of a "typical" bacterial cell (*E. coli*) growing rapidly (40 minutes doubling time) in the laboratory. Data from Neidhardt et al. (1990).

Component	% of dry weight	Weight per cell 10^{-15} g	Number of molecules per cell Total	Unique molecules
Protein	55.0	155	2,360,000	1050
RNA	20.5	59		
23S rRNA		31	18,700	1
16S rRNA		16	18,700	1
5S rRNA		1	18,700	1
tRNA		8.6	205,000	60
mRNA		2.4	1380	400
DNA	3.1	9	2	1
Lipid	9.1	26	22,000,000	4
Lipopolysaccharides	3.4	10	1,200,000	1
Peptidoglycan	2.5	7	1	1
Glycogen	2.5	7	4360	1
Subtotals				
Macromolecules	96.1	273		
Soluble pool	2.9	8		
Inorganic ions	1	3		
Total dry weight	100	284		
Water (70%)		670		
Total weight		954		

found in the laboratory-grown bacterium. The difference is consistent with the smaller size of the natural bacterium, as pointed out in Chapter 1. The biggest difference is the number of mRNA molecules or transcripts in the two types of bacteria. The natural bacterium has only about 200 transcripts, ten-fold fewer than found in the typical laboratory-grown bacterium, reflecting the small size and slow growth of these natural organisms. Protists in the ocean also have a low number of mRNA molecules (Liu et al., 2017).

Both fast and slow growing bacteria have far fewer transcripts than they have proteins (more than 1000-fold fewer) or genes (three-fold fewer for the laboratory bacterium and over ten-fold for the natural bacterium). Bacterial metabolism can carry on with so few transcripts because they are synthesized and degraded ("turned over") very quickly; the life span of a transcript within a cell is only a few minutes and is not thought to vary much with growth rate (Moran et al., 2013). In contrast, genes have to last the lifetime of the cell, and proteins nearly so too; protein turnover in bacteria is limited. This strategy seems the best for minimizing costs while still maintaining cellular functions. Although a bacterium cannot get rid of a gene or protein without losing a function, a transcript is not needed after translation and synthesis of the protein, so it can be degraded. If needed, bacteria can quickly turn mRNA synthesis on, giving them the potential to respond rapidly to changing environmental conditions.

Variation in elemental ratios and biochemical composition

In addition to variation among different microbes, both the elemental ratios and biochemical composition of the same microbe can vary due to changes in growth conditions. One example is the variation in the major macromolecules making up the bacterium *Klebsiella aerogenes*, growing at different rates (Figure 2.5). These data help to explain variation in elemental ratios observed for other bacteria in natural environments. Although some of the macromolecules and elemental ratios change, the relative amount of protein decreases only slightly even when growth rate varies by about ten-fold. This fairly constant protein content helps to explain the lack of large variation in C:N ratios in bacteria and also in many other

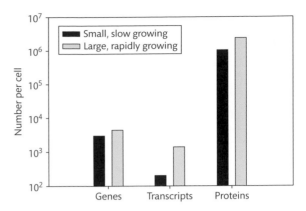

Figure 2.4 Number of genes, transcripts (mRNA), and proteins for a typical bacterium growing slowly in a natural environment with low organic carbon concentrations, and for a bacterium growing rapidly in the laboratory with high concentrations. Data from Table 2.4 and Moran et al. (2013).

heterotrophic organisms (they are "homeostatic" in this case), although there are important exceptions (Persson et al., 2010). In contrast, autotrophs are generally more flexible ("plastic") and can vary their composition more so than heterotrophs.

Relative nucleic acid content changes more as the growth rate increases than does protein, because of changes in both DNA and RNA (Figure 2.5). The relative contribution of DNA to total cell mass decreases, because cell mass increases while genome size does not change with growth rate. The biggest change, however, is the large increase in RNA with increasing growth rate. RNA also increases with growth rate in eukaryotic microbes, including protists and fungi (ter Kuile and Bonilla, 1999; Grimmett et al., 2013) and even in larger organisms like mangrove trees (Reef et al., 2010). As already discussed, rRNA scales with growth because fast-growing cells need more rRNA for more ribosomes and higher rates of protein synthesis. Indeed, fast-growing bacteria can have a thousand-fold more ribosomes than slow-growing bacteria. Rather than the 10,000 or more ribosomes of a fast growing bacterium, slow growing microbes have only the bare minimum. A bacterium in an aquifer has only 42 (Luef et al., 2015), and an archaeon growing slowly in an acid mine drainage community has 92 ribosomes (Comolli et al., 2008). Most ribosomes are involved in protein synthesis (80% in *E. coli*), and rates of protein synthesis

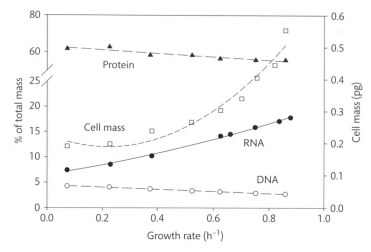

Figure 2.5 Variation in macromolecules (% of total cell mass) and total mass as a function of growth rate for a heterotrophic bacterium (*Klebsiella aerogenes*) growing in pure culture in the laboratory. Data from Mandelstam et al. (1982).

per ribosome do not change. Given the high costs of synthesizing ribosomes, a cell will have only the ribosomes needed for protein synthesis. The increase in ribosome numbers leads to more space within the cell taken up by ribosomes, which in turn leads to an overall increase in cell size with growth rate (Figure 2.5). Ribosomes account not only for a large fraction of total RNA (80%, as mentioned before), but also of dry weight (20–40%).

Because of the connection with P-rich ribosomes and growth, C:P ratios can vary with growth rate. Under P-limitation, the C:P ratio varies from 116:1 to 869:1 for some freshwater bacteria (Godwin and Cotner, 2015). This effect of growth rate has been seen for many other microbes and other organisms (Elser et al., 2003). The "coupling" or covariation among growth, RNA content, and biomass P levels is at the heart of the "growth rate hypothesis," which posits that the relative phosphorus content of organisms increases with growth rate due to higher numbers of ribosomes. Other P-rich macromolecules may also scale with growth and cell size for some microbes (Garcia et al., 2016). Regardless of the macromolecule, C:P ratios can vary greatly, not only for bacteria but other organisms as well. The hypothesis has been supported by work in environments with low P availability for bacteria, algae, and other protists. The coupling among growth and biomass P levels becomes less tight, however, when growth is set by nitrogen.

Variation in C:N and C:P ratios complicates biogeochemical models that assume constant ratios. But this variation reflects fundamental changes in the biochemical composition of microbes responding to environmental conditions. It can provide unique insights into the controls of microbial growth and their contribution to biogeochemical processes.

Architecture of a microbial cell

Arguably the most important physical characteristic of a microbial cell is its size, as already alluded to in Chapter 1. Cell size of many bacteria and other microbes varies with growth rate and other growth conditions. For example, protists get smaller with higher temperatures, perhaps to compensate for the decrease in gas solubility that would compromise O_2 and CO_2 exchange (Atkinson et al., 2003). However, variation in cell size within one type of microbe is small compared to the variation between all microbes. The diameter of microbes ranges at least 100-fold, from about 0.5 μm for bacteria to over 50 μm for some eukaryotes. One of the smallest free-living bacteria, found in an aquifer, is only about 0.3 μm in diameter (Luef et al., 2015), while candidates for the largest protists include slime molds and a type of foraminifera, *Syringammina fragilissima*, both of which can be about 20–30 cm in diameter (Hughes and Gooday, 2004). The diameter of a fungal cell

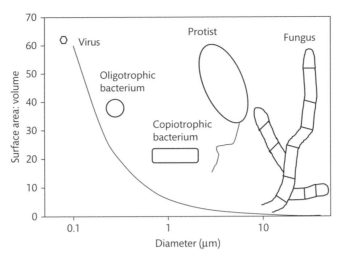

Figure 2.6 Surface area to volume ratio as a function of organism size. Because surface area depends on r^2, while volume scales with r^3, where r is the radius of a cell, the ratio decreases as a function of $1/r$. A bacterium growing in oligotrophic environments with very low nutrients may be half or less than the size of a copiotrophic bacterium living with high nutrient concentrations. Protists can vary from a micron to tens of microns in diameter. The cell diameter of a fungal cell is 2–10 μm, but several cells may be linked in a hypha stretching several meters.

is 2-10 μm, but a string of fungal cells making up "hyphae," which is a common morphology for these organisms, can stretch for meters in length. Some bacteria, cyanobacteria, and algae form chains superficially similar to hyphae, or they bundle together in colonies consisting of several cells arranged in two- or three-dimensional shapes, occasionally reaching sizes big enough to see without the aid of a microscope. Formation of chains or colonies and cell size in general have many implications for predation, competition, and other ecological interactions, as mentioned already in Chapter 1. Some of these interactions are affected by an important aspect of cell size, the surface area to volume ratio (Figure 2.6). This ratio has implications for competition for dissolved material and for thinking about biochemical composition of a microbe.

Membranes of microbes and active transport
All cells, whether microbial or those in macroscopic organisms, must have a barrier, a membrane that separates cellular components in the cytoplasm from the outside environment. This cell membrane, which is about 8 nm thick, keeps cytoplasmic components from leaking out, while preventing the entry of unwanted chemicals from the environment. The overall structure of membranes is

remarkably similar for bacteria and eukaryotes. In both bacteria and eukaryotes, the glycerol and hydrocarbon chain are linked by an ester bond (Figure 2.7), whereas in archaea, the two are linked by an ether bond. In archaea, these hydrocarbon chains are often highly branched with saturated isoprenes and complete rings. Ether lipids are more stable at high temperatures than ester lipids and may give archaea a selective advantage in hot environments (Valentine, 2007).

Some small hydrophobic molecules and gases may pass the lipid bilayers, but hydrophilic or charged compounds cannot. Ammonia (NH_3), for example, readily diffuses across membranes, but ammonium $\left(NH_4^+\right)$ does not. Very few compounds needed by microbes for growth are both small and without a charge. To facilitate the transport of molecules across membranes, all cells have membrane proteins that span the phospholipid bilayer. Some of these membrane proteins are non-specific "porins" that allow into the cell all compounds below a particular size. Several transport proteins are designed to transport specific compounds across the membrane. If membranes consisted only of phospholipids bilayers, a cell would soon starve and die.

Microbes can rely on diffusion to bring only a few compounds into the cell. For most compounds, concentrations

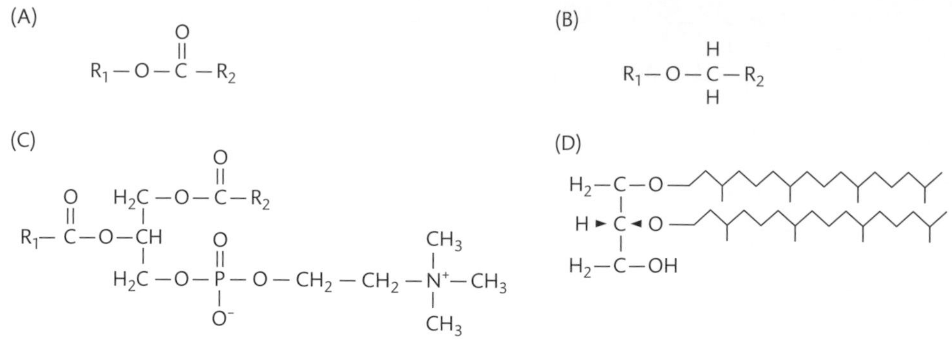

(A)

$$R_1 - O - \overset{\overset{\displaystyle O}{\|}}{C} - R_2$$

(B)

$$R_1 - O - \overset{\overset{\displaystyle H}{|}}{\underset{\underset{\displaystyle H}{|}}{C}} - R_2$$

(C)

$$R_1 - \overset{\overset{\displaystyle O}{\|}}{C} - O - \overset{\overset{\displaystyle H_2C - O - \overset{\overset{\displaystyle O}{\|}}{C} - R_2}{|}}{\underset{\underset{\displaystyle H_2C - O - \overset{\overset{\displaystyle O}{\|}}{P} - O - CH_2 - CH_2 - \overset{\overset{\displaystyle CH_3}{|}}{\underset{\underset{\displaystyle CH_3}{|}}{N^+}} - CH_3}{|}}{CH}}$$

(D)

$$\begin{aligned} &H_2 - C - O \sim \\ &H \blacktriangleright C \blacktriangleleft O \sim \\ &H_2 - C - OH \end{aligned}$$

Figure 2.7 Types of lipids found in microbes. (A) General structure for ester-linked lipids, which are found in bacteria and eukaryotes. (B) General structure for ether-linked lipids, which are found in archaea. (C) An ester-linked lipid (phosphatidylcholine); and (D) an ether-linked lipid. The number of carbon atoms and bond types represented by R$_1$ and R$_2$ vary among microbes.

are higher inside the cell than outside, and diffusion will not work. For these compounds, cells must use energy-requiring, active transport systems to transport the compound against the concentration gradient, from outside the cell to inside. The form of the energy driving the system defines the type of active transport. Simple transport relies on the proton motive force, whereas transport by an ABC (ATP binding cassettes) system is fueled by ATP hydrolysis. Transport by ABC systems may be particularly relevant for bacteria in nature, because this transport mechanism is able to bring in compounds or substrates found in very low concentrations in the external environment. In nearly all cases, several membrane and often cytoplasmic proteins are involved in the transport process. These transport systems are specific for a particular compound (glucose, for example) or a class of related compounds (branched-chain amino acids, for example). Many microbes have more than one system for a particular compound that differ in affinity and energetic costs. Although the transport systems seem redundant, microbes can switch among systems, depending on concentrations, to maximize transport while minimizing costs.

Cell walls in prokaryotes and eukaryotes

In addition to membranes, many cells have a wall that confers a more rigid structure and shape to the cell than is possible with only a membrane. As the word implies, cell walls offer some protection while also helping to prevent the cell from breaking apart due to self-generated

Table 2.5 Cell wall constituents of some microbes.

Microbe	Cell wall	Main constituents
Bacteria	Peptidoglycan	N-acetylmuramic acid, N-acetylglucosamine, amino acids
Archaea	Protein, pseudomurein	Amino acids, N-acetylglucosamine, N-acetyltalosaminuronic acid
Some algae	Cellulose	Glucose
Diatoms	Silica	Silicate
Fungi	Chitin	N-acetylglucosamine

turgor pressure. In bacteria, for example, the concentration of chemicals dissolved in the cytoplasm creates a turgor pressure of about 2 atmospheres, equivalent to that of an automobile tire. Higher plants, fungi, and most prokaryotes have cell walls, whereas animals and some protists do not.

Cell walls vary greatly in eukaryotic microbes (Table 2.5). Yeasts and fungi have chitin, a β 1, 4-linked polymer of N-acetylglucosamine, in their cell walls. Chitin is also found in insects and crustaceans. Similar to higher plants, the cell wall of some algae, like dinoflagellates, a complex group of heterotrophic and autotrophic protists (Chapter 9), is composed of cellulose, a β 1,4-linked polymer of glucose, whereas the polysaccharides of other algal cell walls consist of other sugars. Diatoms, another algal group important in freshwaters, coastal oceans, and soils, are encased in a glass house or "frustule," consisting mostly of silica. The composition of the cell wall of many other protists is not known.

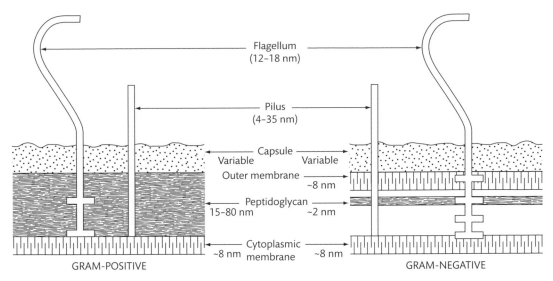

Figure 2.8 The two organizations for cell wall and membranes of bacteria. Taken from Neidhardt et al. (1990) and used with permission of the publisher.

Peptidoglycan, or more specifically murein, is the main component of the cell wall in bacteria. The backbone of peptidoglycan is a polysaccharide (glycan), consisting of another β 1,4-linked polymer, this time N-acetylglucosamine alternating with N-acetylmuramic acid. (Note that this linkage (β 1,4) is found in the three major polysaccharides of cell walls and exoskeletons: cellulose, chitin, and peptidoglycan.) The glycan strands of peptidoglycan are cross-linked by peptide chains of a few amino acids, usually L-alanine, D-alanine, D-glutamic acid, and either lysine or diaminopimelic acid. D-amino acids are unusual because they are not used in proteins. Variation in these few amino acids making up the peptide chain leads to subtle changes in structure, which is the main reason why there are some one hundred types of peptidoglycan.

There are two basic designs for the organization of the wall and membranes found in bacteria: Gram-positive and Gram-negative (Figure 2.8). The names refer to how bacteria react to a stain (the Gram stain) devised by the Danish physician H. C. J. Gram (1853–1938). Later work found that the reaction of a bacterium to the Gram stain depended on its cell wall and membrane architecture. Gram-positive bacteria have a thick cell wall consisting of peptidoglycan and no other external membrane, although they may have capsules and other less well-defined extracellular coverings. Gram-negative bacteria also have these

extracellular coverings and a cell wall, although it is thinner than that of Gram-positive bacteria. In addition, Gram-negative bacteria have an outer membrane containing lipopolysaccharide (LPS), which is found only in Gram-negative bacteria. The space between this outer membrane and the cytoplasmic membrane is the periplasm or periplasmic space.

Cyanobacteria, which are often abundant and important primary producers (Chapter 6), have a Gram-negative-type cell wall and membrane, with some important differences (Hoiczyk and Hansel, 2000). The peptidoglycan layer of cyanobacteria is much thicker than is typical of Gram-negative bacteria, and the cross-linking between peptidoglycan chains is higher. Like Gram-negative bacteria, cyanobacteria have LPS and an oligosaccharide, the O-antigen. The latter may contribute to the toxicity of some cyanobacterial strains, as it does in Gram-negative pathogens. In addition to LPS and the O-antigen, the outer membrane of cyanobacteria has other constituents not found in typical Gram-negative bacteria, including carotenoids and unusual fatty acids, such as β-hydroxypalmitic acid.

Neither LPS nor muramic acid is present in the cell walls and membranes of archaea. Instead, these prokaryotes have a variety of cell wall types (Mayer, 1999). Some have pseudomurein or pseudopeptidoglycan, an acidic heteropolysaccharide sharing some characteristics with peptidoglycan in bacteria. Pseudomurein has N-acetylglucosamine, but it has N-acetyltalosaminuronic acid instead of N-acetylmuramic acid. Other archaea, such as many halophiles and methanogens, have a protein or glycoprotein coat (S-layer) and no pseudomurein. Still other archaea do not have any cell wall and survive with only a membrane separating the cytoplasm from the external environment.

In summary, if present, the cell walls of different microbes are made of different components, such as polysaccharides, proteins, or mixtures of both. The composition of a cell wall or its absence gives some insights into the environmental pressures facing the microbe. Information about cell wall composition is also used by microbial ecologists and biogeochemists to explore several questions about microbial communities and organic material in natural environments, as will be discussed further.

Components of microbial cells as biomarkers

Compounds known to be specific for microbes can be used to estimate microbial biomass and to understand the sources of organic material found in natural environments (Bianchi and Canuel, 2011). These compounds are called "biomarkers." Nearly all of these biomarkers are associated with microbial membranes or walls. The biomarkers used to estimate bacterial biomass include LPS and muramic acid, both unique to bacteria, as mentioned earlier. Data on muramic acid were important in the early studies of bacteria in sediments (Moriarty, 1977;

King and White, 1977), and LPS concentrations were used to confirm that the cells counted by direct microscopy in the oceans were in fact bacteria (Watson et al., 1977). The LPS and direct count estimates are similar because Gram-positive bacteria, which do not have LPS, are not abundant in marine waters (Chapter 4). Other compounds used for biomass estimates are phospholipid-linked fatty acids (PLFA).

Membrane and cell wall components have other uses in microbial ecology and biogeochemistry. In addition to estimating biomass, PLFAs are used to determine the presence or absence of some microbial taxa, such as sulfate reducers (the PLFA i17:1 and 10Me16:0) and *Actinomycetes* (10Me17:0 and 10Me18:0). Other PLFAs are found in eukaryotic algae (20:5 ω 3). Similarly, archaea can be traced by their unique ether lipids. In geochemistry, the number of cyclopentane rings in archaeal membrane lipids (the TEX_{86} proxy) is used to estimate temperature over geological timescales (Qin et al., 2015).

The biomarker-based approaches can be combined with ^{13}C stable isotope analysis. [13]C occurs naturally in the biosphere, albeit only at about 1% of total C with ^{12}C accounting for nearly all of the rest. The natural abundance of ^{13}C can be used to deduce sources of organic carbon used by an organism because the relative amounts of ^{13}C, expressed as δ ^{13}C, vary among primary producers (Figure 2.9), depending on how much each autotroph favors ("discriminates") CO_2 with the lighter ^{12}C over the heavier [13]C. (The symbol "‰" is interpreted as parts per thousand, analogous to "%" for parts per

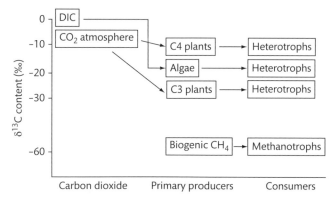

Figure 2.9 The relative abundance of ^{13}C in the atmosphere, dissolved inorganic carbon (DIC), methane, plants, and microbes. Biogenic methane comes from methanogenic archaea, whereas thermogenic methane, which is produced by geothermal heating of organic material, has a much higher (less negative) δ ^{13}C value than biogenic methane. Adapted from Boschker and Middelburg (2002).

hundred or percent. The numbers are negative when there is less ^{13}C compared to a standard.) The δ ^{13}C of C4 plants found on land is about −14‰ whereas it is −21‰ for algae and −27‰ for C3 land plants. Members of a food chain will have δ ^{13}C values similar, differing by only 1‰ or less, to the value of the primary producer at the base of the food chain; the values change by about 1‰ for each step in the food chain. In short, "you are what you eat." For example, a herbivore feeding on C4 plants would have a δ ^{13}C value of about −15‰. The δ ^{13}C of a carnivore preying on the herbivore would be about −16‰.

The same general rules can apply to the ^{13}C value of biomarkers and other individual organic compounds, an approach called "compound-specific isotope analysis." The ^{13}C value of these compounds may differ from the value of the entire organism or carbon source (the "bulk" value) because of fractionation during biosynthesis of the compound. For example, lipids are depleted in ^{13}C by 2 to 6‰ relative to bulk ^{13}C values; if the bulk δ ^{13}C is −26‰, then the δ ^{13}C of lipids may be −28 to −32‰. An example using this technique is a study that examined the flow of carbon through fungi and bacteria in forest soil food webs (Pollierer et al., 2012). Data on δ ^{13}C of lipids indicated that root-derived organic carbon enters into the soil food web via root-associated fungi (Chapter 14), although bacteria contribute more than previously thought.

Extracellular structures

In addition to cell membranes and walls, many microbes have other structures and macromolecules that are attached to the cell but extend beyond the outer cell membrane or wall. These extracellular structures and macromolecules serve a great variety of functions, from propelling microbes through their habitat to keeping them stuck in one place.

Extracellular polymers of microbes

Depending on the environment and growth state, microbes can secrete a complex suite of extracellular polymers, often dominated by polysaccharides. For some bacteria, these polymers can be organized to form a defined layer, the capsule, around the cell

(Figure 2.10A), while other bacteria and other microbes attached to surfaces produce less coherent and more extensive networks of polymers (Figure 2.10B). Some terms used to describe these extracellular polymers include glycocalyx, extracellular polysaccharides, and extracellular polymeric substances (EPS). Sometimes the simple word "slime" is most appropriate. Because of carbon limitation, free-living bacteria in natural environments probably do not form thick capsules, but bacteria attached to organic and inorganic surfaces often produce and are embedded in extracellular polymers.

Regardless of the name, these polymers have several potential functions for microbes. They may serve as a carbon source for heterotrophic microbes when environmental conditions change from being carbon-replete

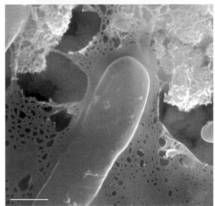

Figure 2.10 Bacteria surrounded by extracellular polymers. Photomicrograph of bacteria surrounded by capsules after negative staining (the capsules are not stained) by India Ink (A). Each cell is about 5 μm long. Taken from Hoffmaster et al. (2004), used with permission. Copyright (2004) National Academy of Sciences, USA. The bacterium *Shewanella oneidensis* surrounded by extracellular polymers (B). The scale bar is 0.5 μm. Picture provided by Alice Dohnalkova from Dohnalkova et al. (2011) and used with permission from the authors and publisher.

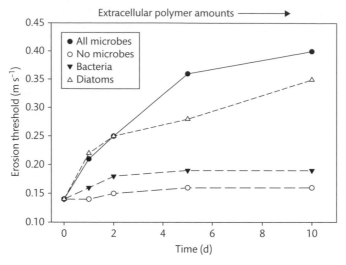

Figure 2.11 Effect of microbes on sediment stability, measured by the current velocity required to disrupt sediments without any microbes, with just bacteria or diatoms, or with all microbes. Much of the effect is due to extracellular polysaccharides (EPS) and other polymers produced by the microbes. The amount of these polymers increases with time, leading to higher erosion thresholds and more stable sediments. Data from Lundkvist et al. (2007).

to carbon-limited. The polymers help to glue attached microbes to surfaces, while also providing some protection against ingestion by grazers and perhaps lysis by viruses. Complex polymers are important components of the symbiosis between bacteria and root nodules of legumes (Chapter 14).

Extracellular polymers can protect pathogenic bacteria and limit the effectiveness of antibiotics in controlling infections. The polymers themselves can contribute to diseases caused by microbes, as in the case of the bacterium *Pseudomonas aeruginosa* and cystic fibrosis. Eukaryotic microbes can also secrete extracellular polymers, and their function is probably quite similar to that for prokaryotic microbes. As with prokaryotic microbes, these extracellular polymers are often dominated by carbohydrates, although protein and other macromolecules can be abundant (Grabowski et al., 2011). The carbohydrate composition of the polymers produced by algae varies greatly among different species, with glucose usually being the most common monosaccharide (Biersmith and Benner, 1998). Benthic diatoms can produce large quantities of extracellular polysaccharides, whereas some planktonic diatoms secrete chitin strands extending several cell lengths away from the diatom. These strands may help protect the diatom from predation or keep it afloat in the water

column. However, these extracellular polymers can also lead to aggregation and sinking of diatoms from the water column.

In addition to being important to microbes, extracellular polymers are important to other organisms and the environment. While polymers can protect microbes from predation by some organisms, metazoans that eat detritus and associated microbes ("detritivores") ingest the extracellular polymers along with the microbes. Polymers released by microbes also contribute to aggregates in aquatic habitats and soils. This organic material is so important to soil quality that artificial polymers have been added to soils to retard erosion and promote retention of water and nutrients. In aquatic habitats, extracellular polymers produced by microbes contribute to the physical stability of sediments even when those microbes are not in thick biofilms (Chapter 3) (Malarkey et al., 2015). Benthic diatoms may have an especially important role in stabilizing sediments, but extracellular polymers from bacteria also contribute (Figure 2.11).

Flagella and cilia; fimbriae and pili
Many microbes are motile and can swim quite rapidly through aqueous environments, including the aqueous

microhabitats of soils. Many motile prokaryotes and eukaryotic microbes are propelled by hair-like structures, the flagella, sticking out from the cell. Some microbes have one or more flagella attached to one end or pole of a microbe (polar flagella), while others have flagella sticking out from all sides (peritrichous flagella). A microbe may have several short versions of flagella or "cilia," which are also involved in motility. Some microbes can move without any flagella by gliding along solid surfaces. These microbes include pennate diatoms (shaped like a cigar), which glide by excreting polymers through a hole (the raphe) in the frustule touching the surface. Some filamentous cyanobacteria also glide by rotating and flexing of cells linked together in the filament. Other bacteria, such as myxobacteria and some members of the *Bacteroidetes* phylum, are well known for gliding, but the mechanisms remain unclear. One member of the *Bacteroidetes*, *Flavobacterium johnsoniae*, appears to glide along surfaces thanks to 5 nm-wide tufts of filaments attached to the outer membrane of this bacterium (Liu et al., 2007).

Other appendages sticking out from bacterial cells include fimbriae and pili (the two terms are often used interchangeably), which are composed of protein-like flagella but are shorter and are not involved in motility (Figure 2.12). Fimbriae and pili can mediate attachment to inert surfaces or to other cells. For example, the bacterium *Bordetella pertussis* uses fimbriae to bind to glycoconjugates of human epithelial cells, an important step leading to the disease whooping cough. One type of pili, the sex pili, is involved in the transfer of DNA from one cell to another, a process called "conjugation." The work on the ecological aspects of fimbriae and pili has focused on detecting their genes in natural microbial communities (Chen et al., 2015b). Bacteria in oligotrophic environments like the open oceans do not seem to have many of these genes (Yooseph et al., 2010).

Still another type of extracellular appendages are "nanowires," used by some anaerobic bacteria in redox reactions (Shi et al., 2016). All other anaerobic bacteria as well as aerobic ones carry out redox reactions intracellularly by transporting the necessary redox compounds into the cell. Some redox compounds, however, are insoluble and cannot cross cell membranes. Examples include minerals containing oxidized iron or manganese

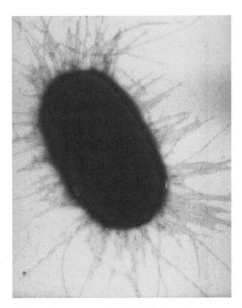

Figure 2.12 A bacterium (*E. coli*) with fimbriae designed to anchor the cell to the mucous membrane of mammals. From Gross (2006) and used under the the the terms of the Creative Commons Attribution License.

in the form of Fe(III) or Mn(IV). To use these, bacteria like *Geobacter metallireducens* and *Shewanella oneidensis* are thought to transfer electrons produced during organic material degradation to Fe(III) or Mn(IV) minerals along protein filaments, the nanowires, sticking out from the cell. Electrons can also be transferred between cells from different species or even from different domains, such as between *Geobacter* species and methanogenic archaea.

Another fascinating example of long-distance electron transfer is cable bacteria (Lovley, 2017; Pfeffer et al., 2012). Bacteria belonging to the family *Desulfobulbaceae* form long chains of cells lined up end to end, extending thousands of cells in length from oxic zones to anoxic zones rich in reduced sulfur. Electrons from the oxidation of the reduced sulfur appear to travel along ridges that run along the length of the filament to cells in the oxic zones bathed in oxygen. There may be interspecies electron transfer from chemolithoautotrophic *Epsilon*- and *Gammaproteobacteria* to the cable bacteria. Nanowired and cable bacteria are likely to be important in degrading organic material in anoxic environments and at the interface between oxic and anoxic habitats (Chapter 11).

Summary

1. The elemental and biochemical composition of microbes has large impacts on the chemistry of natural environments. Information about microbial composition can be used to examine several biogeochemical processes.

2. The elemental and biochemical composition of microbes varies with growth conditions and the type of microbe. This variation can be used to deduce microbial growth conditions.

3. The cell walls of prokaryotes, fungi, and protists differ greatly in composition, but they often contain β 1,4-linked polymers. The cell wall of bacteria is made of peptidoglycan with unusual components, including muramic acid and D-amino acids.

4. The membranes of bacteria and eukaryotes are similar and consist of ester-linked lipids. In contrast, archaea have ether-linked lipids. All membranes have transport proteins for bringing desired compounds into the cell.

5. Compounds in cell walls and membranes can serve as biomarkers for estimating biomass and for tracing carbon sources.

6. Microbes can release extracellular polymers dominated by polysaccharides. These polymers help microbes attach to surfaces and impact the physical environment surrounding microbes.

7. Microbes have various hair-like structures extending out from their cell surface that are used for motility, attachment, and transfer of DNA from one cell to another. Other bacteria use nanowires to transfer electrons to insoluble minerals like Fe(III) oxides.

CHAPTER 3

The physical-chemical environment of microbes

We are familiar with many aspects of the environments where large organisms live and interact. We are accustomed to seeing deer on land and fish in water, lush forests where rain is plentiful, and cactuses where it is not. Many of these familiar environmental factors also affect microbes and microbial processes. For these, the same mechanisms and equations that describe effects on macroscopic organisms are applicable to microbes. Other properties, however, are unique to life at the micron scale. For these properties, our intuition based on macroscopic life breaks down, and we need new ways of thinking about how microscopic organisms interact with their environment.

Some important physical-chemical properties

First, however, let's consider familiar environmental properties well known to affect all organisms and introduce some basic terms and principles about the physical-chemical environment of microbes. We will re-visit many of these properties in other chapters.

Water

The search for life on other planets often focuses on water, because we know that on Earth, where there is water there is life. All cells, whether microbial or metazoan, are about 70% water by weight (Chapter 2), and all organisms require water for growth. Some microbes can survive without water by forming a resting stage, called spores or cysts, depending on the microbial type, but none can grow while completely desiccated. Even in soils, microbes need an aqueous environment at the micron scale to metabolize and grow. Water has a huge impact on the types, abundance, and growth of microbes in soils.

A pioneer in microbial ecology, T. D. Brock (1926–), pointed out that the unusual properties of water explain its "admirable utility as a menstruum for the evolution of living creatures" (Brock, 1966). Among its 63 anomalous properties (Kivelson and Tarjus, 2001), water is polar, has a high dielectric constant, and is a small molecule, making it an excellent solvent for many biologically important compounds. It is also a viscous liquid, an important feature for understanding life at the scale inhabited by microbes. Water is much more viscous than air. This obvious difference between water and air has many fundamental implications for how life at the microbial scale differs from how we and other macroscopic organisms experience it. Water is the medium of microbes even if present only as a thin film in soils.

Temperature

Microbes can live or at least survive temperatures from well below freezing to well above 100 °C. At the lower end, microbes are found in Antarctica, where temperatures dip down routinely to −60 °C, with the record being about −90 °C. While some metabolism may continue in solid ice, liquid water is needed for any substantial microbial activity, which is possible even at very low temperatures. One example is the brine channels of Arctic sea ice where temperatures can be −20 °C. Water

Processes in Microbial Ecology. Second Edition. David L. Kirchman. Oxford University Press (2018). © David L. Kirchman 2018.
DOI 10.1093/oso/9780198789406.001.0001

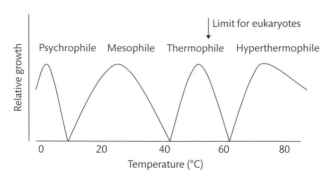

Figure 3.1 Terms used to describe organisms with different temperature ranges.

Table 3.1 Terms used to describe organisms growing under various environmental conditions.

Environmental property	Organism	Growth conditions
Temperature	Psychrophile	<15 °C
	Mesophile	15–40 °C
	Thermophile	45–80 °C
	Hyperthermophile	>80 °C
pH	Acidophile	pH < 5
	Neutrophile	pH = 6–8
	Alkaliphile	pH > 8
Salt	Mild halophile	1–6% NaCl
	Moderate halophile	6–15% NaCl
	Extreme halophile	>15% NaCl
Pressure	Piezotolerant	Survival but no growth above atmospheric pressure
	Piezophile	Growth under moderate pressure (10–80 mPa)
	Hyperpiezophile	Growth under high pressure (>80 mPa)

can remain liquid at this temperature only because salinity is high, reaching 20% or nearly ten-fold higher than that of seawater, as mentioned in Chapter 1. At the other end of the thermometer, microbes proliferate in hot springs on land and in hydrothermal vents found at the bottom of some oceans. Microbes have been found in >150 °C waters, coming out of vents (Chapter 14), but it is not clear whether they are metabolically active; they may come from cooler waters mixed into the hot hydrothermal water. Currently, the temperature record of 121 °C is held by an iron-reducing bacterium isolated from the Juan de Fuca Ridge in the Northeast Pacific Ocean (Harrison et al., 2013). Water can be liquid at temperatures >100 °C only because of the high pressure in the deep ocean. It has been hypothesized that liquid water, rather than temperature per se, sets the limit of life. Figure 3.1 summarizes some of the terms used to describe the microbes growing within various temperature ranges.

Eukaryotic microbes grow in very cold environments, but not in extremely hot ones. Diatoms, other eukaryotic algae, and heterotrophic protists live in the cold brine channels of sea ice, and fungi are commonly isolated from Antarctic soil. Light, more than temperature, limits the growth of phototrophic microbes in sea ice, and the grazing of protists (Chapter 9) on bacteria and other microbes may be physically inhibited by the small confines of brine channels. At the other extreme, the upper limit of eukaryotes is about 65 °C, well below the 121 °C record of prokaryotes. The maximum temperature for growth by eukaryotic phototrophs is even lower. The hottest water in which a eukaryotic alga (*Cyanidium caldarium*) can grow is 55 °C, whereas prokaryotic phototrophs can thrive in waters >70 °C.

Table 3.1 lists the terms for describing the temperature preference of microbes and analogous terms for other environmental properties. Many of these terms end with "phile," which comes from the Greek meaning "loving." A psychrophile, for example, grows below 10 °C, whereas a thermophile is best suited for temperatures above 40 °C. Some thermophilic bacteria appear to have diverged earlier than other bacteria in evolution, suggesting that life arose in hot environments such as hydrothermal vents. Hyperthermophiles, including many archaea, grow at temperatures above about 60 °C. None of these organisms grows well at "normal" temperatures, at least normal by human standards.

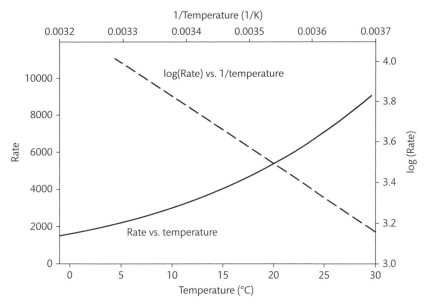

Figure 3.2 Example of how rates vary with temperature according to Arrhenius. The units for the rate are arbitrary. The activation energy for this reaction was set at 40 kJ mol^{-1}, which is roughly equivalent to a Q_{10} of 2 near 20 °C. The top axis is the inverse of temperature, expressed as Kelvin, the units required in the Arrhenius equation.

Of all environmental parameters, temperature has one of the most profound effects on microbial activity, because of its immediate impact on enzymatic reactions being carried out by microbes and on abiotic reactions in the microbial environment. The rate of all chemical reactions increases with temperature following a well-defined rule encapsulated in the Arrhenius equation. This equation describes how a reaction rate (k, with units of per time) varies as a function of temperature (T, expressed in Kelvin):

$$k = Ae^{-E/RT} \qquad (3.1)$$

where R is the gas constant (8.31 kJ mol^{-1} K^{-1}), A is an arbitrary constant, and E is the activation energy, an important defining characteristic of the reaction. The Arrhenius equation predicts that reaction rates increase exponentially with temperature (Figure 3.2).

Although it is accurate for many reactions, microbes sometimes do not follow the Arrhenius equation, even ignoring temperatures exceeding the minimum and maximum limits on growth. One alternative model couples the Arrhenius equation with the equation from Michaelis–Menten kinetics (Chapter 9) that describes how rates vary as a function of substrate concentrations (Davidson et al., 2012). Another approach, the macro-molecular rate theory, builds on the Arrhenius equation and takes into account decreases in enzymatic activity, such as due to denaturation of the enzyme, as temperatures exceed an optimum (Alster et al., 2016). Other approaches are more empirical. In soils, microbial rates can vary in experiments as the square root of temperature. This transformation is simpler than the Arrhenius equation but it lacks a mechanistic foundation. The Arrhenius equation can be derived from first principles governing how molecules interact as a function of temperature.

Another common expression for temperature effects is Q_{10}, the factor by which a rate increases with a 10 °C increase in temperature. Many reactions in biology have a Q_{10} of 2, an important number to remember. For example, a reaction with a rate of 1 mol m^{-2} h^{-1} at 15 °C would be expected to be 2 mol m^{-2} h^{-1} at 25 °C if Q_{10} = 2. Experimentally, Q_{10} is often measured at temperature intervals other than 10 °C. In this case, the Q_{10} can be calculated using the following equation:

$$Q_{10} = (r2 / r1)^{(10/T2-T1)} \qquad (3.2)$$

> **Box 3.1 Arrhenius and the greenhouse effect**
>
> Svante Arrhenius (1859–1927) was awarded the Nobel Prize in 1903 for his work in electrochemistry, his main research field. However, one of Arrhenius's first scientific contributions was to understand the effect of greenhouse gases on our climate. He published a paper in 1896 entitled, *On the Influence of Carbonic Acid in the Air upon the Temperature of the Ground*, in which he argued that global temperatures would increase by 5 °C if CO_2 concentrations increased by two- to three-fold. Arrhenius was trying to explain the comings and goings of ice ages, but his calculations are relevant to understanding climate change issues. Arrhenius's estimate for the sensitivity of the climate system to CO_2 is remarkably similar to current estimates, derived by much more complicated modeling studies and field observations. From the perspective of his home in Uppsala, Sweden, Arrhenius thought a bit of global warming would be a good thing.

where r1 and r2 are the rates measured at two temperatures, T1 and T2, respectively.

Q_{10} is a convenient way to express temperature effects, but it lacks a mechanistic foundation. Q_{10} may not be constant over a large range in temperature, even if the activation energy is constant, implying that the effect of temperature on a process does not vary over the range of temperatures.

While the effect of temperature on a single reaction in the laboratory is easy to understand, temperature effects are often more complicated to untangle in natural environments. One complication is that a process of interest may be the net result of two different, opposing processes with different relationships with temperature. An important example is the relationship between primary production, which fixes CO_2 into organic material, and heterotrophy, which oxidizes that organic material back to CO_2. Any differences in how temperature affects primary producers versus heterotrophic organisms would have huge implications for understanding the impact of climate change on the carbon cycle and the rest of the biosphere. Traditionally, respiration is considered to be

more sensitive than is primary production to temperature (Davidson and Janssens, 2006), and some models examining climate change assume that primary production will not be directly affected by warmer temperatures but heterotrophy and respiration in soils will. Some studies conclude, however, that there is no difference in how temperature affects metabolic rates of heterotrophs and primary producers when cell size is taken into account (Gillooly et al., 2001), but other studies find differences between the two processes (Figure 3.3).

There are even more complications. The temperature effect on seemingly one process, such as organic matter mineralization, may in fact be a net result of several processes. Temperature-stimulated mineralization of soil organic matter to CO_2 may be counter-balanced by the stabilization of that organic matter by other temperature-sensitive microbial processes (Bradford et al., 2016). Another complication is indirect effects of temperature. Higher temperatures can lead to more drought and lower terrestrial primary production (Zhao and Running, 2010), and a warmer surface layer of aquatic habitats could reduce nutrient inputs, resulting again in lower primary production (Falkowski and Oliver, 2007). Even though it has been examined since the days of Arrhenius in the nineteenth century, the temperature effect on microbe-driven processes remains an active research topic today.

pH

The pH has nearly as great an impact on microbes and their environment as temperature. Those microbes able to grow in a select pH range are described by terms analogous to those used for temperature. Acidophilic microbes grow in waters and soils with pH 1–3, while at the other extreme, alkaliphiles prefer a pH of 9–11. In addition to direct effects, pH can indirectly affect microbes and microbial processes via the way larger organisms react to pH and the dependency of chemical processes on pH. Extremes in pH can greatly reduce the number and diversity of larger organisms, if not exclude them completely. Chemical processes affected by pH include the chemical state of key elements and adsorption of key nutrients to solid surfaces.

The pH of most natural aquatic habitats is constant and within a pH unit or two of neutrality (pH = 7). The pH of marine waters is about 8, while many lakes have a

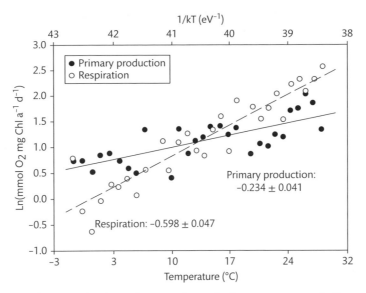

Figure 3.3 Response of gross primary production and community respiration to temperature in the oceans. The top axis of the graph gives values for 1/kT, used in the Arrhenius equation, where k is the Boltzmann constant, and T is temperature in Kelvin. Note that the slope of the regression line for primary production (solid) is much smaller than the slope of the line for respiration (dashed), indicating that primary production is less sensitive to temperature. The amount of variation explained by temperature is also lower for primary production (adjusted r^2 = 0.53 and 0.85 for primary production and respiration, respectively). Data taken from Regaudie-de-Gioux and Duarte (2012).

neutral pH or are slightly acidic. Non-marine geothermal waters are acidic (pH < 5), as are drainage waters from coal and metal mines, with devastating effects on neighboring environments. Rio Tinto in southwestern Spain is highly acidic (pH = 2.3) because its watershed has large iron and copper sulfide deposits which have been mined for millennia. Lakes can become acidic because of "acid rain" caused by pollution from upwind power plants (Box 3.2). Although large organisms cannot survive extremely low pH levels, some eukaryotic microbes can. Some bacteria, such as some that rely on iron oxidation for energy (see Chapter 13), can only flourish in low pH environments.

At the other extreme are alkaline aquatic environments (pH > 10), such as Mono Lake (California), the Great Salt Lake (Utah), and some lakes in the Rift Valley of Africa. These alkaline lakes have very high salt concentrations, ranging from 30 g liter^{-1}, or about the level of seawater, to >300 g liter^{-1}. Like acidic environments, the diversity of biological communities in alkaline lakes is low and consists of only a few metazoans, but potentially several types of microbes. Mono Lake, for example, has no

fish, but is famous for a species of a small brine shrimp (*Artemia monica*) that is food for several migrating birds. The brine shrimp feed on a productive phytoplankton community consisting of a few species of photoautotrophic eukaryotes and cyanobacteria.

The pH of soils varies naturally more than in aquatic habitats, with vast regions of continents being either acidic (pH < 5) or alkaline (pH > 10) (Hengl et al., 2014), depending on the type of soil and moisture content (Slessarev et al., 2016). Soil receiving high rainfall tends to be acidic due to the exchange of cations (Na$^+$, K$^+$, Ca^{2+}, and Mg^{2+}) with H$^+$. Acidic soils are found in the East Coast of the USA, the northeast of Canada, Scandinavia, and Russia. The soils of tropical rainforests in South America and Africa are also acidic. In contrast, semi-arid and arid conditions lead to soils that are alkaline. For that reason, soils in the arid regions of the world, such as many areas of the western USA and northern Africa, are alkaline. As in alkaline aquatic habitats, soils with high pH often but not always have high concentrations of the major cations, notably calcium (Ca^{2+}), along with magnesium (Mg^{2+}),

---

Box 3.2 Acid rain and ocean acidification

"Acid rain" is used in the popular press to describe rain contaminated by sulfur and nitrogen oxides from industrial activity. However, all rain is acidic, even if not affected by human activity. Because it is not well buffered, even pristine water in equilibrium with the atmosphere has a pH of 5.2, due to carbon dioxide and carbonic acid. Regardless, the industrial acids in rain certainly cause harm to freshwaters. A different problem with pH faces the oceans. Increasing carbon dioxide concentrations in the atmosphere have led to more acidity and a lower pH. Although the oceans will not become acidic (pH<7.0) according to even the most dire predictions, the drop in pH is projected to have several negative impacts on marine organisms.

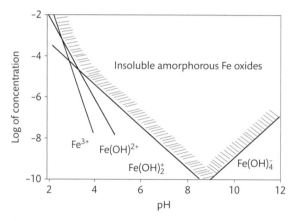

Figure 3.4 Solubility of iron as a function of pH. Iron is soluble (present as Fe^{3+} and other oxide species) at very low pH (pH < 3). At pH >3, iron is present mostly as solid (insoluble) amorphous iron oxides (the area about the hatches). Data taken from Stumm and Morgan (1981).

potassium (K^+), and sodium (Na^+). Many plants have difficulty growing in soils with high pH because it reduces the availability of critical nutrients. Overall, pH has a huge impact on the availability of nutrients for plants and on microbial diversity in soils (Chapter 4).

The pH also affects the chemical state of several important compounds and elements. For example, the most abundant form of iron in oxic environments, Fe(III), is

soluble in acidic environments (pH < 3) but forms insoluble iron oxides at the pH of most environments (Figure 3.4). The adsorption of essential nutrients such as phosphate and nitrate to soil and sediment particles is governed by pH, and is determined by the charge of cations in these particles. The charge of other compounds varies also as a function of pH. One example is ammonium, a key source of nitrogen for many microbes. The exchange between ammonia (NH_3) and ammonium (NH_4^+) is described as

$$NH_3 + H^+ \Leftrightarrow NH_4^+ \qquad (3.3).$$

Because the pKa of this reaction is 9.3, NH_4^+ predominates in most surface environments, except those with a high pH. The simple addition of a single proton changes an uncharged molecule (ammonia), which can easily pass through cell membranes, to a charged molecule (ammonium) that requires specialized transport mechanisms in order for it to be used by microbes. The effect of pH on ammonia versus ammonium has implications for understanding how a key reaction in the nitrogen cycle, nitrification (Chapter 12), may be affected by ocean acidification.

Salt and osmotic balance

The salt concentration of microbial environments varies from distilled water levels to brines near saturation (35% for NaCl). Microbes vary in their capacity to survive and grow in these environments. Halophilic microbes require at least some NaCl if not other salts for growth, whereas other microbes cannot survive with any appreciable salt. Salt curing preserves meat by inhibiting microbial growth, although some extreme halophiles can grow even under these conditions. Extreme halophiles, which include some interesting archaea, can dominate alkaline lakes and evaporation ponds, coloring the water a brilliant red with their pigments. Soils in arid environments such as in parts of Australia are naturally saline because deposition of salt from dust and precipitation exceeds its leaching and removal. Irrigation can increase the salinity of soil if drainage is inadequate.

The problem facing microbes as well as larger organisms is not salt per se, but the relative amount of water—more precisely, water activity—in the cell relative to the environment. With the exception of some extreme halophiles, water activity is lower and solute concentrations

Dimethylsulfoniopriopionate (DMSP): $(CH_3)_2S^+CH_2CH_2COO^-$

Glycerol: $C_3H_5(OH)_3$

Glycine betaine: $C_5H_{11}NO_2$

Glutamate: $C_5H_9NO_4$

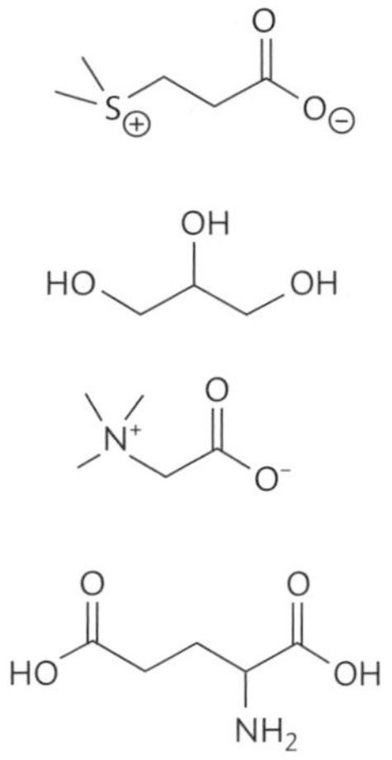

Figure 3.5 Examples of compatible organic solutes found in microbes.

are higher in a cell than in the external environment, resulting in the net flow of needed water into the cell. This gradient is relatively easy for cells to maintain in low salinity environments. However, as salinity increases and thus water activity decreases, cells face the problem of retaining water. To do so, they need to raise their internal solute concentrations by either pumping in inorganic ions (such as K^+) or by synthesizing organic solutes. Whether inorganic or organic, these solutes, called the "compatible solutes," must not disrupt normal cellular biochemical reactions. The compatible organic solutes include glycine betaine, proline, glutamate, glycerol, and dimethylsulfoniopropionate (DMSP); the importance of DMSP is discussed in Chapter 6. Figure 3.5 provides examples of organic compatible solutes.

There are advantages and disadvantages for cells using organic versus inorganic compatible solutes (Oren, 1999). Cells using inorganic solutes have to have enzymes and other proteins specially adapted to high salt concentrations, whereas these adaptations are not needed for cells using organic solutes, because they are either uncharged or zwitterionic at the physiological pH. Consequently, most microbes use organic compatible solutes, while only a few microbes, such as some extreme halophiles, use inorganic ones. The problem with organic solutes is that it is energetically expensive to make them. Energetics may explain why bacteria and archaea relying on low energy-yielding metabolisms, such as methanogenesis

(Chapter 11) and ammonia oxidation (Chapter 12), have not been isolated from high-saline environments. One organic solute, glycerol, can be synthesized cheaply but is used only by some eukaryotes, perhaps because membranes have to be modified to retain this small, uncharged molecule.

Oxygen and redox potential

All metazoans and nearly all eukaryotic microbes (except a few fungi and protists) require oxygen for survival and growth. Many bacteria and archaea are also obligate or strict aerobes, meaning they require oxygen. Many other prokaryotes, however, can grow in the absence of oxygen and either are facultative or strict anaerobes, depending on whether they can or cannot tolerate oxygen. Other chapters discuss the effects of oxygen on microbial community structure (Chapter 4) and the biogeochemical process in anoxic environments (Chapters 11 and 12).

Related to oxygen content is the redox potential of an environment, or the tendency of a solution to gain (positive potential) or lose (negative potential) electrons. Because oxygen is a strong oxidant and can attract electrons from more reduced compounds, oxic environments have a positive redox potential, while anoxic ones have a negative potential defined relative to hydrogen (H^+/H_2). The redox state of water and soils can be measured with a platinum electrode relative to the half-potential of hydrogen (H^+/H_2). The redox potential (E_h) for an individual redox reaction is defined by the Nernst equation:

$$E_h = E° - 0.05916 / n \times \log([\text{reductants}] / [\text{oxidants}]) \quad (3.4)$$

where $E°$ is the standard half-cell potential, n the number of electrons transferred, and [reductants] and [oxidants] are the concentrations of reduced and oxidized compounds, respectively.

Table 3.2 gives the oxidized and reduced forms of some compounds important to microbes and the biosphere. By definition, oxidized compounds can take on electrons and become more reduced, whereas the opposite is the case for reduced compounds. Oxidized compounds, most notably oxygen, are abundant in oxidizing environments (positive E_h), while reduced compounds are more common in a reducing environment (negative E_h).

Table 3.2 Some examples of oxidized and reduced forms of key elements in microbial environments. The E_h is for the half-reaction, relative to hydrogen.

Element	Oxidized	Reduced	E_h (mV)	Comments
Hydrogen	H^+	H_2	0	$E_h = 0$ by definition
Oxygen	O_2	H_2O	+600 to +400	Oxygenic photosynthesis
Nitrogen	NO_3^-	NH_4^+	+250	Many other reduced forms, including organic compounds
Manganese	Mn(IV)	Mn^{2+}	+225	Mn^{3+} in some environments
Iron	Fe(III)	Fe^{2+}	+100 to −100	Speciation also depends on pH
Sulfur	SO_4^{2-}	S^{2-}	−100 to −200	Sulfide (S^{2-}) usually occurs as HS^-, depending on pH
Carbon	CO_2	CH_4	< −200	Many other reduced forms, including organic compounds

From many possible examples, let us consider again iron and the form it takes as a function of the redox potential. It is convenient to express this potential in a form analogous to pH:

$$p\varepsilon = -\log(e^-) \qquad (3.5),$$

which is related to E_h at 25 °C by the following:

$$p\varepsilon = E_h / 0.0591 \qquad (3.6).$$

Figure 3.6 illustrates how Fe^{2+} dominates in reducing environments with low $p\varepsilon$, whereas Fe(III) is the main form of iron in oxidizing environments with high $p\varepsilon$. Exchanges between Fe^{2+} and Fe(III) and indeed all redox reactions are governed by thermodynamics. Some of these reactions are controlled by microbes, whereas others occur abiotically without direct involvement of microbes. The relative importance of microbial versus abiotic processes is still unknown for some redox reactions.

Light

Light can be used or "harvested" to provide energy for organic carbon synthesis by photoautotrophic and mixotrophic microbes (Chapters 6 and 9) and for growth and survival of photoheterotrophic microbes (Chapter 7). Although it can be harvested safely by some microbes, light energy can damage intracellular macromolecules and it affects organic and inorganic compounds used by microbes. The effect of light varies with wavelength (color) and thus its energy. The most energetic light is in the ultraviolet (UV) range with very short wavelengths. UV-C (200–280 nm) is absorbed in the atmosphere, but UV-B (280–315 nm) and UV-A (315–400 nm) reach Earth's surface with several impacts on natural environments. Even

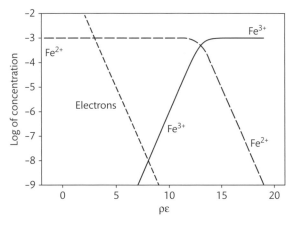

Figure 3.6 Relationship between redox potential (pε) and the concentrations of different Fe species. Data taken from Stumm and Morgan (1981).

visible light with short wavelengths can affect microbes and chemicals.

Light can directly damage DNA and other key macromolecules in microbes. One effect is the cross-linking between adjacent pyrimidine bases of DNA (cytosine and thymine), forming pyrimidine dimers. UV light can also cause the formation of reactive oxygen species, such as peroxide (H_2O_2) and super-oxide radicals (O_2^-), which oxidize DNA, proteins, and other macromolecules in cells. Light energy can be transformed into heat, which can also damage cellular components. Damage to DNA is especially important. If left unrepaired, DNA damage causes mutations and changes in the genetic make-up of the affected microbe.

Nearly all microbes have various mechanisms for preventing or correcting the damage caused by light. Some microbes have sunscreens, pigments such as carotenoids that absorb light and minimize its damage. These pigments

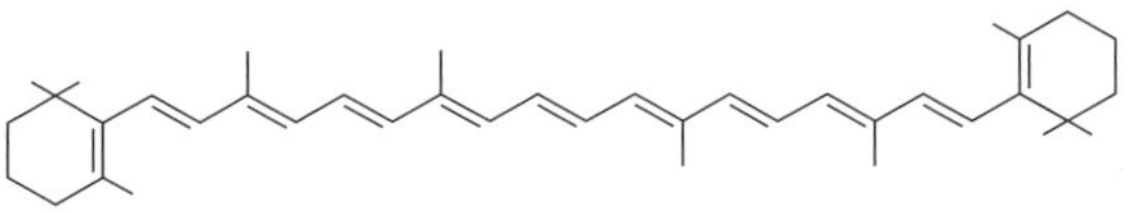

Figure 3.7 Example of a double-bond-rich compound (beta-carotene) common in microbes. This and similar compounds protect microbes from harmful light and give color to even heterotrophic microbes.

have alternating double-bonds (Figure 3.7) that enable light absorption and give them their characteristic color. Microbes also have enzymes, such as RecA, to repair damage done by UV, and other enzymes, such as peroxidases and super-oxide dismutase, to disarm reactive oxygen species produced by photochemical reactions with oxygen. These enzymes are common in aerobic microbes but not in anaerobes in anoxic habitats.

Pressure

The largest biomes on the planet include the deep ocean and the deep subsurface environment—the geological formations deep below the Earth's surface. These biomes are usually thought of as being extreme environments where exceptional microbes live. However, non-extreme environments—as defined by where humans live—may be the exceptional ones, given that the volume of our environment is much smaller than that of the deep ocean and deep subsurface. More of life is subjected to high pressure than to atmospheric pressure.

Pressure inhibits the activity of microbes that are normally found at atmospheric pressure, but even some of these microbes may tolerate high pressure and are able to resume growth when the pressure returns to normal (Fang et al., 2010; Tamburini et al., 2013). Piezophiles, also called barophiles, grow only at high pressures, and hyperpiezophiles are adapted to grow at >60 megapascals (MPa). These microbes are found at the bottom of the oceans, the deepest spot (11 km) being the Pacific Ocean's Mariana Trench, where the pressure is about 110 MPa or over a thousand-fold higher than atmospheric pressure.

Some fish and other metazoans can survive the high pressure of the deep ocean, although no large organisms are found in deep subsurface environments. Piezophiles are thought to have evolved from low-pressure psychrophiles found in high-latitude environments (Lauro et al., 2007). Both types of microbes have similar adaptations

for dealing with their respective extreme environments. The lipids of both piezophiles and psychrophiles are highly unsaturated, and high pressure and low temperatures lead to similar alterations in protein and DNA structures.

An accident provided one of the first glimpses into the effect of pressure on microbial activity. On October 16, 1968, the research submersible Alvin sank off the coast of Massachusetts and came to rest on the sea floor, 1540 m below the surface (Jannasch et al., 1971). Although the crew of three escaped safely, their lunch of bologna sandwiches and apples was left behind. When Alvin was retrieved from the bottom eight months later, the lunch seemed still edible. In contrast, bologna sandwiches kept at the deep sea temperature (3 °C) in a refrigerator on land spoiled within weeks. The implication of this "experiment" is that decomposition of the starch and protein making up the sandwich bread and meat is inhibited by pressure, not just the cold temperatures found at the sea floor.

The consequences of being small

The physical factors affecting microbes discussed so far are all similar to those that affect organisms in the macroscopic world. But smallness itself imposes constraints on what microbes can and cannot do. It would be trivial to say that microbes are small, were it not for all the consequences of being small. Cell size affects transport of limiting nutrients, predator–prey interactions, macromolecular composition, and many other aspects of microbial biology and ecology.

To help with visualizing life at the scale of a microbe, Figure 3.8 scales up the size of organisms in the microbial world to organisms and other things in the macroscopic world. It shows why a 100 μm ciliate or alga is huge in the microbial world. Some interactions among microbes and between microbes and the environment can also be scaled up or simply enlarged to our more familiar macroscopic world. For example, predator–prey relationships in the microbial world are like those in the macroscopic world. However, other aspects of life at the microbial scale are radically different from what is encountered in a macroscopic world. A challenge of microbial ecology is to understand how things happening at the scale of microns and molecules can have such huge impacts on the biosphere and the entire planet.

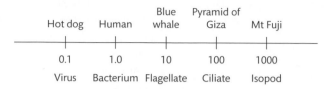

Figure 3.8 Scaling of sizes in the macroscopic and microscopic worlds. The organisms in both worlds vary greatly in size (length) and even more so in volume and mass. Even hot dogs vary from dainty to foot-long.

The world of small cells is fundamentally different from that of macro-organisms. One measure of that difference is the Reynolds number (Re), a dimension-less parameter that is the ratio of inertial forces to viscous forces and is defined by

$$Re = Dv\rho/\mu \qquad (3.7)$$

where D is the characteristic length scale, v the velocity, ρ the density of the fluid, and μ the dynamic or absolute viscosity. The Reynolds number for the world inhabited by humans is huge (10^4). If values from the bacterial world are plugged into the equation (D = 1 μm; v = 30 μm s^{-1}; ρ = 1 g cm^{-3}; μ = 10^{-2} cm^2 s^{-1}), the resulting Reynolds number is low (<1). For this reason, microbes are said to live in a "low Reynolds number environment." Unlike our world, viscous forces dominate over the inertial forces in the low Reynolds number environment of microbes. As the physicist E. M. Purcell once pointed out (Purcell, 1977), for humans, a low Reynolds number environment

would feel like swimming in molasses. A person pushed while in molasses would glide much less than 10 nm. Figure 3.9 gives actual motility rates as a function of organism size in various Reynolds number worlds.

One consequence of being in a low Reynolds number environment is that mixing of molecules is governed by diffusion (a gradient-driven process), whereas in the macroscopic world, mixing is dominated by turbulence (an inertia-driven process). In a diffusion-dominated world, mixing is the result of countless, random collisions between molecules. A measure of how readily this diffusion-driven movement and mixing occurs is diffusivity (D). Movement or flux (J) of a compound as function of distance (z) follows Fick's first law:

$$J = -D_c \, dC/dz \qquad (3.8)$$

where D_c is the diffusion constant for a particular compound with a concentration of C. In words, the flux due to diffusion is a product of the gradient in the concentration

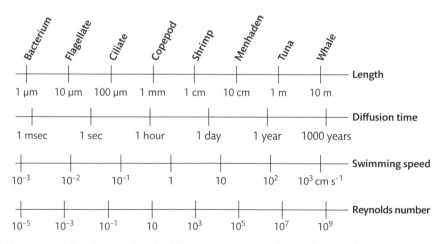

Figure 3.9 Diffusion and mobility for organisms in different Reynolds number worlds. Based on Jørgensen (2006).

(dC/dz) and the diffusion constant (D_c). The flux is always from high concentrations to low, hence the negative sign in Equation 3.8.

The diffusion constant varies with the phase (water, air, or solid), temperature, and the chemical itself. All things being equal, small compounds diffuse more quickly than large compounds, and diffusion of uncharged molecules is faster than of charged ones. Proteins have higher diffusivity than polysaccharides because they tend to be more hydrophobic and more compact (Table 3.3). For gases, diffusion increases with temperature and decreases with pressure. For aqueous solutions, diffusion also increases with temperature, but this is partially offset by a decrease in the viscosity of water. For example, the diffusion constant for oxygen in water is 0.157×10^{-4} and 0.210×10^{-4} cm^2 s^{-1} at 10 and 20 °C, respectively, only a 57% increase for a doubling of temperature.

The Stokes–Einstein relationship (Equation 3.9) relates the time (T) and length scales (L) of diffusion for a compound with a given diffusion constant (D):

$$T = L^2 / 2D \tag{3.9}$$

Table 3.3 Diffusion coefficients of some chemicals and a virus in water at 25 °C. Data from Logan (1999).

Compound or virus	Molecular weight (Daltons)	Diffusivity (cm^2 s$^{-1} \times 10^{-8}$)
Ammonia (NH$_3$)	17	1700
Glucose	180	690
Dextran	60,200	35
Serum albumin	70,000	61
Tobacco mosaic virus	31,400,000	5.3

Table 3.4 Time and distance scales for oxygen and glucose moving by diffusion in water at about 20 °C. Data from Jørgensen (2006).

	Time scale for:	
Distance	Oxygen	Glucose
1 μm	0.34 ms	1.1 ms
10 μm	34 ms	110 ms
100 μm	3.4 s	10 s
1 mm	5.7 min	19 min
1 cm	9.5 h	1.3 d
10 cm	40 d	130 d
1 m	11 y	35 y
10 m	1090 y	3600 y

As one example, consider oxygen and glucose in water at 10 °C (Table 3.4). These two compounds move the length of a bacterium (about 1 μm) within fractions of a second, but they take several seconds to move 100 μm, and years before oxygen and glucose spread over a meter. Diffusion sets an upper limit for how fast a compound can be taken up by a microbe. If every molecule arriving at a cell surface is taken up, then the flux (J) to a spherical cell with a radius r is

$$J = 4\pi DrC \tag{3.10}$$

where C is the concentration in the bulk solution infinitely far from the cell. This flux has units of mass per unit surface area per unit time. In the next chapter, we build on Equation 3.10 to discuss why small cells are better at taking up dissolved compounds than large cells, which in turn explains why microbes in oligotrophic environments, like the open oceans, are small.

Microbial life in natural aquatic habitats

The rules of a low Reynolds number environment apply to all microbial habitats, from soils and sediments even to pure cultures in the laboratory. These habitats perhaps share more similarities than differences, given the importance of size in structuring them. However, of course, there are important differences between the water column of aquatic environments and soils and sediments. The water column has far fewer solids than in soils or sediments, and consequently the microbial environment may seem simple and sparse, as illustrated by the calculations that follow. However, we will soon see that even aquatic environments may be more complex than appearances suggest.

A typical milliliter of water from a natural aquatic environment contains about 1 million bacteria (Chapter 1). At this density, each bacterium would be surrounded by an empty sphere of 10^6 μm^3 assuming an even distribution of microbes (Figure 3.10). If so, then the distance between bacteria is on the order of 60 μm, much greater than the 10 μm for bacteria growing in rich media in the laboratory, where cell densities can exceed 10^8 cells ml^{-1}. The distance to the nearest bacterium of the same species is even greater, depending on its relative abundance. The great distances between bacteria in freshwaters and seawater explain why free-living aquatic microbes appear to lack mechanisms for sensing the presence of one another ("quorum sensing"—see Chapter 14) and are "uncommunicative" (Yooseph et al., 2010).

Heterotrophic microbes are also far from sources of organic material they need to support growth. The abundance of detritus and phytoplankton varies greatly but is 10^3–10^4 particles or cells per milliliter. At these densities and again assuming even distributions, a heterotrophic bacterium would be >100 μm away from these potential organic carbon sources. Even the number of many dissolved molecules is low in the immediate neighborhood of a microbe. As has already been mentioned and will be emphasized again and again, concentrations of all but a few essential compounds are very low in natural environments.

These low concentrations mean that only a few molecules are near to an average microbial cell. For example, consider a dissolved amino acid with a concentration of 100 nmol l^{-1}. Taking advantage of Avogadro's number, we can calculate that only about 30 molecules of this amino acid occur in a 0.5 μm^3 sphere of water surrounding a cell. The number of molecules for many compounds would be even lower. The same is the case for many other organic and inorganic compounds in natural microbial environments.

Motility and taxis

The calculations just discussed assume an even distribution of cells and other particles that does not change over time, but in fact distributions do change. Many microbes are motile (see Box 3.3), resulting in a more

uneven, dynamic distribution of cells than depicted in Figure 3.10. Those protists that swim by using flagella are referred to as "flagellates," and those using the shorter and more numerous cilia are called "ciliates" and belong to the Ciliophora phylum. These eukaryotic microbes swim through water to increase their chances of encountering prey. Many bacteria also use flagella for propulsion, although the structure of a bacterial flagellum differs greatly from its eukaryotic counterpart. Some bacteria and diatoms glide along surfaces, propelled by the secretion of polymers. Microbes can swim incredibly fast, with speeds ranging from 1–1000 μm per second. This speed sounds more impressive when scaled up to our size. If a one meter-tall person could swim as many body lengths as a microbe can (e.g. 100 μm per second), their speed would be over 300 km per hour.

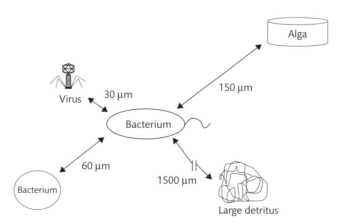

Figure 3.10 The spatial distribution of microbes in aquatic environments assuming all particles are evenly distributed.

Analogous to the use of motility by microbial predators, other motile microbes swim to increase uptake of inorganic and organic nutrients. If the microbe is big enough, the mere act of moving helps to break down limits to uptake set by diffusion. Some motile microbes may use "chemotaxis" to swim toward sources of essential dissolved compounds. To do so, microbes have to sense and follow up the concentration gradient toward the nutrient source. One mechanism for sensing this gradient is the "run and tumble" strategy (Figure 3.11). The duration of runs, or swimming in a straight line, increases when the microbe swims against the concentration gradient. If concentrations decrease, implying that the microbe is going in the wrong direction, then it tumbles and heads randomly in another direction. The net result is chemotaxis towards the source of the desired compound. Alternatively, the opposite may happen for negative chemotaxis if a microbe is to avoid an inhibitory compound.

Microbes are attracted to or repelled by other environmental clues besides dissolved compounds. Phototropic microbes may sense light and use phototaxis, while other microbes rely on aerotaxis to search for oxygen. Magnetotactic bacteria use intracellular magnets, composed of magnetite (Fe_3O_4) or greigite (Fe_3S_4), to align along the Earth's magnetic field. This mechanism in combination with aerotaxis enables these bacteria to find the depth in sediments with optimal concentrations of oxygen and other dissolved compounds (Chapter 13).

Given the many advantages of motility and moving to more suitable micro-habitats, it may be surprising to learn that not all microbes are motile. The estimates for the proportion of motile oceanic bacteria, for example, range from 5–70%, depending on the season and location (Mitchell and Kogure, 2006). Two of the most abundant bacteria in the ocean, *Pelagibacter* and *Prochlorococcus*, are not motile (Zehr et al., 2017). Some microbes may not be active enough to afford the energetic expense of motility, and others may be just too small to gain any benefit. Theory suggests that microbes have to be at least 3.7 or 8.5 µm (depending on the calculations) in order to use motility to escape limitations set by diffusion (Dusenbery, 1997). Just as it is advantageous for some microbes to be motile, it is advantageous for others to remain non-motile.

Submicron- and micron-scale patchiness in aqueous environments

While not all microbes are motile, certainly enough are for us to revisit the picture sketched before about the spatial distribution of microbes and detrital particles suspended in a milliliter of water from a natural aquatic habitat. Remember that the calculations using typical abundances and concentrations gave the impression of everything being isolated and widely dispersed. But we now know that some microbes swim at top speed to track down nutrient-rich sources. These sources, such as an algal cell or decaying detrital particle, may be surrounded by swarms of chemotactic bacteria, which in turn attract grazers in search of easy prey.

Even the distribution of organic material in a milliliter of natural water is complex and far from being a well-mixed, homogenous soup. Because of how dissolved and particulate materials are separated (see Box 3.4), the "dissolved" pool contains many "things" that are closer to being particulate than to being dissolved. These "things" include colloids (any 1–500 nm particle), inorganic and organic aggregates, and gels. Gels spontaneously form by coagulation of smaller organic and inorganic components and can range in size from less than one to several microns. Patchiness of microbes at larger scales is better

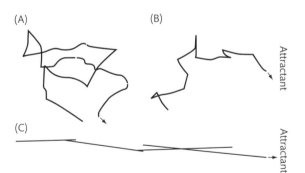

Figure 3.11 Trajectories of bacteria moving randomly (panel A) and moving by positive chemotaxis toward an attractant (B and C). The classic "run and tumble" (biased random walk) is shown in panel B. The bias towards the attractant is only about 1% and has been exaggerated here to show the effect. Another mechanism for chemotaxis, "run and reverse," is common in aquatic bacteria (panel C). In this mechanism, long runs towards the attractant are interrupted by short reversals. The angle between the paths is only a few degrees and has been exaggerated in panel C for clarity. Figures provided by Jim G. Mitchell and used with permission.

Box 3.4 Dividing line between dissolved and particulate

Filters made of glass fibers are used to examine dissolved and many particulate components from natural environments. These filters can easily be cleaned of contamination, most drastically by burning ("combusting") the filter at about 500 °C to remove all organic compounds. Glass fiber filters that retain the smallest particles are Whatman GF/F filters ("**g**lass **f**iber **f**ine"). Anything that passes through a GF/F filter is, by definition, dissolved, while anything retained is particulate. The GF/F filters are advertised to retain particles of about 0.6 μm, but things smaller than this size can be trapped or stick to the glass fibers, while particles larger than 0.6 μm may slip through and appear to be components of the dissolved pool. Delicate cells and detrital particles may be broken up during the filtration process and the pieces may pass through the filter. "Size fractionation" approaches in aquatic microbial ecology rely on filters composed of different materials, such as polycarbonate or cellulose nitrate, and with various pore sizes, ranging from 0.1 to 10 μm.

---

understood than at smaller scales. Microbe communities attached to large particles (tens of microns) differ from the "free-living" community, or those microbes not on large particles. Microbial ecologists have looked at patchiness by obtaining water samples using small pipettes separated by millimeters (Long and Azam, 2001). Both total abundance and the type of microbes differed in these samples. For practical reasons, microbial ecologists usually sample liters or several grams of sediment or soil at a time, yet often the community detected in these large samples can be understood only by considering the microscale environment actually inhabited by microbes.

Microbial life in soils

The immediate physical environment of a microbe in soil is similar in many respects to the aqueous environment just described. Active soil organisms are found in a film of water covering soil particles. These organisms have been called "terrestrial plankton" by one soil ecologist (Coleman, 2008), who borrowed the term "plankton" used by aquatic ecologists to describe free-floating biota. Redox state, pH, and temperature all affect soil microbes as much as they do aquatic microbes, and soil microbes also live in a low Reynolds number world.

Of course, soils differ from the water column of aquatic habitats in several important ways. The soil environment is defined by inorganic and organic particles separated by open channels (pore space) through which air and water pass. The physical environment for microbes, including key properties such as oxygen concentration and redox state, can differ drastically between locations separated by microns, much more so than seen in the water column of aquatic environments.

The three-dimensional physical arrangement of particles in the soil microenvironment has been explored using computerized X-ray tomography, analogous to CT scanning used by physicians to visualize potential tumors in the human body (Voroney and Heck, 2015). Another approach is secondary ion mass spectrometry (SIMS). These approaches have revealed much about total pore space and the size, shape, and connections between pores, all important for understanding microbial life in soils. These properties vary with the type of soil (Table 3.5). Mineral soils are 35–55% pore space by volume, while organic soils are 80–90%. In pores larger than about 10 μm (macropores), air and water readily move by diffusion and drainage. Macropores can be created by plant roots or movement of earthworms and other non-microbial soil organisms. Pores smaller than about 10 μm (micropores) retain water and can limit the movement of soil organisms.

Water content of soils
The extent to which pore space is filled with water has a huge impact on soil chemistry and microbial life. One of the largest impacts is the diffusion of oxygen and other gases; it is much slower in water than in air. For example, the diffusion coefficients at 20 °C for oxygen are 0.205 $cm^2 s^{-1}$ in air and 0.0000210 $cm^2 s^{-1}$ in water. Consequently, oxygen penetrates water-logged soils very slowly, often much slower than in its use by heterotrophic organisms,

Table 3.5 Some properties of the three major inorganic constituents of soils. Particles are assumed to be spherical.

Property	Sand	Silt	Clay
Porosity	Large pores	Small pores	Small pores
Particle size (mm)	0.05–2	0.02–0.05	<0.002
Permeability	Rapid	Low to moderate	Slow
Number of particles per gram	10^2–10^3	10^7	10^{11}
Water holding capacity	Limited	Medium	Very large
Soil particle surface ($cm^2\ g^{-1}$)	10–200	450	8×10^6
Cation exchange capacity	Low	Low	High but varies with mineral

resulting in anoxia. Water content depends on soil type and texture. Because clay has more micropores than sand, soils rich in clay are often more poorly aerated than sandy soils.

Water in soils can be described by two fundamental properties: water content and water potential. Water content is simply the amount or volume of water per amount or volume of soil. It can be measured by weighing soils before and after drying. Water potential is the potential energy or the amount of work potentially done by water moving without a change in temperature. Water potential is the sum of four components: matrix, osmotic, gravitational, and atmospheric pressure. The matrix component consists of adsorption of water to soil constituents, leading to a negative water potential. Osmotic effects, which are also negative, are due to the solutes dissolved in water. Both contribute to the retention of water in soils, while gravitational and atmospheric pressures, which are usually positive, pull and push water out of soils. The units of water potential are the same as for pressure: pascals (Pa) or, more commonly, kilopascals (kPa).

To see how these different components of water potential interact, consider a field that has been saturated with water, such as after a heavy rain or the spring thaw. At first, the negative matrix and osmotic effects on water potential are balanced by positive pressure effects, giving a water potential of 0 kPa. Once the rains have stopped or all snow has melted, gravity takes over, draining the soils of water until the matrix and osmotic forces are large enough to retain water in the soil. At this point, the soil is at "field capacity." The amount of water left, the field capacity, varies with soil type. Loam soils have a soil water potential of –33 kPa, while in sandy soils it is –10 kPa. The water potential of –50 kPa corresponds to water contents of 10% in sandy soil and 45% in clay soils.

Water potential can be used to describe how much water is needed for microbial activity. Even terrestrial microbes need water to grow and be active, as previously mentioned. For microbes to move, soils have to have a water potential of about –30 to –50 kPa, which corresponds to a water film thickness of 0.5 to 4.0 μm on soil particles. Bacterial metabolic activity becomes limited in soils with a water potential of –4000 kPa or a <3.0 nm film of water, although experimentally dried-out soils still have some metabolic activity (respiration) even at -1.0×10^5 kPa (Figure 3.12). There is some evidence that different types of bacteria respond differently to water content, and fungi grow better in dry soils than bacteria. Fungal hyphae can traverse dry spaces in soils better than even filamentous bacteria. Some soil microbes may form resting stages, spores, or cysts, to survive periods of desiccation (Box 3.5).

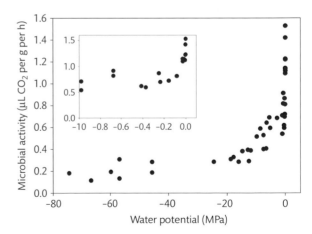

Figure 3.12 Microbial activity (here respiration) as a function of experimentally manipulated water potential. The insert uses the same data but only for high water potential. Data taken from Orchard and Cook (1983).

Box 3.5 Tough nuts

Some Gram-positive soil bacteria, including *Bacillus* and *Clostridium*, form spores, said to be the most resilient biological structure in the biosphere. The resilience of spores is in part due to the number and nature of protein coats surrounding the core proto-plast consisting of the bacterium's DNA. The coats contain unusual compounds not found in "regular," vegetative cells, including dipicolinic acid in com-plexes with calcium ions. Spores can remain viable for at least decades, and there are controversial reports of spore-forming bacteria being recovered from 24–40 million-year-old bees trapped in amber and from 250 million-year-old salt crystals (Vreeland et al., 2000).

Interactions between temperature and water content in soils

Temperature affects aquatic and soil microbes equally, with both groups having a Q_{10} of about 2 over the typical temperature range in nature. However, unlike aquatic habitats, temperature can have an additional impact on microbial activity in soils via its effect on water content. Warmer temperatures due to climate change would increase microbial activity, but could also lead to drier soils and eventually lower microbial activity. Extensive field work has demonstrated that temperature is an excellent predictor of soil respiration until a threshold above which soil moisture comes into play.

Water potential and other factors can greatly compli-cate attempts to look at the relationship between micro-bial activity and temperature. The apparent Q_{10} is often substantially higher than the canonical value of 2 and is more variable when estimated from the change in microbial rates over the seasons as temperatures naturally rise and fall. In soils, moisture and other soil properties also change with the seasons, complicating efforts to understand temperature effects (Figure 3.13). As tem-perature rises, which would increase microbial activity, moisture may decrease, which would slow microbial activity; the balance between the two opposing factors is difficult to predict when each is considered separately. One analysis of experiments done in many soil habitats found that the decrease in moisture with warming canceled out the temperature effect such that in the end soil respiration was not significantly altered (Carey et al., 2016b). Understanding how microbes in soils and aquatic

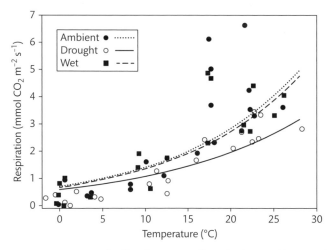

Figure 3.13 Interactions between temperature and soil moisture on soil respiration. In this example, precipitation was experimentally manipulated to give ambient (natural rainfall), wet (50% above natural levels), or drought (50% below) conditions for ambient soil temperatures as they varied through a year. The experiment was repeated monthly with old field soil in the northeast USA. The lines are exponential fits to the data. Compared with ambient precipitation, both wet and drought conditions led to lower soil respiration and lower responses to temperature (lower Q_{10}). Data taken from Suseela et al. (2012).

environments respond to temperature is critical for predicting the feedback of the biosphere to global warming.

The biofilm environment

This chapter has discussed a few interactions between microbes and solid surfaces, and many more will be explored in other chapters. We'll conclude here with a few words about a special type of surface-associated environment, that created by biofilms, complex communities of microbes attached to surfaces. The term is usually used for communities on millimeter-sized or larger surfaces that are also inorganic, such as rocks or stones in a stream, a ship hull in the ocean, or teeth in an animal's mouth. But biofilms also grow on living tissue of plants and animals. As these examples suggest, biofilms can form on many types of surface and cause problems in many industrial and biomedical settings, in addition to being important in natural environments. Biofilms can also be part of the solution, such as in removing dissolved compounds from waste water in sewage treatment plants. Entire research institutes are focused on biofilms. Microbes other than bacteria, such as diatoms and cyanobacteria, can be important in biofilms growing attached to submerged surfaces exposed to sunlight. However, most studies have focused on heterotrophic bacteria.

A biofilm forms any time a surface is immersed in water or moist soil. It starts with the colonization of a surface by planktonic bacteria, perhaps attracted to organic compounds at the surface or as a way to escape predation (Figure 3.14). These initial colonizers are joined by other free-living bacteria, and all then divide and multiply to form several layers of cells over time. The timescale of this process differs depending on the envir-

onment, but the initial colonization phase may last hours to a couple of days. More complex biofilm structures may take weeks to months to form. Along with the addition of new cells by colonization and growth, biofilm bacteria and other microbes secrete extracellular polymers, mainly polysaccharides, as mentioned in Chapter 2. These polymers help to anchor cells to the surface, and they store carbon, protect against predators, or perhaps fend off competitors; they also keep extracellular enzymes close to the cell (Flemming and Wingender, 2010). The polymers often constitute a larger fraction of a mature biofilm's total mass than do the cells and often determine the role of a biofilm in applied and basic environmental problems.

Bacteria in biofilms differ from their free-living counterparts in several respects. First, the species composition of biofilms differs from the free-living community, just as the species make-up of microbes attached to detritus differs from its surrounding planktonic counterpart. Second, the metabolism of an initially planktonic bacterium changes after it colonizes a surface and after the biofilm matures with time. In contrast with the isolated existence of a planktonic cell, a biofilm bacterium is surrounded by other microbes, which may or may not be daughter cells, with limited exchange with the outside environment. A microbe at the biofilm's outer boundary may experience the same dissolved compounds as if it were in the bulk fluid. In contrast, a microbe buried deep within a biofilm may never see some compounds from the bulk phase and may be immersed in metabolic by-products from the biofilm.

A microbe embedded deep in a biofilm may have little contact or exchange with the outside world, but not because the biofilm restricts diffusion. Because a biofilm

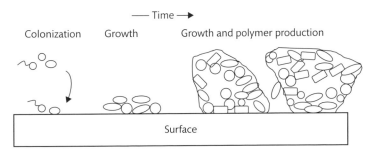

Figure 3.14 Development of a biofilm over time, starting with the colonization phase and culminating in extracellular polymer production.

is mostly water, diffusion within a biofilm is still 60% of that in the bulk fluid (Stewart and Franklin, 2008). The problem facing a deeply buried microbe is consumption of compounds by other biofilm microbes. Oxygen is an important, well-studied example. One study of a mature biofilm found that oxygen concentrations were completely depleted within a 175 μm layer of a 220 μm-thick biofilm. In this example, concentrations even at the biofilm outer surface were only 40% of the bulk fluid. So, anoxic niches and anaerobic microbes and processes can occur in biofilms immersed in an oxic environment.

Microbial ecologists once thought that a mature biofilm resembled tiramisu and consisted of layer upon layer of microbes evenly covering a surface. Confocal microscopic studies (see Box 3.6) have demonstrated that, rather than being two-dimensional, biofilms are complicated three-dimensional structures with channels of fluid flowing over bare surface between towers of microbes. This complex three-dimensional structure helps

---

Box 3.6 Microbial life in 3-D

Confocal microscopy has been instrumental in understanding biofilm microbes and structure. In regular fluorescence microscopy, a single plane of focus is excited by light and the resulting emitted light is analyzed, giving only a two-dimensional view of the sample. Anything not in the plane of focus is not seen. This approach is adequate for many applications in microbial ecology. In contrast, confocal microscopes are capable of taking images at several planes of focus that are then compiled to reconstruct a three-dimensional image of the sample. These 3-D images have provided new insights into biofilm structure and function.

---

to explain variability in chemical and biological composition and several other biofilm properties.

Summary

1. Different microbes are able to survive and grow in environments that vary greatly in temperature, pH, salt, pressure, and other physical properties. Terms building on the suffix "philic" are used to describe these microbes, including psychrophilic, thermophilic, acidophilic, halophilic, and piezophilic.

2. Reaction rates in microbes often increase by two-fold when temperatures increase by 10 °C ($Q_{10} = 2$). Temperature affects several aspects of a microbe's environment, which complicates efforts to understand the impact of global warming.

3. Microbes live in a low Reynolds number environment in which the movement of compounds is affected more by diffusion than by turbulence. Diffusion limits large cells more than small cells in the uptake of dissolved compounds.

4. The physical structure of aquatic habitats is potentially quite sparse for microbes because of relatively low numbers of cells and other particles and because of very low concentrations of many dissolved compounds. However, chemotaxis and the presence of particles that vary in size and composition create a patchy environment at the microbial scale.

5. Soil microbes live in pores of various dimensions in between soil particles. Pore sizes vary with soil type and determine water content, which in turn affects many soil properties and microbial activity.

6. Biofilms are complex structures of microbes living so close together that the availability of oxygen and other dissolved compounds is restricted. In addition to cells, many properties and practical roles of biofilms are shaped by extracellular polysaccharides and other polymers.

CHAPTER 4

Community structure of microbes in natural environments

This chapter will discuss the types of microbes seen in natural environments and the diversity and "structure" of these microbial communities. The term "community structure" refers to the list of organism names, their phylogenetic relationships, and abundance in an environment. One motivation for exploring community structure is to gain insights into biogeochemical processes and other "functions" mediated by microbes. However, even if there is no connection or only a weak one between structure and function, we should want to know the names of the most abundant organisms on the planet, the microbes. The topics covered in this chapter are a large part of modern microbial ecology.

This chapter will focus on bacteria and archaea, with a few observations about fungi. Protists are discussed in Chapter 9 and viruses in Chapter 10.

Taxonomy and phylogeny via genes: introduction to 16S rRNA-based methods

Chapter 1 pointed out that few microbes can be grown in pure cultures on agar plates or using other, traditional cultivation-dependent approaches. In fact, there is a spectrum of culturable microbes in a sample, from those that can be easily cultivated by traditional methods to those that are dead and cannot be revived by any method (Figure 4.1). Many microbiologists believe that all living microbes found in nature can be cultivated, if the right conditions are found, and that there is no such thing as an unculturable microbe, a microbe that cannot be grown by itself in the laboratory under any circum-

stances. It is true that since Robert Koch's work with agar plates in the nineteenth century, new approaches have been developed to isolate difficult microbes and to grow them in pure cultures in the laboratory (Overmann et al., 2017). However, there are candidates for truly unculturable microbes, such as symbiotic microbes and those living in complex consortia. More debatable are seemingly independently growing, free-living microbes that have not yet been cultivated. Perhaps some of these could be cultivated if we knew the right approach, but others may remain uncultivated regardless of the approach.

There are several problems with exploring the ecology and biogeochemical role of an uncultivated microbe, starting with its identification and even recognizing that it is in a sample. To identify cultivated microbes, the traditional approach was to look at its appearance (for large protists and fungi) or its response to a battery of biochemical tests (bacteria and archaea): Gram staining, the capacity to degrade key compounds, other enzymatic activity, and so on (Box 4.1). However, for uncultured microbes, even electron microscopy is only informative for a few, large protists and hardly at all for small ones and prokaryotes. The biochemical characteristics used for traditional identification are not generally observable for an uncultivated microbe co-occurring with others in complex communities. The problem then is how to identify the vast majority of microbes that have not been cultivated.

The solution is to use a gene. Simply put, organisms with similar sequences of this gene are more closely related than organisms with different sequences. In

Processes in Microbial Ecology. Second Edition. David L. Kirchman. Oxford University Press (2018). © David L. Kirchman 2018.
DOI 10.1093/oso/9780198789406.001.0001

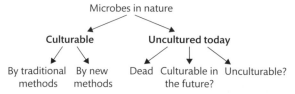

Figure 4.1 Distinctions between microbes that are culturable, uncultured, or unculturable. Among the microbes that are uncultured today, some are unculturable by traditional methods, but can be grown as pure cultures with extraordinary effort and an innovative approach. The microbes resisting even these efforts may really be unculturable. Inspired by a figure by Madsen (2016).

Box 4.1 Bible of bacterial taxonomy

One important repository for bacterial taxonomic information is *Bergey's Manual*. First produced by a committee chaired by David H. Bergey, the manual was published in 1923 under the auspices of the Society of American Bacteriologists, now called the American Society for Microbiology. Nine editions of *Bergey's Manual of Determinative Bacteriology* followed, with the last published in 1994. It is now an online manual, *Bergey's Manual of Systematics of Bacteria and Archaea* (http://www.bergeys.org/).

---

addition to its use in microbiology, the basic idea is also used for exploring the taxonomy and phylogeny of larger organisms. The Barcode of Life Project uses sequences of the cytochrome *c* oxidase subunit 1 (COI) gene and other genes found in mitochondria to identify and classify invertebrates and vertebrates. As mentioned in Chapter 1, the gene most commonly used for prokaryotes is that for 16S rRNA, and for eukaryotic microbes it is the 18S rRNA gene. Both are small subunit (SSU) rRNA genes. Another favorite target is the internal transcribed spacer (ITS) region. This region in prokaryotes is between the 16S rRNA and 18S rRNA genes, whereas eukaryotes have two ITS regions: ITS1 is between the 18S and 5.8S rRNA genes and ITS2 is between the 5.8S and the 28S rRNA genes (26S in plants). These genes and others have been essential for exploring natural microbial communities in an environment independent of cultivation, but they have also been essential in deciphering the taxonomic

and phylogenic relationships even among easily cultivated microbes.

Chapter 1 pointed out the work of Carl Woese, who first used 16S rRNA gene sequences for exploring the taxonomy and phylogeny of cultivated bacteria and archaea. Why the 16S rRNA gene? There are several reasons:

- The 16S rRNA gene is found in all bacteria and archaea. Even eukaryotes have it in mitochondria and chloroplasts.
- Different regions of the gene have different levels of variability, ranging from highly conserved regions that are very similar in all organisms to other regions that are highly variable and differ greatly between distantly related organisms. Both types of regions are needed. The highly conserved regions are very useful for finding all 16S rRNA genes in complex samples, while variable regions are essential for distinguishing one microbial taxon from another.
- Phylogenies derived from the 16S rRNA gene agree well with phylogenies derived from other conserved genes, and are therefore likely to represent the evolutionary history of the organism as a whole. The gene is not thought to be transferred horizontally (Chapter 5).

As mentioned in Chapter 1, approaches that do not rely on cultivation are collectively called cultivation-independent methods. Some authors use "to culture" and its grammatical relatives instead of "to cultivate," but the meaning is the same. The methods developed for 16S rRNA genes in bacteria have been adapted for other genes in all microbes.

Now that we have the 16S rRNA gene to base our cultivation-independent method on, the problem becomes, how do we retrieve it from complex communities consisting of many organisms with many other genes? The problem is solved by using the polymerase chain reaction (PCR). Many other textbooks and web sites describe how PCR works, so here is just a brief overview. In addition to sample DNA and an enzyme, a heat-resistant DNA polymerase (Box 4.2), PCR requires two oligonucleotides ("primers"), each being about 20 base pairs (bp) in length, designed to match the conserved regions of the 16S rRNA molecule. When incubated in a PCR machine, the DNA polymerase synthesizes lots and lots of the desired gene (the "amplicons") from all

Box 4.2 Needles, haystacks, and photocopying a gene

The problem solved by PCR is analogous to finding a needle in a haystack; the needle is the desired DNA fragment and the haystack is all the other genes and genetic material not of immediate interest. PCR turns the haystack into a pile of needles. PCR is also like a photocopier that makes many copies of the same DNA fragment. Invented by Kary Mullis in 1987 (Nobel Prize, 1993), PCR is one of the most commonly used techniques in molecular microbial ecology. A key ingredient in PCR is a heat-resistant DNA polymerase. The polymerase used by Mullis (Taq) was from *Thermophilus aquaticus*, which was first isolated from a Yellowstone Park hot spring by Thomas Brock, one of the founders of microbial ecology.

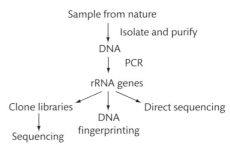

Figure 4.2 Summary of approaches for examining microbial community structure via rRNA genes by PCR-based methods. A "clone library" is a collection of *E. coli* colonies containing the cloned DNA from another organism, in this case the PCR-generated fragments of the rRNA gene. "DNA fingerprinting" refers to a collection of methods, such as denaturing gradient gel electrophoresis (DGGE) and terminal restriction fragment length polymorphism (t-RFLP), that physically separate rRNA gene fragments. The method most commonly used today is direct sequencing, also called tag sequencing.

organisms in the sample; more precisely, what is synthesized is the DNA fragment between the two primer sites from all organisms with DNA sequences sufficiently similar to the primers.

Once the PCR is completed, the problem is then how to separate out and analyze the PCR products. Figure 4.2 outlines the main approaches that have been used over the years. The initial work used clone libraries and the Sanger sequencing approach. It was laborious and costly but yielded high quality, long "reads" or long sequences from large DNA fragments. (Frederick Sanger (1918–2013) was a British biochemist who won two Nobel Prizes, one in 1980 for his DNA sequencing approach and the other in 1958 for his work on protein sequencing and structure.) Long, complete 16S rRNA gene sequences generated by the Sanger approach were used by Woese and others to lay the foundation of modern bacterial taxonomy and phylogeny.

Today the most common approach is direct sequencing of the PCR amplicons using high throughput sequencing. This approach is often called "tag sequencing" because only a short part, a "tag," of the rRNA gene is sequenced and analyzed. ("Tag" also refers to an added oligonucleotide sequence used to label and later separate sequences from different samples that are analyzed together.) The short sequenced part is one or more of the nine variable

regions of the 16S rRNA gene (Yang et al., 2016). Only a short part is sequenced because the average "read length" of today's sequencers is short; currently, the read length is only a few hundred base pairs, far short of the roughly 1500 base pairs of the complete 16S rRNA gene. The read length has been increasing as sequencing technology continues to develop (Singer et al., 2016). Microbial ecologists are more than willing to have short read lengths because the high throughput, also called "next generation," sequencers yield many, many more sequences than the traditional Sanger approach. Once laborious and expensive, sequencing is now easy and cheap, thanks to the development of high throughput sequencing approaches.

When applied to a microbial community, sequencing of PCR amplicons yields a list of difference sequences, each from a different organism. The number of times a sequence appears in the list is assumed to reflect the abundance in the sample of the organism with that sequence. Tag sequencing has largely supplanted DNA fingerprinting and cloning, because of its ease and power in yielding much information about all microbes in complex communities.

However, the cultivation-independent approaches outlined in Figure 4.2 can produce a misleading picture of microbial communities because of problems at each step of the process, from isolation of the DNA to the

Box 4.3 Seeing is believing

Cultivation-independent methods based on PCR are said not to be quantitative because of possible bias with PCR and other problems. One solution is to use a version of PCR, quantitative PCR (qPCR). It is still PCR but it avoids some of the problems of the common PCR approach. qPCR is used to enumerate a few specific OTUs, but it is not used for broad surveys of complex microbial communities. Another solution is fluorescence in situ hybridization (FISH). The approach relies on the binding of fluorescently labeled oligonucleotide probes to the rRNA in ribosomes. Cells with the bound probe become fluorescent and can be counted via epifluorescence microscopy.

sequencing (Tremblay et al., 2015). Box 4.3 discusses another cultivation-independent approach, FISH, which avoids many of these problems, although it too has limitations. Still another solution is metagenomics, discussed in Chapter 5. In spite of the problems, the PCR-based approaches have been extremely powerful and productive in revealing many basic properties of microbial community structure and diversity.

The species problem

The description of tag sequencing mentioned that "different" organisms have different sequences. But what is "different"? The implication is that organisms with the same gene sequence are the same organism. Is that really true? These questions are part of the species problem. The problem is that there is no clear definition of what "species" means for microbes. Without a clear definition, it can be difficult to discuss pathogenicity, the specificity of symbiotic relationships, and perhaps even the role of specific microbes in the environment. The definition of a species is behind several topics in classic ecology; a niche, for example, is defined as the multidimensional space of resources available to and used by a species. These topics and more involve the species concept and how we distinguish between "same" and "different" organisms.

The classical definition, the "biological species concept," is that a species is a collection of individuals that can mate and produce offspring that are themselves capable of reproduction. This definition is meaningless for prokaryotes, many other microbes, and even for metazoans that reproduce asexually. Other definitions have been suggested, but none is universally accepted.

The best definition based on the 16S rRNA gene is that two bacteria or two archaea belong to the same

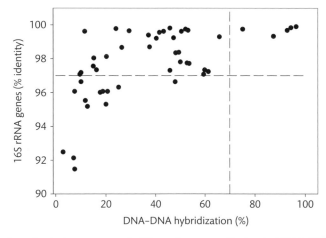

Figure 4.3 16S rRNA genes of one bacterium compared with another versus the level of DNA–DNA hybridization between the same two bacteria. The horizontal line at 97% is the level of identity probably closest to a "species" based on this gene. Points to the left of the vertical line at 70% are different species as defined by DNA–DNA hybridization. Data from Stackebrandt and Goebel (1994).

Table 4.1 Taxonomic levels and corresponding 16S rRNA identity for bacteria. Because other genes are needed to distinguish between prokaryotes and eukaryotes, the degree of 16S rRNA identity is not applicable (NA) at the domain level. Data from Yarza et al. (2014).

Taxonomic level	Number per level*	% identity	Example
Domain	3	NA	*Bacteria*
Phylum	92	75	*Proteobacteria*
Class	5	78.5	*Gammaproteobacteria*
Order	14	82	*Enterobacteriales*
Family	1	86.5	*Enterobacteriaceae*
Genus	42	94.5	*Escherichia*
Species	5	97–99	*Escherichia coli*
Strain	?	>97	*E. coli* O157

* The number of taxa at each level. For example, there are three domains of life, of which *Bacteria* is one. There are 92 phyla of bacteria (the exact number varies among bacterial taxonomists and how candidate phyla are counted), of which *Proteobacteria* is one. *Gammaproteobacteria* is one of five classes of *Proteobacteria*, and so on.

species if their 16S rRNA genes are ≥97% identical. This ≥97% cut-off is based in part on data comparing 16S rRNA similarity versus DNA–DNA hybridization for bacteria in pure cultures (Figure 4.3); DNA–DNA hybridization is expressed as a percentage of how much of the genome from one organism hybridizes or binds to the genome of another organism. The data indicate that two organisms belong to the same species if their DNA–DNA hybridization is at least 70% and they have 16S rRNA genes that are ≥97% similar.

However, other data clearly indicate the problems with this threshold. The problem is that two organisms sharing ≥97% identical 16S rRNA genes may or may not belong to the same species. For example, the 16S rRNA genes of three *Bacillus* species (*B. anthracis*, *B. thuringiensis*, and *B. subtilis*) are >99% identical, yet key features of their physiologies differ greatly, which is why they are treated as separate species. Using a higher threshold, such as 99% or even 100%, does not necessarily solve the problem.

With pure cultures of microbes, microbiologists can examine many more genes than just that for rRNA; now they can look at the entire genome (Chapter 5). This multi-gene approach does not solve the species problem, however, and it is harder for natural communities with uncultured microbes because of difficulties in assembling genomes from metagenomic data (Chapter 5). Even when whole genomes are known there is still a problem in defining a species.

Many microbial ecologists avoid using "species" and instead use other terms, such as "ecotype," "phylotype," and "ribotype," each with a different shade of meaning and implication. Another common term is "operational taxonomic unit" (OTU), which usually is defined as organisms with 16S rRNA genes that are ≥97% similar. Microbial ecologists do use other terms for higher taxonomic levels, such as phyla and class (Table 4.1). "Clade," another common term, is a closely related group of organisms descended from a common ancestor, meaning it is "monophyletic." "Strain" is the word used to describe different isolates or new cultures of the same microbe.

Terms of diversity

Before going over results from 16S rRNA studies of natural microbial communities, we need to discuss two facets of diversity: "species richness" and "evenness."

Species richness is the number of OTUs in a community. One way to characterize richness is with "rarefaction curves" constructed by calculating the number of OTUs that would be found for a particular number of individuals (here, bacteria) or genes sampled (Figure 4.4). As more and more individuals are sampled, more new OTUs are encountered until the rate of discovery slows down, and the curve starts to level off. Rarefaction curves are useful for comparing the diversity of two communities represented by different numbers of samples or individuals. Indices like Chao1 (named after Anne Chao who developed the index) and ACE (abundance coverage estimator) are designed to yield an estimate of OTU richness, but they can be biased by sample size.

The other facet of diversity is "evenness," which takes into account the number of individuals per OTU. A highly even community would have the same number of individuals for each OTU, whereas a highly uneven community would be dominated by a few OTUs represented by many individuals, while the other OTUs would have few individuals. A measure of this aspect of diversity is the Simpson evenness index. A commonly used measure

that includes both richness and evenness is the Shannon index, named after Claude Shannon (1916–2001), a mathematician and electrical engineer who worked on information theory.

Another set of terms, alpha, beta, and gamma diversity, comes from the plant ecologist, R. H. Whittaker (1920–80). Alpha diversity is the diversity within a habitat, whereas gamma diversity is the total diversity over a landscape with several habitats. The example illustrated in Figure 4.5 emphasizes the richness aspect of diversity, but the Shannon or another diversity index could also be used. Beta diversity was originally defined by Whittaker to be

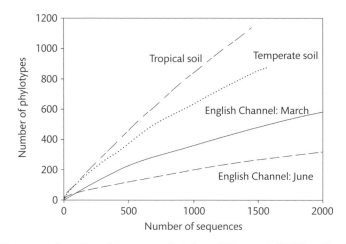

Figure 4.4 Typical rarefaction curves for soils and the oceans. Data from Gilbert et al. (2009) and Lauber et al. (2009).

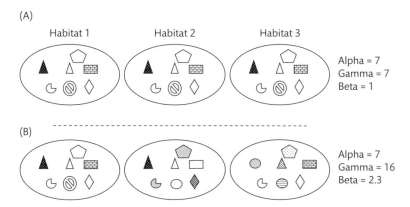

Figure 4.5 Two examples of beta and gamma diversity. In panel A, the number and types of species are the same for three habitats (same alpha diversity), leading to a low gamma diversity and even lower beta diversity. In panel B, species richness and alpha diversity are the same for the three habitats, but the species differ among the habitats, leading to higher gamma and beta diversity than seen in panel A.

the ratio of gamma diversity to alpha diversity, which is the definition used in Figure 4.5, although there are other definitions (Anderson et al., 2011). Beta diversity is also thought of as "species turnover" or the differences among habitats over a landscape.

Patterns of diversity

Let us now discuss some general observations about microbial diversity in natural environments. Many of these observations emerged from early cultivation-independent studies using clone libraries and DNA fingerprinting approaches, but patterns and generalizations became especially clear as results from high-throughput sequencing began to roll in.

Bacterial communities are diverse but very uneven
Early microbial ecologists immediately noticed the diversity of bacterial communities, when they placed environmental samples on agar plates and found growing there large numbers of bacterial colonies with different colors, shapes, and textures. This initial impression of diversity was confirmed by the first studies using cultivation-independent approaches.

It is difficult to assign a precise number to this diversity and to say how many bacterial species are in a sample, but generally bacterial diversity is thought to be quite high. Locey and Lennon (2016) estimated that the entire biosphere of the Earth has 10^{12} different bacterial species based on the correlation between species richness and abundance, much higher than seen for large organisms (Figure 4.6). The precise estimate for the number of OTUs in bacterial communities varies with the environment, but usually it is in the thousands, if not tens of thousands for a particular habitat and time point. Overall, the *Bacteria* domain has more major lineages than do *Archaea* or eukaryotes (Hug et al., 2016).

Some habitats are more diverse than others, providing some clues about what sets the diversity levels and controls the structure of these communities. Soils have more bacterial OTUs than aquatic habitats, evident in the rarefaction curves (Figure 4.4) and also predicted based on the richness versus abundance argument (Figure 4.6). Remember that for the same volume, most soils have about 1000-fold more bacteria than do waters from

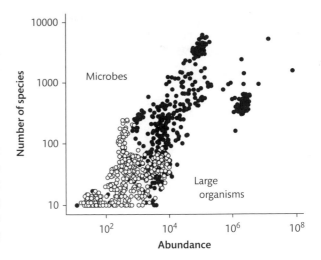

Figure 4.6 Species richness as a function of abundance (number of individuals or sequences) of microbes (closed symbols) or of large organisms (open symbols). Data from Locey and Lennon (2016), provided by Ken Locey.

lakes or the oceans. The higher diversity and microbial abundance in soils is most likely due to their greater spatial complexity and the large number of microenvironments (Tecon and Or, 2017). This greater complexity would permit more microbial taxa to coexist, by minimizing the chance of a superior microbe excluding others by out-competing them for limiting resources. Marine sediments may be the most diverse of all habitats. They have the spatial complexity of soils and even greater chemical complexity because of the range of electron acceptors (Chapter 11) and other physico-chemical properties (Lozupone and Knight, 2007). The water columns of lakes and rivers tend to have more bacterial OTUs than the oceanic water columns (Lozupone and Knight, 2007), again following the richness versus abundance relationship. Freshwaters generally have two- to ten-fold more bacteria than a typical oceanic region. Freshwaters are also affected more by soil microbes from runoff than are even coastal oceans.

Microbial communities are very diverse, but they are not very even. Bacterial communities have a few very abundant OTUs and many rare ones; the 100 most abundant OTUs may account for 80% or more of the entire community, while the remaining 20% is spread out over thousands of OTUs. This collection of rare OTUs has been called the "rare biosphere" (Sogin et al., 2006). The unevenness of microbial communities is illustrated in rank abundance curves, which plot the relative abundance versus the rank of the OTU in the community, starting with the most abundant and ending with the least abundant (Figure 4.7). These curves start off very steep because the communities are dominated by a few, very abundant OTUs, and then they curve sharply down, ending in long tails with the many rare OTUs. In the English Channel, for example, there are a few very abundant types of bacteria, each making up 10% or more of the total community, whereas the other OTUs are less abundant, making up 1% or much less of the total. In contrast, soil communities are more even with fewer highly abundant OTUs.

While aquatic microbial communities are less diverse than soils, they are still quite diverse, so diverse that it has troubled aquatic ecologists. The limnologist G. E. Hutchinson (1903–91) called this the "paradox of the plankton" (Hutchinson, 1961). He pointed out that a large number of aquatic microbial species seem to be competing for the same small number of resources in a physically simple, unstructured environment. These microbial communities should not be very diverse. Competition should result in only a few successful spe-

cies that out-compete and exclude all others, according to the competitive exclusion principle. The paradox is, how can so many species coexist in such a simple environment? Hutchinson had a few suggestions for resolving the paradox, one being that aquatic environments vary substantially over time, thus allowing for coexistence. He was thinking of phytoplankton, but the paradox and Hutchinson's answer also apply to heterotrophic bacteria and other microbes.

There are other ways out of the paradox for heterotrophic microbes. One is that the environment of microbes can be quite complex at the micron scale, even in water and certainly in soils and sediments (Chapter 3). Also, unlike phytoplankton, which use the same carbon source (CO_2) and a few inorganic nutrients, heterotrophic bacteria can use a myriad of organic compounds, which may select for specialization and allow the coexistence of many different types of bacteria. Another resolution to the paradox is that many bacteria are not active and thus are not in direct competition with each other; the genes targeted by cultivation-independent approaches could even come from dead cells. Finally, top-down control by grazers and viruses may allow the coexistence of bacteria that otherwise could not persist together.

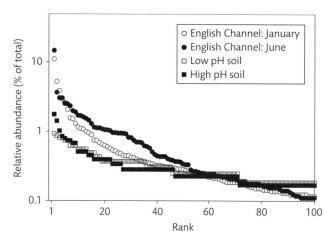

Figure 4.7 Relative abundance of OTUs in the English Channel and two soils. The soils were from a forest with low pH and from scrubland with high pH. The relative abundance is the number of individuals in a particular OTU divided by the total individuals in the community. These data are plotted versus their rank in abundance, starting with the most abundant type in each community. Only the 100 most abundant OTUs are given; each community has thousands of rare bacteria. Data from Gilbert et al. (2009) and Lauber et al. (2009).

Higher mortality caused by grazers and viruses could prevent the superior competitor from excluding less competitive microbes.

There is a bit of the same diversity paradox in soils and sediments. When discussing larger organisms, the problem has been called the "the enigma of soil animal species diversity" (Coleman, 2008). Whether a paradox or an enigma, micro-environments go only so far in explaining the large number of microbial types in soils and sediments. The resolution of the paradox/enigma is similar to that for aquatic environments: temporal variation, diversity in organic material, dormancy, and top-down processes.

Uncultivated bacteria are not the same as easily cultivated bacteria

One of the first questions addressed with cultivation-independent methodology was whether or not cultivated microbes are the same as uncultivated ones. While this question has been examined most intensively for heterotrophic bacteria in oxic habitats, it applies also to other microbes and habitats. The culturability problem means that we cannot use agar plates to enumerate bacteria, but it leaves open the question of whether cultivated bacteria are representative of the uncultivated

bacteria found in nature. Perhaps the bacteria and other microbes that can be cultivated are closely related to those microbes that cannot be cultivated.

Microbial ecologists found that the 16S rRNA genes of bacteria cultivated by traditional methods are not very similar to those assayed by cultivation-independent methodology. In soils, for example, the bacteria most commonly cultivated on agar plates are those in the genera *Streptomyces* and *Bacillus*, whereas these bacteria are not abundant in natural communities, according to clone libraries of 16S rRNA genes or other cultivation-independent approaches (Janssen, 2006). Likewise, in coastal oceanic waters, bacteria in the genera *Pseudomonas* and *Vibrio* are often isolated and grown on agar plates, but their 16S rRNA genes are not common by cultivation-independent approaches. In both soils and aquatic habitats, the differences between cultivated and uncultivated bacteria are evident at higher phylogenetic levels (phylum and subphylum), not just the presence or absence of a few species (Figure 4.8). The types of bacteria cultivated by standard approaches are quite different, at all phylogenetic levels, from the uncultivated microbes found in natural communities.

Like bacteria, few archaea in cultivation are representative of the archaea found in natural environments. Similarly, studies of the ITS region have revealed a large

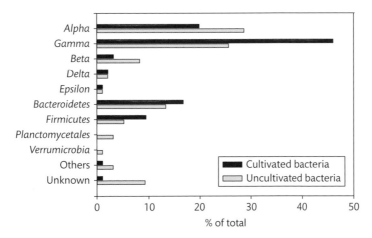

Figure 4.8 Bacterial community structure of a marine environment sampled by cultivation-dependent and cultivation-independent approaches based on 16S rRNA gene sequences. "Unknown" are 16S rRNA genes that could not be classified. "Others" refers to bacteria that could be classified, but individually were <2% of the total community. *Alpha, Beta, Delta, Epsilon,* and *Gamma* refer to classes of *Proteobacteria.* Data from Hagström et al. (2002).

diversity of fungi not seen by studies using cultivation-dependent approaches (Rosenthal et al., 2017). The diversity of protists is also much higher according to cultivation-independent methods than seen in cultures, but as discussed in Chapter 9 in more detail, there is some debate about the ecological relevance of this diversity.

Different bacteria are found in soils, freshwaters, and the oceans

Microbial ecologists are trying to learn about the distribution of various types of microbes among the major habitats of the planet. This is an important question in "biogeography", the study of the geographic distribution of organisms. Although much work remains to be done, we do know of differences between soils, lakes, and the oceans even at the level of phylum and other high phylogenetic groups. Here we focus on oxic environments; the absence of oxygen selects for quite different types of bacteria and other microbes (Chapter 11) than those seen where oxygen is abundant.

Of the >90 known phyla of bacteria in nature, only a few are abundant in any particular environment. Bacterial communities in the oceans, for example, have about ten phyla, of which five each account for at least 5% of the total abundance (Yilmaz et al., 2016). A couple of these phyla are found in many environments, while the abundance of the others varies among the main habitats of the biosphere (Figure 4.9). The *Proteobacteria* phylum is found everywhere, but different proteobacterial classes dominate freshwaters, the oceans, and soils. In freshwaters, *Betaproteobacteria* are most abundant, followed by *Gammaproteobacteria* and *Alphaproteobacteria*. In oxic marine waters, *Alphaproteobacteria* are usually most abundant while *Betaproteobacteria* are much less abundant. Soils also have these three classes of *Proteobacteria*, with *Gammaproteobacteria* being the most abundant of the three.

The other phyla abundant in natural environments include *Bacteroidetes* and *Actinobacteria*. *Actinobacteria*, one of two phyla with Gram-positive bacteria, is abundant in freshwaters and soils, but less so in the oceans. A few groups of *Actinobacteria*, such as those in the acI and acIV clades, can be especially abundant in lakes (Newton et al., 2011). The other phylum of Gram-positive bacteria common in natural environments is the *Firmicutes*. These low G-C bacteria are often retrieved from soils by cultivation-dependent approaches and are abundant in mammalian guts (including our own), but are less abundant than *Actinobacteria* in soils or aquatic habitats when assayed by cultivation-independent methods.

The phylum *Acidobacteria* is quite abundant in soils, but it is rarer in freshwaters and the oceans. This group of bacteria was recognized as a separate phylum only in

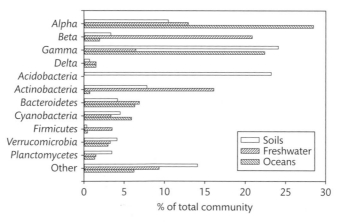

Figure 4.9 Community structure of bacteria in soils, freshwater, and oceans. *Alpha, Beta, Gamma,* and *Delta* refer to classes of the *Proteobacteria*. "Other" includes phyla that individually make up <1% of communities, except for *Gemmatimonadetes*, which accounts for 3.5% of soil communities. Data from Barberan and Casamayor (2010) and Fierer et al. (2012).

1995, because few can be easily cultivated and grown in the laboratory (Kielak et al., 2016). As implied by the name, these bacteria grow best in acidic conditions, and cultivation-independent approaches have demonstrated that *Acidobacteria* are especially abundant in low pH soils, where they can make up a very large fraction, well over 50%, of the total community. As pH increases, their relative abundance decreases, but even for soils with pH values above 6, *Acidobacteria* are still quite abundant, making up 20% of the community (Figure 4.10). While pH is undoubtedly important, other environmental properties, such as concentrations of aluminum, calcium, and magnesium, may also contribute to determining *Acidobacteria* abundance. Most *Acidobacteria* are aerobic heterotrophs while a few are facultative anaerobes (Kielak et al., 2016).

A few bacteria are widely distributed

The differences among major biomes at the phylum and class level are also mirrored at finer phylogenetic levels. Nearly all OTUs are abundant in only a few environments, and few are abundant in many environments. One of the few widespread, ubiquitous bacterial taxa is the SAR11 clade, which was discovered by one of the first cultivation-independent studies of a natural environment, the oceans (Giovannoni, 2017). After its discovery in the Sargasso Sea (explaining the "SAR" of the name), SAR11 has been found in all ocean basins, where it is usually the most abundant clade of bacteria, making up as much as

40% of the total community in surface waters. It is a diverse clade with about ten subclades and a couple of cultivated representatives in the *Pelagibacter* genus. These bacteria are capable of growing aerobically on very low concentrations of only a few organic compounds, making them "oligotrophic" bacteria. This partially explains their success in the oceans. Another key to their success is the size and unique organization of their genomes, as discussed in Chapter 5.

There are few obvious analogues to SAR11 in freshwaters and soils. One reason is that the great heterogeneity and diversity of freshwaters and soils prevents one OTU or even one clade from dominating. There is a freshwater SAR11 clade, subclade IIIb, called LD12 in the freshwater literature (without "SAR"), but it is usually not that abundant. A more abundant group of freshwater bacteria is the acl clade of *Actinobacteria*, mentioned previously. Like SAR11, at least one isolate in the clade has a small genome, only 1.2×10^9 base pairs (1.2 Mbp) (Garcia et al., 2013). However, often lake communities are dominated by *Betaproteobacteria*, in which case the genera *Limnohabitans* or *Polynucleobacter* in the Bet1 and Bet2 lineages are most abundant, depending on whether the lake is acidic or alkaline, respectively (Jezbera et al., 2012). These freshwater bacteria are aerobic heterotrophs.

In soils, one candidate for the terrestrial SAR11 is the class *Spartobacteria* in the phylum *Verrucomicrobia* (Brewer et al., 2016). In contrast with Figure 4.9, other studies

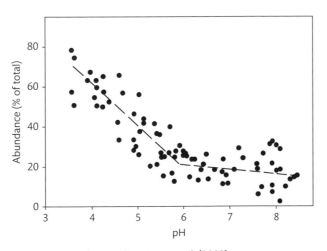

Figure 4.10 Abundance of *Acidobacteria* in soils. Data from Jones et al. (2009).

have found *Verrucomicrobia* to make up over half of the community in prairie soils of the USA, with one taxon, DA101, being especially abundant. DA101 is six-fold more abundant in grassland soils than in forest soils. The only described and sequenced isolate of the *Spartobacteria* is *Chthoniobacter flavus*, a slow-growing aerobic hetero-troph. But this isolate is not closely related to DA101 (92% similar 16S rRNA gene sequence). Analogous to *Pelagibacter*, the genome size of DA101 is much smaller than the typical soil bacterium. A small genome size may be the only feature shared by these abundant bacterial clades, a topic returned to in Chapter 5.

Archaea in non-extreme environments

DNA sequence data from a few isolates in pure cultures were the basis for putting archaea into its own domain of life, quite distinct from bacteria and eukaryotes. Based on these isolates, it was once thought that archaea thrive in only extreme environments and were analogues to early life on the planet, as mentioned in Chapter 1. However, studies using cultivation-independent methods have demonstrated that archaea live in nearly all natural environments, not just extreme ones. Still, archaeal abundance is low relative to bacteria in most natural environments. In soils and surface waters of the oceans, for example, archaea make up <5% of total microbial abundance (Bates et al., 2011; Karner et al., 2001). Also, the diversity of archaea is much lower than bacterial diversity. In contrast with the thousands of bacterial OTUs in a typical soil community, for example, one study found only two OTUs of archaea in the group 1.1b of *Thaumarchaeota*, referred to as *Crenarchaeota* at the time (Bates et al., 2011).

One environment where archaea can be relatively abundant, although still not very diverse, is the deep ocean (Figure 4.11). In waters below about 500 m, archaea can account for as much as 50% of all microbes. Of the two main marine archaeal phyla, *Thaumarchaeota* (once placed in the *Crenarchaeota*) is usually more abun-dant than *Euryarchaeota*. Because the deep ocean makes up 75% of the total biosphere, the total abundance and biomass of archaea in the entire biosphere is substantial. Archaea are also abundant in surface waters in winter near Antarctica and in the Arctic Ocean, whereas in the summer their abundance is low in both polar systems

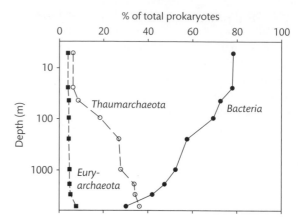

Figure 4.11 Abundance of bacteria and two archaeal phyla in the North Pacific Ocean. Data from Karner et al. (2001).

(Church et al., 2003). Like the deep ocean, polar waters in winter are dark, with low concentrations of organic material and phytoplankton biomass. A worldwide survey found that a member of Marine Group 1 in the *Thaumarchaeota* was the most abundant archaeon and the second most abundant prokaryote on average in the deep oceans (Salazar et al., 2016). In contrast with the results from Karner et al. (2001) given in Figure 4.11, Salazar et al. (2016) found that archaea made up only 2–16% of all microbes in the deep ocean. The difference could be due to methodology or to poorly characterized variability among deep oceanic habitats.

Key to the distribution of archaea in nature is their physiology. As will be discussed in Chapter 12, many archaea in soils and the oceans appear to be chemo-lithoautotrophs that oxidize ammonia to obtain energy. These archaea appear to outcompete and outnumber ammonia-oxidizing bacteria when ammonium concen-trations are low (Stahl and de la Torre, 2012). The abun-dance of ammonia-oxidizing archaea in the deep ocean is high relative to heterotrophic bacteria because the low supply of organic material in deep waters limits the heterotrophs. But in surface oceanic regions and soils where organic material is readily available, heterotrophic bacteria can attain much higher biomass levels, using metabolisms that yield much more energy than ammonia oxidation and other chemolithotrophic pathways. The result is many heterotrophic bacteria and few archaea. It is unclear why heterotrophic archaea are not more abundant in aquatic habitats and soils.

Ecological processes that assemble microbial communities

We have seen that the taxonomic make-up of microbial communities in soils, lakes, and the oceans varies greatly. Some of that variation was explained by differences in environmental properties, such as organic carbon, ammonium concentrations (for chemolithoautotrophic archaea), and pH. We will now discuss in more detail these properties and others and how they shape microbial diversity and community structure. Before doing so, let's step back and ask about the ecological and evolutionary ("eco-evolutionary") processes governing the assembling of microbial communities. Studies of macro-organisms have suggested four general processes (Table 4.2).

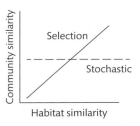

Figure 4.12 Similarity between microbial communities among differing habitats. If selection, a deterministic process, is most important, then different habitats will lead to different microbial communities (solid line). If stochastic processes such as drift and limitations to dispersal are important, then the biotic and abiotic characteristics of the habitat are less important in shaping the make-up of the community (dashed line).

Deterministic versus stochastic processes

If only selection is operating, the presence or absence of a species in a community depends only on its capability to survive, if not thrive, under the abiotic and biotic conditions of the habitat. Selection is a deterministic process in that the success of a species in a habitat is set strictly by how its biochemical, physiological, and ecological capabilities match up with the habitat. The selection hypothesis leads to the prediction that habitats similar to each other should have more similar microbial communities than habitats that differ (Figure 4.12). It also predicts that given enough time (a key point), the structure of a community will always end up to be the same, regardless of the starting point and founding members.

In contrast with selection, diversification and drift are both stochastic processes. Diversification, which is the generation of new genes and genetic mechanisms, depends on random acts of mutation. Drift is also based on chance, not on the microbe's capabilities in survival and growth. The drift hypothesis leads to the prediction that similarity between communities should have nothing to do with similarity in habitat characteristics (Figure 4.12). Studies of large organisms have shown that drift is important when selection is weak and diversity and abundance are both low.

Dispersal, the fourth ecological mechanism assembling communities, has both stochastic as well as deterministic elements. A microbe may be transported to another habitat because of its small size (a deterministic element) but also because, by chance (stochastic), it is in the right place at the right time to be carried by wind or currents or by large organisms moving from one habitat to another. Dispersal may be most important for those microbes able to enter into a resting state, such as a spore, that protects the microbe from desiccation, ultra-violet light, and other harsh environmental conditions experienced during transit. Also, for bacteria, archaea,

Table 4.2 Ecological and evolutionary processes governing the assembling of microbial communities. Based on Nemergut et al. (2013).

Process	Related terms	Definition
Selection	Species sorting, environmental filtering	Capabilities of a taxon determine whether it is present under the abiotic and biotic conditions of the habitat
Diversification	Mutation, lateral gene transfer, speciation	Generation of new genes and genetic controls
Drift	Neutral theory*, stochastic processes	Random changes over time
Dispersal	Mass effect	Movement of organisms over space

* Communities that assemble under neutral processes including dispersal effects as well as drift.

and other asexual microbes, only one cell is needed to establish a population in a new habitat.

Microbial ecologists usually focus on deterministic processes, especially selection or "environmental filtering," and explore how biotic and abiotic factors shape microbial community structure. However, stochastic processes may be more important than usually assumed. One thorough review concluded that environmental properties and distance (dispersal limitation) explained only about 50% of the variation in microbial community structure in soils and aquatic habitats (Hanson et al., 2012a). The large fraction of unexplained variation in community structure raises questions about deterministic processes. On the other hand, the missing 50% may be due to unmeasured environmental properties. More work is needed on this important topic.

Everything everywhere?

Another way to describe some of the deterministic processes that assemble microbial communities is with the aphorism attributed to Lourens G. M. Bass Becking (1895-1963): "everything is everywhere, but the environment selects." That is, environmental conditions determine whether a taxon is abundant or not in a particular habitat; geography and history play no role. As a result of "everything is everywhere," there are many rare OTUs making up the rare biosphere. These rare OTUs, according to the "seed bank" hypothesis (Lennon and Jones, 2011), can respond to new environmental conditions, fulfilling the second part of the aphorism, "the environment selects."

There is evidence for and against the Bass Becking hypothesis. The hypothesis is the best explanation for the occurrence of spores from thermophilic bacteria in Arctic fjord sediments (Müller et al., 2013). It also accounts for why a symbiotic bacterium living in the giant marine ciliate *Zoothamnium niveum*, in the Mediterranean Sea, is the same as the one in the Caribbean Sea (Rinke et al., 2009). It is remarkable that bacterial communities from similar soil environments but at different latitudes are similar (Fierer and Jackson, 2006), bacteria in Arctic lakes appear similar to those in Antarctica, and likewise for archaea in soda lakes of Mongolia and Argentina (Pagaling et al., 2009). Gibbons et al. (2013) concluded that they would have found in their English Channel samples all phylotypes living in all oceans if only they had sequenced sufficiently "deeply" or thoroughly.

Evidence against the Bass Becking hypothesis comes from studies showing that dispersal of microbes is in fact limited; everything is not everywhere at least immediately. These studies examine the similarity (beta diversity) of communities separated by various distances: the "distance–decay" relationship. The slope of similarity versus distance graphs is on average lower for microbes than for larger organisms (or for fungi), but it is not zero, indicating that the greater the distance between two habitats, the more different the microbial communities are (Figure 4.13). Dispersal limitation is one explanation for significant distance–decay relationships. It has been argued that it would take much longer than the age of the planet for all microbes to disperse around the world through the atmosphere (Nemergut et al., 2013).

Testing the Bass Becking hypothesis runs into several complications. To begin, it is impossible to sequence all microbes in a sample and show that a microbe is really absent, which would disprove "everything is everywhere." Also, the evidence previously cited for or against the hypothesis can be questioned. Rather than indicating dispersal limitations, a significant distance–decay relationship could result if habitats separated by greater distances experience increasingly different environmental conditions. The similarity of communities in geographically separate habitats seems consistent with the hypothesis, but Bass Becking would say that the microbes have to be exactly the same. However, whether two microbes are the "same" depends on the phylogenetic resolution of the analysis. Time also matters. A community in a freshly created habitat cannot initially have all of its eventual members, because even unlimited dispersal is not instantaneous; time is needed for new colonizers to arrive.

For biogeochemical processes, rather than the Bass Becking hypothesis, a more relevant aphorism may be, "similar enough things are everywhere, but the environment selects." That is, perhaps even if everything or every microbe isn't everywhere, similar enough ones with the necessary metabolic capabilities are present in every habitat such that a process never depends on new colonizers arriving from elsewhere. Whether this new aphorism is correct depends on the redundancy of microbial communities, a topic discussed in Chapter 7.

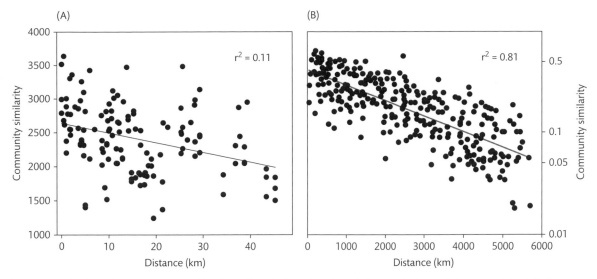

Figure 4.13 Similarity between communities (beta diversity) as a function of distance for bacterial communities in Yellowstone Park lakes (A) and vascular plant flora in Appalachian spruce fir forests (B). The beta-diversity indices were the number of shared OTUs for bacteria and the Jaccard's index for the plant communities. Note that distance explains a low fraction (a low r^2) of the variation in beta diversity for bacteria, but a high fraction for higher plants, indicating that dispersal is less limited for bacteria than for higher plants. The bacteria data were provided by M. Beman from Hayden and Beman (2016) and the plant data are from Nekola and White (1999).

More work is needed to explore this and other concepts related to the Bass Becking hypothesis.

What controls diversity levels and bacterial community structure?

Regardless of the importance of dispersal limitations and other, purely stochastic processes, selection is clearly important in setting microbial diversity and shaping the make-up of bacterial communities. The challenge is figuring out which factors are most important in which environments for a given time frame. These factors potentially include anything that affects growth rates or biomass levels, collectively referred to as "bottom-up" or "top-down" factors, respectively. We are interested in whether these factors affect alpha diversity, in particular species richness, and the composition (structure) of microbial communities.

Oxygen, temperature, salinity, and pH

These factors have huge effects on many properties of microbes, including the abundance of taxa within micro-

bial communities. They directly affect microbes and indirectly affect them via how they control other environmental properties. Oxygen (O_2) is most important. In addition to affecting the speciation of other compounds, oxygen selects for aerobic heterotrophic microbes that use oxygen as an electron acceptor and for other microbes able to tolerate toxic by-products formed from oxygen. The absence of oxygen (anoxia) selects for an entirely different suite of microbes with very different types of metabolism (Chapter 11).

Temperature, salinity, and pH have many large effects on microbial processes and in some cases on the make-up of microbial communities. Temperature, for example, explains a large portion of the variation in bacterial community structure in the oceans and in hot springs, but there is no apparent relationship between diversity and temperature in soils (Fierer and Jackson, 2006). Temperature is a reason, albeit not the only one, why bacterial diversity tends to be lower at the poles than in low latitude oceans, but again this is not the case for soils (Figure 4.14). However, in experiments intended to mimic climate change, long-term warming does affect the composition of bacterial communities in soils (DeAngelis et al., 2015). Salinity may

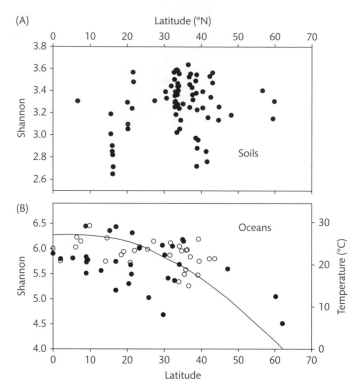

Figure 4.14 Diversity of bacterial communities in soils (A) and the oceans (B) from the tropics to the poles. There is a significant difference in the Shannon index with latitude in the oceans, but not in soils. The difference in diversity between soils and the oceans is due to methodology; other data indicate that soils are more diverse than the oceans. The oceanic data are from both the northern (open symbols) and southern hemispheres (solid symbols). Temperature (the solid line) explains only 15% of the variation in oceanic microbial diversity. Soil data are from Fierer and Jackson (2006). The oceanic data were provided by Guillem Salazar from Sunagawa et al. (2015).

explain the high abundance of *Betaproteobacteria* in lakes and its low abundance in the oceans, and it accounts for large-scale differences in community structure of all bacteria and archaea (Lozupone and Knight, 2007).

Extremes in salinity and in other physical factors lead to low diversity. That is the case for salt pans with very high salt concentrations—near saturation for NaCl. These systems have only a couple of types of bacteria and archaea. Waters with very low pH are also not very diverse. Ponds polluted by runoff from mines can have pH near 1 or lower, and harbor only a couple of prokaryotic taxa, although biomass may be quite high (Méndez-García et al., 2015).

Moisture and soil microbial communities
Water content has large, well-known impacts on microbial activity in soils but complex, poorly understood

effects on diversity. One impact of water on diversity is tied to soil composition and pore connectivity. Soils with high silt and clay content have low water potential and pore connectivity. In these soils, competition is reduced because pores with water and thus active bacteria are separated by dry stretches, creating micro-environments where some bacteria can flourish (Carson et al., 2010). The end result can be higher bacterial diversity in drier soils. Water content also affects oxygen availability. It is low in saturated, water-logged soils because the use of oxygen by respiration exceeds its input by diffusion. The lack of oxygen means that anaerobic bacteria can flourish (Chapter 11), potentially leading to high diversity. Other soil properties, such as salt and nutrient concentrations, connected to water content could also affect diversity. Given these complications, perhaps it is not surprising that bacterial alpha diversity did not change in subtropical soils when water input was experimentally varied, whereas fungal diversity decreased (Zhao et al., 2017). However, the taxonomic composition of both the bacterial and fungal communities varied with water input.

Organic material and inorganic nutrients

One explanation for the high diversity of tropical rain forests and coral reefs is their high rates of primary production. As productivity increases, more herbivores can be supported, which in turn feed more carnivores and so on up the food chain, but only up to a point. At very high levels of productivity, diversity decreases because only a few opportunistic species take over. The end result is a hump-shaped relationship between productivity and diversity. This pattern is also seen for some microbes in some places. Figure 4.15A shows the hump-shaped relationship for fungal diversity, and it has been seen for phytoplankton in the oceans (Vallina et al., 2014), but it may not be the norm for other microbial communities. Overall, negative relationships between diversity and productivity actually may be more common in freshwater and marine habitats (Smith, 2007) and in soils (Figure 4.15B).

The shape of the diversity versus productivity curve has implications for whether top-down or bottom-up factors are more important. The increase in diversity with productivity suggests that top-down control by grazing and viral lysis prevents superior competitors from dominating the community, leading to high diversity, whereas a negative relationship would result if those superior competitors crowd out other taxa, leading to low diversity.

In addition to affecting the diversity (richness), productivity and organic carbon can affect the composition of bacterial communities at phylogenetic levels ranging from OTUs to even phyla. These effects have been observed in studies examining the variation in bacterial communities during phytoplankton "blooms" (large increases in algal biomass; Chapter 6) and in soils differing in organic carbon levels. The effects have also been shown by experiments in which microbial communities are followed over time after the addition of organic material. The additions select for "copiotrophic bacteria," which by definition are capable of using high concentrations of organic material. In contrast, low concentrations select for "oligotrophic bacteria." Chapter 5 discusses genomic data, indicating that oligotrophic bacteria probably dominate most natural environments because concentrations of organic and inorganic compounds are usually quite low. Chapter 8 discusses other data, suggesting that copiotrophic bacteria can grow quickly in response to the occasional high input of organic material.

In addition to amounts, the type of organic compounds in an environment may affect bacterial community structure. We expect selection for bacteria capable of growing quickly on organic compounds that require specific enzymes in order to be used. We do know that addition of one or two organic compounds often leads to dominance of the community by only a few types of bacteria; this is the principle behind enrichment cultures (Chapter 1). The presence of, for example, structural polysaccharides from higher plants, such as cellulose and hemi-cellulose, selects for certain bacteria. These

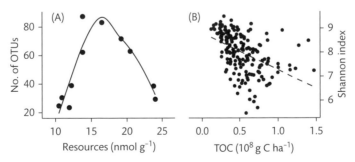

Figure 4.15 Diversity of heterotrophic microbes as a function of organic resources. In Panel A, the "hump" relationship, typical for large organisms, was seen between the number of fungal OTUs and a measure of resource availability, total microbial biomass, as measured by phospholipid fatty acids. Panel B illustrates another common pattern, a negative relationship between bacterial diversity and total organic carbon (TOC) in soils. Panel A data from Waldrop et al. (2006) and panel B data provided by M. Delgado-Baquerizo from Delgado-Baquerizo et al. (2017).

"specialist" bacteria differ from "generalists" able to use a wide variety of organic compounds. The terms are also used for bacteria affected by other environmental properties.

Inorganic nutrients such as ammonium and phosphate contribute to determining the diversity of heterotrophic bacteria in soils, but have less of a role in aquatic habitats. In one study, inorganic nutrient fertilization decreased the richness and changed the composition of soil bacterial communities (Zeng et al., 2016). In contrast, an analogous study did not see any direct effect of phosphate additions to Mediterranean Sea bacterial communities, even though this system is thought to be phosphorus limited (Fodelianakis et al., 2014).

Predation and viral lysis

The bottom-up factors just discussed are only half of the story about what sets microbial diversity and the composition of microbial communities in natural environments. The other half is top-down factors: grazing and viral lysis. Both have the potential for determining the success of specific microbes and thus shaping diversity levels and overall community structure of microbes. Although more data are needed to say for certain, most studies point to viruses having a larger impact than grazers on community structure.

The way grazing affects community structure involves the same factors known to affect grazing rates. Cell size is one such factor, because grazing varies strongly with the size of the prey (Chapter 9). The chemical composition of cellular surfaces is another property of prey that affects grazing and is likely to account for why a particular grazer preys more heavily on one microbial group than another. For those two reasons and others, some bacteria are known to be relatively resistant to grazing, helping to explain their high abundance. Resistance to grazing helps to explain the success of the betaproteobacterial genus *Polynucleobacter*, and of the phylum *Actinobacteria*, in lakes. Similar mechanisms are undoubtedly operating in the oceans and in soils. However, other studies have not been able to detect an impact of grazing on bacterial community structure in marine or soil habitats (Sauvadet et al., 2016; Baltar et al., 2016).

There are theoretical reasons to expect the impact of viruses on diversity to be stronger than that of grazers.

These reasons are based on aspects of viral and protist ecology discussed in more detail in Chapters 9 and 10, so only a brief outline will have to do here. The first reason for a larger viral impact is that viruses substantially outnumber grazers. Viruses are usually more abundant than prospective microbial hosts by about ten-fold; while in contrast, grazers are much less abundant than their microbial prey. Put together, there are about 10^4 more viruses than grazers on average in nature. Another important part of the argument is that viruses infect their host by highly specific mechanisms, often involving transporter proteins synthesized by the microbe to take up compounds from the environment. In contrast, grazers are less selective in what prey they eat (Chapter 9). The end result is a larger impact by viruses than grazers on the diversity of bacteria and potentially other microbes.

The role of viral lysis in shaping microbial communities has been explored via the "kill-the-winner" hypothesis (Figure 4.16). Key to the hypothesis, the microbe that takes up more dissolved compounds grows faster; it is the "winner." But being a "competition specialist" comes with the cost of having poor defenses against viruses, because the same transporter proteins making it a superior competitor also open up these microbes to more attacks by viruses. So the successful competitor attracts more viruses and higher viral lysis, which in turn depresses abundance of the winner: the "kill" part of the hypothesis.

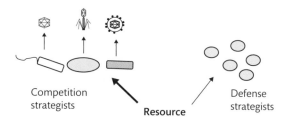

Figure 4.16 Role of viruses in structuring microbial communities, according to the "kill-the-winner" hypothesis. Even though the competition specialists are better at using resources (signified by the thicker arrow from "resource") and grow faster, they do not become abundant because of viral lysis. The viruses also select for diversification among the competition specialists and create more diversity. The defense specialists are abundant in spite of slow growth, according to this theory, because of minimal losses due to viruses. Modified from a diagram provided by T. F. Thingstad. See also Thingstad (2000).

Because of viral lysis, the superior competitor does not crowd out inferior competitors, allowing more species to coexist and ending up with a more diversity community. Viruses also force an arms race within the competition specialists, leading to the splitting up of species into strains and eventually the formation of new species (Thingstad et al., 2014). The opposite, the "defense specialist," grows slowly because it is not efficient at using compounds, but it is abundant because it is resistant to viruses (and perhaps grazers).

The kill-the-winner hypothesis has spawned much debate and research because it encapsulates many important ideas about the interactions between factors thought to affect microbial diversity. The abundant SAR11 clade is an important test case. The clade was initially thought to be abundant because it is a defense specialist. The small cell size of bacteria in the clade would lead to low predation, and its surface properties might limit grazing by some types of predators (Dadon-Pilosof et al., 2017). At first, viruses specific for infecting SAR11 were unknown. Other evidence, however, came in against SAR11 being a defense specialist. FISH-based data indicated that SAR11 bacteria are as big and grow as fast as other bacteria in the north Atlantic Ocean (Malmstrom et al., 2005). Then viruses attacking the clade were discovered to be quite abundant (Zhao et al., 2013). Våge et al. (2013) proposed a hypothesis to reconcile the existence of viruses attacking a defense specialist, but more experimental work is needed. The original discoverer of the clade, S. J. Giovannoni, has argued for SAR11 being a superior competitor for a few dissolved organic compounds (Giovannoni, 2017).

Community structure of fungi

Cultivation-independent approaches are just as important in revealing the diversity of eukaryotic microbes as they are for bacterial communities. Here we discuss a few general aspects of fungal diversity. Chapter 14 has more about mycorrhizal fungi, which form symbioses with higher plants. Fungal diversity has been explored by sequencing the ITS region, although it can give the same picture as the 18S rRNA gene data (Berruti et al., 2017).

Fungal biomass and diversity are much higher in soils than in aquatic habitats (Figure 4.17), and fungi can account for most of the eukaryotic rRNA genes found

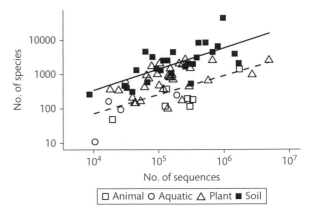

Figure 4.17 OTU richness of fungal communities in four systems as a function of sequencing effort. The solid line is from a regression analysis of fungal communities in soils, while the dashed line is from aquatic fungal communities. Fungi are more diverse in soils and on plants than in aquatic habitats and on animals. Data from Peay et al. (2016).

in soils (Urich et al., 2008). Fungal biomass can equal or exceed that of bacteria in some soils, depending on environmental conditions (Chapter 7). Even so, fungal communities are about half as diverse as bacterial communities in soils; they are 10 to 10,000-fold less diverse in the oceans (Peay et al., 2016). Fungal communities in soils are made up of mostly Ascomycota (58% of all fungal rRNA transcripts) and Basidiomycota (23%) (Choma et al., 2016). The most abundant classes, such as Leotiomycetes (10%) and Eurotiomycetes (8%), are all in the Ascomycota.

The processes assembling bacterial communities are also relevant to thinking about how fungal communities are put together. In particular, like bacterial ecologists, fungal ecologists have discussed the role of dispersal limitation versus selection. More so than seen by bacterial ecologists, fungal ecologists have found that the geographical distance separating similar habitats can result in different fungal communities (Peay et al., 2016). As with bacteria, the comparisons depend on scales of space, time, and taxonomy. Although the major genera of fungi, such as *Russula* and *Boletus*, are found everywhere, enough fungal species are endemic and restricted to specific locations that the source of dust can be deduced by its accompanying fungi. Major environmental properties affect bacteria and fungi differently, a topic returned to in Chapter 8. For example, unlike soil bacteria, the diversity of most soil fungi is highest in the tropics (Tedersoo et al., 2014).

Relevance of community structure to understanding processes

We should want to know about the diversity of microbial communities if only because these organisms are the most abundant on the planet. Understanding the phylogeny of microbes is key to understanding the evolution of all life. Yet for this book, it is relevant to ask about the relationship between community structure and biogeochemical cycles. What, if anything, do we really need to know about microbial community structure if the goal is to understand processes? The answer will vary with the process and with what specifically we want to know.

Some biogeochemical processes are carried out by microbes in well-defined phylogenetic groups. In this case information about the relevant taxa informs our understanding of the process. A good example is the photoautotrophs in oxic habitats. As discussed in Chapter 6, several different types of microbes, ranging from small coccoid cyanobacteria to large diatoms, are involved in primary production in these habitats. A system based on cyanobacterial primary production differs greatly from one based on diatoms. Other examples include methanogenesis and sulfate reduction.

For other processes, the utility of community structure information is less clear cut. An example is the oxidation of organic material by aerobic heterotrophs (Chapter 7). It clearly matters whether fungi or bacteria are the main decomposers of organic material in soils, but the argument for more detailed information about the community structure of just bacteria or just fungi is subtle. The relationship between diversity and organic carbon decomposition is complicated by the presence of many "redundant" taxa with similar metabolic capacities.

Still, for some organic compounds, such as organic pollutants, the type of microbe does matter. Both cultivation-dependent and 16S rRNA gene-based approaches have found known petroleum-degrading bacteria to be enriched following the 2010 Deepwater Horizon oil spill in the Gulf of Mexico (Hazen et al., 2010), and other studies have found that rare biosphere bacteria were more important than the most abundant bacteria in hydrocarbon degradation following the spill (Kleindienst et al., 2016). These studies reveal important insights into the timescale of the microbial response and of Gulf's recovery following this large oil spill.

In all other areas of ecology, it is essential to know which organism is present and active in an environment. It would be remarkable if microbial ecology were different.

Summary

1. To circumvent the problem of culturability, microbial ecologists use cultivation-independent methods to examine specific genes, usually 16S rRNA genes for prokaryotes and 18S rRNA genes and the ITS region for eukaryotes, to deduce the taxonomic composition of communities and the relative abundance of individual taxa.

2. The microbes isolated by standard cultivation approaches are quite different from those observed in nature using cultivation-independent methods.

3. Of the >90 phyla of bacteria found in the biosphere, only a few (<10) are abundant in any particular habitat. Microbial communities are usually dominated by a few OTUs and clades, while most are in low abundance, making up a rare biosphere. Only a few bacteria are widespread within soils, lakes, or the oceans.

4. Microbial communities are assembled by both stochastic and deterministic processes. Among the deterministic processes, selection by both bottom-up (such as salinity, temperature, and organic material) and top-down (viruses) factors determine the diversity and composition of microbial communities.

5. In every other field of ecology, identifying the organisms is essential for understanding their role in the environment. The ecology of microbes is likely to be no different.

CHAPTER 5

Genomes and meta-omics for microbes

Questions about the types of microbes present in natural environments are usually addressed by examining rRNA genes, as discussed in Chapter 4. When interested in a specific biogeochemical process, however, microbial ecologists often turn to other genes. These other genes, the "functional genes," encode a key enzyme of the process being investigated, such as *rbcL* for CO_2 uptake, *pmoA* for methane oxidation, or *nah* for naphthalene degradation. While informative, there are several problems with this functional gene approach. When examined individually, these genes are retrieved by PCR-based methods using primers that target conserved regions of the genes. Any genes too dissimilar to the primers will not be sampled. Often, several genes are important in a process, meaning several primer sets have to be used, and several PCR reactions run. Also, functional genes often cannot be used to deduce which microbe is carrying out the process, for reasons to be discussed in this chapter.

This chapter will describe some genomic approaches to circumvent these problems. We'll also discuss other "omic" approaches, those that focus on RNA (transcriptomics) and protein (proteomics); still another omic approach is "metabolomics," which is the study of the metabolites or the low molecular weight compounds in cells. When an omic approach is applied to an entire community, the prefix "meta" is added; metagenomes refer to the genomes of all microbes in a community, metatranscriptomes for the transcripts expressed by the entire community, and so on. From the viewpoint of microbial ecology, omics can be considered simply as a suite of approaches to examine the evolution, physiology,

and biogeochemical role of microbes in nature. However, genomic-based fields are much more than collections of methods. The genome and the other "omes" of a cell are all major, defining features of an organism. To know those features for a microbe is to know that microbe, a big step forward in understanding its roles in nature. Rather than focusing on those roles or on processes, however, this chapter will provide the basics for understanding the application of omic approaches in microbial ecology. Genomics and other omics are large, dynamic parts of the field.

What are genomics and environmental genomics?

In contrast with the study of a single gene or even of several genes simultaneously, the field of genomics uses data about entire genomes of organisms: the complete sequence of all genes in the right order and organization in each chromosome (but see Box 10.1). The first genome, that of a virus (ϕX174), was sequenced in 1977. It is extremely small, only 5386 nucleotides, just enough for only eight genes. The genomic field really got its start in 1995 with the complete sequencing of two bacteria, *Haemophilus influenzae* and *Mycoplasma genitalium* (Loman and Pallen, 2015). These two bacteria were chosen for sequencing, in part because they are pathogens, but perhaps more importantly because they have small genomes—small for bacteria, but still much larger than the genome of ϕX174 and other viruses. The genome sizes are 1.8 and 0.58 Mb for *H. influenza* and *M. genitalium*, respectively, where "Mb" is a million base

Processes in Microbial Ecology. Second Edition. David L. Kirchman. Oxford University Press (2018). © David L. Kirchman 2018.
DOI 10.1093/oso/9780198789406.001.0001

pairs of DNA. The small size and simplicity of these bacterial genomes were important at the time because genomic sequencing approaches were just being worked out. The bacteria were sequenced by a shotgun cloning approach, a radical idea when it first appeared (Figure 5.1). The sequence of the first eukaryote, *Saccharomyces cerevisiae* (baker's yeast), was published in 1996, followed by publication of a draft of the human genome in 2000.

Now it is routine to sequence prokaryotic genomes and nearly routine for eukaryotes, including humans. It has become routine because the cost of sequencing has greatly decreased since φX174 was sequenced. In the 1970s, sequencing a gene cost nearly $1000 per base pair, and $2.3 billion was spent in the late 1990s sequencing the human genome. Today, it costs a few dollars to get a draft sequence (Box 5.1) of a bacterial genome, and soon it will be possible to sequence a human genome for less than $1000 thanks to the development of new high-throughput sequencing approaches. Knowing your own genome will be the starting point of evaluating your health and physical well-being. New, high-throughput sequencing technology has been driven by the promise

---

Box 5.1 Complete or draft genomes?

Some sequencing projects may gather all of the sequences from the automated part of the sequencing process but not attempt to put them together into a "closed" genome with no gaps. Instead of being complete or closed, the genome is said to be a "draft." Nearly all of an organism's genome may be determined, but the remaining 1–10% is left undone. Assembling the sequences, filling in the gaps, and finding all of the missing pieces are time consuming and expensive. The time and money saved by not finishing the genome can go into doing more genomes, which is important for many ecological questions. The disadvantage is that a gene missing from a draft genome may in fact be in the unsequenced genetic material.

---

of high profits in biomedical applications, but the new approaches are being used in all areas of biology, including microbiology and microbial ecology. The expensive and time-consuming part of genomic work is now not the sequencing but the analysis and interpretation of the sequence data.

The number of ecologically relevant organisms that have been sequenced is still small compared to the number of pathogens, but these numbers are increasing, nearly exponentially so. Now over 75,000 genomes of bacteria and archaea are available in one form or the other (https://gold.jgi.doe.gov), as are the genomes from over 19,000 eukaryotes, including over 6800 fungi. "Environmental genomics" or "ecological genomics" are terms used for genomes of organisms important in the environment. Even though our main focus here is on natural communities of uncultivated microbes, there is much to be learned from the genomes of cultivated microbes, even if these microbes are often only distantly related to the most abundant or ecologically important microbes in nature. Some generalizations based on cultivated microbial genomes will be discussed. A huge advantage of looking at these genomes is that the function of the genes can be experimentally determined and the overall genomic structure can be tied to the biology of the cultivated microbe.

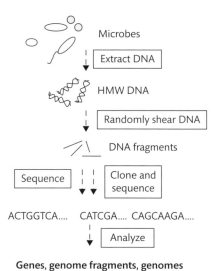

Figure 5.1 Shotgun sequencing of genomes. The approach was first used for bacteria in pure culture, but now it is used for microbes in complex communities. Initially, DNA fragments had to be cloned before sequencing was possible, but high-throughput sequencing techniques have eliminated the need for cloning. "HMW" is high molecular weight DNA.

Turning genomic sequences into genomic information

Once the sequence of a genome has been determined, the real work begins. While some analyses can use the raw sequence data, most questions require that the sequences be given some meaning, if not a name and a solid description of the function. The first step is to find open reading frames (ORFs), sequences of DNA that may encode a protein and begin with a start codon (usually ATG, which encodes methionine) and end with a stop codon (TAA, TGA, or TAG), and are possible "protein-encoding genes." Next, bioinformaticians try to determine whether the ORF is actually a gene by comparing the sequence with others already archived in databases such as GenBank. If the gene is similar enough to known genes in the databases, a function can be assigned. A frequently used tool for comparing sequences is "BLAST" or basic local alignment search tool. BLAST analysis and other steps in the annotation process are automated, with most of the work done by sophisticated computer programs. People doing "manual annotation" are also necessary for some genes and parts of the genome.

The BLAST analysis can turn up significant "hits" or genes in GenBank that are similar in sequence to the unknown gene in the genome being examined. Ideally, the function of the known gene in GenBank has been established experimentally, in which case it is likely that the unknown gene has the same function. Often, however, the function of the known gene has not been directly characterized, and it too is assigned a function based on similarity to another known gene. Enzymes identified by gene sequence similarity alone are often called "putative," to emphasize the uncertain nature of their assigned function. The new gene may turn out to be most similar to another known gene in GenBank without any assigned function. These are "conserved unknown" genes. Finally, sequencing often turns up truly unknown ORFs, sometimes called ORFans, without a significant similarity (homology) to known genes. Some of these ORFans may have come from viruses or other mobile genetic elements (Yu and Stoltzfus, 2012).

Even well-characterized organisms have many genes with unknown function (Figure 5.2). When one of the best characterized organisms, *E. coli* K12, was first sequenced, nearly 40% of its 4288 protein-encoding genes could not be assigned a function (Blattner et al., 1997). The fraction of unknown genes has dropped, but it is still surprisingly high even after years of work. Calling a gene "unknown" depends on one's definition of "unknown" and standards for the amount of data needed before concluding that the function of a gene is truly known. Regardless of semantics, there is still much to be learned about many genes in genomes, even from *E. coli* and other intensively studied organisms. The number of unknown genes is even higher for microbes and other organisms that have not been examined extensively.

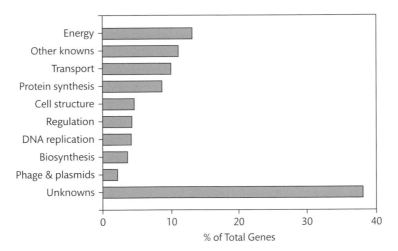

Figure 5.2 Annotation of genes in *E. coli* K12 when it was first sequenced. "Other knowns" is the sum of several, less abundant functional categories that individually make up <2% of the genome. The number of unknown genes has decreased by at least 50%, but it is still high. Data from Blattner et al. (1997).

Lessons from cultivated microbes

The number of poorly characterized and completely unknown genes is one of the first lessons to be learned from examining genomes of cultivated microbes. There are several others, all useful for applying genomic approaches to and thinking about uncultivated microbes in nature.

Similar rRNA genes, dissimilar genomes

The first phase of genomic studies focused on microbes from different genera and even domains; an archaeon (the methanogen *Methanococcus jannaschii*) was sequenced soon after the first bacteria. The second phase focused on organisms that appeared to be very similar, and even strains of the same microbial species were compared. These organisms were usually pathogens or microbes, such as *E. coli*, whose physiology and molecular biology had been examined extensively over the years.

It was startling when early comparative genomic studies found that closely related organisms can have quite different genomes. For example, when the first three strains of *E. coli* were sequenced, it was surprising to find that they had in common only about 40% of all of their genes (Welch et al., 2002). Similarly, strains assigned to the plant pathogenic bacterium *Ralstonia solanacearum* are only 68% similar at the genome level. Another example is *Prochlorococcus*, the most abundant photoautotroph in nature (Chapter 6). Although the 16S rRNA genes from the high- and low-light ecotypes of this cyanobacterium differ only slightly, whole genome sequencing revealed many differences between the two, even in terms of genome size (Rocap et al., 2003). The high-light ecotype has a smaller genome than the low-light ecotype (1716 versus 2275 genes). Diversity within these bacterial species is greater than the difference between humans and pufferfish (Philippot et al., 2010). Because of this diversity, two bacteria sharing nearly identical 16S rRNA genes can have very different genomes.

Core genomes and pangenomes

Although many isolates and strains within a species have differences in some parts of their genomes, they do have a common set of genes found in all members of that species. This common set is the "core genome." These genes tend to be "house-keeping" genes, essential in central metabolism for keeping a cell running smoothly, and other genes encoding functions that help to define the species. All genes not in the core genome make up the "pangenome." The size of the core versus the pangenome varies among bacteria. One way to visualize this variation is to examine how many new genes are added to the pangenome as more and more isolates are sequenced (Figure 5.3). For some bacteria, sequencing more isolates does not add many new genes, and the pangenome is only slightly bigger than the core genome. These bacteria are said to have a "closed pangenome." The classic example is *Bacillus anthracis*, the cause of anthrax, a disease of livestock and occasionally of humans. It took only about four isolates of this Gram-positive bacterium to find all the genes of its pangenome (Medini et al., 2005).

In contrast, for a bacterium with an "open pangenome," sequencing new strains reveals more and more new genes, resulting in the open pangenome being much larger than the core genome. *E. coli* is an example of an open pangenome; even after sequencing thousands of *E. coli* isolates, new genes continue to turn up. The initial analysis of *Prochlorococcus* ecotypes did not reveal as much diversity as seen for *E. coli*, as previously discussed, but new data suggest that this cyanobacterium also has a diverse, open pangenome; only about half of the genes in a single ecotype are in the core genome for this cyanobacterium (Biller et al., 2015).

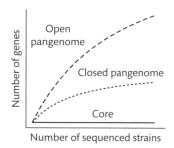

Figure 5.3 The size of closed and open pangenomes relative to the core genome. Here, the core genome for bacteria with a closed pangenome or open pangenome is assumed to be the same size, but that is not necessarily the case.

We expect species with open pangenomes to be members of complex communities in diverse environments and to have many opportunities to exchange genetic material (Medini et al., 2005). The mammalian intestinal tract, the natural environment for *E. coli* with its open pangenome, is complex with many opportunities for gene exchange. The environment of *Prochlorococcus*, another bacterium with an open pangenome, is diverse in terms of light, nutrient availability, and biotic interactions. *Prochlorococcus*, SAR11, and their freshwater and soil counterparts may be ubiquitous in part because they have large pangenomes that serve as reservoirs for traits adaptive for different environments (Barberán et al., 2014).

Data about the pangenome also help to understand processes. One specific, practical example is the causative agent of cholera, *Vibrio cholerae*, which is found in brackish waters and can contaminate drinking water in developing countries. The presence of "virulence adaptive polymorphisms (VAPs)" in the pangenome of this bacterium enables the transition of harmless *V. cholerae* to pathogens via horizontal gene transfer, a process to be discussed further (Shapiro et al., 2016).

Genome size

The genome size of an organism can be estimated without sequencing it, so some differences among organisms were apparent before the start of the genomic age. Still, why genome sizes vary among organisms became better understood as genomic data accumulated. We will discuss some of those reasons, but let us start by discussing the size of genomes for prokaryotes versus eukaryotes.

Genome size varies greatly among prokaryotes and between prokaryotes and eukaryotes. The genome size of sequenced bacteria ranges from about 0.18 Mb for the intracellular symbiont *Carsonella ruddii*, to 13 Mb for the soil bacterium *Sorangium cellulosum* (Koonin and Wolf, 2008). As suggested by these two genomes, obligate symbiotic and parasitic bacteria have very small genomes, whereas some of the largest are found in soil bacteria. The average genome size for archaea is about 2 Mb. There is about a ten-fold variation in genome size among free-living prokaryotes, but this is much less than the variation among eukaryotes; genome size for eukaryotes varies by over 65,000-fold, from 2.3 Mb for a parasitic fungus to about 150,000 Mb for a small flowering plant native to Japan (Elliott and Gregory, 2015). Most fungi have genomes of about 35 Mb, as do protists.

There may be a bimodal distribution in genome size for bacteria. Many bacteria have genomes of around 2 Mb, and there is a second, smaller group of bacteria with genome sizes of around 5 Mb (Koonin and Wolf, 2008). However, it has been argued that the bimodal pattern in the bacterial genome data is caused by the large number of sequences from closely related bacteria; when these are removed, there is no statistical support for two peaks in genome sizes (Gweon et al., 2017). Still, the bimodal distribution is consistent with data from flow cytometry (to be discussed further) of bacterial communities in lakes and the oceans. In graphs of DNA content (fluorescence from a DNA stain) versus side scatter (related to cell size) from flow cytometry data (Figure 5.4), there are

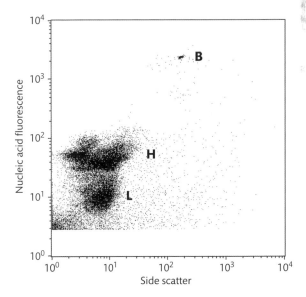

Figure 5.4 Bacteria from coastal waters assayed by flow cytometry. "Side scatter" reflects cell size. Bacteria having low (L) or high (H) amounts of DNA determined by fluorescence from a DNA stain. "B" indicates beads of known size added as a control. Data from Gasol and Moran (2016), provided by J. Gasol and used with permission of the publisher.

often two clouds of points: one due to low nucleic acid-containing cells, the other high nucleic acid-containing cells. The two clouds are interpreted as being from bacteria with small or large genomes, respectively, although some of the fluorescence is also due to rRNA.

The genomes of most eukaryotic microbes are much larger than those of prokaryotes. The genome of the diatom *Thalassiosira oceanica*, for example, is 79.5 Mb and the white-rot fungus *Heterobasidion annosum* has a 40 Mb genome, both 20-fold to over 60-fold larger than that of the bacterium *Pelagibacter ubique*, a member of the ubiquitous SAR11 clade. An exception to the rule for eukaryotes is the green alga *Ostreococcus tauri* (Derelle et al., 2006). Its genome of 12.6 Mb is actually smaller than the genome of the soil bacterium *Sorangium cellulosum* (14.8 Mb), although the alga has 20 chromosomes, while the bacterium has only one. In addition to being small, the genome of *O. tauri* has some other bacteria-like features, such as a large number of genes relative to the total genome size and minimal non-coding DNA, perhaps because this eukaryote is very small, only about one micron.

The genomes of eukaryotes are bigger in part because they have more protein-encoding genes than prokaryotes (Figure 5.5), but they do not have as many as expected for the size of their genome. For example, the white-rot fungus would be expected to have over 36,000 protein-encoding genes for its genome size if its genes were packed into a genome similar to the bacterium *Pelagibacter*. In fact, it has only about 12,000 of these genes. The diatom with its 79.5 Mb genome should have over 85,000 protein-encoding genes, way more than its actual 15,727 genes. Perhaps not surprisingly, the difference between expected and actual number of protein-encoding genes is not as great for the very small green alga, *O. tauri*: it has 8166 protein-encoding genes, not too different from the expected 13,567 genes.

Organization of eukaryotic versus prokaryotic genomes
Eukaryotic and prokaryotic genomes differ in many other aspects in addition to size (Table 5.1). We have just seen that eukaryotic genomes have more protein-coding genes than do prokaryotes, but not as many as

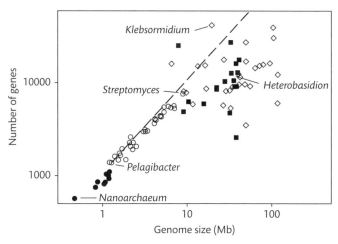

Figure 5.5 Genome size as a function of the number of protein-encoding genes in some bacteria (circles), fungi (closed squares), and protists (diamonds). The bacteria include obligate symbionts and parasites (closed circles) like *Nanoarchaeum equitans* and free-living taxa (open circles) such as the oceanic bacterium *Pelagibacter ubique* and the soil bacterium *Streptomyces coelicolor*. *Heterobasidion annosum* is a white-rot fungus, while *Klebsormidium flaccidum* is a terrestrial alga. The long dashed line is the number of protein-encoding genes predicted from *Pelagibacter* for increasing genome size. Data are from Giovannoni et al. (2005) and Elliott and Gregory (2015).

would be expected based on their genome size. Eukaryotic genomes are large because the fraction of their genomic DNA not devoted to coding for proteins is larger than that seen in prokaryotes (Table 5.1). For fungi, the fraction is about half of the total genome, whereas it is even higher in protists; both are much higher than in bacteria and archaea where nearly all of the genome codes for protein. This non-coding DNA was once called "junk DNA," but now it is thought to be essential for regulation. Another difference is that eukaryotic genes are often interrupted by stretches of DNA (introns) that do not encode for any amino acid and which after transcription are cut out of the resulting mRNA molecule before translation to a protein. Introns take up a large fraction of the genomes for both protists and fungi. All of these and other differences in genome structure have many implications for the regulation of metabolism and thus for the ecology of prokaryotes and eukaryotic microbes.

The modes of regulation in prokaryotes also differ from those in eukaryotes. Genes are often regulated for prokaryotes at the transcription level—the point at which mRNA is synthesized. Gene regulation is more complex in eukaryotes. Regulation can occur at transcription, but eukaryotes also modify mRNA molecules (post-transcriptional regulation) as just discussed, control protein synthesis (translation regulation), or modify proteins (post-translation regulation). The protein-coding genes help define an organism but so too do regulatory genes and mechanisms. Together, they explain differences among organisms. Perhaps because of our inflated self-regard, it was a surprise to discover that *Homo sapiens* has only about 22,000 genes, no more than many vegetables. We differ from turnips and other mammals in part because of how genes are regulated.

Bacteria do have some non-coding regions between genes that are worth highlighting. These are "clustered regularly interspaced short palindromic repeats" (CRISPR, pronounced "crisper"), which are key components of an immune system for bacteria (Barrangou and Horvath, 2017). Next to CRISPRs are *cas* (CRISPR-associated system) genes, encoding endonucleases that work with CRISPRs to fight against infection by viruses. There is tremendous interest in using CRISPR/Cas for editing genomes in many organisms for biotechnological applications and perhaps even human health.

Table 5.1 Genome structure of bacteria and eukaryotic microbes. The values are the mean, and the range in parentheses. Except for rRNA, the data are from finished and permanent draft genomic sequences (https://img.jgi.doe.gov) for the bacteria, or are from Elliott and Gregory (2015) for the eukaryotes. The rRNA data for bacteria are from Roller et al. (2016). The rRNA data for fungi (29 species or strains) and protists (57) come from several sources given at www.oup.co.uk/companion/kirchman.

Property	Bacteria	Fungi	Protists
Genome size (Mb)	4 (0.1–16.4)	33.0 (2.3–215)	38.9 (6.5–1500)
Number of genes	3908 (106–16,692)	9681 (1833–28,232)	9947 (3490–59,681)
Chromosomes	1	8 (3–23)	14 (3–40)
Arrangement of related genes	Operons common	Few operons	Few operons
% protein encoding DNA	87.0 (1.5–98.6)	53.0 (17.1–88.1)	63.5 (57.7–97.6)
% introns*	<1	6.85 (0–13.9)	6.53 (0–40.9)
% repeated sequences*	<0.1	7.52 (0–56.9)	10.15 (0.7–70.4)
rRNA genes per genome	3 (1–15)	85 (20–220)	43,504 (1–400,000)

* % of total genome

Horizontal gene transfer

Genomic sequence data of microbes have revealed a new mechanism in evolution not envisioned by Darwin and his successors. The traditional mode of evolution is that genes are handed down from generation to generation, from parent to offspring, with some genes persisting in offspring that survive, while others disappear when offspring die before reproduction. This "vertical" transfer of genes is captured by traditional phylogenetic trees of rRNA gene sequences (Chapter 4). Similar trees can be constructed with other genes. The problem comes when the trees don't agree. These discrepancies first became apparent when even just a few genes were compared, but they became even more evident as whole-genome data accumulated. One reason for the discrepancies is that rates of evolution for various genes diverge. Another reason is "horizontal gene transfer," also called "lateral gene transfer."

Horizontal gene transfer is the movement of genes from one organism to another, unrelated organism, in contrast with the vertical passing of genes from parent to offspring (Figure 5.6). This transfer can be effected by viruses ("transduction"; see Chapter 10) or by the uptake of free DNA ("transformation"). Genes have a greater chance of remaining in the recipient organism if the donor and recipient are related, but there are many examples of genes being exchanged between unrelated organisms, even between bacteria and archaea, and between the two prokaryotic domains and eukaryotes, including humans.

In addition to discrepancies between a gene's phylogenetic tree and an rRNA tree, the effect of horizontal gene transfer is seen in the relatedness of various genes next to each other in a microbial genome. While most of the genes are most similar to those from a close relative sharing the same ancestor, genes brought in by horizontal transfer are most similar to those from a distantly related organism. The GC content and codon usage of the recently transferred gene may also differ from that of the rest of the genome because these features vary among phylogenetically distant taxa.

Horizontal gene transfer calls into question the idea that any single gene can be used to follow the evolution of an organism. It is certainly true that the tree metaphor for describing evolution must be modified to include many intertwining branches due to instances of horizontal gene transfer. However, those genes involved with information processing, such as DNA and protein synthesis, tend not to be subjected to horizontal gene transfer (Daubin et al., 2003). In particular, the SSU rRNA genes appear to follow an organism's phylogeny, determined by comparing whole genomes in "phylogenomic" analyses (Wu et al., 2009). The SSU rRNA genes probably do not undergo substantial horizontal gene transfer, because the gene products cannot be easily accommodated into the recipient's existing molecular machinery without fatal consequences. In the case of 16S rRNA, a foreign rRNA molecule cannot easily fit into the complex structure of a ribosome, which consists of several other rRNA molecules and more than 50 proteins. In contrast,

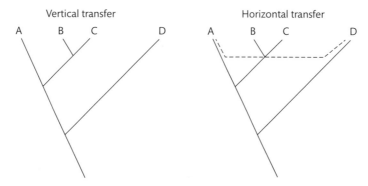

Figure 5.6 Vertical and horizontal gene transfer. The dashed lines in the tree on the right indicate the transfer of a gene from Phylotype D to Phylotype A, independent of other genes. Analysis of only this gene would indicate a closer relationship between Phylotypes A and D than indicated by the genes transferred vertically.

functional genes encoding enzymes often act alone and can tolerate more variation. A foreign enzyme from a horizontally transferred gene may function just fine in its new home and may even provide a new capacity for the recipient.

Horizontal gene transfer has consequences for trying to determine which organism a particular gene came from, an issue that arises when sequence data for only that gene from natural microbial communities are available. In this case, often it is impossible to link the gene to its source organism with any confidence. The enzyme used to hydrolyze chitin (chitinase) is one example (Figure 5.7). Chitinases from various types of bacteria do not fall into the same group as defined by their 16S rRNA genes, probably because of horizontal gene transfer. The presence of chitinases in viruses infecting insects, which have chitin exoskeletons, is further evidence that these genes

are exchanged horizontally. The genes retrieved from marine waters in this example seem to come from vibrios, but we cannot say for sure. This is the problem in interpreting short stretches of DNA sequences, such as retrieved by PCR-dependent approaches, which have parts of functional genes without any sequences from a phylogenetic marker gene.

The discussion so far has emphasized the problems caused by horizontal gene transfer for understanding phylogenetic relationships, but it is also an exciting development in the field of microbial evolution. Horizontal gene transfer is a novel mode of evolution not seen in large organisms. Most of the work has focused on bacteria and archaea, but horizontal gene transfer does occur with microbial eukaryotes. There are also transfers between domains, although some are more common than others. Bacteria to archaea transfers are at least five-fold more

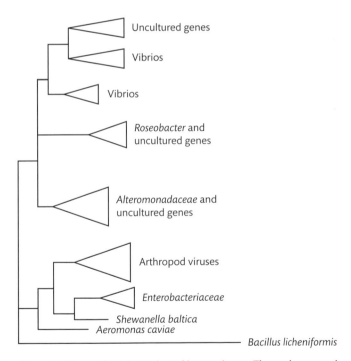

Figure 5.7 Neighbor-joining tree of chitinases from bacteria and insect viruses. The wedges contain several related genes. "Uncultured genes" refers to chitinase genes retrieved by cultivation-independent methods. All genes with taxonomic names come from cultivated organisms. There are several indications of horizontal gene transfer in this tree, such as the presence of virus chitinases among the bacterial genes and the position of *Roseobacter* (member of the *Alphaproteobacteria*) sequences among the vibrios and *Alteromonadaceae* (both in the *Gammaproteobacteria*). Data from Cottrell et al. (2000).

likely than vice versa (Spang et al., 2017). Horizontal gene transfer is more frequent in some taxa than in others. Among archaea, for example, *Thaumarchaeota* and the methanogenic phylum are particularly affected.

Genomes and growth strategies for bacteria

The genome contributes in several ways to a microbe's strategy for survival and growth in natural environments. Of course, functional genes such as those for sulfide oxidation or anaerobic respiration are essential for microbes trying to live by chemolithotrophy or in an anoxic environment. Cataloging those genes is an important part of analyzing genomes and metagenomes. Rather than discussing functional genes, however, here we will discuss general features of genomes that relate to growth. A starting point is genome size.

Small genomes are thought to be one reason why bacteria grow faster than other organisms, at least in nutrient-rich conditions. However, there is no correlation among bacteria between genome size and minimal growth rates (Vieira-Silva and Rocha, 2010), for at least two reasons. First, the energy required for synthesizing DNA is lower than for protein synthesis; the latter accounts for as much as 80% of energy expenditures, overshadowing any advantage in reducing DNA synthesis costs. Another reason is that there are advantages in having a big genome, as will be discussed further.

While there is no relationship between genome size and growth, other genomic features are important in thinking about microbial growth. Our discussion starts with observations about easily cultivated bacteria that grow relatively rapidly when nutrient concentrations are high.

Specific genomic features and growth

In contrast with genome size, there are several predictable relationships between growth and genomic features connected to protein synthesis. Perhaps of most interest to microbial ecologists is the 16S rRNA gene. There is a fairly high correlation between the maximum growth rate of a bacterium and the number of 16S rRNA genes in its genome (Figure 5.8). The genome of a bacterium capable of high growth rates can have several 16S rRNA genes (multiple "copies"). The presence of several rRNA genes enables faster rRNA synthesis and more ribosomes.

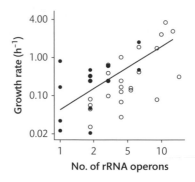

Figure 5.8 Relationship between maximum growth rate and rRNA copy number in bacteria. rRNA copy number explains about 45% of the variation in maximum growth rate (log–log regression analysis). The closed symbols are for bacteria with genome sizes less than 2 Mb, while the open symbols are for >5 Mb genome bacteria. The scatter of both small- and big-genome bacteria around the line is indicative of the lack of a significant relationship between growth rate and genome size. Data from Roller et al. (2016), provided by Ben Roller.

Having more ribosomes translates into faster protein synthesis and ultimately faster growth rates. At the other extreme, bacteria with only one 16S rRNA gene have slow maximal growth rates.

Interestingly, other genes connected to protein synthesis are not necessarily present as multiple copies in fast-growing bacteria (Vieira-Silva and Rocha, 2010). Rather, these genes, such as those for RNA polymerase, ribosomal proteins, and tRNA, tend to be closer to the origin of replication. By being close to where the bacterial chromosome starts to replicate, these genes in effect are present in higher numbers in a rapidly dividing cell than genes further away from the origin of replication.

Another growth-related feature of genomes is the preference for use of one codon for an amino acid over another, referred to as "codon usage bias" (Figure 5.9). Remember that several amino acids can be encoded by more than one triplet of DNA bases. Isoleucine, for example, can be encoded by ATT, ATC, or ATA. Slow-growing bacteria tend to use these various codons equally, whereas fast-growing bacteria have high codon usage bias (von Dassow et al., 2008). The favoring of some codons over others enhances translation efficiency and thus protein synthesis and growth. Interestingly, psychrophilic bacteria tend to have higher codon usage bias than the average bacterium for a given growth rate, while thermophilic bacteria tend to have lower bias. This

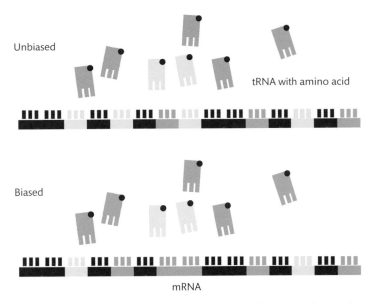

Figure 5.9 One type of codon use bias. Translation is more efficient when the frequency of the codons used in the mRNA match the frequency of the tRNA. The numbers of gray and white tRNA symbols do not match their frequency in the mRNA in the top panel (unbiased) but do in the bottom panel (biased). Based on a figure in Quax et al. (2015).

feature of genomes may allow bacteria to compensate for temperature effects. Psychrophilic bacteria can grow at high rates in spite of cold temperatures slowing down chemical reactions, while thermophilic bacteria can maintain control of their metabolism in the face of high temperatures pushing reactions faster than can be sustained over the long run.

Codon usage bias can also lead to lower nitrogen content because the nitrogenous bases guanine (G) and cytosine (C) have an extra nitrogen atom compared to adenine (A) and thymine (T). Lower G + C therefore means less N and P are required for DNA synthesis, a bonus in the many environments limited by those two elements. Bacteria with low G + C content often live under oligotrophic conditions. Similarly, N-limited bacteria can benefit by switching to amino acids (such as lysine to arginine) with similar chemical properties but with lower N content.

Streamlined genomes

Some bacteria were known to have small genomes even before they were sequenced, but the genomic data revealed interesting features of these genomes. One group of small-genome bacteria are symbionts and parasites that live in very close relationships with eukaryotes. These organisms can survive with small genomes because the eukaryotic host provides many needed compounds, so there is selection against having unnecessary genes.

This explanation does not apply to many other bacteria with small genomes that are free-living. The genomes in these bacteria are said to be "streamlined." While size is important, other characteristics of these genomes are arguably even more important. Streamlined genomes have a very high portion devoted to encoding proteins (Figure 5.10), because they have low amounts of intergenic spacer DNA (the space between genes), few pseudogenes (genes that do not work), and few regulatory systems. They also have low numbers of paralogs (duplicated genes). The examples discussed next are oceanic bacteria, but bacteria with streamlined genomes may be abundant in soils as well (Brewer et al., 2016).

Some numbers about one abundant bacterium, *Pelagibacter ubique*, illustrate how bacteria with streamlined genomes differ from those with a more standard organization (Giovannoni et al., 2014). The non-coding fraction of genomes is low for bacteria in general (about 15%) as previously mentioned, but it is even lower (by three-fold)

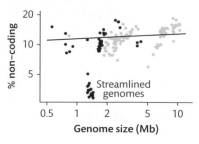

Figure 5.10 Relative amount of non-coding DNA expressed as a fraction of the total genome. The black points are for *Pelagibacter* and *Prochlorococcus* strains, of which some are known to have streamlined genomes. The gray points are from select soil bacteria. The regression line was drawn with data from 6995 species. The slope of the regression line is statistically different from zero, but genome size explains a very small fraction (<1%) of the variation in non-coding DNA. Data from Integrated Microbial Genomes (https://img.jgi.doe.gov), downloaded on March 7, 2017.

Table 5.2 Genomic and other characteristics of copiotrophic and oligotrophic bacteria. Based on Lauro et al. (2009) and Kirchman (2016). + or – indicate the presence or absence of a property.

Property	Copiotrophic	Oligotrophic
Genome size	>2 Mb	<2 Mb
Streamlined genome?	No	Yes
Regulatory systems	Several	Few
rRNA gene copies	>3	1
Cell size	>1 μm³	<0.1 μm³
Motility	+	–
Chemotaxis	+	–
Extracellular hydrolases	Several	Few
Metabolic versatility	High	Low
Growth strategy	Feast or famine	Steady-state
Maximum growth rate	>1 d⁻¹	<0.2 d⁻¹

for *Pelagibacter*. Its median intergenic space is only three base pairs versus as much as 300 base pairs for other bacteria. Finally, *Pelagibacter* has fewer regulatory systems than many other bacteria. One way to count regulatory systems is by looking at σ factors, which are proteins that combine with a DNA-dependent RNA polymerase to initiate transcription at a specific start region, the promoter, for a gene. *Pelagibacter* again has few of these σ factors (less than five), half or fewer than the number seen for other bacteria even when normalized for genome size.

The end result is that a bacterium with a streamlined genome cannot do much because it does not have the genes found in microbes with larger genomes and it cannot respond very well to any change in its environment that would favor faster growth, such as higher nutrient concentrations, because it lacks the regulatory systems to increase synthesis of necessary enzymes. Such a genome sounds like a huge liability, and it may well be the case when nutrient concentrations are high. But in oligotrophic environments where concentrations are rarely high, bacteria like *Pelagibacter* and *Prochlorococcus* with streamlined genomes win the ecological competition and dominate microbial communities, as discussed next.

Oligotrophic versus copiotrophic bacterial genomes
We can now organize many of the genomic features discussed into two different types of bacteria with different

growth strategies (Table 5.2). As discussed in Chapter 4, copiotrophic bacteria grow in high nutrient concentrations, while oligotrophic bacteria do best in low concentrations, by definition. The high concentrations select for fast growth, which requires several rRNA gene copies, along with the other genomic features (such as the position of the protein synthesis genes in the chromosome) that make high growth rates possible. In natural environments, these high concentrations come as complex mixtures of compounds, selecting for bacteria with sensing mechanisms, hydrolases, transporters, and many other proteins and enzymes that need to be encoded by a large genome. Any cost in replicating this DNA is more than compensated by the gains in being able to take advantage of the high concentrations of a complex mixture of compounds.

Oligotrophic bacteria have a different ecological strategy. Rather than being prepared to take advantage of the rare patch of high concentrations, these bacteria are adapted to dealing with the low but steady supply of a few compounds. The limited repertoire of compounds used by one oligotroph, *Pelagibacter*, was discussed in Chapter 4. Oligotrophic bacteria save much energy and limiting elements (C, N, and P) by not synthesizing the many enzymes and regulatory proteins needed only in nutrient-rich environments. They make do with one rRNA operon to carry out their slow rates of protein synthesis and growth. Their streamlined genomes minimize the energetic costs of DNA synthesis and the need for limiting elements.

Genomes from uncultivated microbes: metagenomics

Genomes from cultivated microbes are invaluable for understanding the ecology of uncultivated microbes, but as mentioned repeatedly throughout this book, there are many differences between most of the cultivated microbes grown in the laboratory and uncultivated microbes in nature. Another problem is that there are too many microbes in nature to cultivate them all, even if we could. Fortunately, using metagenomic approaches, genomic information can be accessed directly from microbes without growing them in the laboratory. Although a complete genome is sometimes obtained if the community is simple or lots of sequencing is done, metagenomic approaches often yield only fragments of genomes from several organisms. But the payoff is worth it. The end result is an enormous list of sequences for a large number of genes from many organisms retrieved without using PCR or cultivation.

Metagenomics can provide many clues about the physiology and thus potential biogeochemical role of uncultivated microbes. The sequences of rRNA genes and other phylogenetic markers say much about the types of microbes present in nature, but we cannot link those data easily to a particular biogeochemical function. Metagenomics is one way to make those links.

Other fields of science have used the metagenomic approaches first developed by microbial ecologists to explore, for example, the many uncultivated microbes inhabiting the human body and how they affect our health and well-being. Less obvious is the use of metagenomic data to explore protein structure (Ovchinnikov et al., 2017). Protein biochemists take advantage of the size and diversity of metagenomic data to deduce the most likely three-dimensional structure of a protein; often a difficult problem to solve.

Metagenomic approaches and linking structure with function

Metagenomic approaches have changed greatly since the first study was published in the 1990s. The early studies had to clone the DNA extracted from microbial communities before sequencing was possible (Box 5.2), and because sequencing was so expensive back then, only a few clones or only the ends of several clones were sequenced. Today, most studies using metagenomic approaches rely on the tremendous sequencing power of high-throughput approaches which do not require cloning. Although the sequences can be used directly, often several sequences are pieced together ("assembled") to form a "contig," a much longer, continuous sequence (Figure 5.11). The DNA pieces are stitched together by finding overlapping regions having the same sequence. One potential artifact is that two unrelated DNA pieces may be joined together inadvertently, forming a "chimera," but these can be recognized and removed. The reconstructed DNA fragment, the contig, is assumed to originate from a single microbe.

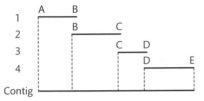

Figure 5.11 Assembling a contig from overlapping sequences. The DNA fragments (labeled here as 1 through 4) can be joined together because they have sequences in common at their ends (here, A through E), forming a contig.

Box 5.2 Cloning and cloning vectors

Cloning consists of putting foreign DNA (the insert) into a cloning vector, which replicates itself along with the insert, inside a host, usually an *E. coli* strain. Vectors are highly modified versions of plasmids and phages once found in nature. Different types of vectors are used depending on the size of the DNA insert. Plasmids are used for cloning PCR products and constructing small-insert metagenomic libraries, while fosmids and bacterial artificial chromosomes (BACs) handle 40 to >100 kb inserts. Cloning is infrequently used today for metagenomic approaches, although fosmid cloning still can be a powerful approach (Haro-Moreno et al., 2017). It yields long sequences with fewer errors than seen with the high-throughput sequencers designed to generate long sequence reads.

Metagenomics can yield an inventory of the functional genes present in a microbial community without the many drawbacks of PCR-based approaches. Perhaps even more important, metagenomics can link functional genes with 16S rRNA genes or other phylogenetic marker genes—the linking of function and structure. The discovery of proteorhodopsin is a good example.

Chapter 7 describes how proteorhodopsin harvests light energy for many heterotrophic bacteria, making them photoheterotrophs. Genes for this chromophore-protein complex were discovered in a metagenomic BAC library (see Box 5.2) of DNA from coastal seawater (Béjà et al., 2000). Screening of the library for 16S rRNA genes turned up a 130 kb clone with a 16S rRNA gene most similar to one from the SAR86 clade of *Gammaproteobacteria*, originally discovered in the Sargasso Sea. Sequencing the clone revealed several other genes (Figure 5.12), including one most similar to a rhodopsin-encoding gene from halophilic archaea discovered using conventional methods years before metagenomic approaches were possible. The presence of the bacterial 16S rRNA gene on the same BAC clone as the rhodopsin gene proved that the rhodopsin was in fact from a bacterium. The rhodopsin protein could then be synthesized from the gene, which was important for showing that its function is similar to that of the archaeal rhodopsin, as predicted by the sequence data.

The story of proteorhodopsin illustrates how metagenomics can reveal previously unknown functions and metabolisms in natural environments. These functions potentially change our ideas about how biogeochemical cycles work and how they are regulated. Metagenomics can also reveal that a known function or metabolism may be carried out by unsuspected microbes. Examples include the oxidization of sulfide and carbon monoxide

in oxic environments by heterotrophic bacteria. Sulfide and other reduced inorganic sulfur compounds are not expected in oxygen-rich habitats, far from anoxic systems where these reduced sulfur compounds are produced (Chapter 11), so it was a surprise to see in metagenomic sequences genes encoding enzymes for oxidizing these compounds in oxic environments (*sox* genes). Metagenomics was also essential in showing the presence of genes for carbon monoxide oxidation (*cox* genes) in unexpected environments and microbes. Both observations suggest that many bacteria may be "mixotrophs" and use more than one mechanism for generating energy. A final example is ammonia oxidation by archaea. Although ammonia oxidation by bacteria had been known for decades, it took metagenomic work in both the oceans and soils to reveal the role of archaea in carrying out this important reaction of the nitrogen cycle.

Single-cell genomics

Metagenomics solves the problem of obtaining genomic data from uncultivated microbes, but it creates another problem: it jumbles together all genes from all organisms in the community. Figuring out which gene came from which microbe is like solving hundreds of jigsaw puzzles whose pieces have been thrown together. To further complicate matters, some of the pieces may be missing or their edges too frayed to fit together with confidence. As mentioned before, it is usually not possible to retrieve large genomic portions of one microbe unless it is very abundant and the community very simple. Increasing the number of sequences and read length (getting more base pairs for a sequence) helps, but current high-throughput sequencing technology cannot solve all of these problems.

Another approach, "single-cell genomics," avoids jumbling all microbes together in the first place. The strategy is to physically separate a single microbe away from all others and to sequence it directly. This powerful approach is used to examine individual cells from multicellular organisms, such as cancer cells in humans (Tanay and Regev, 2017) as well as from diverse microbial communities. A common method for separating microbes is flow cytometry (Figure 5.13). Because current sequencing methods require more DNA than present in a single cell,

Figure 5.12 Map of a BAC clone with the genes for proteorhodopsin and 16S rRNA, indicating the genomic material was from a SAR86 bacterium. The arrows indicate the direction of transcription for each gene. Figure used with permission from Oded Béjà, based on Béjà et al. (2000).

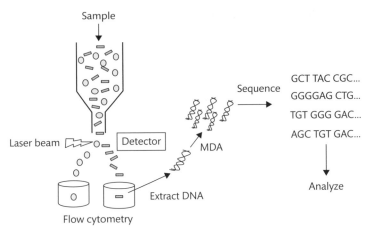

Figure 5.13 Single-cell genomics in which the DNA from a single microbe is amplified (MDA) and sequenced after separation by flow cytometry (shown here) or by micro-fluidic devices.

genomic DNA extracted from the single cell has to be multiplied or amplified by an approach called "multiple displacement amplification" (MDA). The current amplification approach is not perfect as only about two-thirds of the original genome is found in the amplified version, but it has made single-cell genomics possible. Sequencing of the single amplified genomes (SAGs) has yielded many interesting findings.

One important finding from single-cell genomics is that oligotrophic bacteria with streamlined genomes are abundant in the oceans (Swan et al., 2013). A study of various oceanic regions with low nutrient concentrations found that most of the retrieved SAGs were from oligotrophic bacteria with features of streamlined genomes; the genome size, proportions of non-coding DNA, intergenic regions, and numbers of paralogs were all small or low. The G + C content was also lower than for bacteria isolated and grown on rich media. Remember that bacteria need less nitrogen by using codons with A + T rather than G + C. The streamlined genome SAGs had the mixotrophy genes (proteorhodopsin, *sox*, and *cox*) found by earlier metagenomic work. Overall, the genomic data suggested that these bacteria are specialists in using the limited energy and nutrient resources available in oligotrophic environments.

Among the SAGs that did not have streamlined genome features were representatives in the *Verrucomicrobia* and *Bacteroidetes* phyla. Unlike the streamlined genome SAGs, these SAGs had several genes for extracellular hydrolases, consistent with the hypothesized role for these two phyla in biopolymer degradation (Chapter 7).

Single-cell approaches have also been used to examine uncultivated protists. Some of these are large enough to isolate, albeit painstakingly, by pipetting individual cells identified under a light microscope. The smaller but more numerous protists have been examined by the flow cytometry-based approach. Single-cell approaches have reaffirmed the incredible number of rRNA genes in an individual protist cell, over 300,000 in one ciliate (Gong et al., 2013). This high number is thought to be one reason why single-cell approaches can recover more diversity among protists than PCR-based approaches (Heywood et al., 2011), because protists with large numbers of rRNA genes will dominate the PCR-based results. Biases due to PCR primers and other problems are also possible. The single-cell approach gets around the rRNA copy number and PCR-bias problems provided that a high number of protist cells are examined. That is increasingly feasible as sequencing costs decline.

Metatranscriptomics and metaproteomics

The genes from a natural microbial community indicate the potential for a particular function to be carried out, but those genes may or may not be expressed. Biogeochemical approaches can be used to determine whether the function is actually occurring in the environment, but they do not say anything about which organisms are carrying out the process, and they do not detect all functions, such as light harvesting by proteorhodopsin. One way to address these questions is to examine the expression of genes (mRNA synthesis) for metabolic functions connected to a biogeochemical process of interest. Especially for prokaryotes, the presence of mRNA for a particular process from a particular microbe is a strong indication that that microbe is carrying out the process. The presence of a protein is even stronger evidence. For both mRNA and proteins, sequence data indicate the function while also providing clues about the microbe producing the mRNA or protein.

Transcriptomic approaches are harder to do than genomic ones, and proteomics is even more challenging. mRNA is easily degraded, and for prokaryotes the methods for separating mRNA from the ten-fold more abundant rRNA are imperfect; separating mRNA from rRNA is less of a problem for eukaryotes because of a tail of adenines (poly A tail) on eukaryotic mRNA that can be used to enrich for mRNAs. Initially, the mRNA molecules were followed by micro-array technology, but currently the most common technique is "RNA-seq" or the sequencing of mRNA after its conversion to DNA ("complementary" DNA or cDNA) with the help of reverse transcriptase.

It is much more difficult to sequence and identify proteins (Figure 5.14), in part because protein amounts cannot be increased in a test tube like DNA or RNA. The initial studies relied on gel electrophoresis, but most studies now use mass spectrometry, often more than one mass spectrometric technique used in a series (tandem mass spectrometry or "MS/MS"). Unlike DNA or cDNA sequencing which yields sequence data directly, proteins are sequenced by matching ion fragments produced by tandem mass spectrometry to a database. One challenge in metaproteomics is whether the database for deducing the amino acid sequence is sufficient for the task. Ideally, the database is built from genomic or metagenomic sequences from the same system. A database based on other information may lead to misidentification, because important protein sequences are missing from the database or genomic data from an irrelevant organism happens to be sufficiently similar in spite of being incorrect.

The following gives a few examples of using metatranscriptomic and metaproteomic data to identify microbes carrying out a particular ecological process.

Ammonia oxidation is an important step in the nitrogen cycle, potentially carried out by both chemolithotrophic bacteria and archaea (Chapter 12). Metatranscriptomic data as well as reverse transcriptase PCR (PCR that works on mRNA) indicate archaea carry out more ammonia oxidation than do bacteria in some soils (Leininger et al., 2006). Other metatranscriptomic work indicates that some archaea in the deep ocean may be heterotrophs (Li et al., 2015), in contrast with other approaches indicating that most are chemolithotrophs. Metaproteomic studies have explored the use of inorganic and organic compounds by microbes in natural environments. These

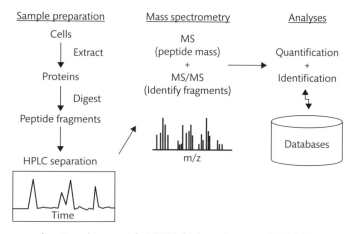

Figure 5.14 Overview of a general proteomic approach. HPLC is high performance liquid chromatography, and MS is mass spectrometry.

studies have found that the most abundant transporters are from only a few bacterial taxa, suggesting the importance of those microbes in organic material degradation (Dong et al., 2014). Some of the more abundant transporters are ABC systems (defined in Chapter 2) and TonB-dependent transporters. Located in the outer membrane of Gram-negative bacteria, TonB-dependent transporters are involved in transport of large compounds like complexed iron and vitamins. Transport proteins provide clues about the types of compounds used by heterotrophic microbes, even some unexpected ones like terrestrial biopolymers in marine waters (Colatriano et al., 2015).

Transcriptional response of oligotrophic and copiotrophic bacteria

Another use of metatranscriptomics is to explore how microbes respond to changes in their environment. One complication is that a microbe has relatively few mRNA molecules and these do not last long in a cell (Chapter 2); once an mRNA has done its job, a cell quickly degrades it and recycles the much needed organic carbon, nitrogen, and phosphorus. As a result, an mRNA molecule lasts only minutes in a cell (Moran et al., 2013), much shorter than a protein which can last as long as a cell lives: hours to days, depending on the organism's growth rate. The short half-life of mRNA and low mRNA levels in each cell complicate the use of metatranscriptomic approaches to

explore the role of microbes in biogeochemical cycles, especially in systems where environmental conditions change rapidly.

Still, transcriptomics is informative about how a microbe responds—or not—to changes in its environment. The response at the level of transcription, which reflects transcriptional control, is quite different for copiotrophic versus oligotrophic bacteria. As mentioned before, the streamlined genomes of oligotrophic bacteria do not have many σ factors, indicating less control at the transcriptional level. This minimal control is evident in examining transcription in an oligotroph as it switches from a fast growth phase to a slow one (Figure 5.15). Transcription of only a few genes changes during this transition, whereas transcription for many genes either increases or decreases when a copiotrophic bacterium shifts its growth rate. Metatranscriptomic studies of natural environments suggest minimal transcriptional control in oligotrophic bacteria in natural environments (Gifford et al., 2013; Ottesen et al., 2013).

Metatranscriptomes of eukaryotic microbes

Metatranscriptomic approaches have been used to examine the types of active fungi and protists in soil and aquatic habitats. The approaches target active microbes because mRNA and rRNA degrade quickly once an organism dies. Metatranscriptomic studies focusing on community structure can take a phylogenomic

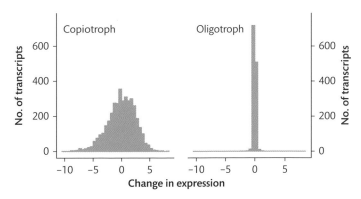

Figure 5.15 Changes in expression in a copiotrophic bacterium (*Ruegeria pomeroyi*) and an oligotrophic bacterium (*Pelagibacter ubique*) switching from fast to slow growth. Some of the changes were negative (lower number of transcripts in the slow growth state), while others were positive (higher number in the slow state). The y-axis gives the number of transcripts with the indicated change in expression. For the copiotroph, several genes changed greatly, whereas for the oligotroph, many of the transcripts did not change at all; the change in expression was at or near zero. Data from Cottrell and Kirchman (2016).

approach and focus on transcripts from several genes to identify microbes, or they can focus on one or two phylogenetic markers such as genes for 18S rRNA, represented in the metatranscriptome.

For example, metatranscriptomics was used to explore the community composition of protists based on SSU rRNA sequences in soils from different habitats (Geisen et al., 2015). The rRNA data suggested that the super-group Amoebozoa was much more abundant in these soils than would have been found by a PCR-based approach. The supergroup along with Rhizaria dominated forest and grassland soils, while Alveolata was most abundant in peat soils. Curiously, transcripts from foraminifera and choanoflagellates, which are aquatic protists, were also abundant in these soils.

Metatranscriptomics arguably is even more necessary for exploring functional genes in uncultivated microbial eukaryotes. It is easier to find the protein-encoding regions of eukaryotic genomes by looking at mRNA rather than looking at the genome or metagenome directly. Sequences from mRNA are not complicated by introns, regulatory regions, and junk DNA. Metatranscriptomic studies have turned up unusual activity in microbial eukaryotes, such as the expression of a proteorhodopsin-like gene in dinoflagellates, a type of eukaryotic alga (Lin et al., 2010), and cellulose-hydrolyzing enzymes, thought to be restricted to fungi, in non-fungal soil eukaryotes (Damon et al., 2012). Metatranscriptomic studies also look at the expression of functional genes in response to perturbations. One such study found that nitrogen fertilization led to lower expression of lignocellulose-degrading enzymes from some soil fungi (Figure 5.16).

One major obstacle for using metatranscriptomics for eukaryotic microbes remains, however: very little is known about the genomes and even the protein-encoding genes of eukaryotic microbes. One metatranscriptomic study found that over 50% of transcripts from soil eukaryotes could not be identified (Damon et al., 2012), and another study also in soils found that 32% of the metaproteome was of novel hypothetical proteins from unknown microbes (Bailly et al., 2007). Even the source of many of the genes could not be identified. Protists accounted for over 10% of the 18S rRNA genes recovered by PCR-based approaches from different soils

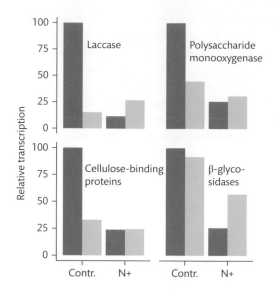

Figure 5.16 Expression of plant polymer-degrading enzymes from two phyla of fungi, *Basidiomycota* (black) and *Ascomycota* (light gray). Transcription levels were normalized to values for *Basidiomycota* in the control treatment ("Contr"). "N+" is the treatment exposed to fertilizer. Laccase is a copper-containing enzyme acting on phenolic components of lignin. Cellulose-binding proteins are fungal proteins involved in cellulose degradation. The general class of β-glycosidases includes several enzymes involved in polysaccharide degradation. The data show that fertilizer caused a decrease in expression of the four proteins, although the response varied with the fungus and the protein. Data from Hesse et al. (2015).

but <5% of the transcripts were assigned to protist taxa, according to a metatranscriptomic approach (Damon et al., 2012). The low representation of protists in the metatranscriptomic data is probably not due to the low number of protist transcripts in these soils, but to the paucity of protist sequences in genomic databases used to identify them. The problem is not just with soil protists. Transcriptomic studies of marine protists have also revealed large number of genes that are not similar to known genes in current databases (Caron et al., 2017).

These technical problems point to the need for more research about the great diversity of organisms in the microbial world. While databases and technologies need to be improved, metatranscriptomics and metaproteomic approaches have made possible new discoveries about the role of specific microbes in biogeochemical processes.

Summary

1. Genomic studies of microbes in pure cultures have revealed new insights into regulation and growth strategies, even though the function of a large fraction of genes remains unknown even for well-studied organisms.

2. Several types of metagenomic approaches have been used over the years. The most recent versions have been greatly aided by advances in inexpensive, high-throughput sequencing technologies. Genomic approaches applied to single cells avoid some of the problems with metagenomics.

3. Metagenomic approaches have been used to identify organisms carrying out specific steps in biogeochemical processes, while also suggesting new functions that were not obvious using microbiological or biogeochemical approaches.

4. Gene and protein expression examined by metatranscriptomic and metaproteomic approaches yields information that more closely reflects actual biogeochemical processes and reveals aspects of microbial communities not gleaned by metagenomic or biogeochemical techniques.

5. A large fraction of transcripts from protists cannot be identified based on sequence analysis of metatranscriptomes.

CHAPTER 6

Microbial primary production and phototrophy

This chapter is devoted to the most important process in the biosphere. Primary production is the first step in the flow of energy and materials in ecosystems. The organic material synthesized by primary producers supports all food chains in the biosphere, and sets the stage for the cycle of carbon and of all other elements. Food web dynamics and biogeochemical cycles depend on the identity of the primary producers, their biomass, and rates of carbon dioxide assimilation and biomass production. In this chapter, we focus on the most important form of primary production, that driven by light energy. Another form, driven by chemical energy, is essential for supporting some habitats such as at hydrothermal vents (Chapter 14) but is rarer and is a small fraction of global primary production.

Primary production by microbes is very important on both global and local scales. Mainly because of their abundance in the oceans, microbes account for about half of all global primary production, while the other half is by terrestrial higher plants. In contrast with life on land, in most aquatic habitats, primary production is mainly by microbes: the eukaryotic algae and cyanobacteria. In terrestrial systems, primary production by microbes is important where larger plants cannot live but a few microbes can. Photosynthetic microbes can grow on rocks (epilithic) or even within rocks (endolithic) and in the top surface layer of soils if enough light is available.

The main primary producers are photoautotrophs that carry out oxygenic photosynthesis, meaning they produce oxygen during photosynthesis, in contrast with anoxygenic photosynthesis which does not evolve oxygen (Table 6.1). Because microbes account for about half

Table 6.1 Use of light energy by microbes. The main pigment used in energy production is given here, although these microbes may have other pigments for light harvesting. Chlorophyll a = Chl a, and bacteriochlorophyll = Bchl.

Metabolism	Purpose of light	Pigment	C source	Role of O_2	Organisms
Oxygenic photosynthesis	ATP and NADPH production	Chl a	CO_2	Produces O_2	Higher plants, eukaryotic algae, cyanobacteria
Anaerobic anoxygenic photosynthesis	ATP and NADPH production	Bchl	CO_2	O_2 inhibits photosynthesis	Bacteria
Photoheterotrophy*	Energy production	Bchl or rhodopsin	Organic C	Consumes O_2	Archaea, bacteria
Mixotrophy*	ATP and NADPH production	Chl a	CO_2 or organic C	Produces and/or consumes O_2	Protists
Heterotrophy	Sensing	Rhodopsin	Organic C	Consumes O_2	Eukaryotes, bacteria, archaea

* Photoheterotrophy and mixotrophy are basically the same metabolism. The first term is used for bacteria, while the second is commonly applied to protists.

Processes in Microbial Ecology. Second Edition. David L. Kirchman. Oxford University Press (2018). © David L. Kirchman 2018. DOI 10.1093/oso/9780198789406.001.0001

of global primary production, they also account for half of the oxygen in the atmosphere. Oxygenic photosynthesis is used by a diverse array of eukaryotes and cyanobacteria. There are no known photoautotrophic archaea, although some halophilic archaea use light energy by a mechanism based on rhodopsin, quite different from that used by photosynthetic bacteria and eukaryotes. Photoautotrophic microbes that are eukaryotic are called "algae" (alga is the singular) or "microalgae," to distinguish them from the macroalgae such as kelp. Algae and cyanobacteria that are free-floating in aquatic habitats are phytoplankton, as mentioned in Chapter 1. Many of the photoautotrophic microbial eukaryotes are protists. Light-using microbes other than photoautotrophs are discussed in Chapter 7 (photoheterotrophs) and Chapter 9 (mixotrophy).

Basics of primary production and photosynthesis

Light-driven primary production is based on photosynthesis (Figure 6.1) and can be summarized as

$$CO_2 + H_2O + nutrients + light \rightarrow biomass + O_2 \quad (6.1).$$

Nutrients include nitrate, phosphate, and many others, the raw material needed for the cell to synthesize proteins, nucleic acids, and the other cellular components of biomass. Photosynthesis is defined here as the fixation of carbon dioxide fueled by light energy (Figure 6.1):

$$CO_2 + H_2O + light \rightarrow CH_2O + O_2 \quad (6.2).$$

"CH$_2$O" refers to organic material, not a specific compound. So, primary production is photosynthesis coupled with the uptake of nutrients needed for biomass synthesis.

The first part of photosynthesis, the "light reaction," generates reducing power (NADPH), energy (ATP), and, in oxygenic photosynthesis, oxygen (O$_2$). Oxygenic photosynthetic organisms use light energy to "split" or oxidize water:

$$2H_2O + light \rightarrow 4H^+ + 4e^- + O_2 \quad (6.3).$$

The four electrons (4e$^-$) produced by the light reaction are used to reduce NADP$^+$ to NADPH and to add a high-energy phosphate bond to ADP to produce ATP. In the second part of photosynthesis, the "dark reaction," the electrons from NADPH reduce CO$_2$ (whose carbon has an oxidation state of 4+) and, together with ATP, form organic carbon (oxidation state of 0):

$$CO_2 + 2NADPH + 2H^+ + 3ATP \rightarrow CH_2O \\ + H_2O + 2NADP^+ + 3ADP + 3Pi \quad (6.4)$$

where "Pi" is inorganic phosphate. This process is also called carbon fixation, because the C in the gas carbon dioxide (CO$_2$) is "fixed" to a nongaseous form of C, an organic compound.

Light and algal pigments

A key step in photosynthesis is "light harvesting," the absorption of light by various pigments in the photoau-

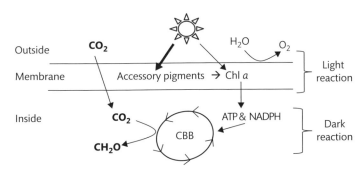

Figure 6.1 Summary of oxygenic photosynthesis. Light energy harvested by "accessory pigments" is transferred to chlorophyll *a* in the reaction center where water is "split" in order to synthesize ATP and NADPH, producing oxygen in the process during the light reaction. The ATP and NADPH are then used to fix CO$_2$ and synthesize organic material ("CH$_2$O") in the dark reaction by the Calvin–Benson–Bassham (CBB) cycle.

totroph. For terrestrial plants and green algae, the light-harvesting pigments are chlorophylls *a* and *b* and "accessory pigments," but chlorophyll *a* is the dominant one. More than 99% of the chlorophyll *a* molecules in phytoplankton are used for light harvesting. This light energy is transferred to special chlorophyll *a* molecules that lie at the heart of the reaction centers of photosynthesis. It is in the reaction centers that light energy is converted into chemical energy. Because the reaction center chlorophyll *a* is essential, all oxygen-producing, photosynthesizing organisms have it.

Phototrophic microbes have a greater diversity of accessory pigments than seen in higher plants on land. These accessory pigments enable phototrophs to harvest the wavelengths of light found in lakes and the oceans, especially green light (roughly 450–550 nm), the main wavelengths penetrating to deep waters. The wavelengths adsorbed by chlorophyll *a* and *b* (<450 nm and >625 nm) become less available in water deeper than a few meters or in microbial mats deeper than a few centimeters. Some common pigments include fucoxanthin (a carotenoid), which is found in diatoms and some other eukaryotic algae, peridinin (another carotenoid), which is found in dinoflagellates, and phycoerythrin, which is made by cyanobacteria and by red algae and cryptomonads. These pigments adsorb in the green part of the light spectrum (Figure 6.2). Because these pigments are more abundant than chlorophyll *a*, phototrophic microbes can be colored yellow, red, brownish hues, or shades of green not seen in higher plants.

Ecologists use pigment data to address two questions about phototrophic microbes in natural habitats: how much biomass or "standing stock" is present? And what is the taxonomic composition of this biomass? The most common use is to estimate biomass from chlorophyll *a*. Ecologists don't often try to measure phototroph biomass directly because it is very difficult to separate microbes from other organic material. Chlorophyll *a* is found only in phototrophic microbes and is easy to detect because of its color. Studies often use concentrations of chlorophyll *a* as an index of phototroph biomass, without converting the data to actual total biomass concentrations. If necessary, biomass (µg C per sample) can be estimated by multiplying chlorophyll *a* concentrations (µg chlorophyll per sample) by an assumed ratio of biomass C per chlorophyll *a*. A commonly used ratio is

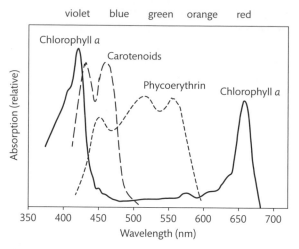

Figure 6.2 Absorbance by some of the main pigments found in photosynthetic microbes. Chlorophyll *a* is found in all oxygenic photosynthetic organisms, ranging from higher plants to cyanobacteria. Phycoerythrin is found in some cyanobacteria and red algae. The names of some colors are given above the top axis.

50:1, although it can vary substantially, up to ten-fold between algal classes and due to light intensity and temperature. Pigments are also used to identify phototrophic microbes, albeit only to the class level, such as diatoms and dinoflagellates. The composition at or close to the species level can be explored using sequences of 18S rRNA genes for eukaryotic algae or 16S rRNA genes in cyanobacteria and in chloroplasts of eukaryotic algae (Needham and Fuhrman, 2016).

The carbon dioxide-fixing enzyme
Another approach to exploring the taxonomic composition of the phototroph community and more generally of all autotrophs is to examine a key enzyme involved in CO_2 fixation. The main pathway for carbon dioxide fixation in oxic environments is the Calvin–Benson–Bassham (CBB) cycle, which is found in higher plants, eukaryotic algae, cyanobacteria, and many chemolithoautotrophic microbes. Other microbes use other fixation pathways with different enzymes and different requirements for ATP and NADPH (Hanson et al., 2012b). These alternative chloroplasts and fixation pathways may have been more important during the early evolution of photosynthesis

on the planet (Fuchs, 2011). In today's environments, however, the CBB cycle is the most common physiological basis of primary production, accounting for 99% of global primary production. A key CBB enzyme examined by microbial ecologists is ribulose-bisphosphate carboxylase/oxygenase (Rubisco) that catalyzes the reaction

$$\text{Ribulose 1,5 - bisphosphate} + CO_2 \atop \rightarrow 2 \ 3\text{ - phosphoglycerate} \qquad (6.5).$$

This enzyme is so important to autotrophs that it sometimes comprises up to 50% of cellular protein, making it one of the most abundant proteins in nature (Tabita et al., 2007).

Molecular approaches targeting genes or transcripts for Rubisco are used by microbial ecologists to explore the contribution of various autotrophs to primary production (Ward and Van Oostende, 2016). Because of the diversity of Rubisco-encoding gene sequences, it is possible to characterize autotrophic taxa to a finer taxonomic level, such as species and even subspecies, than is possible by other approaches. The identification, however, depends on what is known about the Rubisco genes for those autotrophs and on the molecular approach. In addition to identifying the presence of specific autotrophs, transcripts for Rubisco give insights into the contribution of those autotrophs to CO_2 fixation in nature. Rubisco transcription has also been used to explore the response of algal taxa to CO_2 levels and other environmental properties (Endo et al., 2016) and the potential importance of soil algae in carbon storage in terrestrial environments (Yuan et al., 2012).

Primary production, gross production, and net production

The rate of primary production is one of the most important parameters for describing an ecosystem and for understanding microbial and biogeochemical processes. How we measure this rate affects how we interpret the data and how we assess the implications of those data for understanding other processes.

In the light-driven ecosystems discussed here, Equation 6.2, for photosynthesis, provides clues on how to measure primary production. Equation 6.2 suggests that to estimate primary production, we could measure the movement of two elements (C or O) or changes in concentrations of O_2, CO_2, or CH_2O. All of these possibilities are used for various purposes by ecologists. A common method is to add $^{14}CO_2$ (actually $NaH^{14}CO_3$) to a sample and trace the ^{14}C into organic material (CH_2O). The advantages of this method are that it is easy, quick, and sensitive, and the instrument to measure radioactivity (liquid scintillation counter) is relatively inexpensive and common.

Changes in dissolved O_2 concentrations are also relatively easy to measure with the modified Winkler method, O_2 electrodes, or by gas-inlet mass spectrometry. One of the first approaches, the "light–dark bottle method," for estimating production was to measure changes in O_2 concentrations over time in light and dark bottles. Oxygen decreases in the dark bottle due to respiration (R), but it increases in the light bottle if photosynthesis exceeds respiration. The change in oxygen in the light bottle is a measure of net primary production (NPP) (Figure 6.3). Gross primary production (GPP) then is

$$GPP = NPP + R \qquad (6.6).$$

In words, gross primary production is the production of O_2 before respiration takes its toll, while net primary production is gross production minus respiration. To estimate biomass production and CO_2 fluxes, the oxygen-based rates from the light–dark bottle approach can be converted to carbon units by assuming a ratio of O_2 produced to CO_2 fixed, the photosynthetic quotient (PQ). The PQ is usually assumed to be 0.9, but it can vary substantially (Romero-Kutzner et al., 2015). The hidden assumption with the light–dark bottle approach is that O_2 consumption in the light bottle is the same as respiration in the dark bottle. Stable isotope studies with ^{18}O can help to test that assumption by providing another way of estimating primary production.

Investigators who have compared ^{14}C-based and ^{18}O-based measurements of primary production have concluded that the ^{14}C method is measuring something between net and gross production. The ^{14}C method gives a rate that is smaller than gross production, because several processes result in the loss of ^{14}C after fixation of $^{14}CO_2$ into organic material. Any respiration of ^{14}C organic carbon back to $^{14}CO_2$ during the ^{14}C incubation would go unnoticed and would lead to an estimate lower than

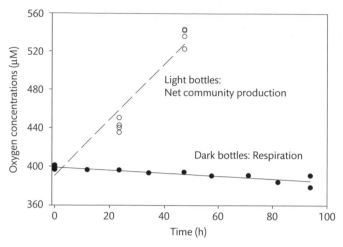

Figure 6.3 A light–dark bottle experiment to measure net community production and respiration. Net community production is the change in oxygen concentration in the light bottles over time (dashed line), while respiration is the decrease in the dark (solid line). Data taken from Cottrell et al. (2006).

the actual rate of gross production. Another problem is excretion or release of dissolved ^{14}C organic material during the incubation. Some of the dissolved ^{14}C organic material can also be taken up by small heterotrophic microbes that are not sampled by standard approaches. Even given these problems, the ^{14}C method is still a powerful and frequently used tool for estimating primary production.

Microbial ecologists and biogeochemists are concerned not only with gross primary production (GPP) and net primary production (NPP) but also with net community production (NCP), which takes into account the respiration of bacteria, protists, and small animals. These different processes are summarized in the following equation:

$$NCP = GPP - R_A - R_H = NPP - R_H \qquad (6.7)$$

where R_A is respiration by autotrophs, and R_H is respiration by heterotrophs. Another term, "net ecosystem production," includes processes in the entire system, such as respiration by benthic organisms.

The magnitude and even the sign of net community production have several important implications for carbon and oxygen fluxes, other biogeochemical processes, and the biota in an ecosystem. When net community production is positive, organic matter can accumulate in an ecosystem as dissolved organic matter and detritus,

or leave the system in sinking particles. Our atmosphere has oxygen because photosynthesis exceeds respiration and the consumption of oxygen. Notably, positive net ecosystem production draws down CO_2 concentrations and lowers the partial pressure of CO_2 (pCO_2) in aquatic systems. When the dissolved pCO_2 is less than the atmospheric pCO_2, there is a net flux of CO_2 into the water. Many regions of the world's oceans are net sinks for atmospheric CO_2. In fact, about a third of all CO_2 released by the burning of fossil fuel ends up in the oceans (Khatiwala et al., 2013). There would be even more CO_2 in the atmosphere if it were not for this flux into the oceans.

Net community production can also be negative for short periods of time or in regions supplied by organic carbon from more productive waters or from land. Many lakes are heterotrophic, with negative net production because of organic inputs from terrestrial primary production (Figure 6.4). In these systems, pCO_2 is higher in the water than in the atmosphere, leading to the release of CO_2, or outgassing, to the atmosphere, a non-trivial flux in global budgets (Tranvik et al., 2009). These heterotrophic lakes may still have large build-ups of algal biomass due to high nutrient inputs. In spite of the algal bloom, oxygen production by the algae may still be lower than respiration fueled by inputs of terrestrial organic material.

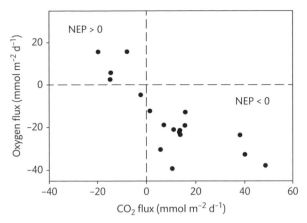

Figure 6.4 Net ecosystem production (NEP) in four lakes over four years in northern Michigan (USA), as determined by oxygen and CO_2 fluxes. Positive values indicate net fluxes from the lake to the atmosphere. Net ecosystem production (NEP) was usually negative in these lakes, as indicated by the negative oxygen fluxes and positive CO_2 fluxes. Data provided by J. J. Cole, taken from Cole et al. (2000).

Primary production by terrestrial higher plants and aquatic microbes

Using the methods for examining biomass and primary production just discussed, ecologists and biogeochemists found that higher plants and photoautotrophic microbes

each account for about half of all global primary production, as mentioned before. But a closer look at the data reveals profound differences in how these photoautotrophs contribute their 50% of the total. Of the many differences between terrestrial and aquatic primary producers, size is the most obvious and probably the most important. The difference in size ends up having huge consequences for how terrestrial and aquatic ecosystems are organized and structured. A comparison of aquatic and terrestrial environments also illustrates important principles about how per capita rates and standing stocks contribute to fluxes.

The averages given in Table 6.2 illustrate the huge differences in biomass and primary production rates among terrestrial and aquatic ecosystems. The biomass of even the most nutrient-rich ("eutrophic") lake or small pond is very small compared with an equal area on land, with the exception of barren deserts. The oligotrophic open oceans have even less biomass, evident from their clear blue waters. Yet rates of primary production in aquatic ecosystems can rival those found on land. The oceans account for a large fraction of global primary production, not only because of their large surface area, but also because rates per square meter can be high for some regions.

The reason why low biomass systems can have high primary production rates is because of high growth rates.

Table 6.2 Photoautotrophic biomass and growth in the major biomes of the planet. "NPP" is net primary production. Turnover time was calculated by Biomass/NPP. Data from Valiela (2015).

	Location	Area (10^6 km²)	Biomass (kg C m⁻²)	NPP (g C m⁻² y⁻¹)	Turnover time (y)
Aquatic					
	Open oceans	332	0.003	125	0.02
	Upwellings	0.4	0.02	500	0.04
	Continental shelves	27	0.001	300	0.00
	Estuaries	1.4	1	1500	0.67
	Wetlands	2	15	3000	5.00
	Lakes	2	0.02	400	0.05
Terrestrial					
	Tropics	43	8	623	12.6
	Temperate	24	5.5	485	11.3
	Desert	18	0.3	80	3.8
	Tundra	11	0.8	130	6.2
	Agriculture	16	1.4	760	1.8

To see why, note that the relationship between production (P), growth rate (μ), and biomass (B) is

$$P = \mu \cdot B \qquad (6.8)$$

where B has units of mass per unit area (gC m^{-2}, for example), and μ has units of per time, such as d^{-1}. Consequently, production has units of mass per unit area per time (g C m^{-2} d^{-1}, for example). The growth rate has units of per time, such as d^{-1}, and does not directly involve biomass. (Both growth rate and production will be discussed in more detail in Chapter 8.) Growth rates can be measured directly for photoautotrophic microbes by following the incorporation of $^{14}CO_2$ into pigments or by the Landry–Hassett dilution approach (Landry and Hassett, 1982). Here, we estimate it simply by using Equation 6.6 and dividing P by B. This method has many experimental and theoretical problems, but it does give a rough idea of how fast plants and photoautotrophic microbes grow. To compare with terrestrial ecosystems, the inverse of μ is calculated to give the turnover time.

This calculation indicates that photoautotrophic microbes have growth rates of about 0.1 to 0.2 d^{-1} and turn over in 4 to 7 days. More accurate measurements indicate growth rates of about 1 d^{-1} (Kirchman, 2016). These growth rates are one hundred- to one thousand-fold faster than growth rates of land plants (Table 6.2). Consequently, although biomass per square meter is much less in aquatic systems than on land, the difference is nearly cancelled out by the much higher growth rates in freshwaters and the oceans. The result is similar rates of primary production per square meter in aquatic systems as on land. This is the first of several examples presented in this book that illustrate the contribution of turnover (with units of time) and standing stocks (units of mass per unit volume or area) to determining a rate or flux (mass per unit volume or area per unit time).

The spring bloom and controls of phytoplankton growth

The numbers given in Table 6.2 give some hints about the variation in rates and standing stocks of primary producers among various ecosystems on an annual basis. However, these also vary over time, on scales ranging from hours to years. Primary production varies from zero every night when the sun goes down to high rates on bright, sunny days. In temperate lakes and oceans, primary production and algal biomass also vary with the seasons. They increase from very low levels in winter to high levels in spring (Figure 6.5). This large increase is called an algal or phytoplankton "bloom." During blooms, net production is high, and growth of phytoplankton exceeds mortality due to viral lysis and grazing. Blooms are large biogeochemical events that affect many ecosystem processes while they occur and long after their demise. If we understand blooms and how they end, we have gone a long way towards understanding the controls of the phytoplankton community and of phytoplankton growth in aquatic ecosystems. The "bottom-up" factors affecting phytoplankton include light and nutrient concentrations, while the "top-down" factors include grazing and viral lysis. Here, we will concentrate on the bottom-up factors.

Why is phytoplankton biomass so high in the spring? Likewise, why is it low in the winter? Temperature may

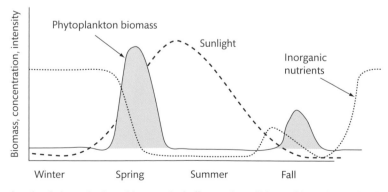

Figure 6.5 The seasonal cycle of phytoplankton biomass, including peaks or "blooms" in spring and in fall, in a typical temperate aquatic habitat. The primary inorganic nutrients include nitrate, phosphate, and, for diatoms, silicate.

be one factor that comes to mind (Kremer et al., 2017), and it is certainly important (Chapter 3), but phytoplankton can form blooms even in the very cold waters of the Arctic Ocean and Antarctic seas, where water temperatures hover near freezing. So, temperature is not the complete answer.

The key to explaining low phytoplankton biomass in the winter is the availability of light. The amount of solar radiation reaching the Earth's surface varies greatly with the seasons. Also, in winter at mid to high latitudes, much of the light does not penetrate the surface of a lake or the ocean because of reflection at the air–water interface. Both the quality and quantity of light have a large impact on the taxonomic composition of the phototrophic community and rates of biomass production. Quality is the wavelength of light, as previously mentioned, and quantity the light intensity. These two parameters of light vary greatly among aquatic ecosystems and with water depth.

Light intensity declines exponentially with depth in water and in soils. Light intensity (I_z) as a function of depth (z) is described by

$$I_z = I_0 e^{-kz} \qquad (6.9)$$

where I_0 is light intensity at the surface (z = 0), and k is the attenuation coefficient. This coefficient is small for open ocean water and large for a murky pond or soils.

The decrease in light intensity with depth leads to a corresponding decrease in photosynthesis (P), but not in a simple linear way, as illustrated in Figure 6.6. Of the many equations proposed to describe the general curve given in Figure 6.6, one of the simplest is

$$P = P_{max} \tanh(\alpha \cdot I/P_{max}) \qquad (6.10)$$

where P_{max} is the maximum rate of photosynthesis, and α is the slope of the initial part of the curve in Figure 6.6. In words, photosynthesis increases with light intensity until it reaches a maximum value. One complication not included in Equation 6.10 or Figure 6.6 is inhibition of photosynthesis at high light intensities, which often occurs in surface waters and bare soils. Because of photoinhibition, primary production is often highest below the surface.

The amount of light experienced by phytoplankton in nature depends not only on the level of irradiance at the surface but also on mixing through the water column. How mixing affects light availability and thus phytoplankton growth is encapsulated in the "critical depth" hypothesis, developed in 1953 by a Norwegian oceanographer Harald Sverdrup (1888–1957). The hypothesis gets its name from the depth at which the availability of light and nutrients is high enough for net phytoplankton growth. This depth is set by mixing due to winds blowing across the sea or lake surface and by convective overturning. Convective overturning occurs when surface waters lose heat to the colder atmosphere, thus decreasing in temperature and increasing in density; the water sinks as

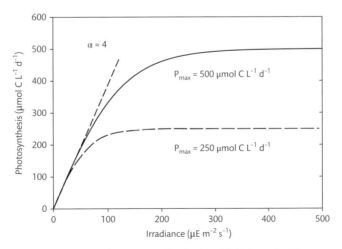

Figure 6.6 Relationship between photosynthesis (primary production) and light intensity. The two cases have the same α but different P_{max} values.

a consequence. Although mixing brings up nutrients to the surface layer from deep waters where concentrations are high, blooms are not possible in the winter because phytoplankton are limited by light. Winter mixing sends phytoplankton deep into the water column where they spend too much time in poorly lit deep waters to grow. When mixing is too deep—below the critical depth—phytoplankton growth is less than losses due to respiration. As winter gives way to spring, the upper layer of the water column warms up and becomes less dense. This warm, less dense surface layer floats on top of colder, denser deep waters. The water column overall becomes more stable and mixing is not as deep, yet the surface layer still has high nutrient levels, similar to those seen in deep waters. Consequently, phytoplankton remain in the light with plenty of nutrients, enabling them to grow fast and to produce more biomass than losses due to respiration. Although not perfect (Behrenfeld, 2010), the theory still provides a useful framework for thinking about light, mixing, and nutrient availability in regulating phytoplankton growth.

Phytoplankton blooms are important, but they do not occur everywhere, and their absence from an aquatic habitat can be important for other reasons. Blooms do not occur in the tropics that do not have the strong seasonal cycles in light and temperature found in the temperate zone. In marine systems, phytoplankton blooms also do not occur in "high nutrient–low chlorophyll" (HNLC) oceans, such as the subarctic Pacific Ocean, the equatorial Pacific Ocean, and the largest HNLC regime, the Southern Ocean. In these waters, concentrations of iron are extremely low and limit rates of primary production (Hutchins and Boyd, 2016). The HNLC regions are possible regions where draw-down of atmospheric CO_2 could be artificially enhanced to counteract the release of CO_2 by the burning of fossil fuels and other human activities.

Major groups of bloom-forming phytoplankton

The main type of phytoplankton making up a bloom varies among different aquatic systems. Which type is dominant has implications for how the bloom affects food web dynamics and the carbon cycle and other biogeochemical processes. These phytoplankton types can be organized by taxonomy, ranging from the algal genus *Phaeocystis* to the phylum *Cyanobacteria*. Another way to organize these organisms is by functional groups. Members within one group may or may not be phylogenetically related, but they do share similar cell wall composition, cell size, or the production of specific compounds (Table 6.3). Variation in these properties explains why different photoautotrophs have different impacts on their habitat. All of the functional groups listed in Figure 6.3 can form phytoplankton blooms, except the last group, the picophytoplankton. These small microbes dominate the phytoplankton community in oligotrophic waters where blooms do not occur.

Diatoms

This algal group, the class Bacillariophyceae, is often the major photoautotroph making up spring phytoplankton blooms in lakes and coastal oceans. Diatoms are also common in algal mats and in the surface layer of soils exposed to sunlight. In aquatic habitats, diatoms do well in the spring, probably because they are better than other algae at using high concentrations of nitrate (the dominant form of inorganic nitrogen in spring) and phosphate, because they grow faster than other algae in the low water temperatures of spring, and because they are able to cope with the huge variation in light that is

Table 6.3 Functional groups of phytoplankton. Diatoms, diazotrophs (except *Trichodesmium*), and picophytoplankton are found in freshwaters and marine waters, whereas coccolithophorids and *Phaeocystis* are only in the oceans.

Functional group	Function	Example
1. Diatoms	Silicate use. Blooms in lakes and coastal waters	*Thalassiosira, Asterionella*
2. Coccolithophorids	$CaCO_3$ production	*Emiliania huxleyi*
3. *Phaeocystis*	Dimethylsulfide production	*Phaeocystis*
4. Diazotrophs	N_2 fixation	*Anabaena, Trichodesmium*
5. Picophytoplankton	Accounts for large fraction of biomass and production in oligotrophic waters	Coccoid cyanobacteria (*Synechococcus, Prochlorococcus*)

also typical of the spring. Diatoms are successful in spite of the fact that they require a nutrient, silicate, not needed by other algae or cyanobacteria; silicate is used for the synthesis of a cell wall, the frustule, found only in diatoms among the algae (Figure 6.7A). Silicate concentrations are high in the spring but decrease as the bloom progresses. However, it is nitrogenous nutrients, specifically nitrate, that end up limiting diatom growth at the end of spring blooms. The depletion of nutrients, among other factors, allows other photoautotrophs to dominate the phytoplankton community as spring transitions to summer.

Spring blooms of diatoms and other algae are critical for ensuring the success of higher trophic levels (larger organisms), and in some sense of the entire aquatic ecosystem. These blooms fuel the growth of herbivorous zooplankton, and the zooplankton in turn are prey for larvae of invertebrates and fish. The spawning of larvae is timed to coincide with spring blooms. Playing on a line from the poem *Leaves of Grass* by Walt Whitman (who in turn borrowed it from Isaiah 40:6), the oceanographer Alfred Bigelow opined that "all fish is diatoms." Herbivores graze on diatoms probably simply because diatoms dominate algal blooms. In fact, diatoms may inhibit reproduction of some types of zooplankton (Carotenuto et al., 2014). Diatoms are also abundant in benthic habitats, microbial mats, and even soils, as will be discussed further.

Coccolithophorids and the biological pump

In the oceans, another group of phytoplankton, the coccolithophorids, can also form dense blooms. Quite unlike other algae, coccolithophorids are covered with calcified scales (coccoliths) made of calcium carbonate ($CaCO_3$) (Figure 6.7B). This algal group is not abundant in freshwaters where calcium concentrations are too low for coccolith synthesis. In addition to being primary producers and fixing CO_2, coccolithophorids contribute to the carbon cycle by forming $CaCO_3$ (Chapter 13) and by exporting carbonate to deep waters and sediments. Carbonate rocks buried in oceanic sediments make up the largest reservoir of carbon on the planet (Chapter 13; see Box 6.1).

Coccoliths make up one component of the "biological pump" in the ocean, which is the sinking of organisms and detritus from the surface layer to deep waters and sediments. $CaCO_3$-containing structures from coccolithophorids and other organisms make up the "hard" part of the pump. The "soft" part consists of organic material originally synthesized by phytoplankton which can be repackaged into fecal pellets by zooplankton and into other organic detritus that sinks to the deep ocean. The sinking of both hard and soft parts "pumps" carbon from the surface layer into the deep ocean and some as far as the sediments. The amount of carbon exported as $CaCO_3$ is generally <10% of total export (Sarmiento and Gruber, 2006). Albeit small, it is an important percentage

(A)

(B)

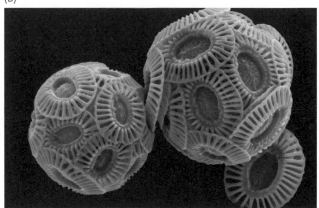

Figure 6.7 Electron micrographs of two common eukaryotic algae: a centric diatom, about 100 μm in diameter (A); the coccolithophorid, *Emiliania huxleyi*, about 4 μm in diameter (B). Pictures used with permission from Ken Bart, University of Hamilton and Alison R. Taylor, University of North Carolina, Wilmington, Microscopy Facility.

Box 6.1 Microbial fossil beds

The preserved cell walls of diatoms and coccolitho-phorids are studied by paleo-oceanographers and limnologists for examining the C cycle over geological times. The amount of these microbial fossils is an index of primary production in the distant past, and the types of diatoms and coccolithophorids can be used to distinguish different water masses in ancient oceans. The ^{13}C content of alkenones, which are produced by coccolithophorids, is used by pale-oceanographers to explore temperature in past geological eras. More dramatic is another type of coccolithophorid fossil. The famous White Cliffs of Dover, England are made of coccoliths ("chalk") that were deposited about 140 million years ago during the Cretaceous period, when southern England was submerged under a tropical sea. The sediments then became exposed as the sea retreated during the ice ages. After the ice ages, the rising sea cut through these soft sediments, leaving behind the English Channel and exposing the cliffs of coccoliths.

---

because of the contribution of $CaCO_3$ to carbon storage in sediments for millennia or longer (Chapter 13). $CaCO_3$-containing structures also act as ballast, speeding up the sinking of organic carbon. Most of the carbon never makes it to the sediments, however, because hetero-trophic organisms oxidize the organic carbon back to CO_2, and $CaCO_3$ structures dissolve back to soluble ions. Even so, the CO_2 regenerated in the deep ocean does not reach the surface layer and remains out of contact with the atmosphere for hundreds of years.

The biological pump is a target of efforts to counteract the release of CO_2 by fossil fuel burning and other human activity. In the absence of the biological pump, atmos-pheric CO_2 concentrations would be 425 to 550 ppm (depending on various assumptions), which is higher than present levels (about 400 ppm in 2017). At the other extreme, if the biological pump were operating at 100% efficiency, atmospheric CO_2 would only be 140–160 ppm (Sarmiento and Toggweiler, 1984). The pump is not work-ing at 100% in the HNLC oceanic regions, resulting in their high concentrations of nitrate and phosphate. It has

been proposed that iron should be dumped into these regions to stimulate phytoplankton growth and to draw down atmospheric CO_2. Adding iron to small areas in a HNLC region does stimulate phytoplankton growth (Tagliabue et al., 2017; Boyd et al., 2007), but large-scale enrichment, an example of "geoengineering," to slow down the increase in atmospheric CO_2 is controversial. Even assuming it would work, many negative side effects are possible.

Phaeocystis *and dimethylsulfide*

A genus of phytoplankton, *Phaeocystis*, which belongs to the Prymnesiophyceae, is another bloom-forming photoautotroph and makes up its own functional group. It has an unusual life cycle. Although it can occur as a solitary, flagellated cell, unknown environmental factors trigger the formation of colonies. Some species of *Phaeocystis* form blooms in coastal waters and cause serious water quality problems. *Phaeocystis* can excrete large quantities of extracellular polymers, so large that unsightly foam several meters deep builds up on the beaches of northern Europe and in the Adriatic Sea. Other species are abundant in Antarctic seas, especially in the Ross Sea where *Phaeocystis* colonies are not grazed on by zooplankton. When growth conditions deteriorate and the bloom ends, *Phaeocystis* colonies can sink rap-idly from the water column.

The main reason why *Phaeocystis* makes up its own functional group is due to its role in the production of dimethylsulfide (DMS), the major sulfur gas in the oceans. DMS is produced during the degradation of another organic sulfur compound, dimethylsulfoniopropionate (DMSP) (Figure 6.8). *Phaeocystis* and some other algae use DMSP as an osmolyte, as mentioned in Chapter 3, or perhaps as an antioxidant or a stress response to anoxia (Sunda et al., 2002; Omori et al., 2015). Initially, ocean-ographers thought that algae make DMS directly from DMSP, but later work showed that DMSP is released from algae, perhaps by zooplankton grazing on the algae, and is then degraded to DMS by heterotrophic bacteria. Some heterotrophic bacteria, the most intensively stud-ied ones being in the *Roseobacter* clade, are known to produce DMS during cleavage of DMSP, whereas other bacteria demethylate DMSP to produce 3-methiolpropi-onate (Moran et al., 2012).

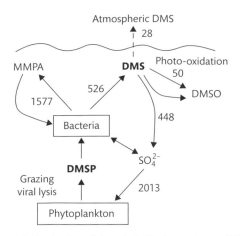

Figure 6.8 Production of dimethylsulfoniopropionate (DMSP) by phytoplankton and its conversion to 3-methiolpropionate (MMPA), or to dimethylsulfide (DMS) which can be degraded to dimethylsulfoxide (DMSO) or sulfate. The numbers are Tg S y^{-1}. Based on Moran et al. (2012).

These organic sulfur compounds are important in the sulfur cycle. DMSP, for example, can supply nearly all of the sulfur used by heterotrophic bacteria (Kiene and Linn, 2000). Even more significant is the possible role of DMS in affecting the world's heat budget. Because it is supersaturated in the upper ocean, DMS outgasses to the atmosphere where it is oxidized to sulfate and contributes to aerosol formation. These aerosols scatter sunlight and act as cloud condensation nuclei, hot spots of cloud formation. The extent of cloud cover has impacts on the amount of light and heat reaching the Earth's surface. This interaction between the plankton and global climate via DMS production is one example of the "Gaia hypothesis," proposed by James Lovelock in 1972 (Stolz, 2017). According to the Gaia hypothesis, negative feedbacks between the biosphere and the rest of the planet maintain the Earth's climate and biogeochemical processes at homeostasis. While extensive research over the years has revealed flaws in the Gaia hypothesis (Tyrrell, 2013), it is still true that DMS is an important compound and that microbes have many important roles in shaping the Earth's climate.

Cyanobacteria and filamentous diazotrophs

Lakes and reservoirs sometimes experience large blooms of cyanobacteria in summer. The cyanobacteria making up blooms often occur as cells strung together in long filaments reaching several millimeters in length (Figure 6.9). Filamentous cyanobacteria are also important in algal mats and in soil microbial communities. Although these microbes have some unique features and thus unique roles in the environment, cyanobacteria carry out primary production like eukaryotic algae. In fact, cyanobacteria share many physiological traits with eukaryotic algae and higher plants, because they are the ancestors of chloroplasts according to the endosymbiosis hypothesis (Chapter 9). All of these organisms use the same

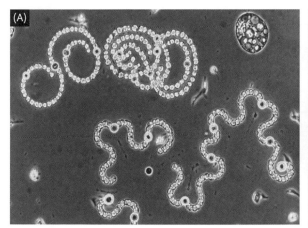
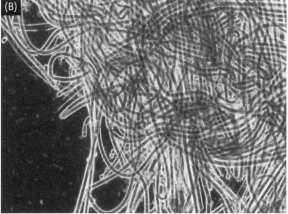

Figure 6.9 Some filamentous cyanobacteria. Chains of *Dolichospermum* (formally *Anabaena*) from Lake Rotongiao, New Zealand. The large cells in each chain are heterocysts (A). *Lyngbya aestuarii*, from a benthic mat near Beaufort, North Carolina (B). Each cell is 5 to 10 μm in diameter. Images courtesy of Hans Paerl. Color versions are available at http://www.oup.co.uk/companion/kirchman.

mechanism to convert light energy to chemical energy, and they fix inorganic carbon into organic carbon by the same pathway, the Calvin–Benson–Bassham cycle.

But in every other respect, cyanobacteria are firmly in the *Bacteria* domain. Cyanobacteria do not have chloroplasts or any organelle, their genome is usually a single circular piece of DNA, and they have cell walls like Gram-negative heterotrophic bacteria, with components such as muramic acid and lipopolysaccharides that are found only in bacteria. The composition and organization of the light-harvesting pigments of cyanobacteria also differs from eukaryotic algae and higher plants. These distinct pigments, the phycobilins, enable cyanobacteria to flourish in low light environments and occasionally outcompete eukaryotic algae.

Cyanobacteria used to be called "blue-green algae," and the chromatic terms of the old name still accurately describe the color of some cyanobacteria; the green is due to chlorophyll *a*, which absorbs violet and red light, allowing only green to reach our eye. One phycobilin, phycocyanin, gives these microbes a blue tinge by removing yellow light. Isolated phycocyanin is a brilliant blue. In contrast, another type of cyanobacterium common in the oceans, *Synechococcus*, has large amounts of a blood-red pigment, phycoerythrin. Dense liquid cultures of this cyanobacterium are pink even though the microbe also contains some phycocyanin and chlorophyll *a*. Table 6.4 lists some cyanobacteria found in nature.

Summer blooms of filamentous cyanobacteria in freshwaters can be caused by high water temperatures which favor these prokaryotes over eukaryotic microalgae. Of even more importance is the supply of phosphate. Because phosphate is often the nutrient limiting primary production in freshwater ecosystems and some oceans,

eukaryotic algal groups with superior uptake systems for phosphate outcompete filamentous cyanobacteria when phosphate concentrations are low. However, phosphate pollution of freshwaters removes the competitive edge of eukaryotic algae and allows cyanobacteria to bloom. These massive outbreaks negatively affect water quality and the general "health" of the ecosystem (Paerl and Otten, 2013). There is evidence that harmful cyanobacterial blooms have become more common over the years because of climate change and other human-related causes (Visser et al., 2016).

The success of some filamentous cyanobacteria in freshwaters with high phosphate concentrations is largely due to their ability to fix dinitrogen gas (N_2) to ammonium (NH_4^+). Organisms capable of fixing N_2 are called "diazotrophs." Some diazotrophic cyanobacteria have "heterocysts," which are specialized cells where N_2 fixation is carried out. When phosphate is plentiful, nitrogen nutrients become limiting, potentially allowing N_2-fixing cyanobacteria to proliferate. While N_2 fixation is carried out by many prokaryotes, not eukaryotes, only some cyanobacteria contribute substantially to both N_2 fixation and primary production. In freshwater ecosystems, filamentous cyanobacteria with heterocysts are the dominant diazotrophs, whereas in the oceans, other diazotrophic cyanobacteria without heterocysts are more important (Chapter 12).

In addition to N_2 fixation, several other traits of filamentous cyanobacteria contribute to their success in freshwater systems. Some of these traits help to minimize losses due to grazing by zooplankton. Individual filaments of some cyanobacteria are often too large to be grazed on effectively by filter-feeding zooplankton. To make matters worse for zooplankton, some filamentous cyanobacteria

Table 6.4 Some important cyanobacterial genera.

Genus	Morphology	Habitat	Noteworthy ecology
Anabaena	Filament	Freshwater	N_2 fixation
Microcoleus	Filament	Soils, desert crust	Tolerates harsh conditions
Microcystis	Filament	Freshwater	Produces toxins
Trichodesmium	Filament	Marine	N_2 fixation
Richelia	Rod	Marine	Endosymbiotic N_2 fixation
Synechococcus	Coccus	Marine and freshwater	Primary production*
Prochlorococcus	Coccus	Marine	Primary production*

* Although all of these cyanobacteria carry out oxygenic photosynthesis like eukaryotic algae, *Synechococcus* and *Prochlorococcus* are especially important as primary producers and account for a large fraction of primary production and biomass in oligotrophic oceans.

form large colonies or aggregates that are even harder for zooplankton to graze on. These aggregates may be visible to the naked eye as green scum on the surface of lakes and reservoirs. Some cyanobacteria float to the surface by regulating buoyancy with gas vacuoles. The end result of high growth fueled by N_2 fixation, assisted by warm summer temperatures, plus low losses due to zooplankton grazing, is a dense bloom of filamentous cyanobacteria.

Some cyanobacteria also have chemical defenses against grazing. We know the most about toxins produced by freshwater cyanobacteria such as *Anabaena*, *Aphanizomenon*, *Microcystis*, and *Nodularia* species. Microcystins, a suite of toxins produced by toxic strains of *Microcystis*, cause liver damage in humans and affect the heart and other muscles of zooplankton (Figure 6.10). These toxins also deter, if not kill off, zooplankton and herbivorous fish grazing on cyanobacterial mats. Because of these toxins and other secondary metabolites, cyanobacterial blooms can lead to a decrease in water quality; drinking water from reservoirs with dense cyanobacterial populations may taste poor. Worse, cyanobacteria-tainted water can be toxic to humans, domestic pets, livestock, birds, and fish.

Toxin production is just one of several negative impacts of cyanobacterial blooms. The switch at the base of aquatic food webs, from primary production by eukaryotic algae to that by filamentous cyanobacteria, can negatively affect the rest of the food web. Herbivorous zooplankton suffer if their only prey are inedible cyanobacteria. Carnivorous zooplankton and larvae feeding on the herbivores are affected next.

In addition to being ugly, cyanobacterial scum floating on the surface of lakes and reservoirs prevents light from reaching other phytoplankton living deeper in the water column, leading to a reduction in oxygen production below the surface. To make matters worse, consumption of oxygen (respiration) is high in scum-filled water, fueled by respiration of organic compounds released from living and dying cyanobacteria. The end result is anoxia or hypoxia just underneath the luxuriant floating mat of cyanobacteria, which still may be actively photosynthesizing and producing oxygen. Among many negative effects of anoxia and hypoxia, hydrogen sulfide (rotten egg smell) and other noxious compounds produced by microbes in low-oxygen waters add to the toxins produced by cyanobacteria.

In addition to noxious cyanobacteria, several eukaryotic algae can cause problems when their numbers are high, leading to "harmful algal blooms" (HAB) in coastal marine waters. HABs include red tides caused by some species of dinoflagellates and brown tides caused by the

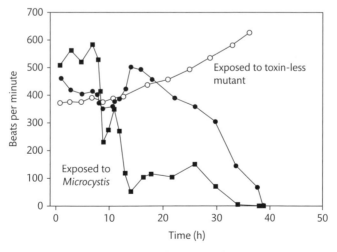

Figure 6.10 An example of the effect of a cyanobacterial toxin (microcystin) on the physiology of a freshwater zooplankton, *Daphnia*. Individual animals were glued in place and fed either a toxic strain (solid symbols) or a toxin-less mutant (open circles) of *Microcystis*. Heart beat (circles), leg motion (squares), and other physiological responses were recorded by videotaping. Data from Rohrlack et al. (2005).

chrysophyte *Aureoumbra*. These HAB species release toxins that can kill mussels, clams, and other invertebrates as well as marine mammals, fish, and sea birds. The diatom *Pseudo-nitzschia* produces the neurotoxin domoic acid, which is known to affect the behavior of California sea lions (Cook et al., 2015). As with cyanobacterial blooms, HABs may be increasing over time due to coastal eutrophication.

After the bloom: competition for limiting nutrients

As the spring bloom progresses and biomass levels build up, microbes strip nutrients from the water column, reducing concentrations from micromolar to nanomolar levels or lower. The result is that algal growth becomes limited by nutrients, explaining why growth slows down. The bloom stops as phytoplankton are removed by grazers or sink out of the surface layer. As the spring bloom ends and large diatoms and other large algae decline, other photoautotrophic groups start to dominate the phytoplankton community.

The other photoautotrophs taking over from diatoms and other large taxa include cyanobacteria and protists in the nanoplankton (cell diameter 2–20 μm) and picoplankton (0.2–2 μm) size classes. The nanoplankton include dinoflagellates, cryptophytes, and other, poorly characterized algal groups. The picoplankton include coccoid cyanobacteria and small eukaryotic algae. These small cells are more abundant and account for more primary production than large phytoplankton in oligotrophic habitats such as the open ocean, where concentrations of nutrients are extremely low (<10 nM). Small cells somehow outcompete large cells for the limiting nutrient when concentrations of that nutrient are low. Why is that?

Physics, specifically diffusion, gives a small cell an advantage over a big cell when concentrations are low. In the microbial world, diffusion is the main process by which nutrients are brought to the cell surface. A stationary cell cannot take up more nutrients than are supplied to it by diffusion. As discussed in Chapter 3, the diffusive flux per unit surface area (J) is a function of the diffusion coefficient (D), the nutrient concentration away from the cell (C), and the cell radius (r):

$$J = 4\pi DrC \qquad (6.11).$$

To calculate the flux per unit volume or biomass of the cell (J_v), Equation 6.11 can be divided by the volume of a spherical cell (V = 4/3 πr^3) to yield

$$J_v = 3D \cdot C/r^2 \qquad (6.12).$$

Equation 6.12 indicates that the upper limit for uptake set by diffusion decreases as cell size increases. As cell size and the radius increase, nutrient concentrations must increase proportionally to achieve the same flux to the cell. For example, a 1.0 μm cell could grow at 1 d^{-1} with about 15 nM of nitrate, while a 5.0 μm cell would require > 100 nM to grow at the same rate (Chisholm, 1992). The story is complicated by cell shape and motility, but the basic physics remains the same. Big cells are more likely to be diffusion-limited in oligotrophic waters than small cells.

In addition to physics, biochemistry and physiology work against large cells in oligotrophic waters. To understand this, it is necessary to look at how uptake varies as a function of nutrient concentrations. This relation is described by the Michaelis–Menten equation, the same used to describe enzyme kinetics:

$$V = V_{max} \cdot S/(K_s + S) \qquad (6.13).$$

Here, uptake (V) is a function of the nutrient concentration (S) and two parameters of the uptake system, the maximum uptake rate (V_{max}) and the half-saturation constant (K_s). The half-saturation constant is equivalent to the concentration at which the uptake rate is half of V_{max}. Over a full range of substrate concentrations, V increases until it reaches V_{max}, resulting in the curve illustrated in Figure 6.11. Note, however, that when S is low relative to K_s, then $K_s + S \approx K_s$ and Equation 6.10 reduces to

$$V = (V_{max}/K_s)S \qquad (6.14).$$

Equation 6.14 says that at very low nutrient concentrations, such as after the spring bloom, the uptake rate depends on the ratio of V_{max} to K_s, the "affinity constant." So, the prediction is that small cells must have either a low K_s or high V_{max}, or both, if they are to outcompete large cells for limiting nutrients. In fact, there is evidence that small cells have a lower K_s and a higher V_{max} normalized per unit of biomass than do large cells, giving small cells an additional advantage when nutrients are low.

So, size explains why small phytoplankton dominate systems with very low nutrient concentrations. Size is also

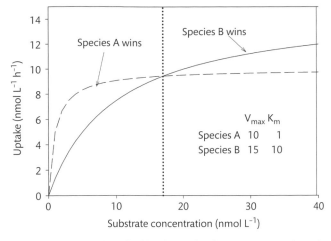

Figure 6.11 Uptake of a dissolved compound as described by the Michaelis–Menten equation. The example has two species competing for the same dissolved compound. The winner is the one with higher uptake rates for a given range of substrate concentrations.

very important in thinking about the interactions between bacteria, phytoplankton, and fungi. Again, the prediction is that smaller bacteria would outcompete larger organisms for inorganic nutrients, which is often the case.

Primary production by coccoid cyanobacteria

One phytoplankton group that is common after blooms is coccoid cyanobacteria. These microbes are especially important in the surface waters of oligotrophic oceans where nutrient concentrations are extremely low (<10 nmol liter^{-1}). Biological oceanographers first recognized the importance of coccoid cyanobacteria when they used filters with different pore sizes to examine the size distribution of primary producers ($^{14}CO_2$ uptake) and of phytoplankton biomass (chlorophyll a concentrations). An example is given in Figure 6.12. These studies demonstrated that in the open oceans, such as the North Pacific Gyre, as much as 80% of all $^{14}CO_2$ uptake and of chlorophyll a is associated with organisms in the <3 μm size fraction (Rii et al., 2016). Subsequent microscopic examination indicated that many of the photoautotrophs among the picoplankton are coccoid cyanobacteria. Given the vast coverage of the open oceans, these data imply that roughly 25% of global primary production is by these cyanobacteria.

There are two diverse groups of coccoid cyanobacteria important in aquatic habitats. The phycoerythrin-rich cyanobacteria, *Synechococcus*, were first examined by marine microbial ecologists in part because these microbes could be counted by epifluorescence microscopy (Chapter 1). But this method missed the second type, *Prochlorococcus*. These microbes are very difficult to see by epifluorescence microscopy, so they were not observed until another technique, flow cytometry (Chapter 5), was used on samples from the oceans (Chisholm et al., 1988). Both *Synechococcus* and *Prochlorococcus* are most abundant in low-latitude open oceans, while freshwater species of *Synechococcus* also occur in oligotrophic lakes. Table 6.5 summarizes the main traits of these two cyanobacteria.

The abundance of *Synechococcus* and *Prochlorococcus* varies differently with depth. Though both cyanobacterial

Table 6.5 Comparison of the two major coccoid cyanobacterial genera found in lakes and oceans.

Property	*Synechococcus*	*Prochlorococcus*
Size (diameter)	1.0 μm	0.7 μm
Chlorophyll a	Yes	Modified
Chlorophyll b	No	Yes
Phycobilins	Yes	Variable
Distribution	Cosmopolitan	Oceanic gyres
Nitrate use	Common	Few*

* Nitrate uptake has been demonstrated for only a few isolated strains, but as much as 50% of the taxa in natural communities can take up nitrate (Berube et al., 2016).

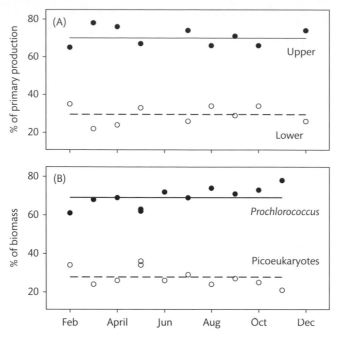

Figure 6.12 Primary production and biomass by *Prochlorococcus* and picoeukaryotes in the North Pacific Ocean. Percentage of primary production by picoplankton (0.2–3 μm, mainly *Prochlorococcus*) in the upper (0–45 m) and lower (75–125 m) layers of the water column (A). Biomass of *Prochlorococcus* and photosynthetic picoeukaryotes, expressed as a percentage of the total biomass (B). Data from Rii et al. (2016).

groups are adapted to the intensity and quality of light found deeper in the water column, *Synechococcus* is generally found higher in the water column than *Prochlorococcus*. *Prochlorococcus* is divided into two types ("ecotypes") which are found at different depths: a high-light ecotype with a low ratio of chlorophyll *a* to chlorophyll *b*, and a low-light ecotype with a high chl *a* to chl *b* ratio. As expected, the low-light ecotype is usually found deeper in the water column than the high-light ecotype. The 16S rRNA genes from isolates of these two ecotypes differ only slightly (average similarity of >97%), less than the level of similarity usually used to separate species (Chapter 4). However, whole genome sequencing has revealed many genetic differences between the two ecotypes, as mentioned in Chapter 5 (Rocap et al., 2003).

Anaerobic anoxygenic photosynthesis

This chapter so far has focused on the dominant form of primary production, with oxygenic photosynthesis as its physiological base, which produces oxygen as a by-product.

There is another form of photosynthesis that does not produce oxygen: "anoxygenic" photosynthesis, a process carried out only by bacteria. It can be represented by a more general version of Equation 6.2:

$$CO_2 + 2H_2A + light \rightarrow CH_2O + 2A + H_2O \quad (6.15)$$

where H_2A is a reduced compound, and A is an oxidized compound with element A. The equation is from the Dutch microbial ecologist C. B. van Niel (1897–1985), who worked for most of his career at the Hopkins Marine Laboratory affiliated with Stanford University. H_2A and A would be H_2O and O_2 in oxygenic photosynthesis, whereas anoxygenic photosynthetic microbes use compounds other than water (H_2A) to provide the electrons needed to reduce CO_2 and synthesize organic carbon (CH_2O). Today, anoxygenic photosynthesis does not contribute much to primary production on a global scale (<1%), although it can be a significant source of organic carbon in some environments.

Anoxygenic photosynthesis is important for other reasons. The variety among anoxygenic photosynthetic mechanisms helps us to understand the biophysics and biochemistry of photosynthesis and its evolution on early Earth. Now restricted to only a few habitats, anoxygenic photosynthetic bacteria once were the only phototrophic primary producers on the planet for eons, from soon after life arose about 3.5 billion years ago to about 2.8 billion years ago, when cyanobacteria with their superior oxygenic photosynthesis evolved. Microbial ecologists and biogeochemists are interested in the anoxygenic process occurring today because of its importance in elemental cycles other than carbon, most notably the sulfur cycle. Anoxygenic phototrophs are major sinks for H_2S and other reduced sulfur compounds. These organisms and processes will also be discussed in Chapter 11 because of their role in oxidizing reduced end products of the sulfur cycle.

A key feature of anoxygenic photosynthetic bacteria is their pigments, mainly bacteriochlorophylls. Although their structures are similar to that of the chlorophylls used by oxygenic photosynthetic organisms, the anoxygenic photosynthetic chlorophylls and other pigments absorb light in different parts of the spectrum than chlorophyll a and other pigments in oxygenic photosynthetic microbes (Figure 6.13). Most notably, bacteriochlorophylls adsorb light with longer (infrared) wavelengths. These pigments enable anoxygenic photosynthetic bacteria to flourish in light environments that do not support the growth of oxygenic photosynthetic microbes. Microbes carrying out anoxygenic photosynthesis do so in anoxic environments.

The possible electron sources, the H_2A in Equation 6.15, used by anoxygenic photosynthesis include H_2, Fe^{2+}, nitrite (NO_2^-), arsenic (As^{3+}), and even some organic compounds (Schott et al., 2010), but the most common and arguably the most important electron sources are H_2S and other reduced sulfur compounds (Table 6.6). The phototrophic users of reduced sulfur compounds include purple sulfur and green sulfur bacteria, which are found in some lakes and marine basins, sediments, and sulfur springs. They can flourish at the intersection between two opposing gradients: light, which decreases with depth, and reduced sulfur compounds, which increase with depth through the water column and sediments. Oxygenic phototrophs cannot absorb the light available at this intersection, and they are inhibited by sulfides.

In microbial mats, purple sulfur bacteria related to *Chromatium* are abundant at depths reached only by near infrared light, because other light is absorbed by overlying oxygenic phototrophs in shallower depths. Use of the long-wavelength light by *Chromatium*-like microbes is made possible by their bacteriochlorophyll b (Jørgensen and Des Marais, 1986). Because growth is possible only when both light and reduced sulfur are available, anoxygenic phototrophs make vividly colored bands a couple

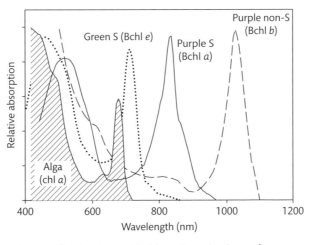

Figure 6.13 Absorption spectra of an alga (*Isochrysis sp.* with chl a as its main pigment), a green sulfur bacterium (*Pelodictyon phaeoclathratiforme* with Bchl e), a purple sulfur bacterium (*Chromatium okenii* with Bchl a), and a purple non-sulfur bacterium (*Blastochloris viridis* with Bchl b). The alga carries out oxygenic photosynthesis, while the three bacteria carry out anoxygenic photosynthesis. Based on data from Stomp et al. (2007).

Table 6.6 Some characteristics of anoxygenic phototrophs. BChl = Bacteriochlorophyll; Alpha = *Alphaproteobacteria*; Beta = *Betaproteobacteria*; Gamma = *Gammaproteobacteria*; L = chemolithoautotrophy; P = photoautotrophy; RPP = reductive pentose phosphate cycle; RCA = reductive citric acid cycle; 3HP = 3-hydroxypropionate cycle; AAP = aerobic anoxygenic phototrophic bacteria. Based on Canfield et al. (2005) and Overmann and Garcia-Pichel (2006).

Group	Taxonomy*	Electron source	Pigments	Autotroph mode	Facultative aerobic	C fixation pathway
Purple S	Alpha	H_2S, S^0, $S_2O_3^{2-}$, H_2	Bchl *a, b*	P, L	Yes	RPP
Purple non-S	Alpha, Beta	H_2S, S^0, $S_2O_3^{2-}$, H_2	Bchl *a, b*	P, L	Yes	RPP
Green S	*Chlorobiaceae*	H_2S, S^0, $S_2O_3^{2-}$, H_2	Bchl *a, c, d, e*	P	No	RCA
Green non-S	*Chloroflexales*	H_2S, H_2, organic C	Bchl *a, c, d*	P	Yes	3HP, RPP
Heliobacteria	*Firmicutes*	H_2S, organic C	Bchl *g*	None	No	None
AAP	Alpha, Beta, Gamma	Organic C	Bchl *a*	None	Yes	None

* Many microbes in the *Alphaproteobacteria*, *Betaproteobacteria*, and *Firmicutes* do not carry out anoxygenic photosynthesis.

of meters thick in the water column or millimeters thick in sediments. In some lakes with high sulfur concentrations, primary production by anoxygenic phototrophic bacteria in these plates can exceed that by oxygenic phototrophic microbes in the much thicker surface layer.

Many anoxygenic phototrophic bacteria use gas vacuoles to maintain their position where growth conditions are optimal, while others are motile (Overmann and Garcia-Pichel, 2006). They either swim using flagella in the water column or glide on surfaces in sediments, moving toward both light and reduced sulfur compounds. Some purple sulfur bacteria appear to migrate a couple of meters every day in response to diel changes in light and reduced sulfur. In microbial mats, which we will discuss in more detail next, the purple sulfur bacterium *Chromatium* carries out diel migration to take advantage of the intracellular sulfur produced from the oxidation of H_2S by anoxygenic photosynthesis during the day. It migrates up at night to the surface and the overlaying water, using positive aerotaxis. Once in this oxygenated microenvironment, *Chromatium* switches to chemolithoautotrophy and oxidizes the stored elemental sulfur.

Several anoxygenic phototrophic bacteria do not use CO_2 as a carbon source and must rely on organic carbon. They use this carbon for biosynthesis and oxidize some of it for energy, augmented by energy gained from phototrophy, a type of metabolism called "photoheterotrophy." One example of photoheterotrophic bacteria is the family *Heliobacteriaceae*, which have an unusual pigment, bacteriochlorophyll *g*. These Gram-positive bacteria have been found mainly in soils and waterlogged rice paddy fields (Stevenson et al., 1997). Another example is the aerobic anoxygenic phototrophic bacteria, which

have bacteriochlorophyll *a*. Microbes with this form of photoheterotrophy are in the *Alpha*- and *Betaproteobacteria*. These bacteria can be quite abundant in lakes and the oceans, but they are outnumbered by yet another type of photoheterotrophic bacteria, those that harvest light energy with rhodopsin, often called proteorhodopsin, to distinguish it from the pigment bacteriorhodopsin first found in archaea (Béjà et al., 2000). Rhodopsin can be found in nearly half of all bacteria in some habitats. These photoheterotrophs are discussed in more detail in Chapter 7.

Sessile algae and microbial mats

The eukaryotic algae and cyanobacteria as well as many of the anoxygenic phototrophic bacteria discussed previously are found in the water column of oceans and lakes. Relatives of these microbes also live attached to surfaces in aquatic habitats and on land. In shallow waters of lakes and oceans, where enough light reaches the bottom, benthic photoautotrophs can be abundant and active enough to contribute substantially to primary production. The benthic photoautotrophic community is often dominated by diatoms, dinoflagellates, and cyanobacteria. Although not substantial on a global scale, the organic carbon produced by benthic photoautotrophs can be locally important in supporting benthic food webs of invertebrates and benthic-feeding fish in shallow sediments.

Photoautotrophs are also found on many surfaces exposed to light on land, as mentioned at the beginning of the chapter. Cyanobacteria are especially ecologically important in alkaline soils in arid and semi-arid regions around the world, where growth of higher plants is limited (Overmann and Garcia-Pichel, 2006). These microbes have evolved strategies of withstanding desiccation and of growing rapidly when water becomes available. Cyanobacterial "desert crusts" contribute to the physical stability of arid soils, akin to the role of microbes in stabilizing sediments in aquatic habitats (Chapter 2), and may even warm up soils due to the production of a microbial sunscreen, scytonemin (Couradeau et al., 2016). Cyanobacteria also grow on the surface of rocks or even in micro-crevices just below the rock surface. Endolithic cyanobacterial communities are usually dominated by the genus *Chroococcidiopsis*. In soils, yellow-green algae (*Xanthophyta*), red algae (*Rhodophyta*), and diatoms, as well as the anoxygenic phototrophic bacteria, purple sulfur, and purple non-sulfur bacteria can be abundant and active in CO_2 fixation (Yuan et al., 2012).

On some surfaces in aquatic habitats, dense, complex communities build up to form microbial mats (Figure 6.14). These communities include phototrophs together with many other metabolic types of bacteria and archaea. Microbial mats can grow in the intertidal sands of marine habitats, saltmarsh sediments, hydrothermal hot springs, and hypersaline ponds. Mats are luxuriant at places like the hot springs of Yellowstone National Park because extreme temperature and pH exclude invertebrates and other animals. Mats are not more common elsewhere probably because of grazing and mixing of sediments by animals. Mats can also be based on the primary production by chemolithoautotrophs in absence of light, such as in the Black Sea, at deep-sea hydrothermal vents, and in the sediments under highly productive waters off the coast of Chile and Peru. However, the best-studied mats are exposed to light and are supported by both oxygenic and anoxygenic phototrophic primary producers. Anoxygenic photosynthesis in mats is especially common where reduced sulfur concentrations are high, such as at hot springs.

One reasons to study microbial mats existing today is that they can provide insights into early life on the planet when a type of mat, "stromatolites," covered vast areas of primordial seas in the Proterozoic (Chapter 13). But microbial mats have intrigued microbiologists and microbial ecologists over the years for other reasons. Mats have large numbers of many different types of microbes, packed tightly together, and intensively interacting over

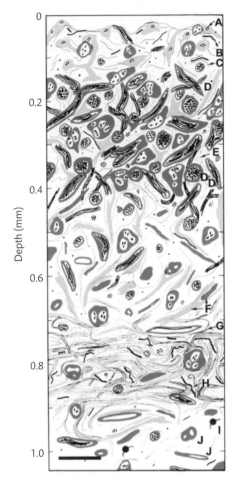

Figure 6.14 A slice through a microbial mat, reconstructed from several transmission electron micrographs. The letters refer to mat microbes: A, diatoms; B, *Spirulina* sp. (cyanobacterium); C, *Oscillatoria* spp. (cyanobacterium); D, *Microcoleus* (cyanobacterium); E, non-photosynthetic bacteria; F, fragments of bacterial mucilage; G, *Chloroflexus* spp. (green non-sulfur bacteria, capable of anoxygenic photosynthesis); H, *Beggiatoa* spp.; I, unidentified grazer; J, abandoned cyanobacterial sheaths. Taken from Des Marais et al. (1992) and used with permission of the publisher.

micron to millimeter scales. They are both intellectually and technically challenging to study. In addition to the molecular methods and other approaches discussed in Chapters 4 and 5, much has been learned about mats from the development and application of microelectrodes for measuring micron-scale gradients in oxygen, H_2S, and other compounds.

Microelectrodes and other approaches have revealed that mats are highly vertically stratified. A few millimeters of a mat can contain many of the organisms and all of the metabolisms discussed in this chapter, as well as those to be covered in Chapter 11, devoted to anaerobic processes. The top layer of a mat may consist of diatoms able to withstand a high intensity of light during the day. Below this is a layer of filamentous cyanobacteria where most of the mat's primary production occurs, evident in a peak of oxygen concentrations during the day. As cyanobacteria decrease with depth, oxygen also decreases because of its use by heterotrophic microbes and by chemolithoauto- trophs and anoxygenic phototrophic bacteria as they oxidize reduced sulfur compounds. Those compounds, most notably H_2S, are produced by sulfate-reducing bacteria abundant in deep, anoxic layers. There are many exchanges of metabolites among members of the mat community. For example, a metatranscriptomic study of a mat in Elkhorn Slough, California, found that organic acids produced by fermentation carried out by the most abundant cyanobacterium were used by the filamentous green non-sulfur bacterium *Chloroflexus* (Burow et al., 2013). This is one of many possible complex interactions and processes occurring in microbial mats.

Summary

1. Microbes account for about 50% of global primary production; the other half is by land plants. Cyanobacteria alone are responsible for 50% of aquatic primary production or about 25% of total global primary production.

2. Rates of primary production are governed by light intensity and quality, along with nutrients such as ammonium, nitrate, phosphate, and iron. One important type of photoautotrophic microbe, diatoms, also requires silicate.

3. Several eukaryotic microbes can dominate the algal community at different times and in different places, with various effects on the ecosystem. These microbes include diatoms (common in blooms), dinoflagellates (some are toxic while others are photoheterotrophic), coccolithophorids ($CaCO_3$ formation), and *Phaeocystis* (DMS producer).

4. Filamentous N_2-fixing cyanobacteria are common in freshwaters, whereas coccoid cyanobacteria (*Synechococcus* and *Prochlorococcus*) are abundant in marine waters, particularly oligotrophic open oceans where they can account for a large fraction of phytoplankton biomass and primary production.

5. Cyanobacterial blooms can cause problems with the quality of water in freshwater reservoirs and ecosystem function in small ponds and lakes. Some eukaryotic algae are harmful to aquatic life and humans.

6. Anoxygenic phototrophic bacteria are important in the sulfur cycle and as primary producers in environments with high sulfur concentrations. These bacteria are also important for understanding the evolution, biochemistry, and biophysics of photosynthesis.

7. Microbial mats are vertically stratified, complex communities of oxygenic and anoxygenic phototrophs, chemolithoautotrophs, and heterotrophs. They serve as models for microbial communities that dominated the biosphere during the Proterozoic.

CHAPTER 7

Degradation of organic matter

The previous chapter discussed the synthesis of organic material by autotrophic microbes, the primary producers. This chapter will discuss the degradation of that organic material by heterotrophic microbes. These two processes are large parts of the natural carbon cycle. Globally, nearly all of the 120 gigatons of carbon dioxide fixed each year by primary producers is returned to the atmosphere by heterotrophic microbes, macroscopic animals, and even some by autotrophic organisms. Consequently, heterotrophy as well as primary production, contributes to determining how much atmospheric carbon dioxide is taken up by the global biosphere. The balance between primary production and organic matter degradation sets many other properties of the ecosystem.

As with all biogeochemical cycles, the carbon cycle consists of reservoirs (concentrations or amounts of material) connected by fluxes (time-dependent rates) driven by both natural and anthropogenic processes (Figure 7.1). The natural rates of exchange between carbon reservoirs are much larger than the anthropogenic ones. In particular, the natural production of carbon dioxide by heterotrophs is much higher than the anthropogenic production, due to the burning of fossil fuels and other human activities. The problem is that because the anthropogenic production of carbon dioxide is not balanced by carbon dioxide consumption, concentrations in the atmosphere are increasing and our planet is warming up (Chapter 1). Nearly all of the natural processes in the carbon cycle are huge and variable, complicating the efforts of biogeochemists to understand how human activity is affecting the carbon cycle and the implications for climate change.

The carbon cycle has several reservoirs of both inorganic and organic material (Figure 7.1). The largest reservoirs are dissolved inorganic carbon (DIC), mostly bicarbonate, in the ocean and calcium carbonate (a major mineral in limestone) on land and in oceanic sediments. The amount of carbon in organisms and in non-living particulate organic material is small. Aquatic ecologists call this dead material "detritus," while terrestrial ecologists also use the terms "plant litter" or simply "litter" when discussing material that is still recognizable as coming from higher plants. Another large reservoir is dissolved organic carbon (DOC), and there is also much organic carbon in sediments of the oceans.

The largest reservoir of organic carbon, however, is in soils and in other terrestrial compartments. These organic reservoirs are as large (oceanic DOC) or larger (soil organic material, SOM) than the atmospheric reservoir of CO_2. In terms of fluxes, two of the largest natural ones are the draw-down of atmospheric CO_2 by primary producers (Chapter 6) and the release of CO_2 by heterotrophic organisms, the subject of this chapter. Not depicted in Figure 7.1 are geochemical reactions such as weathering of carbonate rocks that also take up or release CO_2, topics to be discussed in Chapter 13. Microbes are very important in nearly all of these fluxes.

In this chapter, we discuss aerobic degradation of particulate detritus, litter, and dissolved organic material (DOM) in oxic environments, leaving anaerobic degradation in anoxic environments to later chapters. Here,

Processes in Microbial Ecology. Second Edition. David L. Kirchman. Oxford University Press (2018). © David L. Kirchman 2018.
DOI 10.1093/oso/9780198789406.001.0001

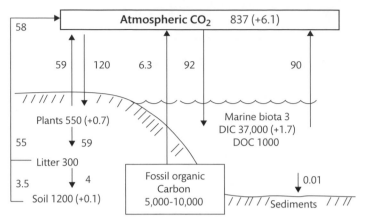

Figure 7.1 Global carbon cycle. The units for the numbers next to the reservoir names are petagrams of carbon (1 Pg = 10^{15} g) and next to the arrows are Pg C y^{-1}. The numbers in parentheses are the yearly changes. Some features not shown here include the CO_2 produced by land-use change (0.9 Pg C y^{-1}) and the biggest carbon reservoir, carbonate rocks (Chapter 13). Based on data presented in Houghton (2007), except the atmospheric CO_2 data from May 2017 (https://www.esrl.noaa.gov/gmd/).

"degradation" means the breaking down of organic material to smaller compounds. "Mineralization" is the degradation of the organic compound to its inorganic constituents, such as carbon dioxide, ammonium, and phosphate. During mineralization in oxic environments, the only element in the organic material whose oxidation state changes is carbon. It is oxidized from its state in organic carbon (zero) to carbon dioxide (4+):

$$CH_2O + O_2 \rightarrow CO_2 + H_2O \qquad (7.1)$$

where again CH_2O symbolizes generic organic material, not a specific compound. In oxic environments, the complete degradation of organic matter is due to aerobic respiration, which consumes oxygen and produces carbon dioxide and water. But degradation involves more than just carbon because organic material always has several other elements. Consequently, organic matter degradation releases several other inorganic or mineral nutrients, such as ammonium and phosphate, in addition to CO_2. Some authors use "remineralization" and "regeneration" to highlight the never-ending cycle of uptake and release of compounds containing essential elements like N and P.

Who does most of the respiration on the planet?

In Chapter 6, we saw that microbes were responsible for about half of all global primary production due to photosyn-

thesis by eukaryotic phytoplankton and cyanobacteria in the oceans. Microbes are also mainly responsible for degrading the organic carbon made by primary production as measured by respiration. Microbes account for much more than half of total global respiration, although the precise percentage is difficult to estimate. The global estimate may be less important than the percentages for individual ecosystems. These percentages indicate the importance of microbes in structuring the flow of energy, carbon, and other elements in these ecosystems.

In aquatic environments, respiration by microbes can be estimated by incubations in which large organisms are removed by filtration, leaving only microbes in the water. The consumption of oxygen is then measured over time in the dark (to stop photosynthesis and oxygen production), sometimes along with estimates of DOC consumption. Simultaneously, respiration by all organisms is estimated from changes in oxygen in other, dark incubations with unfiltered water.

This experiment has shown that nearly all of the respiration is by organisms smaller than 200 μm and nearly half is by organisms that could pass through a filter with 0.8 μm pores (Figure 7.2). The exact percentage varies with the environment, but usually it is very high, 50% or greater. Other analyses show that these organisms are mostly bacteria. Several other methods and approaches support the conclusion that roughly half of the total respiration in aquatic ecosystems is by bacteria. These data also indicate that fungi do not account for much respiration in most

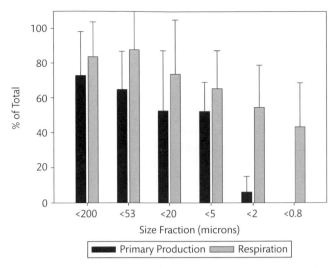

Figure 7.2 The size distribution of respiration and photosynthesis, expressed as a percentage of rates in unfiltered samples. The data are mostly from those aquatic environments where primary production by cyanobacteria and small eukaryotes is low. See Chapter 6. Data from Williams (2000).

freshwaters and marine systems. Some fungi are indigenous to aquatic systems, and they may be abundant on large particles and detritus in some habitats (Bochdansky et al., 2017). Still, their overall biomass is very low compared with bacteria in aqueous environments. Other data indicate that archaea are also not very important in organic matter degradation in aquatic habitats.

Like the water column of aquatic habitats, microbes also account for a large fraction of respiration and organic matter degradation in sediments and soils, although it is difficult to get good estimates. One approach is to look at respiration by large organisms and assume the difference with total respiration is due to microbes. To estimate respiration by large organisms in sediments, one way is to combine data on abundance in the field and respiration rates determined in laboratory experiments. These studies indicate that large organisms account for 5–30% of total respiration in freshwater and coastal marine sediments, implying the remaining 70–95% is due to microbes. For soils, respiration has been measured before and after removing plant roots. These studies found that roughly half of total respiration is by roots (called "autotrophic respiration") and associated microbes in the rhizosphere (Hanson et al., 2000), and nearly all of the rest is by microbes not associated with roots. Little respiration in soils is by large organisms, such as nematodes, earthworms, and insect larvae, because they make up a small fraction (<5%) of total biomass.

Bacteria and fungi both account for a large fraction of respiration and of organic material degradation in soils, in contrast with aquatic ecosystems. Here, we focus on fungi living on dead organic material (saprophytic fungi) and leave discussion of root-associated fungi (mycorrhizal fungi) to Chapter 14. According to experiments using antibiotics and other inhibitors, bacteria and fungi account for about 35 and 65% of microbial respiration in soils, respectively (Joergensen and Wichern, 2008). These percentages may be inaccurate due to inefficiencies in stopping activity with inhibitors, and the contributions by fungi and bacteria certainly vary among soils, depending on environmental factors such as pH, water content, and temperature. Like their contribution to respiration, the biomass of fungi is much higher in soils than in aquatic habitats. On average, fungi make up 35 to over 50% of microbial biomass in soils (Joergensen and Wichern, 2008; van Groenigen et al., 2010), depending on the soil type, geographical location, and method for estimating microbial biomass. As with aquatic systems, heterotrophic archaea are not abundant enough in soils to be important in organic matter degradation.

Bacteria and saprophytic fungi have the same ecological role in nature, but the abundance of fungi and their contribution to total degradation in aquatic habitats versus in soils is quite different. Why? Bacteria win out in the water column of aquatic habitats because their small size makes them superior competitors for dissolved compounds, the same mechanism that explains why small photoautotrophs win out over big ones, as discussed in Chapter 6. This competitive edge is less important in soils, unless they are waterlogged. In terrestrial environments, the hyphae life form taken on by fungi allows them to cross dry gaps between moist microhabitats and to access organic material not available to water-bound bacteria. Some bacteria also grow as filaments, but the resemblance to fungal hyphae is superficial. Unlike bacteria, the cytoplasm of fungi moves within the rigid hyphae to take advantage of favorable growth conditions. Fungi can also move nutrients along their hyphae to support degradation in organic carbon-rich but nutrient-poor microhabitats (van der Wal et al., 2013). The hyphal body form goes a long way to explaining the success of fungi in soils.

Fungi and bacteria use similar organic compounds in soils, but there are differences with implications for understanding rates of growth and carbon cycling (Rinnan and Bååth, 2009). In soils, bacteria tend to use labile organic compounds, while fungi degrade refractory material, the most important being ligno-cellulose, as will be discussed further. Experimental studies have found that bacteria are stimulated more so than fungi by the addition of labile organic material, and bacterial rRNA (an indicator of active cells) is higher on labile leaf litter than on refractory material, whereas the opposite is the case for fungal rRNA (Fabian et al., 2017; Wymore et al., 2013). These differences in organic carbon use probably explain why bacteria generally grow faster than fungi, as discussed in Chapter 8 in more detail. Reflecting the differences in growth rates and the types of organic matter used by these microbes, bacteria are thought to mediate a fast carbon cycling pathway, while fungi are responsible for a slow pathway. Although not perfect (Rousk and Frey, 2015), the slow versus fast pathway model is useful for thinking about the role of fungi versus bacteria in organic carbon degradation.

Detritus and detrital food webs

We have just seen that microbes account for a large fraction of respiration in the biosphere. Because the organic carbon supporting that respiration must have come from primary producers, the respiration data indicate that a large fraction of primary production is somehow routed through microbes. That observation is in contrast with

how ecologists used to picture carbon flows and food webs. The classic food chain consists of plants being eaten by herbivores which in turn are eaten by carnivores, and so on up the food chain. Ecologists have long known that microbes degrade dead organic material and release inorganic nutrients for primary producers, but that seemed necessary only to clean up after the large organisms who carry out herbivory and carnivory. However, the respiration data and other studies indicate that the classic food chain is not the main fate of primary production in most ecosystems. The main fate somehow involves heterotrophic microbes.

The question then becomes, how does organic material in primary producers get to microbes? While some pathogenic and symbiotic microbes get organic compounds directly from living plants and algae, a larger pathway is via detritus and other forms of non-living organic material. This material is produced when algae, higher plants, and animals senesce and die. Detritus can also be a by-product of herbivore grazing, fecal material from metazoan grazers, or viral lysis. The detritus produced by these different mechanisms differs in composition and rates of degradation.

The fate of primary production—herbivory or detritus—varies with the ecosystem and the type of primary producer (Figure 7.3). On land, where higher plants dominate, only a small fraction of primary production is consumed

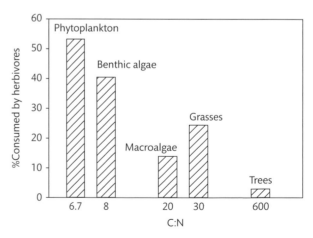

Figure 7.3 Fraction of primary production directly consumed by herbivores. The primary producers range from protein-rich algae with low C:N ratios (6.7) to trees with little protein, resulting in C:N ratios exceeding 100. Data mostly from Cebrian (1999).

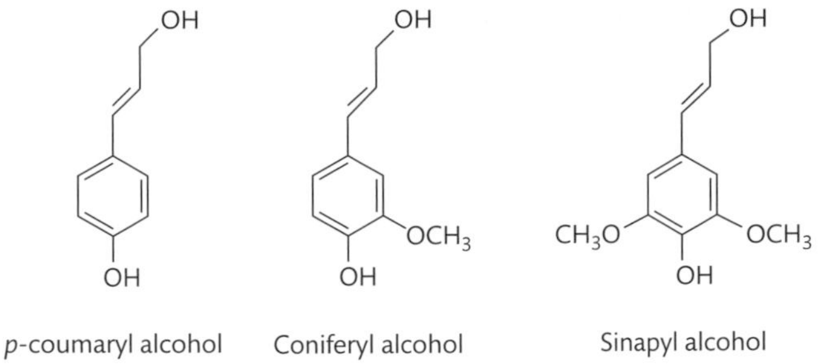

p-coumaryl alcohol   Coniferyl alcohol   Sinapyl alcohol

Figure 7.4 The structure of common subunits of lignin, the main structural element of wood. The amounts of these and other subunits vary with the type of lignin.

directly by herbivores, while the fraction is higher in aquatic systems where primary production is almost entirely by algae and cyanobacteria. But even in lakes and the oceans, as much as 50% of primary production is routed through the dead organic matter pool by various mechanisms. The reason for the difference is the composition and ultimately the size of the primary producer.

A small fraction of higher plant primary production is eaten by herbivores, in part because of the biopolymers and structures plants need to succeed on land. Terrestrial plants need stems, branches, and trunks in order to grow up and out into the air away from soil and towards the light. Trees need bark to fend off herbivores and pathogens. These structures have lots of cellulose, related complex carbohydrates, and lignin. Cellulose is a polymer of glucose linked by β1,4 bonds, whereas lignin is a very complex, ill-defined structure consisting of several phenolic or aromatic groups (Figure 7.4). Lignin is a major component of wood, and its strength explains why trees can grow so high and why they are so hard for herbivores to eat. Although some phytoplankton and other aquatic primary producers have cellulose in their cell walls, they do not make lignin. Suspended by water, phytoplankton and macroalgae do not need lignin and woody structures to survive. Consequently, phytoplankton are relatively rich in protein, much more so than terrestrial plants, because they lack many of the carbohydrates and all of the lignin required for life on land (Table 7.1). Likewise, the particulate detritus in aquatic environments is protein-rich, whereas detritus on land reflects the carbohydrate make-up of terrestrial plants.

Detritus is mineralized by bacteria and fungi to CO_2 and other inorganic constituents, but some is also eaten by some metazoans. The end result is a "detritus food

web" of bacteria, fungi, protists, and metazoans, living directly or indirectly on dead organic material rather than on live plants or algae (Figure 7.5). The detritus-eating organisms are called "detritivores." Marine benthic

Table 7.1 Biochemical composition of plant detritus and organisms in terrestrial and aquatic ecosystems. Data from Canfield et al. (2005) and Randlett et al. (1996).

	% of total				
	Lignin	Carbohydrate	Protein	Lipid	C:N ratio
Terrestrial					
Straw	14	81	1	2	80
Tree leaves	12	77	7	12	50
Pine wood	27	72	0	1	640
Aquatic					
Kelp	0	91	7	<1	50
Diatom	0	32	58	7	6.7
Zooplankton	0	14	46	<1	6.7

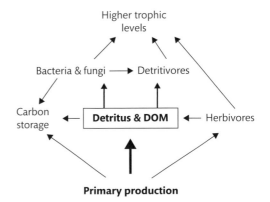

Figure 7.5 Detritus food web. "DOM" is dissolved organic material.

ecologists use the term deposit-feeders because detritus eaten by these animals is deposited onto sediments from plankton production in overlying surface waters. Deposit-feeders and detritivores feed indiscriminately in sediments and soils where detritus and plant litter are abundant. In all cases, microbes, which otherwise are too small to be grazed on by these animals, are included with the detritus as it is ingested. Some examples of detritivores include nematodes and polychaetes in sediments and oligochaetes, such as earthworms, and springtails (*Collembola*) in soils. Detrital food webs are especially important in detritus-rich habitats, such as salt marshes, many estuaries, and bogs. These food webs are even more important in the subsurface environments of soils. Nearly all primary production by trees is routed through detrital food webs, while roughly half is in grasslands (Cebrian, 1999).

Detritivores are very important in effecting the degradation of particulate detritus in both terrestrial and aquatic ecosystems, even though their direct contribution to detritus mineralization is small (Gebremikael et al., 2015). Rather than accounting for much respiration, the more important role of detritivores is to physically break up detritus and plant litter, which decreases the size of detrital particles and as a result increases the surface area where microbes can adhere and degrade the exposed organic compounds (Figure 7.6). Macrofauna, which are organisms larger than 2 mm, can have additional effects on the microbial environment. In soils, these animals break up aggregates and increase aeration and water flow. Earthworms in particular constitute a "geomorphic force" ten-fold stronger than other, purely physical processes (Chapin et al., 2002). In addition to physically disturbing soils, earthworms secrete along their burrow walls polysaccharides that bind to clays, forming stable organic aggregates. They may also enhance degradation by releasing compounds that help microbes further degrade organic detritus, a type of "priming effect," which will be discussed further (Bernard et al., 2012). In sediments, burrows of macrofauna allow penetration of oxygen into otherwise anoxic environments, greatly affecting sediment chemistry. In both soils and sediments, macrofauna disrupt the orderly layers, horizons, and gradients in geochemical properties that would otherwise form in a world without animals. The end result is that detritivores and other macroscopic organisms help to speed up the degradation of detritus, even though most of the actual mineralization is done by bacteria and fungi.

DOM and the microbial loop

In addition to particulate detritus, plant and algal organic material becomes available to microbes through the release or excretion of dissolved organic matter (DOM). In soils and sediments with rooted plants, this release is part of "below-ground production," in contrast with the more visible "above-ground production." Difficult to estimate, below-ground production can be a very large

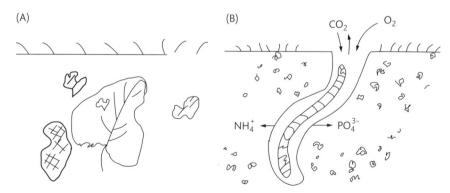

Figure 7.6 Effect of macrofauna, like worms, on the degradation of organic material and on the structure of soils and sediments. Panel A illustrates a worm-less world in which large pieces of detritus are not broken down. In contrast, in the environment depicted by panel B, worms and other macrofauna help to break up detritus and facilitate the mineralization of the organic material to inorganic nutrients like ammonium and phosphate. The burrows of these large organisms also allow faster diffusion of gases in soils and of dissolved compounds in aquatic sediments.

fraction, as high as 50%, of total primary production by higher plants (Högberg and Read, 2006). This released organic material fuels soil microbial activity while by-passing herbivores.

Like excretion by plant roots, DOM is released directly or indirectly by phytoplankton in aquatic ecosystems. This release can be measured by tracing $^{14}CO_2$ into phytoplankton cells caught on filters with 0.2 μm pore size and then into DOM, the radioactivity passing the 0.2 μm filters; any remaining dissolved $^{14}CO_2$ is removed by acidification. This experiment indicates that as much as 50% of primary production can be released as DOM, although the overall average is closer to 10%. Some of the ^{14}C-labeled DOM comes directly from phytoplankton cells, while other components may be released inadvertently by herbivores eating ^{14}C-labelled phytoplankton cells ("sloppy grazing") or by viral lysis (Chapter 10). These and other data indicate that fluxes through the DOM reservoir are quite large, equal to roughly half of primary production, and can support much microbial growth and respiration.

DOM cannot be taken up by many large organisms, and even protists usually do not compete effectively with heterotrophic bacteria for dissolved compounds. Once taken up by bacteria, however, the carbon, nitrogen, and other elements in DOM are now available to protists eating the bacteria. The protists are then eaten by metazoans, and so on up the food chain. The DOM-based pathway, consisting of primary production → DOM → microbes → grazers, is called the "microbial loop" (Figure 7.7). The term was coined by aquatic microbial ecologists, but the concept is applicable to terrestrial ecosystems as well (Bonkowski, 2004). Bacteria and fungi turn indigestible organic material, such as ligno-cellulose, into food for soil metazoans. The microbial loop concept highlights that in addition to being mineralizers, bacteria and fungi can be large components of food webs in natural ecosystems.

However, not all of the carbon taken up by microbes is available for grazers and higher trophic levels. Some of it may be respired as CO_2 and thus is lost from the system. The remaining carbon would be used for biomass production and would be available as food for grazers. Figuring out which of these two fates of carbon—respiration or biomass production—is most important has been called the "sink or link" question (Pomeroy, 1974).

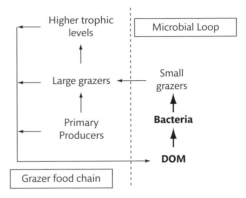

Figure 7.7 The microbial loop, which consists of the production of dissolved organic material (DOM) and its uptake by heterotrophic bacteria. The key concept is that bacteria use a form of organic material (dissolved, in this case) not available to other, larger organisms. To apply to soils, "bacteria" would include fungi, and DOM would be replaced by any organic material used only by bacteria and fungi.

Bacterial growth efficiency and carbon use efficiency

It is possible to determine experimentally whether organic carbon taken up by the microbial loop is mostly respired (it is a "sink") or passed on to higher trophic levels ("link") (Ducklow et al., 1986), but another way to answer the question is to examine the "bacterial growth efficiency" (BGE), called the "carbon use efficiency" (CUE) in the soil literature. This parameter is the ratio of biomass production (P) to the sum of production and respiration (R):

$$BGE = P / (P + R) \cdot 100 \qquad (7.2).$$

The growth efficiency is very important in setting the relationship between growth and other metabolic functioning, such as nutrient mineralization, to be discussed further, for all heterotrophic organisms as well as heterotrophic bacteria.

When the sink–link question was first posed, microbial ecologists thought that the growth efficiency of bacteria was high, on the order of 50%. Growth efficiencies of fungi were also thought to be high. However, results from new experiments with natural microbial communities indicated that the growth efficiency is usually less than 50% (Figure 7.8). For the data given in Figure 7.8, the growth efficiency varies from 20% in lakes, estuaries, and the oceans, to 49% for soil communities. The numbers vary

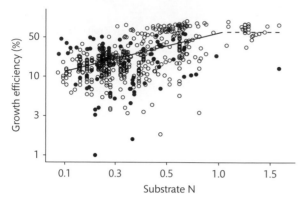

Figure 7.8 Growth efficiencies for natural terrestrial (open symbols) and aquatic (solid symbols) communities as a function of the nitrogen content of the substrate relative to the microbial N content (N:C_s/N:C_b expressed in mass units). The solid line is based on a log–log regression analysis ($r^2 = 0.12$; $p < 0.001$), while the dashed line represents nitrogen-replete systems with maximum efficiency. Data provided by Stefano Manzoni from Manzoni et al. (2017).

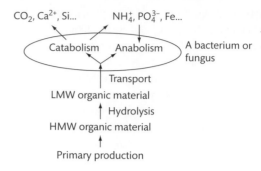

Figure 7.9 Mineralization of organic material by heterotrophic bacteria and fungi. LMW and HMW refer to low and high molecular weight material, respectively. Catabolism is the energy-producing parts of microbial metabolism, whereas anabolic reactions lead to synthesis of cellular components and eventually growth. Some inorganic compounds are potentially used by heterotrophic microbes (NH_4^+, PO_4^{3-}, and Fe), whereas others (CO_2, Ca^{2+}, and Si) are not.

greatly within each system (Lee and Schmidt, 2014). One of several properties thought to affect growth efficiencies is the amount of nitrogen in the organic material used by these heterotrophic microbes. There is a significant but weak relationship between growth efficiency and the relative nitrogen content of the organic material, but clearly other factors are at work (Figure 7.8). The availability of inorganic nitrogen, like ammonium, which is not included in "substrate N" plotted in Figure 7.8, would lead to increases in growth efficiency, as has been shown by carbon use efficiency studies in soils (Spohn et al., 2016b).

Growth efficiency has been examined extensively over the years, and we are still far from understanding how it varies and what controls it. In any case, growth efficiencies of less than 50% mean that most of the carbon is released as CO_2 and little remains in biomass to be eaten and passed on to higher trophic levels. So, the low growth efficiency estimates indicate that the microbial loop is mainly a sink.

Mechanism of organic matter degradation

Equation 7.1 implies that the mineralization of organic material is a simple process. In fact, several biochemical steps are necessary to degrade organic material and to oxidize the organic carbon part to carbon dioxide. Fortunately, here we need to discuss only a few basic steps to understand the ecology of the microbes using this material (Figure 7.9). These basic steps are applicable to both fungi and bacteria. Both microbial groups must take compounds in the external environment across a membrane before using them intracellularly for catabolism (energy generation) or anabolism (biomass production).

Hydrolysis of high molecular weight organic compounds

Even after detritus and plant litter is broken up by metazoans, high molecular weight (HMW) organic compounds need to be reduced in size even further before use by microbes. Compounds larger than about 600 Da must be degraded to compounds small enough (<600 Da) to be transported across cell membranes. The 600 Da cut-off is largely set by the capacity of membrane transport proteins. The mechanism that reduces in size a biopolymer, made up of monomers linked together, differs from the mechanism for lignin. We will first discuss biopolymer degradation and then move on to lignin degradation.

The first step in biopolymer degradation is hydrolysis of the polymer to monomers; hydrolysis, which means "lysis by water," is the breaking of bonds that link monomers together into a polymer. For example, hydrolysis of protein releases amino acids and oligopeptides, but not CO_2 nor NH_4^+. Hydrolysis is often said to be the rate-limiting step or the slowest reaction in the degradation pathway, one piece of evidence being that concentrations of polymers are higher than those of monomers in natural environments. Several types of enzymes, collectively called "hydrolases," are needed to hydrolyze the >600 Da polymers found in the HMW pool. Specific enzymes are necessary for each biopolymer, with the enzyme name usually containing the polymer name, such as the enzyme cellulase that hydrolyzes cellulose. For most polymers, effective hydrolysis requires enzymes that work on different parts of the polymer chain. The breakdown of protein is a good example (Figure 7.10).

Protein must first be hydrolyzed by exoproteases, which cleave off amino acids or dipeptides (two amino acids) at the ends of the polypeptide, and endoproteases, which cleave the peptide chain far from the ends and produce smaller polypeptides. Exoproteases can be further divided into those that work at the N terminus (aminopeptidases) or at the C terminus (carboxypeptidases) of the peptide chain. Any oligopeptides must then be hydrolyzed further by peptidases, although this

hydrolysis step may be inside the cell if the peptide is <600 Da, roughly a pentapeptide. Finally, the monomers can be used to synthesize new polymers or catabolized to provide energy. Note that only during catabolism of monomers, the final step in biopolymer degradation, is organic carbon oxidized to CO_2 and nitrogen mineralized to NH_4^+.

Enzymes that catalyze the initial hydrolysis of HMW biopolymers must be located outside the outer cell membrane, hence their name "extracellular enzymes" or "ectoenzymes." These enzymes can remain tethered to the outer membrane or they can be released into the external environment. Studies with analogs that produce easily measured fluorescent by-products (Table 7.2) have shown that most ectoenzyme activity is associated with cells, but that is not always true for all enzymes. The release option may be an effective strategy for bacteria and fungi in biofilms, attached to particulate detritus, or inside a soil aggregate. In these cases, the released enzyme has a good chance of reaching the targeted HMW compound, and in turn, the LMW by-products cannot diffuse away before uptake by the cell originally releasing the enzyme. It is less clear how releasing the ecotenzyme works for free-living microbes in aquatic

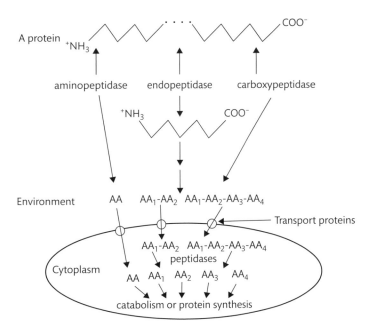

Figure 7.10 The enzymes needed to degrade protein, an important class of high molecular weight organic material. AA is a free amino acid, and AA_1-AA_2 and AA_1-AA_2-AA_2-AA_3 are peptides composed of different amino acids.

Table 7.2 Some polymers, associated hydrolases, and fluorogenic analogs used to study hydrolase activity.

Biopolymer	Hydrolase	Analog*
Proteins	Leucine aminopeptidase	Leu-MCA
Chitin, glycoproteins	N-acetyl-β-D-glucosaminidase	MUF-N-acetylglucosamine
Peptidoglycan	Lysozyme	MUF-N-tri-N-acetyl-β-chitotrioside
Chitin	Chitinase	MUF-N-tri-N-acetyl-β-chitotrioside
Organic phosphate	Phosphatase	MUF-phosphate
Cellulose	Cellulase	MUF-β-D-cellobioside
Polysaccharides with alpha-linkage	α-D-glucosidase	MUF-α-D-glucoside
Lipids	Lipases	Various

* MCA = methylcoumaryl; MUF = methylumbelliferyl.

systems where by-products may never return to the microbe producing the hydrolase and the enzyme itself may be degraded by other microbes. One solution would be to tightly couple hydrolysis and uptake so that few LMW by-products can escape and be used by others.

A large number of different microbes are capable of hydrolyzing biopolymers like protein, cellulose, and chitin. For now, we focus on bacteria because more is known about them, although one study using metaproteomics found that extracellular hydrolytic enzymes for degrading leaf litter were all from fungi (Schneider et al., 2012). Nearly half of all bacteria analyzed so far have genes for at least one hydrolytic enzyme, but few have several (Zimmerman et al., 2013); there are no "super bugs" capable of hydrolyzing all biopolymers. Genomic analyses have found that genes for hydrolytic enzymes are not evenly or randomly distributed among bacterial taxa, but even closely related bacteria may not have the same hydrolytic enzymes, implying that bacteria from different classes or phyla may or may not share similar capacity for biopolymer hydrolysis.

Still, bacteria in the *Bacteroidetes* phylum are well known for having a prodigious number of hydrolytic enzymes (Fernandez-Gomez et al., 2013), and studies using a combination of fluorescence in situ hybridization (FISH) and microautoradiography have found this phylum to be more active than others in using biopolymers in coastal marine waters (Cottrell and Kirchman, 2000). One metatranscriptome study found that cellulases in soils were mostly from *Bacteroidetes* and also *Actinobacteria* (Tveit et al., 2014), while a similar study found that most polysaccharide hydrolases in soils were from bacteria in the *Firmicutes* phylum (Wegner and Liesack, 2016).

Uptake of low molecular weight organic compounds: turnover versus reservoir size

After hydrolysis or the breakdown of large compounds, the next step in organic material degradation is the uptake of monomers and other LMW compounds. These compounds can be released during hydrolysis of biopolymers or directly released by plant roots in soils and by phytoplankton and zooplankton in aquatic environments. We know the most about free amino acids and glucose. These compounds have been examined extensively because proteins and polysaccharides are large components of cells and of the known fraction of organic material. In addition, the use of amino acids and some sugars can be followed easily because their concentrations can be measured by high performance liquid chromatography (HPLC) and they are commercially available labelled with ^{13}C, ^{14}C, or ^{3}H.

If judged by concentrations alone, LMW compounds would not seem very important in fueling microbial growth and in overall organic material degradation. Concentrations of compounds like glucose and amino acids are very low, nanomolar or less. There are more amino acids on your fingertips than in a liter of water or in a few grams of soil. However, in spite of low concentrations, the flux of amino acids and other monomers can be quite high. Flux refers to both production and uptake, which are equal at steady state when concentrations do not change ($dS/dt = 0$). The change in a compound (or substrate, S) over time is

$$dS/dt = P - \lambda \cdot S \qquad (7.3)$$

where P is the production rate, and λ is the turnover rate constant. The units of flux combine the units of concentra-

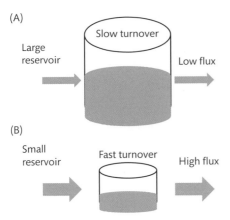

(A)

Large reservoir

Slow turnover

Low flux

(B)

Small reservoir

Fast turnover

High flux

Figure 7.11 Two possible relationship between reservoir size and fluxes. Flux is low in spite of a large reservoir because the turnover rate constant is small (A). In the other case, the flux is high even though the reservoir is small because turnover is high (B). Other cases not shown include a low flux due to a small reservoir and a high flux due to a large reservoir.

tion (mass per unit area, mass, or volume, such as nanomol liter^{-1}) and of the turnover rate constant (per time, such as per day). In spite of very low concentrations, turnover of LMW compounds can be fast enough to result in very high fluxes (Figure 7.11). Microbial ecologists often use the inverse of turnover rate constants, the "turnover time," to quantify the relationship between fluxes and reservoir size. Geochemists use "residence time" for the same concept.

The turnover time of LMW compounds like amino acids can range from minutes to hours. Because of these fast turnover times, fluxes of LMW compounds alone can support a high fraction of bacterial growth in natural environments even though concentrations are very low. Concentrations of these labile, LMW compounds are low because of rapid use by microbes. In some cases, low concentrations signify low fluxes and importance to microbes, but that is not the case for compounds like amino acids and sugars.

Uptake or mineralization of N and P?

Once inside the cell, the LMW organic compound may be used directly for biosynthesis or oxidized to provide energy. This oxidation potentially could release, depending on the chemical make-up of the LMW compound,

ammonium, phosphate, and other inorganic nutrients. This release or "regeneration" is the classic role of heterotrophic bacteria and fungi and is essential in maintaining primary production in the absence of external sources of nutrients. The problem is, heterotrophic bacteria and fungi need some of the N and P in the original organic material for biosynthesis (anabolism) of their own cells. Another complication is that these heterotrophic microbes can take up ammonium, phosphate, and many other inorganic nutrients, thus competing with algae, cyanobacteria, and higher plants. So what determines when there is release or uptake of these nutrients? We will answer this question for N, but the answer can be easily modified to examine the flux of P and other elements.

The short answer is the C:N ratio of microbes versus the C:N ratio of the organic material, plus how much carbon is lost during respiration (Goldman et al., 1987a). The long answer involves a couple of simple equations. Based on the definition of growth efficiency (Y), the amount of C used by microbes for growth is $U_c \cdot Y$, where U_c is the total uptake of the material in carbon units. To calculate N uptake, we convert C uptake to N uptake with the C:N ratio of the original organic material and of microbial biomass; here, it is more convenient to use N:C_s for the organic material (the substrate) and N:C_b for microbial biomass, rather than the more commonly used C:N. So, the total amount of N taken up is $U_c \cdot N:C_s$ and the amount of N used for growth is $U_c \cdot Y \cdot N:C_b$. At steady state, these two are balanced by the uptake or release of ammonium (F_N):

$$U_c \cdot N:C_s + F_N = U_c \cdot Y \cdot N:C_b \qquad (7.4).$$

For net ammonium excretion, $F_N < 0$, whereas for net uptake, $F_N > 0$. In this model, Y is assumed to be constant.

To explore Equation 7.4 in more detail, we can calculate conditions when there is no net excretion or uptake of ammonium, and then use these conditions as a dividing line between the two opposing processes (Figure 7.12). For example, let us use two extreme values for C:N_b, which are averages for bacteria (C:N_b=5.5) and fungi (C:N_b = 8; Chapter 2). Growth efficiency is likely to be between 0.1 and 0.5 in various environments (Figure 7.8). Based on these values, microbes should generally release ammonium except when using very nitrogen-poor

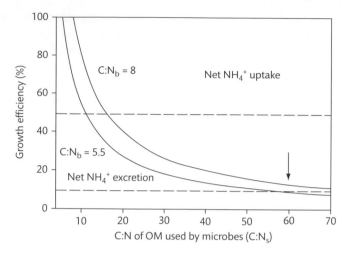

Figure 7.12 Uptake or excretion of ammonium as a function of C:N ratios of the microbial biomass (C:N$_b$) and of the organic material (OM) (C:N$_s$) and the growth efficiency. To determine whether there is net ammonium uptake or excretion, pick a growth efficiency and C:N$_s$ and find where the point is in relation to the solid curves. If it is to the left or below the curve, then there is net ammonium excretion. If it is to the right or above the curve, then there is net uptake. The dashed horizontal lines indicate likely extremes of growth efficiencies. The arrow indicates the C:N$_s$ value (60) above which net ammonium assimilation occurs for a growth efficiency equal to 10% and C:N$_b$ = 5.5.

organic material with C:N$_s$ exceeding 60. That number comes from seeing where the curve for C:N$_b$ = 5.5 crosses the horizontal line set by a growth efficiency of 0.1 (Figure 7.12). Fungi, which need less nitrogen than bacteria (it has a higher C:N$_b$), should also release nitrogen except for extremely high C:N$_s$. These results lead to the prediction that degradation of protein-rich detritus with low C:N ratios, such as from algae, should release ammonium, whereas microbes need to assimilate ammonium when growing on plant litter rich in carbohydrates and other components with high C:N ratios. So the model is consistent with data indicating net ammonium regeneration during degradation of most organic material, except those components with very high C:N ratios.

However, heterotrophic bacteria assimilate more ammonium and nitrate than expected from Equation 7.4. These microbes account for about 30% of the uptake of ammonium and nitrate in aquatic habitats and effectively compete with larger algae for these important N sources. Heterotrophic bacteria and fungi are also successful in competing with higher plants and take up a large fraction of ammonium in soils (Inselsbacher et al., 2010). There are several ways of reconciling inorganic N uptake with the inorganic N excretion predicted by

Equation 7.4. The assumed C:N ratios for the microbes may be incorrect, and different microbes may use organic material with C:N ratios that differ from what was assumed here. Also, the model is about net fluxes and would not preclude uptake of some ammonium even during net mineralization. Finally, Equation 7.4 may be too simple. It assumes Y is constant, yet it can vary with growth rate and environmental factors (Lipson, 2015). However, in spite of its shortcomings, Equation 7.4 is still a useful way of thinking how ammonium fluxes are affected by nitrogen content (C:N ratios) and microbial energetics (Y).

Degradation of lignin and other higher plant compounds

We saw earlier that a large fraction of terrestrial primary production is routed through detritus, because land plants have several biopolymers designed for physical structure and to ward off herbivores. It is worthwhile looking in more detail at how plant detritus is degraded and how degradation of these biopolymers varies. It is an important process in terrestrial systems and an important example of how chemical structure affects degradation rates.

Once a living plant becomes plant litter, the LMW compounds quickly leach out and are easily degraded, as previously discussed. Next to go are simple carbohydrates such as starch, a major storage compound in plants, consisting of α1,4 glucose. Most proteins are also easily degraded. (Some animal proteins, such as keratin found in hair, are not.) Cellulose and hemicellulose are also glucose-containing polymers but with β1,4 glycosidic bonds, making them harder to degrade than starch. Still, they are used more quickly than the many "refractory" or hard to degrade compounds made by large plants (Figure 7.13). These include waxes, cutin, melanins, tannins, and lignin, the main component of wood. Lignin degradation is important for practical as well as ecological reasons. Lignin must be removed to make high-quality paper, yet there is interest in using lignin-rich material for biofuel production. It can be the starting material for the synthesis of chemicals for commercial applications.

In addition to its aromaticity, lignin is hard to degrade because of its many ether and carbon-to-carbon linkages (Bugg et al., 2011). Consequently, it is broken down by a mechanism which is quite different from how other biopolymers are degraded (Figure 7.14), relying on oxidases and abiotic reactions rather than hydrolases. White rot fungi are thought to be the main degraders of lignin in soils with *Phanerochaete chrysosporium*, belonging to the Basidomycota, the best-studied example (Suzuki et al., 2012; Bugg et al., 2011). The name includes "white"

because degradation of the brown, lignin-rich parts has the net effect of bleaching the wood. In contrast, brown rot fungi focus mainly on the white fibers rich in cellulose and hemicelluloses, leaving behind the darker, lignin-rich components.

One key to lignin degradation is the production of hydrogen peroxide (H_2O_2) by a variety of mechanisms, such as from the oxidation of excreted aldehydes by extracellular enzymes. Hydrogen peroxide, a highly reactive compound, then serves as a co-substrate for several enzymes, such as lignin peroxidase and manganese-dependent peroxidase. *P. chrysosporium* makes ten types of lignin peroxidases and five manganese-dependent peroxidases whose roles are not completely known, except that their oxidation–reduction potential varies (Bugg et al., 2011). The reactive oxygen species produced by these microbes can also attack other compounds such as proteins and other biopolymers. This process of "oxidative decomposition" is non-specific in contrast with specific decomposition by hydrolases.

Bacteria were once thought not to be able to degrade lignin at all, and fungi probably are superior degraders of wood and specifically lignin because of their enzymes and hyphal-growth form. Still, bacteria appear to be able to degrade lignin by poorly understood mechanisms (Bugg et al., 2011). Bacteria with lignin-degrading potential have been isolated from soils and sediments (Tian et al., 2016). Bacteria may be more important than fungi in aquatic environments where their sheer numbers give

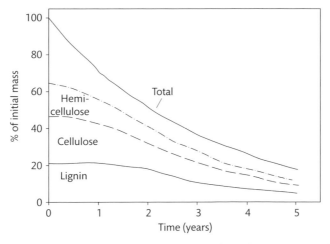

Figure 7.13 Decomposition of various chemical components of litter. Not shown here is the rapid degradation of protein and labile polysaccharides like starch. The example is of litter from Scots pine needles. Data from Berg and Laskowski (2006).

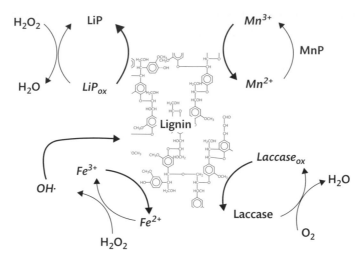

Figure 7.14 Some of the enzymes and mechanisms used by microbes to degrade lignin. LiP = lignin peroxidase and MnP = manganese peroxidase. Lignin decomposition by laccase and the Fenton reaction ($Fe^{2+} + H_2O_2 \rightarrow Fe^{3+}\ OH\bullet + OH^-$) are separate from decomposition by LiP and MnP. An enzyme not shown, the versatile peroxidase, has properties of both LiP and MnP. The enzymes and compounds directly acting on lignin are in italics. Based on Bugg et al. (2011) with input from Tim Bugg.

them an advantage. This question has been examined by following the fate of lignin or cellulose of ligno-cellulose complexes labeled with ^{14}C in incubations with added inhibitors that act against either bacteria or fungi (Benner et al., 1986). A stable isotope probing (SIP) study found that three bacterial genera, *Desulfosarcina*, *Spirochaeta*, and *Kangiella*, were involved in ^{13}C-lignin degradation in coastal wetland sediments (Darjany et al., 2014).

Interactions between organic compounds: protection and priming

One function of lignin is to protect more easily degraded organic components of higher plants from degradation by microbes, insects, and other animals. Degrading the lignin frees up cellulose and hemicellulose for attack by microbes. This relationship between lignin and other plant organic components is an example of how a recalcitrant compound can protect another, labile compound from degradation.

Protection by adsorption or aggregation
Other examples of protection include adsorption to inorganic surfaces and aggregation with inorganic com-

pounds or with organic material or with both. A microbe or even an extracellular enzyme may not have access to an otherwise labile organic compound adsorbed onto mineral surfaces or within nanometer-scale crevices (Castellano et al., 2015). This type of physical protection can occur in the water columns of aquatic systems (Pérez et al., 2011), but it is prevalent in soils, especially those with high mineral content. An organic compound may also be protected by being embedded inside aggregates of compounds like polysaccharides and proteins, or of inorganic material such as the polyvalent cations Al^{3+}, Fe^{3+}, and Ca^{2+}. Regardless of the composition or formation mechanism, adsorption and aggregation help to protect otherwise labile organic carbon from degradation.

Protection by adsorption and aggregation is so important that some soil scientists have argued that models in microbial ecology should not include organic carbon quality—here meaning the chemical composition of the organic pool (Lehmann and Kleber, 2015). They would argue that aggregation and adsorption are more important mechanisms for explaining soil organic material formation. These protection mechanisms are probably less prominent in aquatic habitats where the levels of minerals and other ingredients for these mechanisms are much lower than in soils. Several studies have shown how degradation can vary with the composition of the organic material in aquatic habitats (Kellerman et al., 2015). Chemical structure helps to explain the fate of hydrocarbons released in the Gulf of Mexico following the Deep Water Horizon oil spill (Bagby et al., 2017).

Priming effects

While organic material can help to protect an organic compound from degradation, another type of interaction between organic compounds can stimulate degradation. The "priming effect" is the term used to describe how addition of fresh organic compounds can lead to higher degradation rates of the older organic material. The effect has been examined most extensively in soils, with experiments exploring changes in carbon and nitrogen mineralization following the addition of labile organic compounds.

Natural mechanisms for introducing fresh organic material to soils include excretion by roots in the rhizosphere (Figure 7.15) or by small soil invertebrates or by the introduction of fresh plant detritus to existing soil organic material. Mucus from earthworms, for example, is thought to stimulate degradation of soil organic material, as can the passage of microbes through the earthworm digestive tract (Bernard et al., 2012). Stable

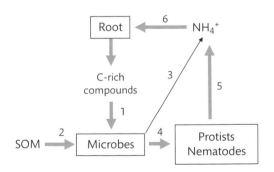

Figure 7.15 An example of the priming effect. Roots can release C-rich compounds (step 1) which stimulates the degradation (step 2) of soil organic material (SOM). Depending on C:N ratios, SOM may stimulate net NH_4^+ mineralization (step 3). Biomass from enhanced microbial production is grazed on by protists and other organisms (step 4), leading to release of NH_4^+ (step 5) that is taken up by plant roots (step 6). By releasing C-rich compounds, the plant may benefit from higher NH_4^+ mineralization rates. Based on a figure by Kuzyakov et al. (2000).

isotope probing experiments with ^{13}C-glucose and ^{18}O-water have shown that several bacterial taxa respond to experimental additions of glucose or other labile organic material (Morrissey et al., 2017). One hypothesis is that the taxa responding to the fresh, labile organic compounds are fast growing r-strategists. Growth of these microbes eventually stimulates the growth of K-strategists and their degradation of the older, more recalcitrant soil organic material. (The terms "r-strategist" and "K-strategist" are discussed in Chapter 8.) Regardless of the hypothesis, priming has large effects on carbon and nitrogen mineralization and organic matter degradation.

There is some debate about whether the priming effect even occurs in aquatic habitats (Catalán et al., 2015), but it seems to in some systems. Organic compounds from phytoplankton do appear to prime degradation of terrestrial components added to freshwaters, but the effect was measureable only for specific organic components, not for the entire organic carbon pool (Morling et al., 2017). Other studies have suggested that labile organic material from algae stimulates the degradation of recalcitrant terrestrial organic material, helping to explain the low concentrations of lignin and other terrestrial compounds in coastal waters (Bianchi et al., 2015). The priming effect is likely to vary with the type of aquatic habitat and microbial community, as is the case for terrestrial habitats.

Photoheterotrophy: energy from organic material and light

So far, the implicit assumption has been that heterotrophic microbes have no use for light and light has no direct role in organic matter degradation. However, it has long been known that many cyanobacteria and eukaryotic algae are capable of using some dissolved organic compounds, probably to access the nitrogen or phosphorus in those compounds. More recently, some bacteria were found to harvest light energy while also oxidizing organic material for energy, a process called "photoheterotrophy." One type of photoheterotroph is aerobic anoxygenic phototrophic (AAP) bacteria.

Anoxygenic photosynthesis carried out by anaerobic bacteria in anoxic environments was discovered early in the history of microbiology. Long after this discovery, obligate aerobic anoxygenic phototrophic bacteria were found in oxic environments. As implied by the name, AAP bacteria require oxygen for growth (they are aerobic), do not produce oxygen during phototrophy (they are anoxygenic), but can use light energy—they are phototrophic—to augment the energy gained from heterotrophy. Unlike anaerobic anoxygenic photosynthetic bacteria (Chapter 6), AAP bacteria do not have Rubisco and require organic carbon for growth. AAP bacteria have bacteriochlorophyll a, the same pigment found in anaerobic anoxygenic bacteria, but not in cyanobacteria or eukaryotic algae. AAP bacteria can be detected by fluorimetric methods and counted by epifluorescence microscopy, because bacteriochlorophyll a fluoresces in the infrared when excited with green light, unlike chlorophyll and other pigments in cyanobacteria and eukaryotic algae. AAP bacteria have other pigments and proteins needed to carry out phototrophy (Figure 7.16A).

At first, AAP bacteria were not considered to be ecologically important, but their discovery in the oceans invigorated new work on these microbes (Kolber et al., 2000). The original hypothesis was that AAP bacteria and other photoheterotrophic microbes would have an advantage in oligotrophic environments. It is true that some of the highest estimates for AAP bacterial abundance have been observed in the South Pacific Ocean (about 20% of total bacterial abundance), one of the most oligotrophic environments on the planet (Lami et al., 2007), but AAP bacteria can be as abundant in eutrophic estuaries. AAP bacteria have been found in soils and just

about anywhere that light reaches. AAP bacteria in the laboratory grow faster and have higher growth efficiencies with light than those grown in the dark, even though these bacteria gain most of their energy from oxidation of organic carbon.

Even more abundant than AAP bacteria are those microbes that have proteorhodopsin (Figure 7.16B). This pigment was first discovered by a metagenomic approach (Chapter 5) in SAR86, an uncultivated clone of *Gammaproteobacteria* (Béjà et al., 2000), which explains why it has "proteo" in its name. Since its discovery, versions of proteorhodopsin have been found in many types of microbes, including SAR11 and *Actinobacteria*. It may be in as many as half of all bacteria in aquatic environments. Archaea use bacteriorhodopsin to harvest light energy; the prefix "bacterio" was applied to archaeal rhodopsins when archaea were classified with the bacteria. Some types of rhodopsin are involved in light sensing in metazoans and other organisms, but proteorhodopsins absorb light for energy production.

However, the amount of energy harvested by proteorhodopsin is small, much smaller than that collected by chlorophyll a- and bacteriochlorophyll a-based systems (Kirchman and Hanson, 2013). Consequently, few bacteria carrying proteorhodopsin grow faster with light. More common is the observation that proteorhodopsin-bearing bacteria survive organic matter starvation longer in the light than in the dark (Figure 7.17). That so many bacteria have proteorhodopsin in spite of its low energy yield is testament once again to the diverse strategies

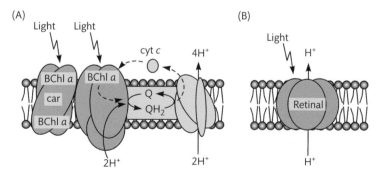

Figure 7.16 Molecular structure of bacteriochlorophyll a-based phototrophy found in AAP bacteria (A) and of proteorhodopsin (B). In part A, BChl=bacteriochlorophyll; car=carotenoid; and cyt=cytochrome. AAP bacteria can harvest more light energy than can proteorhodopsin-bearing bacteria, but synthesis of the several proteins needed by AAP bacteria costs more than synthesizing proteorhodopsin. Taken from Kirchman and Hanson (2013), used with permission of the publisher.

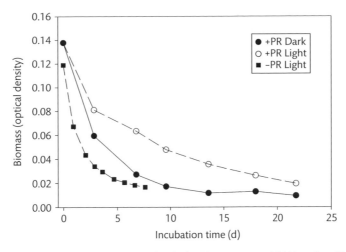

Figure 7.17 Survival of *Vibrio* strain ADN4 with proteorhodopsin (+PR) or a mutant ADN4 strain without it (−PR) in light and dark without nutrients. Data from Gómez-Consarnau et al. (2010).

used by microbes to gain any energy by any means possible in natural environments. Another factor explaining why proteorhodopsin is so common is that it is relatively easy to make, requiring only a few proteins and genes.

Another effect of light is its action on many dissolved compounds in the "colored" or "chromophoric" dissolved organic matter (CDOM) pools (Mopper et al., 2015). CDOM is dominated by aromatic compounds and compounds with alternating double bonds, both more common in terrestrial organic material. High concentrations of CDOM give ponds and small lakes a brownish, tea color. Sunlight can cause photo-oxidation of CDOM to carbon dioxide and carbon monoxide (CO), which is used by bacteria even though it is only slightly more reduced than carbon dioxide. Photochemistry can also lead to the production of labile compounds, such as carbonyl compounds, mainly small fatty acids and keto-acids, and even free amino acids.

Contribution of microbes to ancient organic carbon and SOM formation

This chapter began with discussion about the large amounts of carbon in soil organic matter (SOM) and oceanic DOM. Studies using ^{14}C dating have found that the organic pool in the deep ocean is thousands of

years old, with some components dated to be 12,000 years old (Hansell, 2013). Other ^{14}C-dating studies have found that the age of SOM ranges from about 300 years to over 15,000 years, depending on the extraction method and geological setting (Trumbore, 2009). It is amazing that any organic carbon compound could last even a few months, not to mention millennia when so many microbes are around in need of carbon and energy. Microbes are amazingly effective at degrading organic compounds, even exotic ones made by industrial processes. Yet a very small amount of primary production does in fact escape immediate degradation. This small fraction has built up over geological time, resulting in soils and oceans now having large reservoirs of organic carbon. Geochemists are very much interested in the origin and formation of this ancient organic material.

Obvious sources of this material are the detritus and litter from primary producers such as algae and higher plants. In soils, it was once thought that organic matter from plant litter led to SOM because of "humification," in which known biochemicals from plants underwent abiotic condensation reactions to form large complexes that were resistant to degradation. This material was studied by extraction with alkaline solvents, leaving behind an insoluble "humin" fraction (Lehmann and

Kleber, 2015). It seemed to make sense that humic material, humin, and another fraction extracted by this approach, fulvic acids, would make up the ancient core of SOM, because these complex, aromatic compounds would be hard to degrade. Much of the same terminology and methods have been used to characterize organic material in aquatic habitats.

This picture of SOM formation has changed greatly as new instrumentation and approaches have been used to explore the chemical make-up of organic compounds in soils. It now appears that much of the humic material and related fractions is created during the harsh extraction process (Lehmann and Kleber, 2015) and is not really present in soils, indicating that abiotic condensation reactions are not the main mechanisms for SOM formation. More important are the protection mechanisms discussed before.

In addition, the immediate source of the organic material contributing to SOM formation may not be plant detritus. Although the carbon has to be originally fixed by higher plants (or algae and cyanobacteria in aquatic habitats), several lines of research point to heterotrophic microbes as being the immediate carbon source for SOM and also for many dissolved compounds in aquatic habitats (Figure 7.18). Experiments with ^{13}C-labeled bacteria have traced the ^{13}C into SOM, and scanning electron microscopy has revealed fragments of bacterial cell walls and membranes on soil mineral particles (Miltner et al., 2012). Other studies have found that fungi also contribute organic material to SOM (Kallenbach et al., 2016). Bacteria appear to be important sources of dissolved organic compounds in aquatic habitats (Ogawa et al., 2001).

Regardless of how they are formed, dissolved organic carbon pools in the oceans and SOM on land are large components of the carbon cycle. Even a small change in these large reservoirs could have equally large effects on levels of atmospheric carbon dioxide, with many implications for climate change. It has been well known for years that microbes have huge roles in degrading this organic material, but now we know they have equally huge roles in making it.

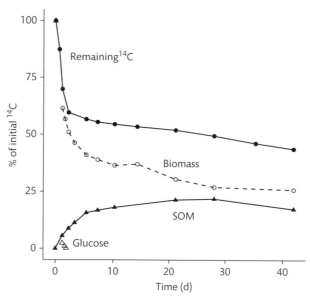

Figure 7.18 The fate of ^{14}C glucose added to northern prairie soil. Glucose disappears within a couple of days, and radioactivity is lost as $^{14}CO_2$ because of respiration, but some ^{14}C remains in biomass and associated with SOM. Data from Paul (2016), originally from Voroney et al. (1989).

Degradation and microbial diversity

Chapter 4 raised the question about the role of organic material in setting the diversity of microbial communities. Here we turn the question around and ask how diversity affects organic matter degradation. The question is connected to concerns about decreasing diversity of large organism communities and even extinction due to climate change and other, less global disruptions of habitats by human malfeasance or indifference. Aside from ethical questions about the loss of species, there are practical concerns about a decline in diversity affecting "ecosystem services," the useful things organisms in natural environments do for us. The ecosystem service relevant here is organic material degradation.

One hypothesis is that a diverse microbial community is necessary to deal with the diversity of compounds in natural organic pools. Direct tests of the hypothesis have returned mixed results (Louis et al., 2016; Banerjee et al., 2016). Experimental studies have manipulated diversity by artificially creating communities with different numbers of cultivated bacteria or by diluting natural communities, resulting in fewer and fewer taxa with more and more dilution. In addition to the drawbacks unique to each approach, with both it is difficult to separate the effect of species richness from the effect of taxonomic composition of the experimental communities. In any case, some studies found faster degradation by more diverse communities, while other studies found no effect. Similar ambiguous results have been found by studies examining connections between degradation and the diversity of fungal communities (van der Wal et al., 2013).

The lack of an effect observed by some studies may be due to large numbers of bacteria and fungi having redundant metabolic capabilities (Figure 7.19). If so, then reducing taxon richness would not necessarily affect degradation, because the remaining taxa would have all of the necessary metabolic functions to carry out organic carbon degradation. One indication of this redundancy is that degradation can remain unchanged even though community structure changes greatly (Leff et al., 2012). This redundancy may become important when the environment changes. A diverse, functionally redundant community may respond more quickly than a less diverse community to

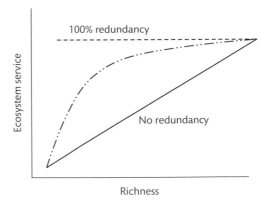

Figure 7.19 Schematic representation of how an ecosystem service may change as a function of taxon richness.

perturbations. Likewise, a diverse community may be more resilient to environmental change, allowing degradation and critical processes to continue uninterrupted. A soil study found this to be the case, although diversity was only important for communities exposed to extremes in temperature, not mercury contamination (Tardy et al., 2014). Diversity may provide "ecological insurance" to help ensure that ecosystem services continue uninterrupted after environmental changes.

Another way in which diversity could matter is illustrated by the response of the microbial community to the 2010 Deepwater Horizon oil spill in the Gulf of Mexico, the largest oil spill to date. The microbial response may have been faster than in other environments because the natural seeps of hydrocarbons in the Gulf select for bacteria capable of oxidizing the hydrocarbons found in the spilled petroleum. However, unexpectedly, rare members of the community, the rare biosphere (Chapter 4), responded to the high concentrations of hydrocarbons from the spill, while the dominant taxa, adapted to the low concentrations steadily leaking from seeps, were unable to cope (Kleindienst et al., 2016). The rare biosphere may provide another type of ecological insurance to help minimize damage when an environment is assaulted like the Gulf of Mexico was by the Deepwater Horizon oil spill.

The loss of a microbe or two would not make the newspaper front pages or be mourned by conservationists, yet microbes are more irreplaceable than a charismatic mammal or bird.

Summary

1. Bacteria and fungi account for 50% or more of total respiration in the biosphere.

2. Bacteria are much more abundant than fungi in aquatic habitats, whereas in soils, both are abundant and important in degrading organic material.

3. Detritus food webs consist of many organisms that feed on dead organic matter and associated microbes as carbon and energy sources. Although directly contributing little to carbon mineralization, detritivores and other eukaryotes break up detritus and increase surface area for attack by microbes, thus speeding up degradation and mineralization.

4. Large organic compounds must be broken down to compounds smaller than about 600 Da before being transported into cells for the final steps of decomposition. Biopolymers are broken down by hydrolysis carried out by a complex suite of enzymes (hydrolases) specific for the biopolymer and sometimes location within the biopolymer. In contrast, lignin is broken down by oxidative decomposition.

5. Compounds like lignin with many different types of chemical bonds and many aromatic components are difficult to degrade, whereas polysaccharides and most proteins are easily degraded by microbes.

6. Bacteria and fungi in soils and aquatic habitats contribute to the production of organic material resistant to degradation. The resulting refractory organic material is a large component of the global carbon cycle.

7. While some microbial taxa specialize in using select organic compounds, many taxa are redundant with similar metabolic capabilities, complicating the relationship between diversity and organic matter degradation.

CHAPTER 8

Microbial growth, biomass production, and controls

In the previous chapter, we learned about the degradation and the mineralization of organic material back to carbon dioxide and the rest of its inorganic constituents. In many aquatic ecosystems and soils, heterotrophic microbes are responsible for half or more of this degradation, and thus they consume an equally large fraction of primary production. Microbes degrade organic material to support their survival and growth, with the evolutionary goal of passing on their genes to future generations. So, to understand organic material degradation and many other biogeochemical processes, we need to understand microbial growth and what controls it. Also, growth and production along with standing stock are fundamental properties of populations in nature.

This chapter will focus on heterotrophic bacteria and fungi in oxic environments, but many of the topics discussed here are also relevant to cyanobacteria and protists in those environments. They are also relevant to microbes in anoxic environments, but electron acceptors have to be considered first because they often control the growth of anaerobic microbes, topics to be discussed in Chapter 11.

Are microbes alive or dead?

The high abundance of bacteria was an important discovery back in the 1970s, when epifluorescence microscopy was first applied to natural samples (Hobbie et al., 1977; Zimmermann and Meyer-Reil, 1974). The question then became, are these cells really active and alive?

It is possible that the observed degradation of organic material is mediated by a small number of live bacteria and fungi and that most of the cells visible by epifluorescence microscopy are dead or dormant. Questions about the metabolic state of bacteria were raised back in the 1950s, in part because it was known that the number of bacteria that grew up on agar plates (the plate count method) was much smaller than the direct count estimate from epifluorescence microscopy (Chapter 1). Although the limitations of the plate count method were recognized early on, it also seemed possible that the difference could be due to dead bacteria. The difference between the plate count and direct count methods suggests that 99% or more of all bacteria are dead.

We now know that the extreme estimate—99% dead—is not correct, but the actual number of alive and dead cells for a given environmental sample is rather hard to pin down. Part of the difficulty is that microbial cells can be in different states of "activity" (Figure 8.1), ranging from truly dead cells, which never could be resuscitated, to microbes that are actively metabolizing and dividing. These metabolic states can be explored by a variety of methods that target different aspects of activity. In the end, the estimate of the number of active or inactive cells depends on the method, what aspect of microbial activity is being examined, and of course the habitat and time.

This part of the chapter will focus on heterotrophic bacteria, in part because there are fewer studies of other microbes, even though it is unlikely that all fungi, cyanobacteria, and protists are equally active. As some

Processes in Microbial Ecology. Second Edition. David L. Kirchman. Oxford University Press (2018). © David L. Kirchman 2018.
DOI 10.1093/oso/9780198789406.001.0001

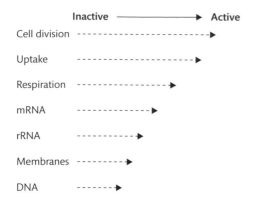

Figure 8.1 Possible activity states of microbes in nature, illustrating how the definition of "activity" depends on what is being measured. Cell properties at a particular level depend on those below it. All actively dividing heterotrophic cells take up organic compounds, respire, and so on. All cells taking up compounds may not be dividing, but are respiring and synthesizing mRNA, and so on down the ladder.

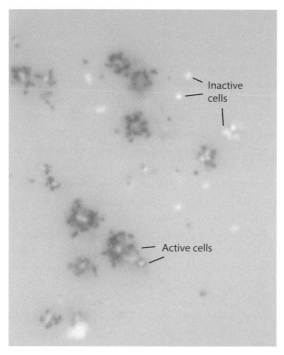

Figure 8.2 Example of microautoradiography showing cells with silver grains that have taken up ^{3}H-amino acids. The white dots are bacteria stained with the DNA stain, 4′,6-diamidino-2-phenylindole (DAPI). The dark areas around active cells are silver grains. Several other methods have been used to examine the metabolic state of bacteria (del Giorgio and Gasol, 2008).

justification for our focus here, eukaryotic microbes might not be able to enter into inactive or dormant states as easily as bacteria (Massana and Logares, 2013); some diatoms and yeasts are among the notable exceptions.

Microautoradiography was one of the first methods used in microbial ecology to explore microbial activity. It is a single-cell method that examines each cell rather than bulk properties of the entire community. Here is how it works. A radiolabeled organic compound, a ^{3}H-amino acid, for example, is added to a sample, incubated for a few hours, and then filtered, or the cells are collected by centrifugation. The microbes are placed into photographic film emulsion. After an exposure time ranging from hours for highly active microbes to days for relatively inactive ones, the film emulsion is developed, the microbes are stained for DNA, and the sample is viewed with epifluorescence microscopy. Cells that have taken up ^{3}H-amino acids have silver grains associated with them (Figure 8.2). These grains are produced by beta particles from the decaying ^{3}H, reducing silver halide in the

---

Box 8.1 Fish macroautoradiography

The most commonly used radioisotopes in microbial ecology, such as ^{3}H, ^{14}C, ^{35}S, and ^{33}P, are all easy to use with only a few safety precautions. The radioisotopes released by above-ground nuclear tests, now banned by international treaties, are an entirely different matter. Back in 1946, the U.S. Navy was finding it difficult to clean up after the plutonium bomb tests at the Bikini Atoll (Weisgall, 1994). The officer in charge of radiation safety, Colonel Stafford Warren, convinced the commanding officer, Vice Admiral William H. P. Blandy, of the dangers of the clean-up by showing him an autoradiograph made by a surgeonfish that had been caught in contaminated reef waters and left on a photographic plate overnight. The resulting X-ray image of the fish was produced by alpha particles from decaying plutonium absorbed by the seemingly healthy animal. That fish macroautoradiograph helped to convince Blandy to call off clean-up efforts to avoid exposing the clean-up crew to more radioactivity.

emulsion to elemental silver, which appears as black splotches around the cells in photomicrographs.

Activity state of bacteria in water and soils
The number of active bacteria detected by microautoradiography in aquatic habitats varies from <10% to 50% or even higher, depending on the environment and radiolabeled compound. This is a large fraction, given that microautoradiography detects only cells synthesizing new biomass, because the radiolabeled compound has to be incorporated into macromolecules which do not leak out of the cell when preserved. A cell taking up the radioactive compound but just mineralizing it without incorporation into a macromolecule would not be scored as being alive. Microautoradiography has also been used to examine which microbes are most important in using the targeted compound. Results from this approach were one reason why microbial ecologists concluded that heterotrophic bacteria dominate the use of dissolved organic material.

It is harder to examine the activity state of bacteria in soils, partly because the physical complexity of these environments creates problems and challenges. There are many practical problems in trying to assay cells within the complex matrix of detritus and inorganic particles, and there are conceptual challenges in dealing with the range of possible activity states of bacteria inhabiting the many microhabitats in a small volume of soil. One study using microautoradiography found that 50% or more of bacteria in sandy and loam soils of New Zealand took up glucose (Ramsay, 1984), but studies using other assays for activity found much lower numbers (<10%) (Blagodatskaya and Kuzyakov, 2013). For unperturbed soils, the number of active bacteria is more than that estimated by the plate count method (<1%) but less than found in aquatic systems (>40%), consistent with the difference in average bacterial growth rates for water and soils, as will be discussed further.

However, the number of soil bacteria ready to become active is much higher than the number active at any one time in unperturbed soils. One piece of evidence is that 50% or more soil bacteria have ribosomes, according to studies using fluorescence in situ hybridization (FISH), the microscopic approach with DNA probes mentioned in Chapter 4. Having ribosomes is one of the minimum

requirements for a cell to be active. More compelling evidence comes from experiments in which water with or without substrates is added to soil. As will be discussed in more detail, microbial growth and respiration respond quickly to these additions, indicating that probably a large fraction of bacteria and fungi are ready to be active when environmental conditions improve.

The initial studies in aquatic and terrestrial systems emphasized that many more bacteria are active than suggested by the number capable of growing on agar plates. In contrast, recent work has emphasized how few bacteria are active in natural environments and how many are potentially dormant. Dormancy helps to explain the high diversity of rare bacteria in the microbial seed bank (Lennon and Jones, 2011), and it may improve biogeochemical models trying to account for organic material mineralization while also reproducing realistic levels of microbial biomass (Wang et al., 2015).

Activity state of individual bacterial taxa
The discussion so far has focused on the entire community of bacteria, even though we know that bacterial communities are very diverse with many taxa. An obvious question is whether all of them are equally active. One method to answer this question combines FISH with microautoradiography, an approach that goes by the acronym of MAR-FISH or CARD-MAR-FISH if the more sensitive version of FISH is used. FISH identifies which cells in the microautoradiography preparation are active in using the supplied radioactive compound. A related version of this approach couples FISH with nanoscale secondary ion mass spectrometry (nanoSIMS) to examine uptake of compounds labeled with stable isotopes. MAR-FISH and nanoSIMS-FISH studies have found, not surprisingly, that activity does vary substantially, as much as ten-fold, among different taxa. What is more surprising is that these differences are evident at high phylogenetic levels, even between the phylum *Bacteroidetes* and classes of *Proteobacteria*, for example (Kirchman, 2016).

The link between activity state and structure has been explored with another approach, stable isotope probing (SIP), which was introduced briefly in Chapter 7. Similar to nanoSIMS, SIP follows ^{18}O, ^{13}C, or ^{15}N-labeled compounds into DNA of active organisms. The heavy DNA is then separated by density ultracentrifugation and analyzed by 16S rRNA gene amplicon sequencing or by metagenomic sequencing. One study used the SIP approach to see which bacteria in pine forest soil incorporated ^{18}O-labeled water or glucose labeled with either ^{18}O or ^{13}C (Morrissey et al., 2016). Like the MAR-FISH work, SIP studies have found different numbers of active bacteria in different phylogenetic groups, ranging from individual OTUs to phyla. The pine forest soil study found that phylogeny explained a significant fraction of the variability in activity (Figure 8.3). It is surprising that phylogeny is important in explaining the uptake of a simple compound like glucose. If bacterial communities are highly functionally redundant (Chapter 7), then glucose uptake would be equal regardless of taxonomic level because many bacteria could take up the compound, and phylogeny would not explain a significant fraction of the variability in uptake. Figure 8.3 suggests that communities may not be as redundant as other studies have concluded.

SIP studies have also tried to link the general activity state of bacteria to degradation processes and to address questions about the relationship between abundance of a bacterial taxon and its contribution to organic material degradation. One SIP study using ^{18}O-labeled water found that abundant bacterial taxa were not necessarily the ones most active (Hayer et al., 2016),

implying that less abundant bacteria have an important role in organic material degradation.

Introduction to growth and biomass production

As microbial ecologists were addressing the question about the number of active cells, it became clear that we did not know how fast bacteria and fungi were growing in natural environments. Microbial ecologists need information about growth rates and biomass production to understand the role of microbes in material and energy fluxes. Many metabolic processes carried out by microbes "scale" with growth rate. When growth is fast, often so too is the process. Parameters about growth and related properties are summarized in Table 8.1.

Growth in the laboratory: batch cultures

Microbes growing as a single species in the laboratory provide two models for growth in nature: batch cultures and continuous cultures. The simplest model, a batch culture, consists of growth in a fresh medium in a closed environment, such as a laboratory flask. When inoculated into new medium, growth begins only after a delay, the "lag phase" (Figure 8.4). Once microbes start to grow,

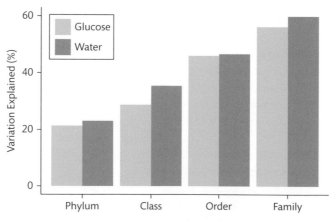

Figure 8.3 Variation in uptake of ^{18}O-water or ^{18}O-glucose explained by phylogenetic level. The variation is the adjusted R^2 expressed as a percentage. Each is significant (p < 0.001). These data show that two organisms are likely to have the same activity if they are phylogenetically related. That likelihood is higher for families than for phyla. Data from Morrissey et al. (2016).

Table 8.1 Terms for basic parameters of microbial biomass and growth.

Parameter	Symbol	Units[a]	Method
Cell numbers	N	cells L^{-1} or cells g^{-1}	Microscopy, flow cytometry
Biomass	B	mg C L^{-1} or mg C g^{-1}	Cell numbers, biomarkers
Growth rate	μ	d^{-1}	From production and biomass
Biomass production	BP	mg C L^{-1} d^{-1} or mg C g^{-1} d^{-1}	Leucine incorporation, acetate-into-ergosterol
Generation time	g	d	From the growth rate
Growth yield	Y	cells L^{-1} or cells g^{-1}	Cell numbers or biomass
Growth efficiency	BGE, CUE[b]	dimensionless	Various

[a] The parameters can be expressed in moles of C instead of mgC. Also, rather than L^{-1} or g^{-1}, the units can include per unit area, such as m^{-2}.

[b] BGE = bacterial growth efficiency; CUE = carbon use efficiency.

they enter into the "log" or "exponential," phase during which abundance increases exponentially. The change in bacterial numbers (N) as a function of time (t) is

$$dN/dt = \mu N \quad (8.1)$$

where μ is the growth rate of the bacterial population; some investigators call this the "specific growth rate" or the "instantaneous growth rate." Growth rates in pure cultures are calculated from the slope of ln(N) versus time; "ln(N)" is the natural log of N or 2.30 × log(N). The change in numbers or biomass (dN/dt) is equal to biomass production. The solution to Equation 8.1 gives an expression for N as a function of time:

$$N_t = N_0 e^{\mu t} \quad (8.2)$$

where N_t is the number of cells at time t, and N_0 is the initial abundance (t = 0). The units for μ are per time; for example, for rapidly growing lab cultures, convenient units are per hour, whereas they would be per day for microbial assemblages growing more slowly in nature.

For some questions, it is adequate to think about growth in terms of cell abundance, but other questions require data on cell size and biomass. Cell size in an exponentially growing pure culture follows a log-normal distribution (Kubitschek, 1969); the biggest cells are those about to divide while the smallest ones are the new, daughter cells resulting from cell division. In spite of this variation in size, growth rates can be calculated using Equation 8.2 and data on total biomass.

Parameters related to the growth rate (μ) include the turnover time of the population (1/μ) and the generation time (g), both of which have units of time, such as hours or days. The generation time is defined as the amount of time required for a population to double. That is,

$$2 N_t = N_t e^{\mu g} \quad (8.3)$$

which yields, after some algebra,

$$g = \ln(2)/\mu = 0.692/\mu \quad (8.4).$$

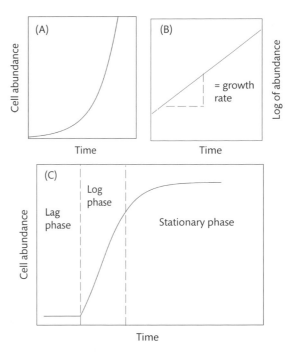

Figure 8.4 Bacterial growth in batch culture. (A) Exponential growth with no lag period or stationary phase. (B) Also exponential growth, but cell numbers are plotted on a log scale. (C) Growth phases of bacteria growing after a lag phase before the log or exponential phase. After the growth-limiting substrate (usually organic carbon) is used up, the culture reaches the stationary phase when cell numbers do not change.

Even in pure cultures, some resource eventually becomes limiting, and net growth slows down and eventually stops completely. At this point, the culture enters the "stationary" phase (Figure 8.4) when abundance is constant over time. The complete curve, without the lag phase, can be described by the logistic equation,

$$dN/dt = rN(1-N/K) \qquad (8.5)$$

where r is the specific growth rate, and K the maximum population size or "carrying capacity" of the environment. (For historical reasons, r is used instead of μ in the logistic equation.) Note that when N is small relative to K, Equation 8.5 becomes the same as Equation 8.1. The symbols r and K are part of terms used to define two types of selection pressures faced by organisms: r-selection and K-selection.

The terms "r-selection" and "K-selection" were originally derived for large eukaryotes colonizing a new habitat. The initial colonizers are r-selected and grow rapidly to take advantage of free space and a new habitat. As the carrying capacity of the new habitat is reached, rapid growth is no longer favored, but rather K-selected organisms with traits for surviving crowded conditions win out. Traits of r-selected organisms or r-strategists allow them to flourish in unstable environments where growth conditions change rapidly, before the build-up of dense populations. In contrast, K-strategists dominate stable environments with invariant growth conditions that promote dense populations. These concepts from large organism ecology can be useful in thinking about microbes in some environments. Some bacteria, the copiotrophs, for example, are adapted to grow rapidly when organic concentrations are high, like r-strategists colonizing a new habitat. In contrast, K-strategists are oligotrophs, adapted to grow slowly on low concentrations in stable environments.

Growth in the laboratory: continuous cultures
The key feature of a batch culture is that it is a closed system with no inputs or outputs; the inoculum is exposed to only one dose of growth substrates at the beginning, and any waste by-products excreted during growth are not removed, except for gases. In contrast with this model of microbial growth, microbes in a "continuous" culture are provided fresh medium continuously and the

old medium—along with waste products and cells—is removed at the same rate. A "chemostat" is a continuous culture in which the concentrations of all chemicals are constant. Although all chemostats are continuous cultures, a continuous culture is not necessarily a chemostat.

Continuous cultures can be quite elaborate and sophisticated, but the basic design is simple (Figure 8.5). To start off, a reaction chamber is inoculated with microbes and is allowed to operate in batch mode initially; at first, there are no inputs or outputs as the microbes multiply. Then new, sterile medium is pumped into the reaction chamber at a fixed rate, and the medium in the reaction chamber is pumped out at the same rate in order to maintain a constant volume within the reaction chamber. Initially, abundance decreases when the pump is turned on, but then it increases as microbes take advantage of the new media and start to grow. These oscillations continue until a steady state is reached when abundance is constant. At this point, it can be shown that

$$\mu = D \qquad (8.6)$$

where D is the dilution rate, defined by

$$D = f/V \qquad (8.7)$$

where f is the flow rate (with units such as L h^{-1}), and V is the volume of the reaction chamber (L). The dilution rate has the same units (for example, h^{-1}) as the growth rate.

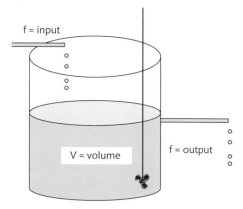

Figure 8.5 A simple continuous culture. The rate (f) at which new medium is added (input) must equal the rate at which medium from the reaction chamber flows out (output). The flow rate can be controlled by simple gravity or by pumps. Air can be introduced to help with circulation and to provide oxygen and other gases used in metabolism.

Equation 8.6 is a very simple but powerful statement about growth. It says that growth is set by the dilution rate, which is under control of the investigator. It also says that growth rates are independent of the supply and concentration of substrate in the continuous culture. The concentration of the substrate, along with the growth efficiency, sets biomass levels.

Continuous cultures provide a different model of growth in nature than batch cultures. Like continuous cultures, microbial abundance is mostly constant over time in nature because growth is balanced by removal: the outflow in the case of continuous cultures, mortality caused by grazing, and viral lysis in nature. The implication is that over some space and time scales in some environments, microbial communities are in a quasi-steady state. On the other hand, if growth conditions do change on relevant timescales, a batch culture may be a more accurate description of microbial growth. It may apply to phytoplankton during the early stages of a spring bloom, for example, when nutrient concentrations are high and mortality is low. Neither batch nor continuous cultures are perfect models for growth in nature. But both provide useful terms and concepts for examining the processes controlling microbial standing stocks and biomass production in natural environments.

Maintenance energy

Early continuous culture studies explored how many properties of bacteria varied with dilution rate and thus growth rate. One property is the rate of substrate consumption. Microbiologists first thought that this rate should increase with the dilution rate and thus the growth rate, and that when the dilution rate is extrapolated to zero, consumption should also be zero. In the actual experiment, the consumption rate did increase with the dilution rate, but it was not zero when the dilution was extrapolated to zero. Bacteria appear to require substrate even when not growing. This substrate uptake at zero growth, first recognized by S. J. Pirt, was hypothesized to be for "maintenance energy" (q_m) needed to keep the cell alive albeit not growing (Hoehler and Jørgensen, 2013). The maintenance energy can be estimated from the intercept of the specific consumption rate versus dilution rate (Figure 8.6).

While the concept seems straightforward and intuitive, in fact maintenance energy is hard to measure and

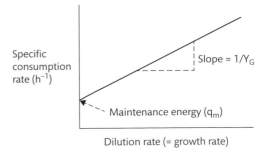

Figure 8.6 Substrate consumption as a function of the dilution rate in continuous cultures. The y-intercept is the maintenance energy. Y_G is the cell yield of growth without maintenance energy.

conceptually more complicated than originally formulated by Pirt. First, it is difficult to run continuous cultures at very low dilution rates and thus low growth rates. When the rate is slow, growth on the culture walls is one of several problems that become significant. A conceptual problem is that maintenance energy may include processes that we would not associate with "maintenance." For example, motility consumes energy not directly tied to growth but is not part of maintenance. To avoid some of the conceptual problems with the term maintenance energy, Hoehler and Jørgensen (2013) suggest replacing it with "basal power requirement."

Whatever the term, the concept touches on several important issues in thinking about the growth and survival of microbes in natural environments. The basal power requirement is the minimum energy a cell needs to counteract entropy and to remain viable, including activities such as repairing mutations and maintaining the integrity of cellular membranes. This minimum sets the growth rate when energy inputs are low and the survival time when microbes are deprived of all energy. It is particularly relevant when thinking of microbes in the deep subsurface and in exploring the conditions under which a microbe can gain enough energy from a chemical reaction to support growth (Chapter 11).

Growth rates and biomass production in nature

Measuring growth in the lab is straightforward, because the rate is calculated directly from the change in abundance or biomass over time in batch cultures or by knowing the dilution rate in continuous cultures. In nature, however, microbes occur in complex communities where a change in abundance could be due to mortality caused by grazing and viral lysis, as well as growth. So, the simple laboratory approach does not work with natural communities unless mortality is minimized somehow.

The commonly used methods for estimating growth (production) in natural environments are based on ^{3}H-thymidine or ^{3}H-leucine incorporation for bacteria and ^{14}C-acetate incorporation for fungi. Conceptually, the methods are similar. Thymidine, which is one of the four nucleosides in DNA, is used to trace DNA synthesis, whereas leucine, an amino acid, is used to trace protein synthesis. Dividing cells must make DNA and thus incorporate thymidine as they grow. Similarly, fast-growing cells make more protein and thus incorporate more leucine than slow-growing cells. The same basic idea is used for estimating fungal growth, except that the starting radiolabeled compound is ^{14}C-acetate, which is followed into a sterol, ergosterol, present in the membranes of all fungi. The acetate-into-ergosterol approach follows membrane synthesis which tracks cell growth like DNA and protein synthesis. Production measured by these methods does not include respiration, nor does it reflect any losses due to grazing or viral lysis.

Production rates are useful for evaluating the general importance of heterotrophic bacteria and fungi in ecosystems and for exploring what controls growth and biomass levels. For aquatic systems, one important observation is that bacterial production usually correlates with primary production: higher primary production leads to higher bacterial production (Figure 8.7). But there is much variation in this relationship. Sometimes the correlation between bacterial and primary production is very high, indicating a tight "coupling" between the two microbial processes, while in other habitats and times, there is no significant relationship (Viviani and Church, 2017). Often bacterial production and primary production are coupled over large spatial and temporal scales but not over small ones.

Another important observation concerns the magnitude of bacterial production compared with primary production and the ratio of the two rates (BP:PP). This ratio is a measure of the importance of heterotrophic bacteria and the rest of the microbial loop in consuming primary production. The BP:PP ratio varies greatly over time and space, but usually it is low in the open oceans, about 0.1, whereas sometimes it is as high as 0.5 or more in lakes and estuaries. The ratio is higher in these aquatic systems in part because of the input of terrestrial organic

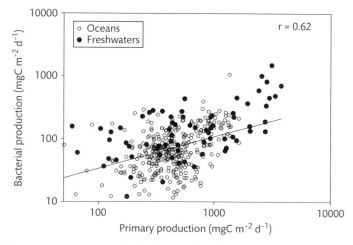

Figure 8.7 Bacterial production versus primary production in a variety of aquatic environments. The correlation coefficient (r = 0.62; n = 441; p < 0.001) and least-squares line are from analyses of the entire data set of oceans and freshwaters, but here only environments with primary production greater than 50 mg C m^{-2} d^{-1} are shown, for clarity. Data provided by Eric Fouilland, taken from Fouilland and Mostajir (2010).

carbon. At the other extreme, the BP:PP ratio for the Arctic Ocean and Antarctic seas is often low (<0.05).

A BP:PP ratio of 0.1 or less may not seem impressive, but its significance becomes clear when it is coupled with the bacterial growth efficiency (BGE). Remembering the definition of BGE given in Chapter 7, we can define total bacterial carbon demand (BCD) as the sum of both production and respiration and express it as a function of BP and BGE:

$$BCD = BP/BGE \qquad (8.8).$$

We can now relate the total use of organic carbon (BCD) by heterotrophic bacteria to primary production by combining data on production rates and bacterial growth efficiency. These data once again indicate the importance of heterotrophic bacteria in processing primary production. Although the open oceans tend to have lower BP:PP ratios, these are offset by low BGE, leading to the observation that about 65% of primary production is routed somehow through dissolved organic material and heterotrophic bacteria. This percentage is roughly the same as that estimated from respiration alone (Chapter 7).

There are not enough data about bacterial and fungal growth in soils to explore relationships between heterotrophic microbial production and primary production in terrestrial ecosystems, although it is known that microbial biomass and community respiration correlate with above-ground primary production (Zak et al., 1994). Previous studies have focused more on how environmental factors like the addition of organic material affect bacterial and fungal growth, as will be discussed further. One study using stable isotope probing with $H_2^{18}O$ found that both microbial growth and total organic carbon decreased along a soil depth profile (Figure 8.8). These and the enrichment experiments to be discussed indicate the importance of organic carbon in setting growth of soil heterotrophic bacteria and fungi.

Growth rates of phytoplankton, bacteria, and fungi
As already mentioned, the growth rate is a fundamental property of any biological population, including microbial ones. Our picture of a population with high abundance but low growth rate would be quite different from one with low abundance and high growth rate. One approach for estimating growth rate of the entire community is with estimates of biomass production and standing stocks (cell abundance or biomass); that is, the growth rate (μ) is production divided by cell abundance or biomass (see Box 8.2). While the approach has its flaws, the estimates still give a good picture of the timescale on which microbes grow in natural environments. The approaches for estimating phytoplankton growth

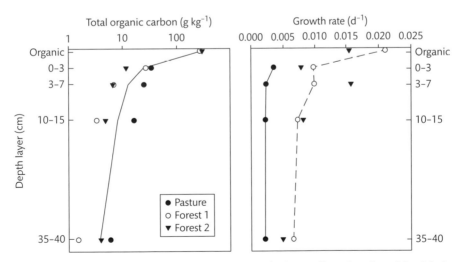

Figure 8.8 Total organic carbon (left panel) and microbial growth rates (right panel) as a function of depth below the surface in three soils. Data from Spohn et al. (2016a).

Box 8.2 Microbial biomass from abundance

Most approaches use data of production and biomass to estimate growth rates of microbes in natural communities. Biomass is easily measured for microbes in pure cultures, but difficult for complex communities. One approach is to convert cell abundance to biomass with an assumption about the amount of carbon per cell. However, cell size varies because of growth rate (Chapter 2), phylogenetic composition, and environmental factors (Straza et al., 2009). Another approach is to measure cell size by microscopy ("biovolume") and convert those data to biomass with an assumption about the amount of carbon per cubic micrometer. Data from these approaches indicate that not only are microbes the most numerous organisms in the biosphere, their biomass can rival that of large organisms in many ecosystems.

and the oceans range between 0.1 and 1 d^{-1} (Figure 8.9), which translates to generation times of between 10 and 1 day, much slower than the hour or less generation times possible in the laboratory. Another striking observation is the slow growth of fungi, at least those in bulk soil away from plant roots where growth is higher (Rousk and Bååth, 2011). Soil fungi grow more slowly on average than any of the organisms in any habitat, including freshwater fungi and soil bacteria; these data support the model of slow and fast carbon pathways mediated by slow-growing fungi and fast-growing bacteria, as discussed in Chapter 7.

Phytoplankton grow much faster than bacteria in the oceans, although rates for both are similar in freshwaters. The data for marine phytoplankton include coccoid cyanobacteria like *Synechococcus* and *Prochlorococcus*, as well as eukaryotic phytoplankton. A closer look at those data reveals that cyanobacteria grow more slowly than eukaryotic phytoplankton but not as slowly on average as heterotrophic bacteria in the oceans (Kirchman, 2016). These rates give a general picture of the timescale over which microbes and microbial processes operate in natural environments.

rates are discussed in Chapter 6. Growth rate data can be used to explore many issues in microbial ecology.

The most important observation has been mentioned before: microbes grow more slowly in nearly all natural environments than in the laboratory under optimal growth conditions. Average growth rates for bacteria in soils, lakes,

Growth rates of individual microbial taxa

The growth rates examined so far have been for entire communities of algae, bacteria, or fungi, yet we know that these communities are made up of many different

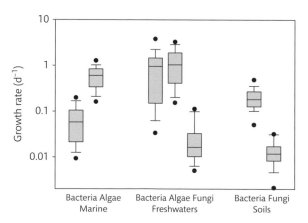

Figure 8.9 Growth rates for bacteria, phytoplankton ("algae"), and fungi in soils and aquatic habitats. Data for freshwater bacteria and phytoplankton are from Fouilland and Mostajir (2010), and data for marine microbes are from Kirchman (2016). Most of the fungal data are from Rousk and Bååth (2011), with additional estimates from Lopez-Sangil et al. (2011) and Kamble and Bååth (2016).

organisms. The next question is whether or not rates vary within each community, which is the same question raised before about general activity. Data on growth rates help us to understand the ecology of these organisms and their contribution to biogeochemical processes. To estimate rates, a couple of different methods have been proposed. One set of methods relies on minimizing mortality, and then measuring the increase over time in the abundance of targeted microbes by following diagnostic pigments, or by FISH using DNA probes for the desired microbial group. Another approach is to take advantage of the relationship between rRNA and growth (Chapter 2); the number of a specific rRNA sequence, relative to its gene abundance (rRNA:rDNA), is an index of the growth rate for the microbe represented by that sequence. Similarly, a few studies have used metatranscriptomics to follow mRNA

for ribosomal proteins in order to track growth (Gifford et al., 2014). Some investigators will not go as far as using rRNA or ribosomal proteins as a "growth rate index," and few will calculate an actual rate because of problems relating rRNA and ribosomes to growth (Blazewicz et al., 2013; Lankiewicz et al., 2016). Rather, they use it as a more generic index of activity. Still, the methods have yielded some general conclusions about bacterial growth. Much less is known about specific fungal taxa.

As expected, different bacteria grow at different rates depending on environmental conditions. Unexpectedly, rates calculated for broad phylogenetic levels, such as order and phyla, also differed, similar to the stable isotope probing results (Figure 8.4). Variation in growth among bacterial groups is evident in rRNA:rDNA data for a marine habitat (Figure 8.10). Ratios were high for the

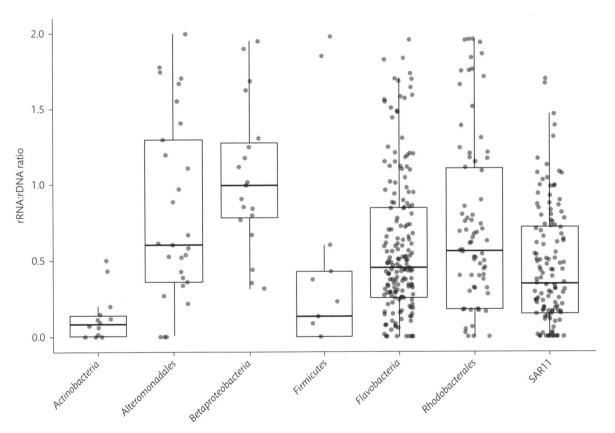

Figure 8.10 Ratio of 16S rRNA to 16S rRNA genes (rRNA:rDNA) for some bacterial taxa ranging from a clade (SAR11) to a phylum (*Actinobacteria*). Each point is an individual bacterial OTU at one particular time. The median and quartiles for the entire taxon are summarized by the box-whisker part of the plot. Data from Campbell et al. (2011). See Kirchman (2016) for a statistical analysis of these data.

gammaproteobacterial families *Alteromonadaceae* and *Rhodobacteraceae*, and low for SAR11 bacteria, implying high and low growth rates, respectively. Consistent with the rRNA:rDNA data, SAR11 grew slowly in experiments that minimized mortality, whereas growth rates for bacteria in the *Alteromonadaceae* and *Rhodobacteraceae* were high (Sánchez et al., 2017).

The differences in growth rates among bacterial taxa can be explored with the help of two interrelated sets of terms: copiotrophs versus oligotrophs, and r-selection versus K-selection. The fast-growing bacteria in the *Alteromonadaceae* and *Rhodobacteraceae* are likely copiotrophs and r-selected to take advantage of high nutrient concentrations. In soils, the r-strategists appear to be in the *Alpha-*, *Beta-*, and *Gammaproteobacteria* and in the phylum *Bacteroidetes* because these respond to the addition of plant litter (DeAngelis et al., 2013). These copiotrophic, r-selected bacteria can increase their growth rate rapidly by taking advantage of the occasional patch or time period of high nutrient concentrations (Figure 8.11). But then growth rates decrease as concentrations decrease. The end result is the "feast or famine" growth strategy mentioned in Chapter 5. In contrast, oligotrophic bacteria maintain the same growth rate and do not have the genetic and physiological mechanisms to respond to nutrient-rich opportunities. They are K-selected organisms. In the oceans, the K-strategists include *Prochlorococcus* and the SAR11 clade (Giovannoni, 2017), while in soils they may be organisms in the *Actinobacteria* and *Deltaproteobacteria* phyla (DeAngelis et al., 2013).

The different growth strategies and variation over time are reasons why the relationship between growth rates and abundance of a taxon is not straightforward.

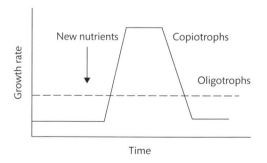

Figure 8.11 Hypothetical growth rates for oligotrophic (dashed line) and copiotrophic bacteria (solid line) over time before and after a nutrient pulse.

The most abundant bacteria are usually oligotrophs, which have low but constant growth rates. Less abundant heterotrophic copiotrophs may have high growth rates if in a patch of high organic concentrations, but low rates or even no growth at all if concentrations are low. The variation in growth rate for a taxon may be more revealing about its growth strategy than the rate at a single time.

Growth in the deep biosphere

The growth rates previously discussed are for microbes in the surface layers of aquatic and terrestrial environments close to sources of energy: light in the case of cyanobacteria and algae, organic material from primary producers in the case of heterotrophic bacteria and archaea. There are, however, microbes far from these energy sources, thousands of meters below the Earth's surface in a "deep biosphere." Prokaryotes have been found 2.5 km below the sea floor in sediments thought to be 100 million years old (Jørgensen and Marshall, 2016), and bacteria, archaea, and perhaps other microbes are also in deep subsurface habitats under terrestrial ecosystems. The growth rate of prokaryotes has been estimated indirectly from data on cell abundance and on metabolic processes such as sulfate reduction; an independent method based on racemization of amino acids has yielded similar estimates. Both approaches indicate that generation times for prokaryotes are on the order of 1000 years in deep subsurface environments.

This incredibly slow rate has many implications for thinking about microbial biology and ecology. The slow rate means that these cells are dormant by the standards set for surface layer communities. However, although some microbes may be in a resting or spore form, it has been argued that a cell needs to be at least somewhat active to repair mutations and cell damage (Jørgensen and Marshall, 2016). Entering into an inactive state as a growth strategy makes sense in environments in which growth conditions vary from unfavorable to favorable, but less so in the deep subsurface which is stable and constant with no chance of growth conditions becoming more favorable. In fact, one study of 219-m-deep, 460,000-year-old sediments off Japan found that over 70% of prokaryotes incorporated ^{13}C or ^{15}N labeled glucose, pyruvate, or amino acids (Morono et al., 2011),

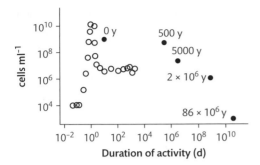

Figure 8.12 Timescale of prokaryotic growth in a pure culture (open symbols), a surface habitat (closed symbol 0 y), and subsurface environments (other closed symbols) that are older and increasingly distant from new energy sources. The years given for the subsurface environments indicate the age of the sediment. Based on Hoehler and Jørgensen (2013).

evidence against dormancy being the norm in this deep biosphere site. In any case, not surprisingly, prokaryotic abundance decreases with depth and thus age of the subsurface environment (Figure 8.12). What is a surprise is that the loss rate calculated from the decrease in abundance with depth is only about 0.4% per generation. The cells that do die provide material and energy to support the growth of the survivors. Grazing is probably minimal, given that eukaryotes have not been observed by microscopy in the deep sea subsurface, and only fungal sequences have been detected by cultivation-independent methods looking for eukaryotes. Even though it would seem unlikely that such slow-growing microbes could support production of viruses, the transcription of virus genes has been detected by metatranscriptomic studies, suggesting viral lysis does occur in deep subsurface environments (Engelhardt et al., 2015).

What sets growth by heterotrophic microbes in nature?

The growth rates of bacteria and fungi in nature are much lower than what can be achieved in laboratory cultures. What then prevents these microbes from growing faster in nature? For photoautotrophic microbes, we saw in Chapter 6 that the answer is fairly simple: light and the supply of inorganic nutrients. For heterotrophic microbes, the answer is more complex. Chapter 4 discussed many factors affecting the taxonomic make-up,

the "structure," of microbial communities. Many of those same factors also affect growth rates, which is a big reason why structure changes; differences in growth rates among microbial taxa lead to differences in abundance and thus community structure. Here we focus on bottom-up factors affecting growth rates of bacteria and fungi at the community level, while remembering that not all taxa within the bacterial or fungal communities respond equally.

Many of these bottom-up factors are "density-independent" because their effect does not vary with microbial abundance. Temperature is a good example. A top-down factor like predation, on the other hand, is a "density-dependent" factor because it varies with predator and prey abundance (Chapter 9). Many abiotic factors are density-independent, but not all. Physical space or room, for example, in a soil micro-environment or in a biofilm, may fill up with microbes and limit growth, making it a density-dependent factor. Soil moisture is a product of both density-independent factors, such as the frequency and intensity of rain events, and density-dependent factors, such as the retention of water by microbially produced extracellular polymers within the soil matrix.

All of these factors and more can affect bacteria and fungi differently (Table 8.2) with implications for carbon cycling and many other biogeochemical processes. Heterotrophic bacteria and saprophytic fungi potentially use the same organic substrates, yet their growth and biomass often vary in opposite directions, suggesting that the two microbial groups are interacting and competing. However, it is also possible that the two are not

Table 8.2 Summary of factors affecting bacterial and fungal growth in soils. The positive effects are indicated by the various numbers of "+," while "−" indicates negative impacts. See main text and also Rousk and Bååth (2011).

Factor	Impact on	
	Bacteria	Fungi
Moisture	+++	++
Temperature	+++	++
Organic carbon	+++	+++
Acidity	−	+
Disturbance	++	+
Metals	−	+

competing but are just responding differently to the same factor. More convincing evidence for competition has come from experiments that follow bacterial and fungal growth after adding or removing fungi or by adding inhibitors of bacterial activity (Rousk et al., 2008). In the latter case, fungal growth increases when bacterial growth is inhibited by the addition of antibiotics such as oxytetracycline and tylosin. These experiments show that bacteria affect fungi in a density-dependent fashion, a strong sign of direct competition between the two microbial groups for the same growth-limiting organic substrates. Some types of bacteria can negatively affect fungi by excreting organic compounds, one example being the polyene nystatin. The soil bacterium *Streptomyces* is famous for producing these antifungal compounds, as well as other antibiotics, which work against other bacteria.

Temperature effects on growth and carbon cycling

Of all bottom-up, density-independent factors affecting growth, temperature is arguably the most important. Chapter 3 discussed how temperature affects all chemical reactions and processes in nature, and microbial growth

is no exception. As a general rule of thumb, the Q_{10} of growth rates is about 2, although it varies, of course. The Q_{10} values for heterotrophic bacteria and fungi may or may not differ, depending on environmental properties. The example given in Figure 8.13 indicates that bacteria and fungi respond similarly to temperature in Swedish arable soil. However, Pietikainen et al. (2005) found that the optimal temperature for bacterial growth in agricultural and forest soils was about 5 °C warmer than that of fungi. This difference could result from bacteria and fungi using different organic compounds whose degradation varies differently with temperature (Kirschbaum, 2013). That mechanism could explain the observation that the ratio of bacterial to fungal biomass increases with temperature of different soil habitats (Chen et al., 2015a).

The example given in Figure 8.13A illustrates the short-term response of bacteria and fungi to different temperatures, some far from the original 15 °C. The short-term response over hours may reflect the physiological response of microbes already in the community; the experiment was too short to allow microbial taxa adapted to the incubation temperature to increase in abundance appreciably. In the study given in Figure 8.13,

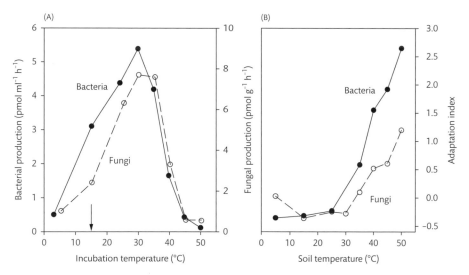

Figure 8.13 Response of bacterial and fungal growth to temperature in soils. Bacterial production (leucine incorporation) and fungal production (acetate-in-ergosterol incorporation) were measured at the indicated incubation temperature. The arrow indicates the initial temperature of the soil (A). In panel B, a temperature adaption index was measured for bacteria and fungi held at the indicated soil temperatures for about a month. After that time, bacterial and fungal production were measured in incubations lasting hours at 5 and 45 °C, and the rates were used to calculate the index (log(rate at 45 °C/rate at 5 °C)). Data taken from Bárcenas-Moreno et al. (2009).

possible adaption was explored by exposing the communities to different temperatures for about a month, then growth was measured at different incubation temperatures. Figure 8.13B shows that only microbial communities exposed to soil warmer than about 30 °C become adapted to the high temperatures (the index was above 0), implying that the temperature response does not change even after a month in the new thermal environment.

The effect of temperature on growth rates should help us understand how biogeochemical processes respond to temperature. Many studies have examined how soil respiration and organic material decomposition may respond to predicted changes in temperature due to global warming, and there is much concern that increasing temperatures would lead to more decomposition and higher CO_2 fluxes back to the atmosphere (Cheng et al., 2017). The problem is especially acute in the Arctic, where warming by only a few degrees would melt permafrost and may not only free up organic carbon that can be mineralized to carbon dioxide but also lead to the release of more methane, a potent greenhouse gas (Yvon-Durocher et al., 2014).

Temperature also affects the growth of bacteria in temperate aquatic environments. In fact, often, bacterial biomass production correlates best with temperature rather than with other properties known to be important in setting growth of these heterotrophs. One example is from Narragansett Bay, Rhode Island (Figure 8.14). In this environment, temperature ranges between −1 and nearly 23 °C over a year, while biomass production varies by over one hundred-fold. The correlation between the two parameters was high during this study (r = 0.70); in contrast, there was no significant correlation with a proxy for the supply of organic carbon, chlorophyll *a*. But the Q_{10} implied by the field data is much higher than the

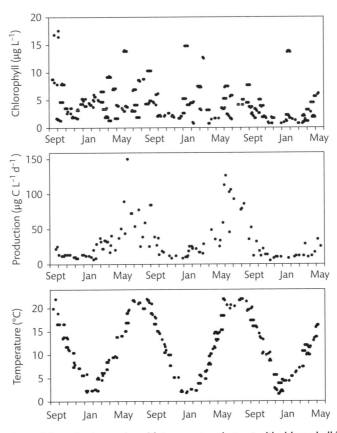

Figure 8.14 Example of how bacterial production varies with temperature but not with chlorophyll in a temperate environment. Data from Staroscik and Smith (2004).

Q_{10} estimated in experiments when only temperature is varied, suggesting that other factors also affected bacterial production and growth. Soil microbial ecologists have also concluded that high Q_{10} values indicate that factors other than temperature are at work (Davidson et al., 2006). One of those other factors is the supply of labile organic material, as will be discussed further.

The Narragansett Bay study is one example of a problem often faced by ecologists who need to use correlations to examine a functional relationship in real natural environments, in this case, between microbial growth and temperature in an estuary. The problem is that a correlation does not necessarily imply causation. In temperate environments, temperature varies greatly along with many other ecosystem properties, such as light, primary production, and biomass, all potentially affecting growth. So, temperature may correlate significantly with heterotrophic bacterial production, in part because temperature co-varies with another, hidden property of the ecosystem that also affects bacterial growth. A correlation analysis is a powerful tool for exploring relationships in microbial ecology, but often other approaches and types of data are needed to really understand what is going on.

pH effects

Chapter 4 mentioned that the diversity and composition of soil bacterial communities is strongly affected by pH, and it should not be surprising that pH has an equally

large impact on fungal communities. It was well known that fungi are relatively more abundant than bacteria in acidic soils, but whether this was due to pH-related differences in growth or mortality was unclear. Work done with soils in the Hoosfield acid strip at Rothamsted Research, United Kingdom provided direct evidence of how pH affects microbial growth. In the Hoosfield fields manipulated by chalk additions to control pH, bacterial production increases with increasing soil pH (Figure 8.15A). Fungal production also increases, but it reaches a maximum in acidic soils, at about pH 5, and then decreases with higher pH to rates as low as in the most acidic soils (Figure 8.15B). This variation is mainly due to variation in growth rates, not changes in biomass levels (Rousk et al., 2009). Low growth of both groups in acidic soils could be due to direct pH effects, although other data suggest those effects are small. More importantly, the input of organic material is low in highly acidic soils due to poor plant growth. Fungal growth is highest at pH 5 because that pH is low enough to inhibit bacterial growth without stopping plant growth and organic material inputs. When bacterial growth is inhibited, fungal growth increases with pH. Like temperature, pH has both direct and indirect effects on microbial growth and organic material degradation.

Soil moisture

The water content of soils is well known to affect many aspects of microbial activity and biogeochemical processes.

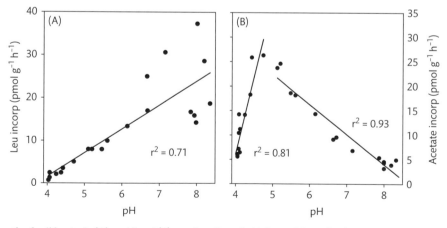

Figure 8.15 Growth of soil bacteria (A) and fungi (B) as a function of pH. Bacterial production was measured by leucine (Leu) incorporation, while fungal production was measured by the acetate-into-ergosterol method. Biomass data indicate that the variation in production is mainly due to changes in growth rates. Data taken from Rousk et al. (2009).

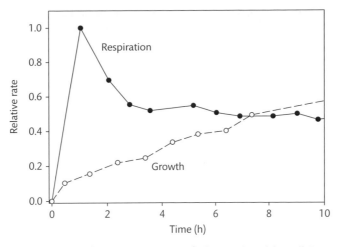

Figure 8.16 Respiration and bacterial growth (leucine incorporation) after wetting of dry soil. Rates were normalized to the maxima. Data from Iovieno and Bååth (2008).

Soil moisture explains more variation in total microbial biomass than any other soil property (Serna-Chavez et al., 2013). Addition of water to dry soil increases bacterial growth within hours, and the stimulation of respiration is even greater and usually quicker in these experiments (Figure 8.16). The time for a measurable response after the water addition and the relationship between growth and respiration vary with the length of dry conditions (Mcisner et al., 2015). We would expect that fungal growth relative to bacterial growth would be higher for drier soils, but this has not been examined. One study, however, found no long-term effect of drought on either bacterial or fungal growth (Rousk et al., 2013a), and growth rates estimated from stable isotope probing with $H_2^{18}O$ were similar for bacteria and fungi following rewetting of soils (Blazewicz et al., 2014). More studies like these would help us predict how changing precipitation and evapotranspiration due to climate change will affect the carbon cycle and other biogeochemical processes in terrestrial ecosystems.

Limitation by organic carbon

The concentration and supply of organic material are often very important factors determining the growth of heterotrophic microbes in both soils and aquatic systems. As mentioned before, concentrations of labile organic components are very low in nature, which explains why growth rates of heterotrophic bacteria and fungi are usually far lower in natural environments than seen in the laboratory. One line of evidence for carbon limitation in aquatic systems is the correlation between bacterial production and primary production in lakes and the oceans (Figure 8.7). The easiest way to explain this correlation is that primary production determines, directly or indirectly, the supply of organic material, which in turn drives heterotrophic bacterial growth. Likewise, there is a correlation between microbial growth and organic material in soils (Figure 8.8) and between soil respiration and primary production (Zak et al., 1994), all evidence for organic carbon limitation of soil bacteria and fungi.

Another line of evidence indicating carbon limitation comes from addition experiments. In these experiments, organic compounds are added to incubations of water or soil, and microbial production is followed over time. Often, bacterial and fungal growth is higher in incubations with the organic compounds than in the no-addition control (Anderson et al., 2016). In soils, the nearly immediate increase in microbial growth and respiration following the addition of an organic substrate is called "substrate-induced respiration" (Reischke et al., 2014).

Both the concentration and the supply rate are important in thinking about limitation by organic carbon and other nutrients. The relationship between concentrations and growth rates is described by the Monod equation (see Box 8.3),

$$\mu = \mu_{max} \cdot S / (K_s + S) \qquad (8.9)$$

where μ is the growth rate, μ_{max} the maximum growth rate, S the concentration of the growth-limiting substrate, and K_s the substrate concentration at which the growth rate is half of the maximum (Figure 8.17). The Monod equation is mathematically the same as the Michaelis–Menten equation (Equation 6.13) used to examine uptake and enzyme kinetics. When concentrations are low, Microbe A with a low K_s and high enough μ_{max} will win out over Microbe B with a higher μ_{max} but also a too high K_s. When concentrations are high, Microbe B wins because of its higher μ_{max}. Oligotrophs have Monod equation parameters similar to Microbe A,

explaining why they outcompete copiotrophs (Microbe B) when concentrations are low.

Limitation by inorganic nutrients

Of the material needed for biomass synthesis by heterotrophic microbes, organic compounds are most important in setting growth rates of heterotrophic bacteria and fungi, as mentioned before. However, the concentration of many inorganic nutrients can be low in soils, lakes, and the oceans, raising the possibility of these compounds limiting growth in some environments. Phosphate does seem to limit primary production and bacterial growth in the Sargasso Sea and the Mediterranean Sea, based on addition experiments (Figure 8.18A), in contrast with the general rule of primary production in marine waters being limited by nitrogen. Another approach has been used to argue for P limitation of soil bacteria. In an analysis of many types of soils, there was a good correlation between microbial respiration and inorganic phosphorus (Hartman and Richardson, 2013). Bacteria can become P limited because of the high phosphorus requirement of ribosomes. That requirement is part of the growth rate hypothesis that links growth rates to the elemental stoichiometry of microbes, as discussed in Chapter 2.

While some studies have found evidence of P limitation, fewer studies have reported N limitation of heterotrophic

---

Box 8.3 Freedom fighter and microbiologist par excellence

The Monod equation is named after Jacques Monod (1910–76), who won the Nobel Prize (along with his compatriots, François Jacob and André Lwoff) for work on the *lac* operon in *E. coli*. This operon was one of the first models of gene regulation at the transcription level. In addition to being a scientist, Monod was a member of the French Resistance that fought against the German occupation of France during World War II.

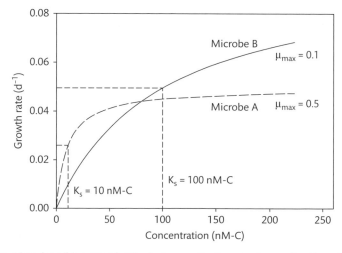

Figure 8.17 The Monod equation describing growth rates by two competing species as a function of limiting substrate concentrations.

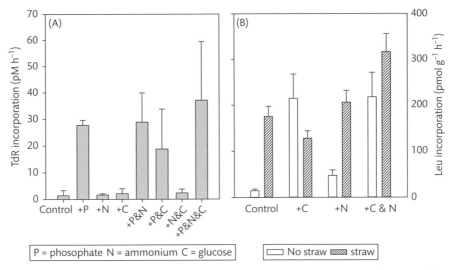

Figure 8.18 Examples of addition studies to explore nutrient limitation of microbial growth. Panel A: Bacterial production in the Sargasso Sea estimated from thymidine (TdR) incorporation was measured over 24–48 hours in bottles receiving the indicated compounds. Panel B: Bacterial production estimated from leucine (Leu) incorporation in soil suspensions with or without straw and without any other addition (Control) or with the addition of glucose (C), ammonium (N), or both. Data from Cotner et al. (1997) and Kamble and Bååth (2016).

bacteria. This work raises two questions: why is heterotrophic growth generally limited by organic carbon and not by inorganic nutrients? And why is P limitation more common than N limitation?

One answer to the first question is that organic carbon is used by aerobic microbes for both biomass synthesis and respiration, but nitrogen and phosphorus are used only for biomass synthesis, so heterotrophs need much more organic C than inorganic N or P. Another answer involves competition for these inorganic nutrients between the heterotrophic microbes and autotrophic microbes in aquatic systems, and between heterotrophic microbes and higher plants in terrestrial systems. In Chapter 6, we learned that small microbes with their high surface area to volume ratios should outcompete large microbes and higher plants for ammonium, phosphate, and other dissolved compounds. However, uptake of inorganic nutrients by heterotrophic microbes eventually would lead to lower growth of autotrophic organisms and lower production of organic material. The net result is limitation by organic carbon, not by inorganic nutrients.

The other question is about why P limitation is more common than N limitation. The answer may be that het-

erotrophic bacteria and even fungi are exceptionally P-rich and have very low C:P ratios, so they need lots of P for growth. Another part of the answer may lie in the biochemicals that contain N and P in microbes. As mentioned in Chapter 2, N is mainly in cellular proteins that last the entire lifetime of a cell; they are not "turned over." In contrast, P is in compounds like ATP and NADPH that turn over rapidly. The ceaseless synthesis of these P-rich compounds would require high amounts of phosphate.

It may seem that soil microbes should be stimulated by inorganic nitrogen because of the low nitrogen content of plant litter, the main organic material supporting the soil microbial loop. Given the very high C:N ratio of wood, lignin-degrading fungi would seem to benefit from higher inputs of inorganic N. This topic touches on the environmental problems caused by N-rich fertilizers and questions raised about whether high anthropogenic N deposition affects organic material degradation. Contrary to expectations, the addition of fertilizer can decrease the rate of litter degradation. It is thought that the decrease results from the fertilizer decreasing expression of lignin-degrading genes and changing the taxonomic composition of the fungal community (Edwards et al., 2011).

Table 8.3 Some cases of co-limitation of microbial growth by at least two bottom-up factors. Based on Saito et al. (2008) and sources cited in the text.

Microbe	Primary factor	Secondary factor	Comments
Photoautotrophs	Light	Nitrate	Nitrate use requires energy
All microbes	Nitrate	Iron	Nitrate use requires iron-containing nitrate reductase
All microbes	Phosphate	Zinc	Alkaline phosphatase requires zinc
All microbes	Urea	Nickel	Urease requires nickel
Diazotrophs	Energy	Iron	Nitrogenase requires iron
Bacteria and fungi	Organic carbon	Temperature	
Soil microbes	Organic carbon	Water	

Co-limitation and interactions between controlling factors

Microbes have adapted to live on very low concentrations of many compounds in natural ecosystems, so it can be overly simplistic to focus on a single limiting factor. We see the consequences of these low concentrations in addition experiments, where often the addition of two compounds stimulates biomass production more than the addition of either compound alone. An example is the experiment illustrated in Figure 8.18B; other data from the study provided stronger support for stimulation of production by the addition of both organic carbon and ammonium. Some authors would conclude that the microbes were under co-limitation by organic C and N, but another interpretation is that N just became the next limiting factor after the addition alleviated organic C limitation.

There are several clearer examples of co-limitation, where the limiting factors are physiologically linked (Table 8.3). For example, microbes may be prevented from using nitrate, and thus are limited by N, because low iron levels limit the expression of nitrate reductase, an iron-containing enzyme essential for reducing nitrate to ammonium, the form needed for biomass synthesis. Nitrogenase, the critical enzyme for N_2 fixation, is another enzyme that requires iron as a co-factor. Several enzymes require other trace metals, such as manganese, nickel, cobalt, copper, and zinc, which occur in very low concentrations, especially in the open oceans. These cases are clear examples of co-limitation, because one compound or element is required for acquisition or use of the other.

Two important examples of co-limitation involving temperature should be mentioned. Growth of microbes in polar environments may be co-limited by organic carbon and temperature. One physiological link between the two factors is that low temperature causes stiff membranes and impedes transport of dissolved compounds. According to this hypothesis, higher organic carbon concentrations are needed for a heterotrophic microbe to grow in cold water at the same rate as in warmer waters. In soil microbial ecology, there has been much discussion about whether the sensitivity of organic material degradation to temperature, as measured by Q_{10}, varies with organic material quality. The other example of co-limitation involving temperature is the interaction between temperature and water content in soils. Warmer temperatures alone would stimulate decomposition and presumably microbial growth in soils, but they also lead to more evaporation and less moisture, which potentially limits microbial activity. As mentioned before, the confounding effects of moisture complicate efforts to estimate Q_{10} for soils and to use Q_{10} in global models to predict the response of terrestrial ecosystems to global warming.

Cooperation between organisms

So far, this chapter has focused on how competition between different microbes has negative impacts on at least one of the interacting organisms. For example, bacteria and fungi likely compete for the same organic material, resulting in lower growth rates of one or the other. But there are many positive interactions, if not actual cooperation, in which one or both of the interacting microbes benefits from the presence of the other microbe. Chapter 12 and 14 will discuss symbiotic relationships between microbes in intimate contact with each other. This chapter will end with some other examples of positive interactions between microbes not necessarily living in close physical contact.

Not all interactions between bacteria and fungi may be negative, at least for bacteria. Those prokaryotes may be able to use labile compounds freed up when fungi degrade cellulose and other structural polysaccharides (Rousk and Bååth, 2011), and similarly, bacteria and fungi hydrolyzing macromolecules may feed low molecular weight by-products to other microbes not carrying the necessary hydrolases. Oligotrophic heterotrophic bacteria specialized to use only a few organic compounds may depend on copiotrophic bacteria and other microbes to produce those compounds (Giovannoni, 2017). These compounds can be carbon and energy sources, but microbes may also take in other compounds that are not degraded for energy but are essential in microbial metabolism. Many microbes are incapable of synthesizing vitamins (they are auxotrophic) and need to obtain them from other microbes (Sañudo-Wilhelmy et al., 2014). Other compounds are also likely exchanged among microbes (Pande and Kost, 2017). Finally, there are positive cooperative, perhaps even symbiotic relationships between specific bacterial taxa and algae (Amin et al., 2015). Positive interactions may be more important in controlling microbial processes than is now commonly believed.

Summary

1. Many, but not all, bacteria and fungi in natural environments are actively metabolizing and growing. The state of activity varies from being dead to actively carrying out cell division and biomass production.

2. Similar to primary production, biomass production of heterotrophic bacteria and fungi can be used to assess their contribution to carbon fluxes. The production data are consistent with data from other approaches indicating the high flux of carbon and energy through heterotrophic microbes.

3. Growth rates of bacteria in nature are much slower than rates in nutrient-rich laboratory experiments. Bacteria appear to grow faster than fungi in soils and in aquatic habitats, consistent with models of slow and fast carbon pathways.

4. Growth of heterotrophic bacteria and fungi is limited by the supply and quality of organic carbon in most oxic environments, although inorganic nutrients, such as phosphate, can be limiting in some habitats.

5. Temperature also has large but probably different effects on bacterial and fungal growth. How temperature affects these microbes has many implications for understanding climate change.

6. In addition to competing for limiting organic and inorganic compounds, microbes can directly interact via the secretion of antimicrobial compounds. However, microbes may help each other out by releasing organic material, or compounds like vitamins necessary for basic metabolism.

CHAPTER 9

Predation and protists

The previous chapter mentioned that growth rates of bacteria, fungi, and algae in natural habitats are generally slow compared to what is possible under optimal conditions in the laboratory. However, even with slow growth, these microbes would quickly fill up the biosphere were it not for some force that kills them off. Some microbes may self-destruct because they lack a limiting nutrient, but many other microbes grow, even if slowly, under the most adverse environmental conditions. Large phytoplankton cells in the upper surface layer of aquatic habitats can sink to their death in deep, dark waters, but many phytoplankton, bacteria, and other small microbes do not sink appreciably, nor do microbes of any size in terrestrial systems. The primary mechanism of keeping microbial populations in check is mortality by predation and viral lysis, collectively referred to as top-down control. How much of mortality is due to predators versus viruses is discussed in Chapter 10. This chapter is about predation and the ecology of the microbial predators, the protists.

Protists are more than just predators, and are diverse in many ways. They range in size from nanoflagellates, nearly as small as bacteria, to some amoebae and myxomycetes over a centimeter in length. The cell shape of protists varies from simple coccoids to elaborate houses. Protists are involved in a diverse suite of biological interactions and take on just about every possible role known in ecology (Table 9.1).

Protists have been known for centuries but by other names, such as "animalcules," a term used by Antonie van Leeuwenhoek in the seventeenth century to describe the microbes he saw with a primitive microscope in samples of his stool and of scum from his teeth. Darwin and his contemporaries called them "infusoria." Protozoa is another term still used by some microbial ecologists today, but "protist" is more appropriate, especially if the microbe is capable of photosynthesis or if its metabolism is unknown (Box 9.1). The metabolism and thus ecological roles of many protists are not known, mainly because they have not been isolated and grown in the laboratory. Just as for bacteria and fungi, cultivation-independent approaches are now being used to figure out what some protists are doing in nature. Here we first focus on the ecological roles of protists, before

Table 9.1 Ecological roles of protists in nature.

Ecological role	Organisms	Comments
Primary production	Phytoplankton and algae	Many autotrophic protists are capable of grazing.
Herbivory	Flagellates and ciliates	Protists are major grazers of phytoplankton.
Bacterivory	Nanoflagellates and amoebae	Many protists are capable of grazing on bacteria.
Mixotrophy	Several	Mixotrophic organisms obtain energy from both phototrophy and heterotrophy (grazing or uptake of dissolved organic compounds).
Carnivory	Ciliates and others	Large flagellates are capable of eating small flagellates.
Parasitism	Flagellates and ciliates	
Symbiosis	Algae and flagellates	

Processes in Microbial Ecology. Second Edition. David L. Kirchman. Oxford University Press (2018). © David L. Kirchman 2018. DOI 10.1093/oso/9780198789406.001.0001

Protozoa, the plural of protozoan, comes from the Greek for "first animal" (*protohi zoa*) and has been used to refer to microbial eukaryotes that graze on other microbes. Some microbial ecologists prefer to use "protists" and argue against using the term "protozoa," because many of these microbes are capable of photosynthesis and have other characteristics quite different from animals. The scientific society devoted to the study of these microbes changed its name in 2005 to the International Society of Protistologists, from the Society of Protozoologists. Still, protozoa and protozoan are useful terms for colorless protists that are not capable of photosynthesis and carry out only heterotrophy.

number was not particularly interesting. About the same bacterial abundance was found wherever and whenever it was measured. Bacterial abundance does vary over time, especially in temperate regions with the seasons (high abundance in summer, low in winter), but it varies less than ten-fold. It is not very interesting to find the same number all the time, but the constancy of bacterial abundance is quite interesting and raises two questions: why is bacterial abundance so constant over time and space? And is there anything special about 10^9 cells per liter or per gram of soil? Why this number and not another, radically different? Part of the answer to these questions is "bacterivory," that is, the eating of bacteria by protists and other organisms.

To find out who is eating bacteria, new methods had to be developed. In contrast with the relative ease of estimating primary production and heterotrophic bacterial production, it is difficult to examine grazing on bacteria and other microbes. Consequently, no single method has emerged to be that of choice, and all have yielded some information on rates and aspects of protist biology. Some of these methods are not just for examining bacterivory. For example, the dilution method is commonly used to estimate grazing rates by protists on small eukaryotes and cyanobacteria (Landry and Hassett, 1982).

ending with some discussion of protist names, taxonomy, and phylogeny.

Bacterivory in aquatic habitats

As data came in about bacterial abundance in natural environments, microbial ecologists discovered that the

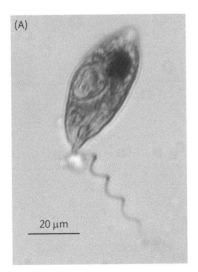

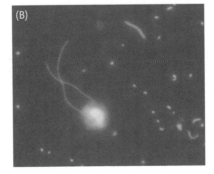

Figure 9.1 Some examples of flagellates able to graze on bacteria and other microbes. Panel A: Heterotrophic euglenoid flagellate from the Bering Sea, viewed by light microscopy. Courtesy of Evelyn Sherr. Panel B: Unidentified flagellate from the Sargasso Sea, stained with a DNA stain and viewed by epifluorescence microscopy. The small cells are bacteria about 0.5 μm. Courtesy of Craig Carlson.

Some of the methods provide information about which organisms are the main bacterivores in aquatic habitats. One simple experimental design uses filters with various pore sizes to remove all organisms larger than the respective pore, and then bacterial abundance is followed over time. An increase in bacterial abundance implies that the main bacterivore was removed by the filtration step. This approach has shown that removing large organisms such as copepods does not lead to immediate changes in bacterial abundance, implying that the main bacterivores are still present and are not large zooplankton. However, filtering out organisms less than 5 μm in size does result in an increase in bacterial abundance over time, indicating that the main bacterivores are usually in that size range. Methods using fluorescent bacteria or fluorescent beads, which mimic bacteria, have revealed that the small grazers are flagellates, often called heterotrophic nanoflagellates, 1–5 μm in length. These microbes earn their name by having one or more flagella that are used for locomotion and for feeding (Figure 9.1). They are very abundant in all environments, reaching 10^7 cells per liter in aquatic systems.

We know that many of these flagellates are heterotrophic, because they lack photosynthetic pigments and because these microbes only survive in laboratory cultures when fed bacteria. Other flagellates are capable of both photosynthesis and of feeding on bacteria and small microbes. Some common flagellates are listed in Table 9.2.

Another group of bacterivorous protists, naked amoebae, are not very abundant in the water column of aquatic habitats. They are less abundant than flagellates, although they can be as abundant as ciliates (Lesen et al., 2010). The low abundance of amoebae in the water column may reflect adaptation for growth on surfaces, which are less common in the water column than in soils and sediments. At times, however, amoebae can contribute significantly to grazing on bacteria and other small microbes. More work is needed on these fragile microbes.

In addition to flagellates and amoebae, several other organisms potentially graze on bacteria and similar-sized microbes in some aquatic habitats. Ciliates, to be discussed further, can prey on bacteria but are more important in grazing on other, larger microbes. Other bacterivores are not protists, but can be important in top-down control of bacteria and other microbes. In freshwaters, non-protist bacterivores include zooplankton belonging to the order Cladocera, such as the genus *Daphnia*. These zooplankton feed by filtering out prey with a mesh of hair-like structures (setae) which are spaced closely enough to capture micron-sized particles, including bacteria. In marine waters, other potential bacteriovores include gelatinous zooplankton, such as larvaceans, salps, and doliolids, all belonging to the phylum Chordata (which includes *Homo sapiens*). Larvaceans, for example, live in gelatinous houses and feed on bacteria and other small microbes by catching them in a fine-meshed, sticky, filtering structure. When the filtering mesh becomes clogged, larvaceans throw it away and build a new one, often several times a day in productive waters. Notably, larvaceans and other bacterivorous gelatinous zooplankton are much larger (millimeters to centimeters) than flagellates (microns). This discrepancy in the size of the prey (bacteria) and predator (gelatinous zooplankton) has several implications for thinking about how material and energy move through food webs.

Table 9.2 Some flagellates common in natural environments. From Sherr and Sherr (2000) and Howe et al. (2009).

Group	Example genus	Habitat	Characteristics
Bicosoecids	*Cafeteria*	Freshwater and marine	Heterotroph with two unequal flagella
Chrysomonads	*Ochromonas*	Freshwater and marine	Mixotrophic
Dinoflagellates	*Peridinium*	Freshwater and marine	Some toxic, others symbionts in corals
Choanoflagellates	*Monosiga*	Freshwater and marine	Heterotroph with one flagellum and collar
Euglenozoa	*Euglena*	Freshwater	Mixotroph
Cercozoans	*Bodomorpha*	Soils	Abundant heterotrophic flagellate
Cercozoans	*Heteromita*	Soils	Abundant heterotrophic flagellate
Parabasalia	*Trichomitopsis*	Termite gut	Hydrolyzes cellulose
Zoomastigophora	*Giardia*	Mammalian intestines	Parasite without mitochondria

Grazers of bacteria and fungi in soils and sediments

Soil protists are not as well studied as protists in aquatic habitats (Geisen et al., 2017), but we know that several types are also the main grazers of bacteria and some fungi in soils (Figure 9.2). Soil protists were traditionally divided into four morphological groups: the flagellates, naked amoebae, testate amoebae, and ciliates. (As with aquatic protists, 18S rRNA gene sequence data indicate that organisms making up these groups can be quite different in terms of phylogeny, but the groups are still useful for introducing the ecological roles of protists in natural environments.) Flagellates are functionally similar to those seen in aquatic habitats and eat bacteria as their main prey. Flagellate abundance ranges from 10^2 cells per gram of desert soils to 10^5 cells per gram in forest soils (Coleman and Wall, 2015).

Unlike in aquatic systems, naked amoebae are very abundant and active in many types of soils, eating not only bacteria but also fungi, algae, and even small detrital particles. Because the lack of rigid cell walls makes them very flexible, naked amoebae are able to explore small crevices and pores in soils where other grazers cannot go. One study found amoebae to be the most abundant type of protist in a soil (Geisen et al., 2014). In contrast with the naked amoebae, the testate amoebae have a rigid external "house" and usually are not as abundant as the naked variety. Microbes in the fourth protist group, the ciliates, also eat bacteria, but they are likely to be less important than flagellates or amoebae as bacterivores because ciliate abundance is much lower than that of other protists, reaching only about 1000 g^{-1} in temperate soils.

Soil fungi were traditionally thought to be eaten only by arthropods and nematodes, while protists were considered to graze on only bacteria (Geisen et al., 2016). However, that view may result from the cultivation methods using bacteria as prey, selecting against "mycophagous" or fungus-eating protists. Recent work has shown the importance of amoebae and ciliates in grazing on fungi. But fungal predators certainly still include nematodes, one of the most abundant and diverse groups of multicellular organisms in the biosphere (Coleman and Wall, 2015). Along with flagellates, ciliates, and rotifers, nematodes live in aqueous films and water-filled pores in soils, where they feed on a variety of prey, including fungi. Some nematodes have a hollow stylet, a dagger-like structure for piercing fungal hyphae, roots, or root hairs. Feeding on fungi has been examined by following the appearance of fungus-specific fatty acids in nematodes (Ruess and Chamberlain, 2010). Other grazers of fungi include arthropods, such as mites and springtails, many of which can graze on prey in air-filled pores.

In sediments of aquatic habitats, flagellates are major grazers of bacteria, but as in soils, some meiofauna (metazoans between 0.45 μm and 1 mm) and macrofauna (>1 mm) can also eat bacteria (Pascal et al., 2009). In a mudflat, the dominant meiofaunal grazers were foraminifers, nematodes, and harpacticoid copepods, while one macrofaunal type, a mudsnail, also ingested bacteria. An important feature governing the relative abundance of these grazers is grain size and sediment structure, determined by the amount of clays, sand, and organic material. Often, grazing seems low compared to bacterial growth, implying that another form of mortality, most likely viral lysis, accounts for top-down control in these ecosystems. However, the many methodological difficulties in working with samples with heavy particle

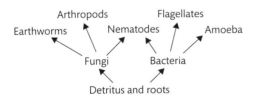

Figure 9.2 Some of the predators of bacteria and fungi in soils. The arthropods here include mites and springtails. Based mainly on Chapin et al. (2002). The diagram does not show all possible predator–prey interactions, such as the feeding of amoebae on fungi (Geisen et al., 2016).

Box 9.2 Role reversal

Nematodes and microarthropods usually feed on fungi, but some fungi can turn the tables. There has been much research on nematophagous fungi that trap and digest nematodes for food, and there is at least one case of a fungus apparently eating collembolans (Klironomos and Hart, 2001). Fungi are suspected to attack nematodes and microarthropods for their nitrogen, as well as for carbon and energy.

loads may lead to underestimation of grazing in soils and sediments.

Mechanism of protist grazing

The mechanism by which protists feed on other microbes gives insights into protist behavior and ecology. Some protists feed by extracting the cytoplasm out of their prey (Box 9.3). Many other protists feed by "phagocytosis" (Figure 9.3), a process by which microbes engulf particles and digest them in a food vacuole. Understanding phagocytosis helps to explain several aspects of protist biology and ecology. While there are many parallels between predation by protists and by metazoans, phagocytosis is fundamentally different from how a macroscopic predator eats its prey.

The first problem faced by a protist grazer is finding and encountering its prey. To understand the first step, it is crucial to remember that protists and their prey live in a low Reynolds number world where viscous forces dominate, quite unlike the world of their macroscopic counterparts. As mentioned in Chapter 3, to imagine life in this world, think of swimming in molasses or hot tar. To feed in this low Reynolds number world, protists have to get water flowing past them in one direction. They achieve this unidirectional flow by moving asymmetrically (Strom, 2000). Flagellates do so by swimming in a corkscrew pattern or by moving their flagella asymmetrically. Ciliates beat their cilia like rowing a boat. Heterotrophic protists can be classified by how they obtain their prey (Montagnes et al., 2008). Filter feeders, such as some

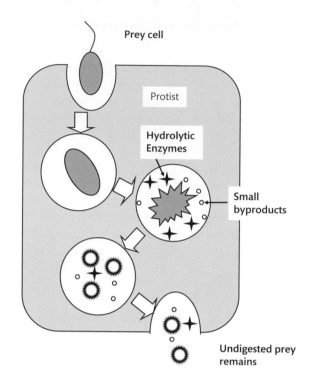

Figure 9.3 Phagocytosis by a protist feeding on another microbe.

ciliates and flagellates, produce feeding currents, while diffusion feeders, such as heliozoans, stick out stiff arm-like structures (axopods) into which prey collide. Raptorial feeders, which include some ciliates, flagellates, and naked amoebae, actively hunt and capture prey. Some protists have specialized feeding apparatus, such as the collar of choanoflagellates, that assist in the capture of small food particles.

Once captured by any of these mechanisms, the prey particle is packaged into a "food vacuole", formed by the protist's outer membrane stretching around the prey particle. This process of phagocytosis is similar to what happens when a mammalian lymphocyte encounters a foreign particle. In protists, the entire process must be very efficient and fast, ranging from minutes to hours, to account for the observed feeding rates at prey concentrations typically seen in nature. Once inside the food vacuole, digestion can begin. This consists of the release of various enzymes, such as proteases and lysozymes (for bacterial prey), into the food vacuole to break down the entrapped prey. The acidity of the food vacuole also

---

Box 9.3 Grisly dining

Some heterotrophic dinoflagellates have a unique way of feasting on their prey. They stick a feeding tube, called the peduncle, into hapless prey and suck its guts out. Dinoflagellates similar to *Pfiesteria* prey on a variety of phytoplankton and can be grown on red blood cells of fish (and presumably humans) (Jeong et al., 2007). Vampyrellid amoebae use an analogous feeding mechanism to prey on fungi and algae. Abundant in soils (Geisen et al., 2016), "vampire" amoebae have also been found in the oceans (Berney et al., 2013).

helps to disable the prey and to assist in digestion, analogous to the mammalian digestive system. The products from the digestion process are carried into the cytoplasm by pinocytotic vesicles, analogous to food vacuoles except that the pinocytotic vesicles are much smaller. During the entire digestion process, the food vacuole moves around the protistan cell until digestion is completed, at which time it fuses with the protist outer membrane; in ciliates, this fusion occurs at a miniature anus, the cytoproct. The undigested contents of the food vacuole are then expelled to the outside environment.

Factors affecting grazing

When confronted by changes in food availability and prey numbers, protistan grazers respond in two general ways. Most of this section will focus on "functional responses," which are how the grazing rate responds to changes in prey abundance over a short timescale. The second way, the "numerical response," is how grazer growth rates change in response to prey abundance. This response occurs over longer timescales (the generation time of the protist, which is roughly a day) than the functional response occurring within minutes to hours. Both types of response are important in thinking about the ecological roles of protists and indeed of all grazers in soils and aquatic habitats. The factors affecting protistan

grazing are prey numbers, prey size, and the chemical composition of the prey.

Prey number and predator–prey cycles

One of the simplest but most important factors affecting grazing is the number of prey. From first principles, we would expect grazing in the microbial world to increase as prey abundance increases, because it increases the chance of a predator encountering a prey. However, the rate cannot increase indefinitely due to the limit set by the rate of phagocytosis and digestion. So, after increasing in response to increasing prey abundance, the grazing rate reaches a maximum (Figure 9.4). The general shape of the ingestion versus prey curve is very similar to what we have seen before, such as the uptake of a dissolved compound as a function of its concentration, described by the Michaelis–Menten equation (Equation 6.10). An equation similar to the Michaelis–Menten equation can be written to describe the ingestion rate as a function of prey abundance. Unlike uptake, however, protists can stop feeding at low prey density. The result is the curve crossing the x-axis in Figure 9.4 at low but positive prey numbers, because the ingestion rate becomes zero even though prey are present at some threshold level. This response is in effect the end result of a cost–benefit analysis by the protist. It may cease feeding when

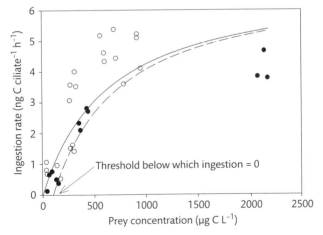

Figure 9.4 Ingestion of algal prey by a ciliate as a function of prey abundance. Two algal prey were used: *Nannochloropsis* (filled circles) and *Isochrysis* (open circles). The solid line was determined by regression analysis of the actual data: I = 8.96 × P/(641 + P), where I is the ingestion rate and P the prey concentration. The dotted line illustrates the effect of a threshold on ingestion, though there is no evidence of this in the actual data. Data from Chen et al. (2010).

energetic costs outweigh the benefits of grazing on scarce prey.

The existence of grazing thresholds is one answer to the question of why bacterial abundance is about 10^9 cells in a gram of soil or sediment or in a liter of water from an aquatic habitat. These abundances may reflect grazing thresholds. Growth brings up bacterial abundance to the threshold levels, but grazing prevents these microbes from exceeding the threshold for long. The thresholds may be set at abundances of 10^9 cells because of fundamental limitations in the effectiveness of feeding behavior and energetics of bacterivores. Grazing thresholds can also account for why protists and other microbes can exist in nature, in spite of grazing by hungry carnivores searching for food. These carnivores may go after other prey when prey numbers drop below the threshold level.

Figure 9.4 illustrates a functional response to prey abundance. The graph of the numerical response of protists to the prey would look very similar to Figure 9.4; that is, protist growth rates also increase with prey concentrations before reaching a maximum, analogous to the response of heterotrophic bacterial growth to organic carbon concentrations, or algal growth to inorganic nutrient concentrations (see Figure 6.11). The equation describing growth rate as a function of prey abundance is exactly the same as that for ingestion rate as a function of prey abundance. But a graph of protist growth versus prey abundance is a static picture of how protists respond to initial prey level. In nature, both predator and prey abundance vary continuously because of one population impacting the other.

Predator–prey interactions have been extensively explored with the Lotka–Volterra model, developed independently by the American biophysicist Alfred Lotka in 1925 and the Italian mathematical biologist Vito Volterra in 1926. This model has been used to examine all sorts of predator–prey relationships, the classic one being snow lynxes and hares in Canada. The same ecological principles can be used in the microbial world. The Lotka–Volterra model consists of two differential equations, the first describing how the prey changes as a function of its growth rate (r) and a grazing rate constant (a), multiplied by the prey abundance (H) and the predator abundance (P):

$$dH/dt = r \cdot H - a \cdot H \cdot P \qquad (9.1).$$

In words, the change in prey abundance over time is equal to prey cell production minus predation on the prey. The second equation is for the predator:

$$dP/dt = b \cdot H \cdot P - m \cdot P \qquad (9.2)$$

where b is the growth rate of the predator, and m is the specific mortality rate for the predator. In words, the change in predator abundance over time is equal to predator cell production minus mortality of the predator. This model assumes that the prey grows exponentially, that rates are proportional to population size, that the rate constants do not vary with population size (in spite of the known numerical response), and that predation is the only ecological process at work. The mathematics and implications of the Lotka–Volterra model have been examined in great depth. Here we concentrate on only a couple of predictions from this model.

The first is that predator and prey abundances oscillate over time (Figure 9.5). Some values for the model parameters lead to unstable solutions, meaning the populations either go extinct or increase to infinitely large levels. The stable solutions imply that the predator population lags behind the prey population and that both vary around each other forever. The model can also be used to predict the period (how long it takes for a population to return to a starting value after increasing and decreasing) and the amplitude (the difference between minimum and maximum abundance) of the oscillation. The period is the same for both predator and prey, while the amplitudes differ.

Although classic predator–prey oscillations can be seen in controlled experiments in the laboratory, they are rarely observed with microbes in nature. One problem is that it would be difficult to see the oscillations illustrated in Figure 9.5 in nature, even if they exist. The amplitude in oscillation of prey abundance is only on the order of 20% in this example, which is small and would be difficult to detect in nature. More problematic is the long timescale of these oscillations. The Lotka–Volterra model predicts that the period of the oscillation should be about 20 days, given rates and population levels typical of natural habitats. To be really convincing, data from two or, better, three cycles, equivalent to 40–60 days of observations, would be needed to test the model. That would be difficult even for a study of soils or of a small lake, and nearly impossible for an ocean, where a long

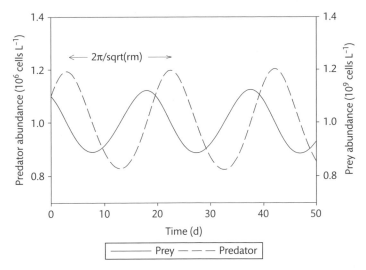

Figure 9.5 Variation in the abundance of predator and prey over time according to the Lotka–Volterra model. This is a hypothetical example but uses realistic values for r and m (0.2 and 0.5 d⁻¹, which are typical microbial growth rates) and for the initial predator and prey abundances. See Murdoch et al. (2003) for more on the mathematics of the Lotka–Volterra model and other, more sophisticated representations of predator–prey relationships.

scientific cruise is 30 days. So, grazing could account for the apparent constancy of bacterial abundance in many natural ecosystems even if the Lotka–Volterra model applies, because the predicted oscillations are relatively small and hard to see.

A more fundamental problem, however, is that the Lotka–Volterra model is not realistic. It does not include many ecological processes, such as switching by the predator to other prey, several predators (and viruses) going after several prey, competition among predators and prey, and bottom-up controls. Many of these other processes would tend to dampen the oscillations and disrupt the timing between predator and prey variations. Still, the Lotka–Volterra model is a useful starting point for more realistic and sophisticated models of predator-prey relationships.

Size relationships of predator and prey
We have seen that the dominant grazers of bacteria, which are about 0.5 μm in nature, are 1–5 μm flagellates. This observation raises a general question about the relationship between sizes of other prey and predators and whether there is a general rule that can predict who is eating whom in the microbial world. The Danish microbial ecologist Tom Fenchel (1940–) suggested an elegant

answer to this question (Fenchel, 1987). He considered a hypothetical spherical protist with a radius R eating an equally hypothetical spherical prey with a radius ρ, and argued that the clearance rate of a predator should vary as a function of ρ/R. He went on to provide empirical evidence that the ratio is on the order of 0.1 (Figure 9.6). That is, predators are generally about ten-fold bigger than their prey.

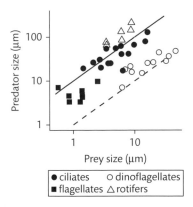

Figure 9.6 Relationship between length of a prey and its predator. The solid line indicates a ratio of predator size to prey size of 10:1 (spherical equivalent diameter), while the dashed line is the 1:1 ratio. The slope of the regression line (log-log transformation) is 0.597 ± 0.13. Data from Wirtz (2012).

As more data came in after Fenchel's initial work, exceptions to the 10:1 rule became evident. Perhaps the most striking exceptions are the dinoflagellates and other flagellates, which are substantially below the 10:1 line in Figure 9.6; dinoflagellates eat prey not much smaller than their own size, including diatom chains extending for >100 μm (Sherr and Sherr, 2009). Less dramatic are the 1–3 μm protists that eat 0.5 μm sized bacteria, again breaking the 10:1 rule. Other exceptions, not in the figure, include microbial predators that are much larger than expected, such as nematodes in soils or gelatinous bacterivorous zooplankton, like larvaceans. Some species of large detritivores and bivalves in soils and sediments are more important grazers of small microbes than expected from the 10:1 rule. Even given these exceptions, the 10:1 rule is still a useful if crude guideline in thinking about predator–prey relationships in the microbial world.

A related issue is how grazing by one predator of a particular size varies as a function of prey size. Fenchel's 10:1 rule leads to the prediction that grazing is low on very small prey and on very large prey, relative to an optimal prey size for a particular predator. Big prey are beyond the capacity of a predator to ingest, while small prey are captured too inefficiently. Somewhere in between the two extremes in prey size is the optimum that results in the highest grazing rate. As a predator increases in size, the optimal prey size also increases.

Experimentally, this size effect is demonstrated by feeding one predator prey of different sizes, as was done to obtain the data in Figure 9.7. Another experiment is to observe the change in size distribution of a bacterial assemblage or a bacterial culture over time with and without a protist. The presence of a grazer can result in the bacterium forming long chains or aggregates that are too big for protists to eat.

So, size is a key factor in determining grazing rates and who is eating whom in the microbial world; it explains many observations. Size is why plastic beads are ingested by some protists at nearly the same rate as microbial prey of the same size. (It is also why microplastic pollution is so worrisome.) Because of this, fluorescent plastic beads can be used as a surrogate food to estimate grazing rates. The size effect also implies that prey of similar size should be eaten by the same predator. For example, heterotrophic bacteria and the cyanobacteria *Prochlorococcus* and *Synechococcus* are probably eaten by the same suite of protists, given that cells in the three bacterial groups are similar in size. Likewise, carnivores that eat heterotrophic nanoflagellates should also eat similar-sized algae.

Chemical recognition and composition

While size is a powerful predictor of grazing behavior, microbial ecologists knew early on that it was not the

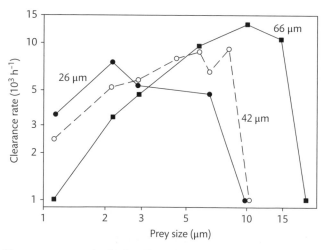

Figure 9.7 Grazing by three ciliates on prey varying in size. The units for clearance rate are in ciliate volume cleared per hour. The number next to each line is the diameter of the ciliate as if it were a sphere (equivalent spherical diameter). Note that the optimal prey size increases with the predator size. Data from Jonsson (1986).

only factor. The chemical composition of the prey also has a role. Early evidence for this conclusion came from the experience of growing protists on different bacterial species in the laboratory. Some bacteria were better food than others, even though they seemed to share the same size and appearance. Also, protists sometimes grow more slowly on heat-killed bacteria than on live food. One explanation is that heating changes prey chemistry, similar to what cooking does to our food. Think of the difference between fresh and cooked eggs. Finally, there is evidence of discrimination against plastic beads and fluorescent-labeled bacteria; grazing rates on these particles can be lower than on natural, unaltered bacteria. These findings indicate that chemical properties of prey, especially the composition of the prey cell surface, affect grazing of protists. It seems that somehow protists can taste their food.

More support for this idea comes from experiments that examine feeding on plastic beads coated with various organic compounds (Wootton et al., 2007). These experiments also suggest a mechanism (Figure 9.8). Wootton and colleagues found that the marine dinoflagellate *Oxyrrhis marina* ingested more plastic beads coated with mannose than with other sugars. Furthermore,

feeding by the dinoflagellate on its regular phytoplankton prey was inhibited when the dinoflagellate was exposed to mannose, but again not by other sugars. These are classic experiments demonstrating that a cell–cell interaction is mediated by a type of cell surface receptor, a "lectin." Lectins are sugar-binding proteins involved in a great variety of cell–cell interactions in organisms ranging from plants to humans. In this case, the dinoflagellate uses a mannose-binding lectin as a receptor to recognize mannose on the prey cell surface. Mannose itself is not why the grazer selects the prey, but rather its presence indicates a desirable food item for the grazer. The general model has been applied to other protists (Dürichen et al., 2016).

Once inside the food vacuole, the cell surface chemistry of the prey can also determine digestion by the protist grazer and therefore their growth. The prey cell surface can affect the efficiency of digestion, that is, the amount of prey carbon incorporated into protist cytoplasm. For example, some bacteria are resistant to grazing because of their unique cell walls (Tarao et al., 2009). Some prey cells are rejected by the protist and are ejected ("egested") from the food vacuole back to the outside environment.

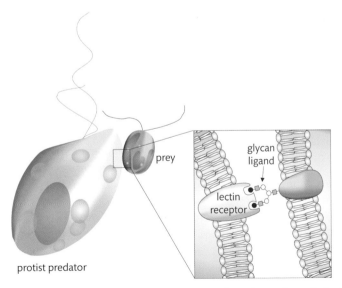

Figure 9.8 Model of how protist predators may be able to recognize prey through protein–carbohydrate interactions. Lectin receptors on the cell surface of the predator bind to specific carbohydrate conformations on the surface of the prey. Lectins are carbohydrate-binding proteins, and glycans are the carbohydrate portions of glycoproteins and glycolipids. Figure provided by Emily Roberts. See Wootton et al. (2007) for more details.

Among several reasons why grazers may select for some prey over others, is that the preferred microbial prey may be more nutritious for the protist. "More nutritious" here means that the chemical composition of the preferred prey leads to faster or more efficient growth of the protist. There is some evidence of protists favoring prey with a low C:N ratio (Montagnes et al., 2008), probably due to high protein content of these prey. The lipid content of the prey also appears to have a big impact. Diets of prey with certain polyunsaturated fatty acids, for example, promote faster growth of protists and metazoans, and the lack of these fatty acids in bacteria may limit flux through food webs based on bacteria (von Elert et al., 2003). Both heterotrophic bacteria and cyanobacteria also do not have the sterols that are needed by protists and other eukaryotes (Martin-Creuzburg and von Elert, 2009). Oddly, lipids unique to bacteria can be retained in the lipids of eukaryotic predators, thus providing evidence that the predator grazes on bacteria (Taipale et al., 2012).

In addition to chemical characteristics of a prey cell, compounds released by microbes may affect protist feeding behavior. Volatile organic compounds produced by six cultivated soil bacteria affected the abundance and grazing by three protists (Schulz-Bohm et al., 2017). Additional experiments with mutant bacteria suggested that terpenes were among the volatile compounds affecting bacteria–protist interactions. Similarly, protists may sense microbial prey by detecting dissolved compounds produced by the prey. The chemical ecology of protist predators and their prey is probably more complex than is currently appreciated.

Grazing by larger protists: ciliates and dinoflagellates

If flagellates 1–5 µm in size are the main grazers of heterotrophic bacteria, coccoid cyanobacteria, and micron-sized eukaryotic algae, who eats the flagellates? The 10:1 rule would predict a carnivore of about 10–50 µm. In soils, these carnivores include amoebae, ciliates, and nematodes. In the water column of aquatic habitats, a complex suite of microzooplankton protists in the 20–200 µm size range are potential predators of flagellates, including big flagellates grazing on small flagellates (Figure 9.9). Some of the first microzooplankton protists

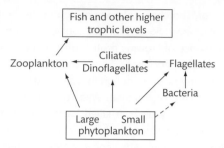

Figure 9.9 Model of a microbial food web to illustrate some of the roles of ciliates and dinoflagellates in aquatic habitats. "Bacteria" here refers to heterotrophic bacteria; coccoid cyanobacteria are included with the small phytoplankton. The dotted line linking phytoplankton and bacteria indicates that the connection between these organisms is indirect and involves dissolved organic material and detritus.

studied by microbial ecologists were tintinnids (Dolan et al., 2013), which are choreotrich ciliates, characterized by their elaborate houses (loricae). The sturdy houses enabled microbial ecologists to collect these otherwise fragile ciliates with fine-mesh plankton nets and to identify them. The houses vary among species of tintinnids. Other collection and fixation methods led to the discovery that ciliates other than tintinnids are generally much more abundant than tintinnids. These other ciliates are sometimes referred to as being "aloricate" or "naked" because of their houseless body form.

Ciliates as predators of bacteria, flagellates, and fungi
In addition to grazing on bacterivorous nanoflagellates, ciliates are important herbivores and graze on many types of phytoplankton, including long chains of diatoms. This top-down control of phytoplankton was previously thought to be by crustacean zooplankton larger than the microzooplankton. Aquatic ecologists now believe, however, that zooplankton such as copepods may be more important as carnivores eating microzooplankton than as herbivores grazing on phytoplankton. In addition to herbivory, small ciliates may graze directly on bacteria and similar-sized algae, such as the coccoid cyanobacteria *Synechococcus* and *Prochlorococcus*.

The most common ciliates in aquatic habitats are oligotrich ciliates in the class Spirotrichea, (the phylum Ciliophora). These microbes are round or oval with a crown of cilia at the oral cavity. Another class abundant in the oceans is Oligohymenophorea, which include a number of symbionts and parasites (Gimmler et al., 2016). Not all ciliates are strict heterotrophs. Many are mixotrophic and use both phototrophy and heterotrophy, while others are mostly autotrophic. Examples of ciliates are given in Figure 9.10 and Figure 1.1D in Chapter 1.

Ciliates are not as ecologically important in soils as they are in aquatic habitats, one reason being that they are restricted to very moist soils (Coleman and Wall, 2015). The abundance of ciliates is lower than that of flagellates in soils, which is the case in aquatic habitats as well. Many ciliates feed on bacteria, but some are capable of feeding on fungi (Geisen et al., 2016), as mentioned before, and still others can graze on flagellates, an ecological role they share with nematodes. Soil ciliates are adapted to life on the surface of soil particles (thigmotactic). Cultivation-independent methods have found that, as in aquatic habitats, Spirotrichea is an abundant order of ciliates, but so too are Colpodea and Oligohymenophorea (Geisen et al., 2015).

In the benthic habitat of aquatic systems, ciliates are important in sandy sediments with large interstitial spaces. Sediments dominated by silt and clay have small interstitial spaces and lower ciliate numbers than sandy sediments. As in soils, sediment ciliates have several adaptations to life on surfaces and in interstitial spaces. Some are long (millimeters in length) with cilia only on one side of the cell. These microbes feed on benthic algae, flagellates, other ciliates, and bacteria. Depending on time of year and location, ciliates can be the dominant consumers of algae and bacteria in sediments of aquatic habitats.

Heterotrophic dinoflagellates

Other prominent members of the protist community include heterotrophic dinoflagellates. We first encountered dinoflagellates in Chapter 6 in the discussion of phytoplankton and primary production. In addition to their contribution to primary production, many dinoflagellates are mixotrophic, while others are strictly heterotrophic. This group of protists is well known to be common in aquatic habitats, but they may be more common in soils than previously thought (Bates et al., 2013). If there is a "typical" dinoflagellate, it is pear-shaped with one groove, the girdle, around the middle of the cell and another groove, the sulcus, going down from the girdle (Figure 9.11). The microbe moves thanks to two flagella, one wrapped around the cell within the girdle, the other beating in the sulcus and off the posterior end of the cell. Some species are armored with thecal plates composed of cellulose, while others are "naked" or "unarmored."

Dinoflagellates vary greatly in size, shape, and in metabolism (Table 9.3). Species in the genera *Amphidinium* are >20 µm, have thecal plates, and are capable of feeding on nanoflagellates using their peduncles, tube-like structures, to penetrate into prey. Other species feed by more conventional phagocytosis on diatoms and ciliates. Still others are referred to as "veil feeders" because they exude

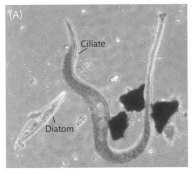
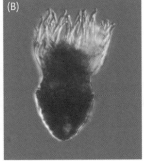

Figure 9.10 Examples of ciliates. Protostomatid ciliate of the genus *Tracheloraphis* found in marine sediments, along with a pennate diatom. The ciliate, which is contracted here, is about 500 µm long. If stretched out, it would be over 1 mm. The large dark blobs next to the ciliate are sand grains. Image by David J. Patterson, used courtesy of micro*scope (microscope.mbl.edu) (A). An oligotrich ciliate, common in the water column of aquatic systems, about 30 µm long. Image courtesy of John Dolan (B). Color version of panel B is available at www.oup.co.uk/companion/kirchman.

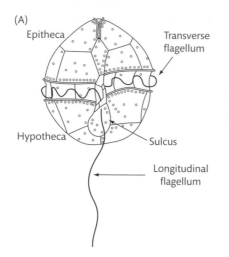

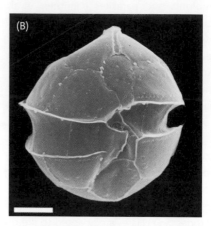

Figure 9.11 The ventral view of a "typical" dinoflagellate in a schematic diagram (A) taken from Jeong et al. (2005) and in a scanning electron micrograph (B). This particular species (*Stoeckeria algicida*) is about 17 µm by 13 µm. Used with permission from Hae Jin Jeong.

Table 9.3 Some common dinoflagellates. Data from Hansen (1991).

Genus	Feeding mechanism	Prey	Comments
Amphidinium	Predation by a peduncle	Nanoflagellates	Armored
Ceratium	Photoautotrophic	None	Armored
Gonyaulax	Photoautotrophic	None	Some toxic species
Gymnodinium	Predation by a peduncle	Diatoms	Unarmored
Noctiluca	Predation by engulfment	Nearly all particles of the right size	Some bioluminescent
Oxyrrhis	Predation by engulfment	Nanoflagellates, diatoms	Easily grown in the laboratory
Protoperidinium	Predation by a feeding veil	Diatom chains, other large phytoplankton	Attaches to prey with a "tow filament"
Symbiodinium	Photoautotrophic	None	Coral symbiont

a pseudopodial cytoplasmic sheet that envelopes even large diatom chains, digesting them extracellularly, and absorbing nutrients before being hauled back into the cell. Some heterotrophic dinoflagellates are bioluminescent, with *Noctiluca scintillans* being the most famous example. This microbe is about 500 µm or bigger and produces light by the luciferin–luciferase system (Chapter 14). In response to unfavorable growth conditions, dinoflagellates can form cysts which sink to and survive in bottom sediments.

Mixotrophic protists and endosymbiosis

The discussion so far has implied that protists are either photoautotrophic (algae) or heterotrophic predators (protozoa), but in fact many protists are in between the two extremes. Some protists, like diatoms, are devoted 100% to photoautotrophy. Diatoms cannot graze on other microbes because of their siliceous cell walls. At the other extreme, the colorless protozoa cannot carry out photoautotrophy because they lack the pigments necessary to harvest light energy. In between the two extremes are mixotrophic protists (Figure 9.12). These microbes can carry out photosynthesis while also being able to capture prey by phagocytosis. Some may also assimilate dissolved organic material. Mixotrophic protists have been divided into two functional groups (Mitra et al., 2016): constitutive mixotrophs, which are photoautotrophs capable of phagocytosis, and non-constitutive mixotrophs, which are phagotrophs that acquire their photoautotrophic capacity by ingesting photoautotrophic prey.

Near the strict phototrophic end of the continuum are the phagotrophic algae. These protists are phototrophic,

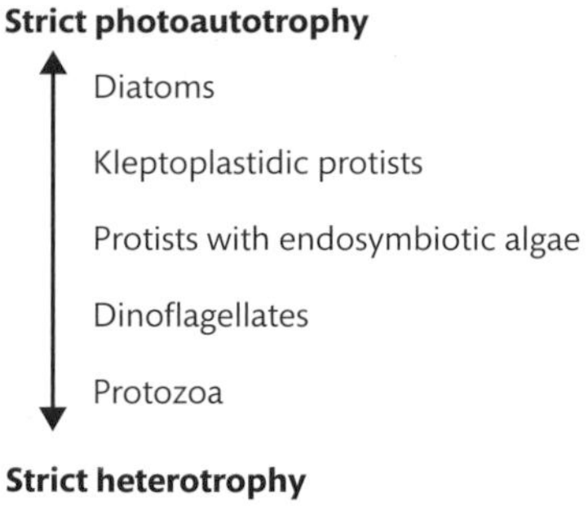

Strict photoautotrophy

Diatoms

Kleptoplastidic protists

Protists with endosymbiotic algae

Dinoflagellates

Protozoa

Strict heterotrophy

Figure 9.12 Protist metabolisms, ranging from strict photoautotrophic organisms incapable of growing without light, to strict heterotrophs incapable of growing without prey. In between these two extremes are the mixotrophs.

with their own functional chloroplasts that are thought to carry out some phagotrophy to obtain essential vitamins, specific lipids, or organic material rich in nitrogen and phosphorus. Prey fed on by phagotrophic algae are rich nuggets of nitrogen and other elements that supplement the elements supplied with dissolved compounds such as ammonium and phosphate, which are often in very low concentrations. In fact, grazing by mixotrophic haptophytes has been shown to be high when phosphate concentrations are low (Unrein et al., 2014). This type of "phytoplankton" can be quite common and may account for up to 50% of the entire community in freshwaters, and are equally important in marine systems as well. Their grazing impact can also be substantial, with rates comparable to those of strict heterotrophic grazers. Mixotrophic protists can occur in the top layer of soils receiving sufficient sunlight.

Another type of mixotrophic protist feeds on algae and completely digests its prey, except for the chloroplasts, at least for a short time. The undigested chloroplasts retained by the protist are referred to as "kleptochloroplasts" or kleptoplasts, with "klepto" coming from the Greek for an irresistible urge to steal. Some examples of protists with kleptochloroplasts include the ciliates *Strombidium* and *Mesodinium*, and the dinoflagellates *Gymnodinium* and *Amphidinium*. Protists with kleptochloroplasts can obtain some organic carbon from photosynthesis carried out in these chloroplasts. These chloroplasts differ in many fundamental ways from those found in a strict phototrophic protist. One difference is that kleptochloroplasts are usually lost over a few

days if the protist does not feed again. These protists are basically heterotrophic and opportunistically take advantage of photosynthesis by chloroplasts from their prey. Chloroplast retention is best known among the ciliates, and species with this form of metabolism can make up nearly half of the total ciliate community in estuaries and oceanic environments. It is also common among dinoflagellates and rare but present in some benthic foraminifera.

It should now be clear that mixotrophy is ubiquitous among protists. The high abundance of mixotrophic protists attests to the selection for evolutionary strategies that extract as much energy and limiting elements as possible from natural environments that are often depauperate in labile organic carbon, nitrogen, or phosphorus. Because of their diversity and high abundance, mixotrophic protists contribute substantially to both primary production and grazing activity. One study, for example, found that a group of mixotrophic protists (haptophytes) account for about half of predation on bacteria in oligotrophic coastal marine waters (Unrein et al., 2014). The mixotrophic dinoflagellate *Noctiluca* can graze on diatoms but also replace them at the base of the food chain, with implications for higher trophic levels that normally depend on diatoms (Stoecker et al., 2017). Microbial ecologists are still working on incorporating mixotrophy into models of trophic dynamics and carbon flow in natural environments (Figure 9.13).

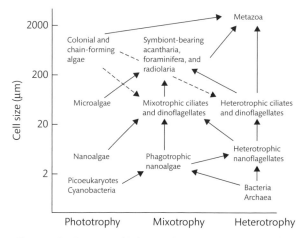

Figure 9.13 Some trophic interactions among phototrophic, mixotrophic, and heterotrophic microbes. The dashed lines indicate the grazing of small predators on large prey. Based on Caron et al. (2012).

The relationship between a chloroplast-retaining phagotrophic protist and its prey is one-sided, but that is not the case with protists that host functional algal cells within their cytoplasm. Instead of digesting the prey, these phagotrophic protists have evolved mechanisms to retain and nurture the alga in an endosymbiotic relationship, playing host to only a single algal species, in contrast with chloroplast-retaining protists which may steal chloroplasts from several types of algae. The symbiotic algae include chlorophytes, prymnesiophytes, prasinophytes, diatoms, and many dinoflagellates (Caron et al., 2012). Some algal lineages seem to be especially common in endosymbiotic relationships. Among the dinoflagellates, for example, *Pelagodinium beii* (formally *Gymnodinium beii*) is found in four species of planktonic foraminiferans, and is a not-too-distant cousin of the coral's symbiotic dinoflagellate, the genus *Symbiodinium*. Several species of radiolarians harbor *Brandtodinium nutricula* (formally *Scrippsiella nutricula*).

The phagotrophic protist and its endosymbiotic algae probably each enjoy several benefits from the relationship. The otherwise heterotrophic protist gains another source of carbon and energy from organic material synthesized and exuded by the phototrophic algae. The host may gain additional material and energy by digesting some of the algae from time to time, which may be necessary to keep the endosymbiont population at a manageable level. The algal symbiont may also absorb ultraviolet light and protect the protist. On the other side of the relationship, the alga benefits by being protected from predation (ignoring the occasional digestion by the protist host) and perhaps from viral lysis. The endosymbiotic alga is also physically close to a ready source of inorganic nutrients, such as ammonium and phosphate, released as wastes by its host. The large number of algae, reaching several thousand cells in some cases, inside of one protist host is an indication that the relationship is benefitting the alga.

In addition to their ecological importance, phagotrophic protists with endosymbiotic algae are important examples in support of the endosymbiotic theory for the evolution of algae (Figure 9.14). One of the first endosymbiosis events was between a heterotrophic protist and a cyanobacterium, which eventually became the chloroplast in algae and higher plants. Additional endosymbiosis events are needed to explain other features of some

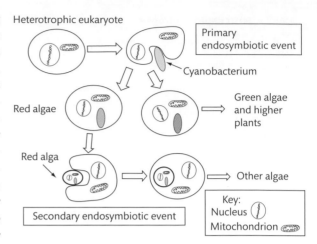

Figure 9.14 The endosymbiotic events leading to algae and higher plants. An important endosymbiotic event not shown is that leading to mitochondria. Based on Delwiche (1999) and Worden and Not (2008).

algae, such as the presence of three to four membranes around chloroplasts in dinoflagellates, haptophytes, and cryptophytes. The plastids in these algae are thought to have arisen by secondary or even tertiary endosymbiotic events, that is, the phagocytosis of a photoautotroph other than a cyanobacterium. In the case of the Chromalveolata, the plastids are thought to have come from a red algal ancestor (Reyes-Prieto et al., 2007). Today, these chloroplasts are fully integrated into the algal cell, a relationship cemented by the transfer of genes from the former symbiont to the host nucleus. The existence of endosymbiotic protists today is strong evidence that these endosymbiotic events occurred during algal evolution millions of years ago.

Protist community structure and the evolution of eukaryotes

Much of the discussion so far in this chapter has been about protists that have distinctive morphologies, such as flagellates, amoeba, and ciliates. Many protist biologists still rely heavily on morphology to classify protists. However, small protists (<10 μm) often cannot be distinguished by morphology, and as seen several times in this book, many of the most abundant and presumably important protists are difficult to cultivate, necessitating the use of cultivation-independent methods analogous to those developed for prokaryotes. Instead of the 16S rRNA gene used for bacteria and archaea, the 18S rRNA gene is key for protists. In addition to using these gene sequences for identifying protists and for exploring the structure of protist communities in natural environments, sequences of the 18S rRNA gene and other genes have also upended ideas about the evolution of all eukaryotes.

The current concept of eukaryotic evolution is quite different from traditional phylogenetic trees and from the popular image of humans and other large organisms at the tree's pinnacle. The five kingdom scheme of Whittaker (1969) had protists ("Protista") at the bottom, eventually leading to the "crown" phyla in three eukaryotic groups: the Plantae, Fungi and Animalia. (The fifth kingdom, Monera, consisting of what we now know to be bacteria and archaea, was below Protista.) In contrast, the best current model of eukaryote evolution places higher plants, fungi, and animals as mere branches off a tree dominated by protists (Figure 9.15). These organisms are divided into five supergroups with a possible sixth (Hacrobia) still being debated (Burki et al., 2016). The supergroup SAR includes the stramenopiles (with diatoms and oomycetes as members, among many others), alveolates (ciliates, dinoflagellates, and Apicomplexa), and rhizaria (foraminifera and the abundant Cercozoa phylum). (This "SAR" is quite different from the bacterial clade SAR11 and the other SAR clades discussed in Chapter 4.) The other four well-established supergroups include the Archaeplastida (land plants and green algae), Amoebozoa, Opithokonts (higher animals, fungi, and choanoflagellates), and Excavates (euglenids). The tree of eukaryotic life is dominated by protists and consists almost entirely of microbes.

Exploring protist community structure

The 18S rRNA gene data has revealed the huge diversity of protists sharing similar morphologies and the problems with classifying these microbes by appearance alone. The gene sequence data indicates that groups defined by morphology contain several unrelated taxa. For example, several protists in unrelated taxa can take on the amoeba morphology. Once placed in the Sarcodines, most "amoeba" are now in two supergroups: Amoebozoa and Rhizaria (Pawlowski and Burki, 2009).

Similarly, "flagellates" are found in several, distantly related taxa in the eukaryotic tree. Only ciliates are monophyletic, all being in Ciliophora in the Alveolata supergroup.

While the gene sequence data answered some questions, new ones were raised that are still being addressed. One is the same question raised about bacteria and 16S rRNA genes: what is a species? Most protists have sexual phases, so the biological species concept could be applied to these organisms. However, the concept is not applicable to those protists that do not have sexual phases and the many protists that remain uncultivated and are known only by their 18S rRNA genes. As with prokaryotes, an operational taxonomic unit (OTU) for protists is defined by organisms sharing ≥97% similar 18S rRNA genes, although some investigators advocate higher relative similarities, close to 100% (Massana, 2015). At the other extreme, some microbial ecologists believe that it is ecologically more meaningful to focus on morphospecies and only large differences in 18S rRNA genes.

The answer is probably between the two extremes and is likely to vary depending on the protist and the process. Some protists with ≥97% similar 18S rRNA gene sequences probably have the same ecological role, while other protists with 98% or maybe even 100% similar 18S rRNA gene sequences may differ in their ecology. We need more work to explore the ecological meaning of small differences in 18S rRNA gene sequences.

We can gain insights into protist ecology with data about the abundance or biomass of protist taxa with different 18S rRNA genes and about how they vary in natural environments. But there is a large problem in getting these data from the relative abundance of different 18S rRNA gene sequences in a sample. The problem is the number of "copies" of the 18S rRNA gene per protist. This is a problem for bacteria but it is a much bigger problem for protists. Although bacteria can have between 1 and 15 copies of the 16S rRNA gene per cell, the most abundant type, the oligotrophic bacteria, have only one or two (Chapter 5). In contrast, this number ranges between 1 and 400,000 among protists (Table 5.1 in Chapter 5). Consequently, a taxon could account for a large fraction of 18S rRNA sequences because it is truly abundant in the sample, or because it is a protist with a high copy number. Conceivably, the copy number problem could

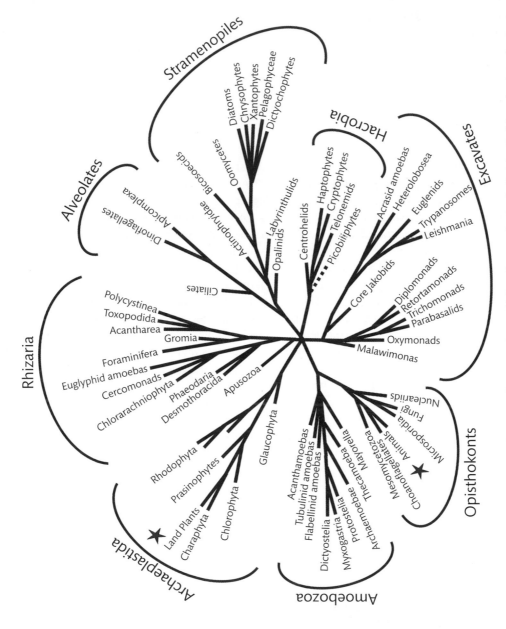

Figure 9.15 Phylogenetic tree of eukaryotes. The stars point out higher plants and animals. Nearly all of the other branches are protists. Modified from a version provided by Ramon Massana, based on Baldauf (2003) with additional information from Okamoto et al. (2009).

Box 9.4 An early protist biologist

The German polymath Ernest Haeckel (1834–1919) made beautiful drawings of many protists and other organisms, published in *Kunstformen der Natur* (*Art Forms in Nature*) in 1904. He used those drawings to describe new species of protists, most notably of Radiolaria. He also came up with many terms still used today in biology, including phylum, phylogeny, ecology, and most relevant here, protista. Although brilliant and creative, his legacy is marred by his writings on social Darwinism and scientific racism, which years later contributed to the rise of Nazism.

be solved by knowing the number of 18S rRNA genes per cell for each protist in a sample. However, that is often not possible because many protists have not been cultivated and/or sequenced enough to know their copy number. Fortunately, there is a correlation between copy number and cell size (Figure 9.16), so an abundant 18S rRNA gene sequence would indicate at least that the microbe contributes substantially to protist biomass. Given the huge variation in copy numbers, it is noteworthy that there is a correlation, albeit weak, between 18S rRNA sequences and direct counts of protists determined by DNA probing (Giner et al., 2016).

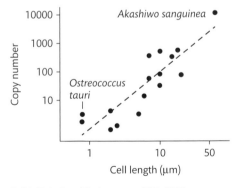

Figure 9.16 Relationship between 18S rRNA gene copy number and the cell size of protists. The smallest protist (*Ostreococcus tauri*) with the lowest number of 18S rRNA genes is an alga (a prasinophyte), while the largest protist with the most 18S rRNA genes is a dinoflagellate (*Akashiwo sanguinea*). The dashed line is from a regression analysis ($r^2 = 0.75$). Data from Zhu et al. (2005).

Biogeography of protists

Questions about the geographic distribution of protists are similar to those already discussed for bacteria and fungi. Protists appear to follow the classic trend with latitude as seen for higher organisms and oceanic bacteria (but not soil bacteria): they are more diverse in the tropics than in polar habitats (Dolan et al., 2016; Wu et al., 2017). Like bacteria, there is a rare biosphere for protists, and the small cell size and high abundance of protists should help ensure that their geographic distribution is not limited by dispersal. However, cultivation-independent evidence suggests that not all protists are cosmopolitan (Grossmann et al., 2016) and that the Baas Becking aphorism "everything is everywhere but the environment selects" does not completely apply to all protists.

The alternative to Bass Becking has been called the "moderate endemicity" hypothesis (van der Gast, 2015); the dispersal of some protists is limited, leading to endemicity, that is, taxa found in only some habitats. Perhaps protists are big enough and not abundant enough to facilitate limitless dispersal, or they may be less resilient to stress during transport from one hospitable habitat to another.

Regardless of the mechanism, protist communities in different habitats differ even at the phylum and supergroup level (Figure 9.17). Overall, the SAR supergroup is the most abundant in all of these biomes, but the abundance of SAR phyla differs among the three biomes. Of the SAR phyla, Cercozoa in the Rhizaria group is the most abundant phylum in soils but is much less abundant in marine and freshwaters. In the data sets used to make Figure 9.17, another SAR supergroup member, the algal class Chrysophyceae, is more abundant in freshwaters than in the oceans or soils. Dinophyceae, the dinoflagellates, are abundant alveolates in the oceans as discussed before. Among the other supergroups, amoebae (Amoeboza) are abundant in soils, but are nearly absent from aquatic habitats, as previously discussed. Two other supergroups, Archaeplastida and Opisthokonts, the latter represented in Figure 9.17 by the choanoflagellates (thought to be the protist ancestor of animals), also differ among the three major biomes, but their abundances are <10%. The final supergroup, Excavata, in the "other" category of Figure 9.17, accounts for about 3% of protists in freshwater lakes and <1% in the oceans and soils.

While there are many parallels between bacteria and protist biogeography, the compositions of the two microbial

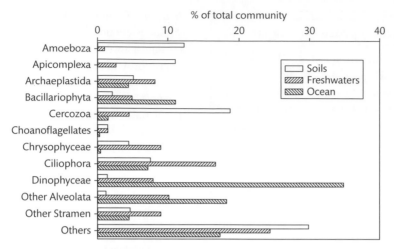

Figure 9.17 Composition of protist communities in lakes, soils, and a coastal ocean. "Other Alveolata" are those taxa not in the Cilophora, Dinophyceae, or Apicomplexa. "Other Stramen" refers to Stramenopiles not in the Bacillariophyta and Chrysophyceae. "Others" includes all taxa contributing <5% (most <1%) of the total. An important exception is Acantharia, which make up nearly 8% of the protist communities in the oceans. Data from Grossmann et al. (2016) and Massana et al. (2015).

communities are controlled by different factors. For example, pH is very important in structuring bacterial communities in soils, but it is less so for protists (Bates et al., 2013), although one study found pH to be important for parasitic protists (Dupont et al., 2016). Generally, soil moisture and other environmental properties are more important in shaping protist community structure (Stefan et al., 2014). In part because environmental factors affect these microbes differently, the alpha diversity of bacterial and protist communities are only weakly correlated (Bates et al., 2013). The lack of a tight coupling between bacterial and protist diversity may also reflect the low specificity of protist predation (many bacterial taxa have the same cell size and may be grazed on the same) and because protists may graze on other microbes. The link between bacterial and protist diversity is also weak because of photoautotrophic and mixotrophic protists that do not depend on bacteria.

Connecting protist communities with processes

We have already discussed the role of protists in grazing bacteria and other microbes in natural environments.

Many of these conclusions were deduced by clever experiments and by using morphological features to identify the protists to broad phylogenetic levels. However, the composition of protist communities can be linked to biogeochemical processes at finer phylogenetic levels by using approaches based on 18S rRNA analogous to those used for bacteria. For example, a version of fluorescence in situ hybridization (FISH) has been used to explore grazing by a mixotrophic protist (haptophytes) on bacteria (Unrein et al., 2014). Prey cells have been labeled with the stable isotopes ^{13}C or ^{15}N or both and used to follow carbon and nitrogen flow through protists in soils and grazing by mixotrophic haptophytes, stramenopiles, and dinoflagellates on cyanobacteria in the oceans (Crotty et al., 2013; Frias-Lopez et al., 2009). RNA sequence data suggest that parasitic and symbiotic protists may be more common than is currently appreciated (de Vargas et al., 2015; Dupont et al., 2016). More of this work could be used to explore whether protists with different 18S rRNA sequences but similar morphology have different ecological roles.

Summary

1. Flagellates and ciliates are abundant grazers in both aquatic and terrestrial habitats, and amoeboid microbes are also important in soils. In aquatic habitats, flagellates 1–5 µm in size are the major grazers of bacteria. They and amoebae are important bacterivores in soils. Nematodes and ciliates eat fungi in soils and sediments.

2. Protists usually feed by phagocytosis, which consists of three phases: encounter and cell–cell recognition of the prey by the protist; engulfment (phagocytosis) of the prey particle; and digestion of the prey in the food vacuole.

3. Grazing rates are affected by prey size, prey numbers, and chemical composition of the prey.

4. Many protists are mixotrophic, capable of both photoautotrophy and predation on other microbes. Some mixotrophic protists have their own chloroplasts, while others have kleptochloroplasts, which are chloroplasts taken from partially digested photoautotrophic prey; and still others have symbiotic photoautotrophs. Mixotrophic protists are abundant and contribute substantially to both predation and primary production in many habitats.

5. Protists can be identified to some extent by morphology and ultrastructure visible under light or electron microscopy, but the phylogenetic relationships among protists have been determined in more detail via sequencing of 18S rRNA genes. This work has found that the most abundant protists are in the SAR supergroup, especially the alveolates and stramenopiles in aquatic habitats and Cercozoa in soils.

6. There are many parallels between the biogeography of bacteria and protist biogeography, but there are differences. One is that some protists may be endemic to a region and are not as ubiquitously distributed as bacteria.

CHAPTER 10

The ecology of viruses

Bacteria, archaea, and protists are the most abundant organisms in the biosphere, but the most abundant biological entities are viruses. There are about 10^{31} of them on the planet, enough to extend past the nearest 60 galaxies in the universe, if lined up and strung end to end (Suttle, 2005). In spite of being inert particles outside their hosts, incapable of catalyzing a chemical reaction, viruses directly and indirectly affect many biogeochemical processes and all food webs.

That ecological view of viruses, however, is not appreciated by all biologists, evident from remarks by two Nobel laureates. Peter Medawar described a virus as "a piece of bad news wrapped up in a protein," while David Baltimore thought that "if they weren't here, we wouldn't miss them" (Ingraham, 2010). We wouldn't miss the diseases caused by viruses, but eventually we would see very different ecosystems if viruses were magically blotted out. This chapter will discuss how viruses play irreplaceable roles in the ecology and evolution of microbes and of all organisms. Arguably, life could not exist without viruses.

A defining characteristic of a virus is that it must infect a host in order to replicate. Because infection can be fatal to the host cell, viruses are a form of top-down control of microbial populations, as already mentioned in previous chapters. In this chapter, we learn more about viruses to understand this top-down control and to explore other ecological roles of viruses in soils and aquatic habitats. Probably all organisms on the planet are potentially infected by at least one virus, but the most common viruses in nature are thought to be those that infect bacteria, because bacteria are the most abundant form of cellular life. These viruses are called "bacteriophages" or simply "phages." This chapter will have lots about phages, but other viruses will be discussed as well.

What are viruses?

In some ways, viruses are very simple, consisting of only nucleic acids (Medawar's "bad news") surrounded by a protein coat, the capsid, and for some viruses by membranes and tails. The protein coat is needed to protect the viral nucleic acids from degradation by microbes, host defenses, and physical forces. Viruses come in several morphologies, ranging from simple geometric shapes to more complicated structures that resemble a lunar landing craft (Figure 10.1). Some of these shapes are determined by how capsid protein subunits are arranged to house the viral genome. Enveloped viruses, such as the human immunodeficiency virus (HIV), take

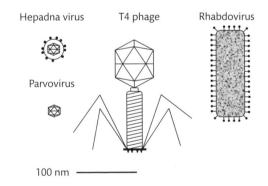

Figure 10.1 Some examples of viral shapes and sizes. Rhabdovirus is encased in a lipid envelope with surface glycoproteins. Some rhabdoviruses are shaped like a bullet.

Processes in Microbial Ecology. Second Edition. David L. Kirchman. Oxford University Press (2018). © David L. Kirchman 2018.
DOI 10.1093/oso/9780198789406.001.0001

Table 10.1 Some types of viruses, classified by nucleic acid. The genetic material can be double-stranded (ds) or single-stranded (ss). Single-stranded nucleic acids can be present in viruses as the positive strand (+) or the negative strand (−). Some of this genomic material is first transcribed to DNA by reverse transcriptase (RT). Genome size is measured in thousands of base pairs (kbp), or in thousands of bases (kb) in the case of single-stranded nucleic acid viruses. The examples given here are biased towards those causing disease in humans or economically important organisms. See Koonin et al. (2015) and the Universal Virus Database of the International Committee on Taxonomy of Viruses (http://www.ictvdb.org).

Type	Genetic material	Example family	Example	Genome size (kbp or kb)
I	dsDNA	*Myoviridae*	T4	39–169
II	(+)ssDNA	*Parvoviridae*	Aleutian mink disease	4–6
III	dsRNA	*Reoviridae*	Rotavirus A	19–32
IV	(+)ssRNA	*Picornaviridae*	Hepatitis C	7–8
V	(−)ssRNA	*Orthomyxoviridae*	Influenza A	12–15
VI	ssRNA-RT	*Retroviridae*	HIV	7–12
VII	dsDNA-RT	*Hepadnaviridae*	Hepatitis B	3–8

on a lipid membrane from their host. Some viruses of archaea have weird shapes, including those of spindles and bottles (Prangishvili, 2013). These structures and regular geometric shapes become important when using electron microscopy to distinguish viruses from detrital particles in samples from natural environments.

More dramatic and fundamental than differences in morphology are the differences in the viral genomic material. In stark contrast with prokaryotes and eukaryotes, viral genomes are not just double-stranded DNA (dsDNA), but occur in every possible variation of nucleic acid: single-stranded DNA (ssDNA), double-stranded RNA (dsRNA), or single-stranded RNA (ssRNA), as well as dsDNA (Table 10.1). A viral genome may be in a circular chromosome like prokaryotes or in a linear one like eukaryotes. Some of the RNA viruses have positive-sense RNA, which is essentially mRNA, and can use it to synthesize proteins immediately after entry into the host cell. The negative-sense RNA viruses must first convert their RNA to the positive sense using an RNA polymerase. Still other RNA viruses, the retroviruses, first make DNA using the viral enzyme reverse transcriptase, and the resulting cDNA is incorporated into the host chromosome. Most of the viruses examined in the laboratory are disease-causing RNA viruses; in contrast, we know the most about dsDNA viruses in nature. There have been few studies of RNA viruses in natural ecosystems.

The size of viruses varies greatly, in part due to the size of the viral genome (Table 10.1). The smallest virus, circovirus, has only two genes (<2000 nucleotides), and its capsid is only 20 nm in diameter. Most viruses need only a few genes because they rely on host genes for reproduction. Other viruses approach the size of a bacterial cell, with genomes bigger than that of some bacteria. Mimiviruses are nearly 1 μm in diameter with a 1.2 Mb genome (Colson et al., 2017). But most viruses in nature are small with capsids about 50 nm in diameter. As a general rule, viruses, even those in laboratory cultures, are ten-fold smaller than bacteria in size and have many fewer genes in smaller genomes.

The number of viruses in natural environments

How do we know that viruses are abundant enough to reach another galaxy if lined up properly? The methods for estimating viral abundance in nature are analogous to those for counting bacteria and other microbes. The first method to be discussed, the plaque assay, is not used much in microbial ecology today because it greatly underestimates viral abundance. Understanding why gives insights into some of the problems that complicate the study of viruses in natural environments.

Counting viruses by the plaque assay

This classic assay is used for counting phages, but it can be applied to viruses that attack any microbe (or cell) capable of growth on solid media such as agar (Figure 10.2). The sample containing the viruses is mixed with a culture of the host microbe, which without any viruses would form a dense, continuous "lawn" of host cells, and is

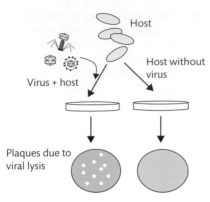

Figure 10.2 The plaque assay for counting and isolating viruses from a sample. Each plaque in the lawn (continuous layer of host cells) corresponds to where one virus infected a host, replicated, and lysed enough hosts to create a cell-free hole.

spread on an agar plate. The plate is then incubated for several hours to days. As the viruses replicate and lyse the host cells, holes or "plaques" in the lawn become visible. It is assumed that one plaque starts off as one virus infecting one cell in the lawn. As the virus replicates and infects more and more hosts, eventually enough cells in the developing lawn are lysed and the plaque becomes visible to the naked eye. So, the number of plaques on the bacterial lawn is proportional to the number of viruses in the original sample. In addition to counting viruses, the plaque assay can be used to isolate a virus from complex mixtures. The plaque can be subsampled and the viruses from it grown in liquid culture with the host cells.

The plaque assay greatly underestimates the total number of viruses actually in a natural sample. In one study of an estuary, for example, only 10 of 36 samples yielded detectable viruses, and the overall estimate of viral abundance was seven plaque-forming units (pfu) per liter (Wommack and Colwell, 2000). In fact, there were about 10^{10} viruses per liter in those samples. The reason for the severe underestimation by the plaque assay is the inability to grow the right host cell on solid media, another consequence of the culturability problem encountered in Chapter 1. As mentioned several times already in this book, nearly all of the bacteria and most of the other microbes known to be present in nature cannot be grown on solid media. Non-traditional cultivation approaches that are successful in growing

these microbes in the laboratory are not easily modified to include the plaque assay or anything similar to it. Another problem is the diversity in hosts. Even if they could be grown on an agar plate, the large number of possible hosts means it is impractical to include enough for an effective plaque assay. Consequently, the assay misses the many viruses that infect uncultivated microbes in nature.

The inadequacy of the plaque assay approach has a huge consequence for examining viruses in nature. It means that most viruses in nature cannot be isolated, identified, and studied in the laboratory by traditional methods, as is the case for nearly all microbes found in nature. If the host cannot be grown in the laboratory, then the virus infecting it cannot be isolated and identified by traditional methods.

Counting viruses by microscopy

The plaque assay indicated that viruses were present in nature but in very low numbers. It was not until transmission electron microscopy (TEM) was used to examine samples from coastal marine habitats that the high abundance of viruses was discovered (see Box 10.1). The first step in the TEM method is to spin down viruses by centrifugation onto a small grid placed at the bottom of the centrifuge tube. After centrifugation, the small grid is

---

Box 10.1 Discovery before its time

Bacteria, algae, and other protists have been known to be important in natural environments since the late nineteenth century, but the ecological importance of viruses was not appreciated until about 1990. Even a study published in 1979 reporting on 10,000 viruses per ml of coastal seawater did not attract much attention (Torrella and Morita, 1979). Although based on TEM, the 1979 estimate was about 1000-fold too low because the investigators examined only >0.2 μm viruses retained by a filter. Finally, ten years later, studies using the ultracentrifugation–TEM approach found not only high viral abundance in seawater, but also evidence of high mortality caused by viral lysis (Bergh et al., 1989; Proctor and Fuhrman, 1990).

taken out, prepared, and viewed by TEM. The TEM pictures reveal many amorphous particles of unknown origin but also particles with shapes and sizes identical to known viruses (Figure 10.3A), including classic ones like the T4 virus. If it is assumed that all viruses in the volume of water above the grid are collected onto the TEM grid, the number of viruses in the original sample can be estimated by counting the viruses in a known area of the grid. Microbes infected by viruses can be seen with or without thin sectioning of the microbes and examination by TEM. While quite powerful, the TEM approach has its problems. It requires an expensive instrument which requires training and skill to operate.

An alternative, now the most common approach, is to enumerate viruses by epifluorescence microscopy. The approach is nearly the same as that used to count bacteria and other microbes, with one critical difference: the sample is stained with a very bright nucleic acid stain such as SYBR Gold. When viewed by epifluorescence microscopy, viruses stained with SYBR Green I look like small pinpoints of green light, while bacteria and other microbes appear huge in comparison (Figure 10.3B). The smallest particles are viruses. Remarkably, estimates of viruses by this approach are similar to those by the TEM method. Still another method for enumerating viruses uses flow cytometry. When used properly, flow cytometry yields similar estimates of viral abundance to epifluorescence microscopy, but with higher precision and greater ease and speed in processing many samples.

(A)

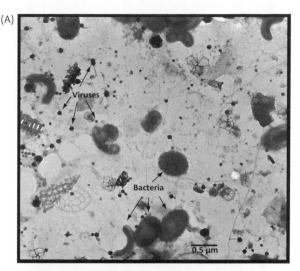

(B)

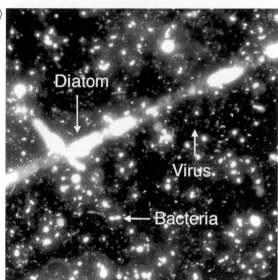

Figure 10.3 Examples of viruses in nature, as revealed by transmission electron microscopy (A) and epifluorescence microscopy (B). Panel A is used with permission from K. E. Wommack, and Panel B is used with permission from M. T. Cottrell.

Virus–bacteria ratio in nature

Because of studies using TEM and microscopic direct count approaches, we can now make intergalactic analogies about viral abundance in nature. These studies have revealed very high numbers of free viruses outside of hosts in virtually all habitats of the biosphere, ranging from about 10^7 per milliliter in surface waters of aquatic habitats to 10^{10} per gram of soils and sediments, in stark contrast with the plaque assay results. Viruses are found everywhere microbes live, usually in much higher abundance than cellular life. An informative way to express viral abundance is relative to bacterial abundance, that is, the virus to bacterium ratio (VBR). The justification for using the ratio is the assumption that bacteria are

probably hosts for nearly all of the viruses counted by microscopy.

The VBR was initially thought to be about 10 because that is the average found by the initial studies on the topic (Figure 10.4). That value has been observed often elsewhere and probably isn't a bad initial estimate for an unstudied habitat. However, VBR clearly varies greatly

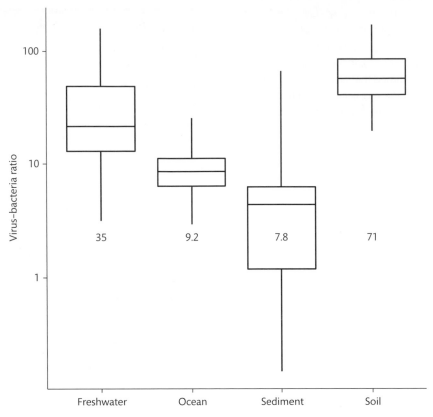

Figure 10.4 The virus to bacteria ratio (VBR) for four environments. The numbers are the mean VBRs for the four environments. Data from Knowles et al. (2016).

among habitats. The lowest reported VBR value of 0.001 was found 200 m below the surface of marine sediments, whereas soils currently hold the record for the highest VBR, of 8200 (Knowles et al., 2016). As elsewhere, most of the viruses counted in soils probably attack bacteria. Although fungi are often quite abundant in terrestrial habitats, fungal viruses or "mycoviruses" would not be counted by standard methods because of the nature of the mycoviral genomic material and the transmission mechanism of mycoviruses from one fungus to another, as will be discussed in more detail.

More insights into virus–host relationships are gained by looking at plots of viral abundance versus bacterial abundance. This plot shows that viral abundance increases with bacterial abundance at a slope equivalent to a VBR of 10; but then viral abundance is lower than values predicted by the VBR = 10 relationship when bacterial abundance is very high (Figure 10.5). These

lower than expected viral abundances have been observed even in the oceans where the VBR = 10 relationship was first seen (Wigington et al., 2016). The relatively low viral abundances have been hypothesized to result from more viruses staying in and replicating with the host genome (the lysogenic phase) where they would not be counted by standard approaches. Why VBR varies so much is not well understood (Parikka et al., 2017).

Viral replication

Viruses are obligate parasites that must invade a host cell and take over its biochemical machinery to replicate more viruses. Understanding the basics of viral replication is essential for understanding viral biology and the dynamics of viral and microbial communities. Viruses like phages have two basic strategies for replication. One

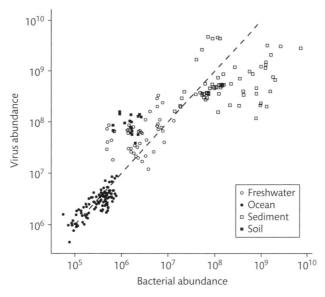

Figure 10.5 Viral abundance versus bacterial abundance in four environments. The units are viruses or bacterial abundance per ml for freshwater and the oceans, or per g for sediments and soils. The dashed line indicates a virus to bacterial abundance ratio of 10. Data from Knowles et al. (2016).

strategy is taken by "virulent viruses" which go through only the "lytic phase." During this phase, the virus immediately begins the process of viral replication inside the host cell after infection. Bacteriophages inject only their genetic material into the host, whereas other viruses, such as those infecting eukaryotes, enter the host cell as a complete virus, including the capsid. For some amount of time, the viral particle is not visible and the host will not appear to be infected. This period is the "latent phase." The virus then directs the host cell to make more viral genomic material and structural proteins. Although the virus genome may encode for some of the enzymes necessary for viral replication, the virus also takes over key genes and enzymes of the host for its own needs. When sufficient genomic material and protein coats are ready, the viral genomic material is packaged with the protein coating, resulting in many fully formed viruses inside the host cell. Now the host cell will appear to be infected if viewed by electron microscopy. The viruses then break out of and thus break up the host cell (lysis), expelling many viruses into the environment. The number of viruses manufactured and released by a single host cell, which is called the "burst size," varies from <50

for oligotrophic bacterial hosts (10 to 40 for *Pelagibacter ubique*) to about 500 for copiotrophic bacteria (Chow and Suttle, 2015).

The "temperate" viruses use a different strategy and have another phase, "lysogeny," before the lytic. During lysogeny, the viral genomic material either integrates into the host genome or forms a self-replicating plasmid. Either way, the viral genome is replicated along with the host genome for an indeterminate time, during the latent phase (Figure 10.6). A microbe in this state is said to be "lysogenic" or is a "lysogen." Each time the lysogenic microbe divides, each daughter cell includes a copy of the viral genome. When the host is a bacterium, the integrated viral genome is called a "prophage." There are prophage analogs (proviruses) in eukaryotic hosts.

At some point, given the right environmental cues, the prophage switches to the lytic phase and starts to make new virus particles. The switch from the lysogenic phase to the lytic phase has been examined in great detail in the laboratory, because it was an early model for the regulation of gene expression. In any case, the end result is the same as for virulent phages: the host is lysed, releasing new viruses into the environment.

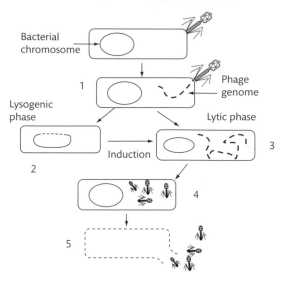

Figure 10.6 Lysogenic and lytic phases of a temperate phase. Step 1 consists of the injection of phage nucleic acid into the bacterium. In the lysogenic phase (Step 2), the phage nucleic acid is incorporated into the bacterial chromosome, as shown here, or forms a self-replicating plasmid (not shown). The incorporated phage DNA, the prophage, is indicated by a dashed line. During induction, the prophage is excised from the bacterial chromosome and replicates itself (Step 3). During Step 4, the phage nucleic acid is packaged into phage heads. Once packaging is complete, the bacterial cell is lysed (Step 5).

The mechanisms by which virulent and temperate viruses replicate are extremes among several other possibilities. Some viruses do not lyse, or otherwise kill their host when ready to re-enter the environment. Rather, the host cell releases these viruses by extruding them through the membrane, or viruses are encapsulated in host membranes and are budded off away from the host cell into the environment. This process may occur over several generations, with the host in the state of chronic infection. In these cases, the host cell is being parasitized by the virus but is not killed by it. Chronic infection in natural microbial communities has not been examined extensively (Howard-Varona et al., 2017).

Temperate viruses and lysogeny in nature

The prevalence of temperate viruses in nature has implications for relationships between viruses and their hosts. Virulent viruses kill the host soon after infection, while that is not necessarily the case for temperate viruses. Consequently, temperate viruses may have a smaller immediate impact than virulent viruses, but a larger effect in shaping host populations over the long term. The number of lysogenic bacteria in a sample from a natural environment is estimated by counting viruses before and after the addition of mitomycin C or by exposure to UV light, or both to induce the switch from the lysogenic to the lytic phase (Figure 10.7). An increase in viruses after the induction treatment indicates the presence of inducible prophages within the microbial community and is a proxy for the number of temperate viruses in that habitat.

Temperate viruses are probably very abundant in natural environments if the prevalence of prophages in cultivated bacteria is any guide; about half of all sequenced genomes have phage genes (Howard-Varona et al., 2017). In theory, the abundance and distribution of temperate viruses could be followed by examining characteristic marker genes, but that approach has not been used extensively to date, perhaps because these marker genes are too specific and the genetic mechanisms for the temperate lifestyle are too diverse. The chemical induction assay for lysogeny has turned up lower percentages for the number of temperate viruses in nature than suggested by the number of prophages in genomes. For example, lysogeny was only observed in less than half the samples taken during a two-year study in the Mediterranean Sea (Boras et al., 2009) and in only 20% of the samples from a year-long study in the Gulf of Mexico (Williamson et al., 2002). The percentages may be higher in soils. One study found a very high fraction (85%) of lysogenized bacteria and temperate viruses (Figure 10.7). Another study found high fractions in temperate soils (22–68%), although lower (4–20%) in Antarctic soils (Williamson et al., 2007). There has been much speculation about how lysogeny varies in environments with various levels of abundance and production (Howard-Varona et al., 2017; Knowles et al., 2017). More work is needed with direct assays of lysogeny.

Host range of viruses

After learning that nearly all natural environments have millions and millions of viruses, it may seem impossible to swim in a lake or thrust a hand into soil without being attacked by viruses. The reason why we are not attacked

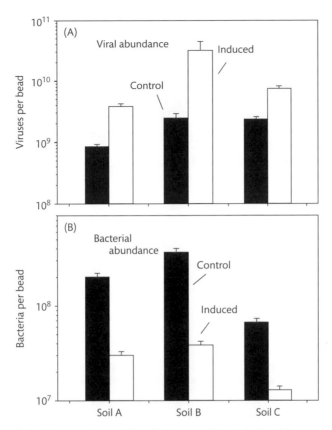

Figure 10.7 An example of an induction experiment with soil viruses and bacteria. Panel A presents the number of viruses in the control (no treatment) and in incubations with mitomycin C, which induces temperate viruses in lysogenized bacteria to switch to the lytic phase, leading to an increase in total viral abundance. The number of bacteria (panel B) decreased at the same time because of viral lysis. The unusual "per bead" units reflect how the viruses and bacteria were isolated from these soil samples. Data from Ghosh et al. (2008).

by viruses in lakes or soils explains why the plaque assay for viruses greatly underestimates viral abundance. The assay counts only those few viruses able to infect the microbe (usually a bacterium) used to make the lawn, because viruses have a limited number of hosts they can infect.

The host range for viruses is usually limited because the virus infects its host by highly specific mechanisms. First, the virus needs to bump into and recognize a specific component, a receptor, in the host outer membrane. Successfully interacting with this receptor initiates attachment by the virus to the host, followed eventually by invasion of the virus or injection of the viral genome into the host. The receptors are not made by the host to encourage viral attack. Rather, these membrane components have some other function of importance to the host. A classic example is a protein encoded by the gene *lamB* for maltose transport by *E. coli*. It is at this membrane protein that the lambda phage attaches to *E. coli*. These receptors on the host surface are often proteins, but they can be the carbohydrate part of glycoproteins or glycolipids.

Another example is receptors used by bacteria to transport iron, an element often occurring in concentrations too low to support high microbial growth. As mentioned in previous chapters, iron at near-neutral pH in oxic environments is insoluble and is often associated with organic chelators. One such class of chelators could be the tails of T4-like viruses (Figure 10.8). These iron-rich tails could be used by the virus to gain entry into a bacterium by interacting with iron-binding receptors. The bacterium does get the iron, but also a deadly virus,

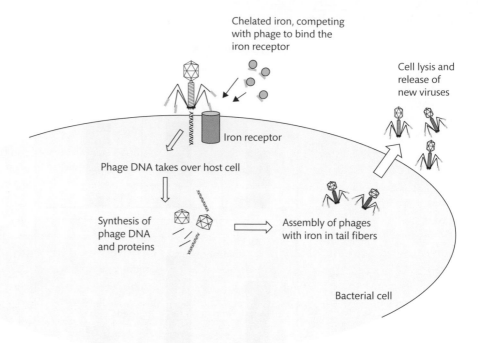

Figure 10.8 A hypothesis about the use of iron receptors by viruses to infect bacteria requiring iron, an example of the specific recognition required by a virus to infect a host. Based on a figure in Bonnain et al. (2016).

a combination that led the authors to propose the name "Ferrojan Horse Hypothesis" for this potential virus–bacterium interaction (Bonnain et al., 2016).

Because of the specific molecular interactions between viruses and hosts, a virus is thought to attack only one type of host, whereas a host is potentially attacked by several types of virus, each targeting a different receptor molecule in the host membrane. It is because of this specificity that we can swim in a lake or weed a garden without any worries of aquatic or terrestrial viruses attacking us. Hosts attacked by the same virus generally belong to the same species or are even more closely related. This specialization, supported by metagenomic analyses (Paez-Espino et al., 2016), is crucial in thinking about how viruses potentially control microbial communities. However, there are examples of viruses attacking more than one host. Metagenomic analyses of viral sequences occurring within bacterial contigs suggest that some viruses may be able to attack bacteria from different phyla (Paez-Espino et al., 2016). Although most viruses in nature are probably specialized for one host, the few viruses capable of attacking several hosts, said to have a

"broad host range," may have an outsized role in virus-mediated processes, such as horizontal gene transfer between distantly related microbes.

Viral production, loss, and mortality of bacteria

One of the main roles of viruses in nature is similar to that of grazers. Like heterotrophic protists, viruses help to control the biomass levels of their hosts. Any increase in the number of a particular bacterial taxon increases the chances of colliding with a virus and thus increases viral lysis, leading to the reduction in that particular taxon. But how important is this control mechanism? How much of total microbial mortality is due to viral lysis versus grazing?

The high number of viruses suggests that viral lysis accounts for a large fraction of bacterial mortality, but that is not necessarily the case. A free virus may be non-infectious and incapable of attacking a host cell. Even if all free viruses are infectious, a rate cannot be estimated from a standing stock measurement (here, the number of viruses) without many assumptions. So, we need a more direct method. There are at least six such methods. Two are discussed here as a way to learn more about viral ecology.

Percentage of infected cells

One of the simplest and most direct methods to estimate viral-mediated bacterial mortality is to count the number of bacterial cells that are infected by viruses (Figure 10.9). Because an infected cell is doomed to eventual death by lysis, the fraction of the bacterial community that is infected is the fraction of bacterial growth being controlled by viruses. The fraction of infected cells, visible by TEM, is low, about 1–5% in nature (Proctor and Fuhrman, 1990), which would seem to imply an equally small effect on bacteria. However, the observed fraction does not take into account infected cells without visible viral particles in the latent phase. Cells could have viruses that have not yet reached the later stages of replication when the intracellular structures of viruses are recognizable. The final stage of viral infection when viral particles are visible takes up only about 10–20% of the entire viral life cycle. So, the fraction of infected cells has to be increased by the 10–20% factor to estimate the full impact of viruses on their host. When thus corrected, the estimate of 1–5% infected cells implies that 5–50% of bacterial mortality is due to viruses.

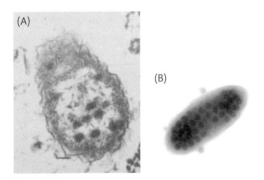

Figure 10.9 Bacteria infected by viruses as viewed by transmission electron microscopy. The relative number of visibly infected cells has been used to estimate mortality caused by viruses. Infected bacterium from the eastern Caribbean Sea (A). Photomicrograph from Proctor and Fuhrman (1990), used with permission of the publisher. Infected bacterium from the Mediterranean Sea (B). Photomicrograph courtesy of Markus Weinbauer, used with permission.

The viral reduction method

As implied by the name, in this approach the number of free viruses is reduced (ideally, to zero) by filtering water through filters with pore sizes small enough to remove viruses while retaining bacteria and other microbes. (As with many methods in microbial ecology, this one does not work with soils or sediments.) Some samples are also treated with mitomycin C to induce the lytic phase of temperate phages in lysogenized bacteria. The very low number of viruses left in the sample ensures that no viruses are lost due to adsorption and no new viruses are produced by new attacks on bacteria. Total viral abundance should increase over time, as both virulent and temperate phages (in the presence of mitomycin C) are released from hosts infected before the experiment began. An example of this method is given in Figure 10.10. Estimating viral mortality with the viral reduction method requires estimates of the burst size; one assumption is 50 viruses released from one host cell. Although far from perfect, this method is one of the best for estimating rates of viral mortality.

Contribution of viruses versus grazers to bacterial mortality

Whether a microbe is killed off by a virus or a grazer has several important implications for food web dynamics and biogeochemical processes. Mortality by viruses, for example, could lower the amount of carbon and energy available for transfer by grazing to higher trophic levels. Other ways in which the impact of viruses and grazers differ are discussed later. We assume here that viral lysis and grazing are the only two forms of mortality for microbes, but another mechanism, called "programmed cell death" or "apoptosis" may be relevant for some microbes. Even this mechanism can involve viruses (Bidle, 2015).

The results from the two methods for measuring virus-induced bacterial mortality just discussed, plus those from all of the other methods, indicate that viruses can account for a large fraction of bacterial mortality. Roughly half of all bacterial mortality can be attributed to viruses, the other half to grazing by various protists, but these percentages vary greatly over time and space. The relatively few habitats examined so far are nearly all marine. Even among these environments, the fractions attributed to viruses and grazers vary greatly.

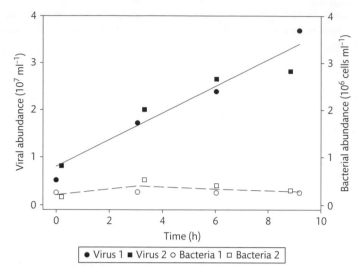

Figure 10.10 Example of data from the viral reduction method for estimating rates of viral production. Viral and bacterial abundance are reduced by filtration and dilution with virus-free water. The increase in viral abundance is due to their release from lysogenized bacteria and from virulent virus-infected bacteria. Bacterial abundance does not change on this timescale. Results from two separate incubations are given. Data from Wilhelm et al. (2002).

Viral lysis is probably more important as a top-down controlling agent in some environments than in others, but which environments is not entirely clear. One hypothesis is that viruses are more important in eutrophic habitats where host abundances are high. Grazing accounted for all of the measured bacterial production in oligotrophic marine environments, but not in eutrophic ones, according to one analysis of several studies published at the time (Strom, 2000), implying that viruses accounted for the mortality not due to grazing. Since that analysis was done, other have studies found that viruses were responsible for about half of all bacterial mortality in the oligotrophic Mediterranean Sea (Figure 10.11) and North Atlantic Ocean (Boras et al., 2010). Still, it makes sense that the impact of viruses would be higher in environments where high nutrients promote more cell production and higher biomass, leading to more contact between viruses and hosts and to larger burst sizes.

Viruses also probably contribute more to bacterial mortality in habitats where protists do not grow well. Examples include habitats with low pH or high temperature, two factors that select against protists and other eukaryotes. Viruses are likely to be the major form of mortality of prokaryotes in anoxic habitats where the lack of oxygen excludes nearly all eukaryotes, although protist grazing can still be substantial in low-oxygen waters (Medina et al., 2017). Some of the highest levels of viruses and rates of viral lysis have been found in the anoxic hypolimnion of a lake (Weinbauer and Höfle, 1998). Protists have not been observed in deep subsurface sediments (Chapter 9), suggesting that viruses are the main agent of mortality in that vast habitat. These hypotheses about the importance of viruses in extreme environments need to be tested by more studies.

Viral lysis and grazing are often viewed as being unrelated processes, but they are potentially linked by the chemical characteristics of the microbial outer surface. The connection probably explains why grazing is higher on the alga *Emiliania huxleyi* when it is being lysed by a virus (Breitbart, 2012). In contrast, changes in the outer surface may enhance one mortality process but not the other. A mutant of the cyanobacterium *Synechococcus* with a modified lipopolysaccharide layer was more resistant to viruses but more susceptible to grazing by heterotrophic nanoflagellates. Viruses and grazers of microbes other than these two photoautotrophs are also likely linked.

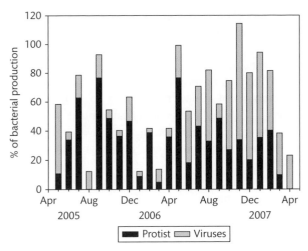

Figure 10.11 Mortality of bacteria due to viral lysis or protist grazing, expressed as percentage of bacterial production, in the Mediterranean Sea. The two mortality processes should add up to 100%. In some cases, the sum is not statistically different from 100%, given the errors in both measurements. If the sum is normalized to 100% (assuming morality is only due to lysis and grazing), then the average percentage of mortality due to protist grazing is 54%, while 45% is due to viral lysis. Data from Boras et al. (2009).

Viral decay and loss

Microbes and other organisms are infected and eventually lysed by viruses, releasing millions of new viruses into the environment every day. Consequently, even though there are already many viruses in most environments, there would be even more if there were no mechanism to remove them. Fortunately, viruses are removed or lose the capacity to infect organisms by several possible mechanisms.

Perhaps the most important mechanism in aquatic systems is sunlight, specifically ultraviolet (UV) light, which damages viral genomic material beyond repair (Suttle and Chen, 1992; Breitbart, 2012). In soils and sediments, a major loss mechanism is the inactivation of viruses by adsorption to colloids and other particles. Drying of soils also leads to inactivation of viruses, according to cultivation-dependent assays, but temperature is often the property that explains most of the variation in cultivation-dependent assays of virus infectivity in soils (Kimura et al., 2008); infectivity decreases with higher

temperature. Heterotrophic bacteria may be able to degrade viruses, treating them as just another nutrient-rich particle, and protists may graze on large viruses. Several studies have examined the loss and inactivation of pathogenic viruses, but more work is needed on these processes for viruses infecting natural microbes and other organisms in natural environments.

Viruses are not grazers

One ecological role of viruses, that of killing off hosts and effecting a form of top-down control, is similar to that of grazers. But this similarity between viruses and grazers is rather superficial; the ecological roles of viruses differ from those of grazers in many ways. As mentioned previously, virus-induced mortality may lower the transfer of carbon and energy to higher trophic levels, whereas grazing-induced mortality enhances the transfer. Viruses and grazers differ in several other ways.

Viral shunt and DOM production

Grazing and viral lysis both kill prey and host cells, but what physically remains after the two processes are completed differs greatly. A grazer can completely consume its prey, oxidize the organic carbon to carbon dioxide, and mineralize the organic nitrogen and phosphorus back to ammonium and phosphate. In contrast, viral lysis releases dissolved and particulate organic material because of the dependence of the virus on a functioning host. A virus needs the host cell to remain viable up to the end of the infection cycle, throughout the synthesis of viral components and assembly of new viral particles. Although hijacked by the virus and transformed to suit the purpose of the virus, the biochemical machinery of the host cell is not destroyed. Consequently, lysis by the virus releases the host's cellular contents into the environment with little oxidation or mineralization. In soils, the released cellular contents may adsorb onto surfaces, whereas in aquatic habitats they may become part of the dissolved organic material (DOM) pool. In all environments, cellular debris from lysis may form particulate organic detritus.

The production of DOM by viral lysis and its subsequent use by microbes is called the viral shunt (Figure 10.12).

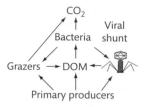

Figure 10.12 The viral shunt and the rest of the microbial food web. This diagram implies that only bacteria and phytoplankton are attacked by viruses, but in fact, viruses infect all organisms, potentially releasing dissolved organic material (DOM) and other detrital organic carbon.

Most of the organic compounds released by viral lysis are thought to be labile and readily used by microbes. Lysis of algae or higher plant cells would make organic material available for bacteria and fungi that otherwise would have gone to a herbivore. Viral lysis may also release material containing potentially limiting elements, like phosphorus and iron. Some evidence suggests that average bacterial growth is faster in habitats with high viral lysis because uninfected bacteria feed off the DOM produced by viral lysis. DOM from viral lysis is potentially a large part of total DOM production (Lønborg et al., 2013).

Population dynamics of a virus and its host
Another similarity between viruses and grazers is that in both systems, one host or prey oscillates out of phase with the virus or predator. Phage–bacteria systems have even served as models for exploring theoretical questions about predator–prey interactions (Kerr et al., 2008). However, there are crucial differences between virus–host interactions and predator–prey interactions.

One difference is that a host can evolve defenses against a virus more easily than prey can fend off predators. A single mutation can make the host impervious to viral attack. In laboratory experiments with a single bacterium and a single phage, a spontaneous mutant of the bacterium often arises that is resistant against the phage. Among several mechanisms for resistance, one is simply not to synthesize the protein or other membrane components used by the virus to recognize the host. Even if the virus gets past the first line of defense by the host, the viral genome may be degraded by bacterial

enzymes ("restrictionases") and inactivated. So, a virus cannot completely kill every microbe in a population. Indeed, co-existence has been demonstrated for a virulent virus infecting an algal species (Thyrhaug et al., 2003) as well as for phages that infect bacteria. Some microbes can evolve mechanisms to avoid grazing, such as forming large chains or clumps that are too big for the grazer to eat. But there are fewer anti-grazing mechanisms than the anti-viral defenses potentially used by bacteria.

Why then are not all microbes resistant against viruses? The answer has two parts.

First, there is a cost associated with resisting viruses. Elimination or even just modification of the virus cell-surface receptor could affect how the host interacts with its environment. For example, an *E. coli* strain without the maltose-transport protein cannot be attacked by the lambda phage, but it also would no longer be able to take up maltose. The cost of the antiviral defense may not always be so obvious; the phage-resistant *E. coli* strain may still grow more slowly than the parent strain even if the main carbon source is not maltose. The cost depends on the type of mutation, the metabolic capacity of the host, and the environment. In any case, because of these costs, virus-resistant microbes often grow more slowly than the original, virus-sensitive strain and do not necessarily become the dominant members of the community. Simple laboratory experiments have shown that virus-resistant and virus-sensitive microbes can coexist (Figure 10.13), with the former being regulated by bottom-up control (organic carbon) and the latter by top-down control (viral lysis).

The other part of the answer is that viruses evolve in response to the microbial host. The complete loss of a host receptor would be difficult to overcome, but more subtle changes in the host receptor would select for mutant viruses able to recognize the mutated receptor. In addition to selection due to virus–host receptor interactions, there may be selection for mutations in the latency period or the burst size. Mutations in the virus would select for more mutations in its microbial host, and so on and so on. The end result is an evolutionary arms race between viruses and their microbial hosts (Schwartz and Lindell, 2017), neither side gaining an upper hand and winning.

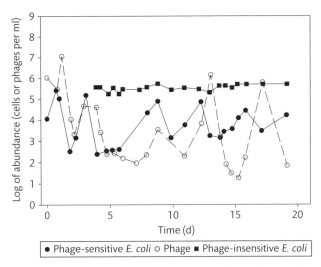

Figure 10.13 Coexistence of phage-sensitive and phage-insensitive mutants of *E. coli*, along with the phage in an organic carbon-limited continuous culture. The phage-insensitive mutant does not take over because it is a weak competitor for organic carbon. Note the predator–prey oscillations between the phage and the phage-sensitive *E. coli* strain. Data from Bohannan and Lenski (1999).

Genetic exchange mediated by viruses

So far, it seems that viruses have only negative impacts on host populations, and it is certainly true that many microbes are killed off daily by viruses. However, viruses also have other potential impacts with many far-reaching implications for the ecology and evolution of microbes and higher organisms. Viruses effect a form of sex for microbes. In more scientific terms, viruses mediate the exchange of genetic material among microbes. We will focus on bacteria here because they have been examined the most extensively.

Bacteria normally reproduce by asexual cell division, meaning that the same genetic material is passed from mother to daughter without the mixing of genes from another cell. However, bacteria can take on genes from other cells (sex) via three mechanisms. The first, "transformation," is the uptake of free DNA from the surrounding environment. The second, "conjugation," involves exchange of DNA from one cell to another via a proteinaceous tube (pilus) connecting the two cells. The third mechanism, "transduction," involves viruses. It is not clear which mechanism is most common or important in nature, but viral infection of microbes is quite common, as should now be clear, and viral genes are abundant in the genomes of microbes and larger

organisms. Genetic exchange mediated by viruses is thought to be very high in nature.

Viruses can mediate the exchange of microbial genes because of "mistakes" during the packaging of the viral genetic material into viral particles (Figure 10.14). While the viral genetic material is being loaded into the capsid, genes from the microbial host genome can be included as well. The host genes are either randomly selected from the entire genome (generalized transduction) or the genes may be specific because the virus inserts itself into specific sites within the host chromosome (specialized transduction). In either case, the newly formed virus will now carry those host genes into a new host after infection. The newly infected host could express the virus-borne host genes and potentially gain a new metabolic capacity, a new version of metabolism it already had, or simply more of the same metabolism.

Well-studied cases of bacteria acquiring new metabolic pathways involve the conversion of a nonpathogenic strain of a bacterium to a pathogen. One example is the conversion of *Vibrio cholera* from an innocuous estuarine bacterium to a cholera-causing pathogen, due to infection by the filamentous phage CTXphi. The phage carries a "pathogenicity island" consisting of the cholera toxin and other genes, including those for pili that facilitate

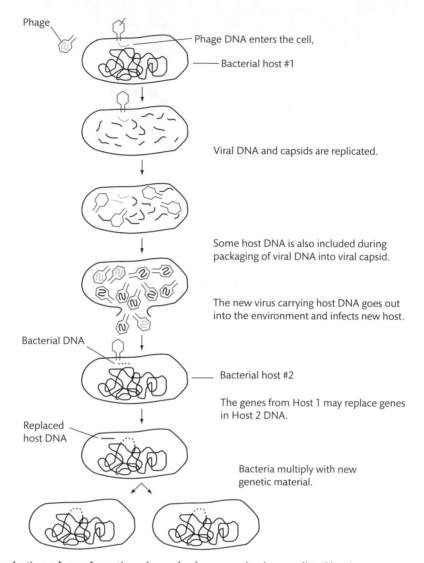

Phage

Phage DNA enters the cell,

Bacterial host #1

Viral DNA and capsids are replicated.

Some host DNA is also included during packaging of viral DNA into viral capsid.

The new virus carrying host DNA goes out into the environment and infects new host.

Bacterial DNA

Bacterial host #2

The genes from Host 1 may replace genes in Host 2 DNA.

Replaced host DNA

Bacteria multiply with new genetic material.

Figure 10.14 Transduction, a form of genetic exchange (sex) among microbes, mediated by viruses.

attachment. Infection by the viruses promotes survival of this bacterium by enhancing attachment to chitin and gut cells and by increasing the capacity of the bacterium to embed itself in biofilms (Faruque et al., 2006).

Viral genetic material, along with genes from prior hosts, can become permanent fixtures in the host genome. At least some of the genes resulting from horizontal gene transfer (Chapter 5) get into the genome by transduction. In fact, the genomes of bacteria, other microbes, and indeed all organisms, including humans, are littered with the remains of viruses. Most sequenced bacterial genomes have at least one prophage or parts of one, as mentioned before (Howard-Varona et al., 2017), and viral genes have also been seen in bacterial metagenomes. Over a third of the single-amplified genomes from bacteria and archaea had virus genes, in one study (Labonte et al., 2015). Viruses also have been detected in single-amplified genomes from protists, but those viral sequences have been interpreted as coming from large viruses that had been eaten by the protist (Yoon et al., 2011).

Viral and other genes brought in microbial hosts by transduction serve both the virus and the host. Being replicated along with host genetic material is one mechanism for a viral gene to reach its selfish goal: to make more copies of itself. That goal can be attained by the gene inhabiting a free viral particle or by being tucked away among the many genes of a host genome. The microbe also potentially wins. It gains new genetic material—genes that may prove advantageous as the microbe is challenged by changing environmental conditions.

Metagenomics of viruses

While much can be learned about viruses by counting them and measuring lysis rates, many questions about their ecology and importance in natural environments can be addressed only with information about the diversity of viruses found in a particular habitat. Similar to microbes, we can only learn so much from virus morphology or from cultivation-dependent approaches. As pointed out before, because few microbial hosts abundant in natural environments can be cultivated in the laboratory by conventional approaches, their viruses cannot be isolated and examined by cultivation-dependent approaches.

Another obstacle is that viruses have nothing analogous to rRNA genes (Chapter 4). There is no marker gene present in all viruses that could be used to identify them, although some studies have examined specific genes conserved within a few viral families, such as the *g*20 and *g*23 genes in T4-like myoviruses and the *RdRp* gene in picorna-like viruses (Brum and Sullivan, 2015). Most of the characterized viruses infecting prokaryotes are in the order *Caudovirales*, with dsDNA and tails (Krupovic et al., 2011).

Early cultivation-independent approaches used DNA fingerprinting methods to explore viral diversity, but viral ecologists quickly switched to metagenomics as the limitations of the fingerprinting methods became obvious and as sequencing became easier and cheaper (Chapter 5). In the most common version of viral metagenomics, most of cellular life is filtered out with a 0.22 μm pore-sized filter, the viruses in the 0.22 μm filtrate are concentrated, and the DNA is purified and cloned or directly sequenced (Figure 10.15). Soil and sediment samples can be suspended in water and then treated like an aqueous sample. Genomic information about viruses can also be gained by looking at infected hosts isolated by flow cytometry or microfluidic devices. Some clever

approaches use fluorescent stains and epifluorescence microscopy or flow cytometry to explore specific virus–host interactions (Brum and Sullivan, 2015).

Viral diversity

One of the most striking observations from early viral metagenomic studies was the high number of unknown genes and unknown viruses in natural environments. One study collated metagenomic results from many environments and found that about 75% of protein-encoding viral genes were not similar to sequences from previously examined viruses mainly infecting cultivated microbes or large organisms (Paez-Espino et al., 2016). Another study of several oceanic regions found that only two of 38 viral clusters were similar to well-known viruses, ten were known only from previous metagenomic work, and another 18 clusters were completely new to science (Roux et al., 2016). Metagenomic studies of microbes also turn up a large number of unknown organisms and genes, but the number of unknowns is even greater for viruses. Part of the problem is the limited database used to analyze viral sequences. The number of sequenced genomes for dsDNA viruses and retroviruses is about 5000, or ten-fold less than the nearly 50,000 bacterial genomes that have been sequenced (Paez-Espino et al., 2016).

While the functions of many genes remain unknown, viral ecologists can still use metagenomic data to trace viral populations and to test hypotheses about their distributions in the environment (Brum et al., 2015). Key to this work was the development of bioinformatic approaches to remove non-viral genes from the metagenomic data so that the analyses focus only on viral genomes. The non-viral genes come from microbial DNA passing through filters intended to remove all cells (Figure 10.15).

Core, noncore, and auxiliary metabolic genes

Genomic and metagenomic studies of viruses have found sets of genes analogous to the core and pangenome of microbes. As for microbes, the core genes are found in all members of a viral group, implying they are essential for viral construction and replication (Krupovic et al., 2011). Examples include genes for capsid proteins and DNA metabolism. For microbes, the other genes not

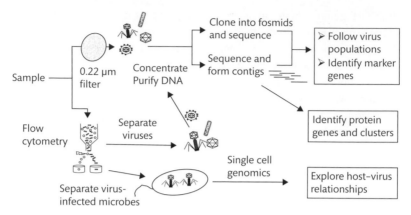

Figure 10.15 Metagenomic approaches for examining viruses. Based on Brum and Sullivan (2015).

found in all strains within a taxon are said to be part of the pangenome (Chapter 5). For viruses, the term is simply the "noncore" genes. Although the function of these genes is often unknown, in general they are thought to help viruses adapt to new environments by providing new genetic material for new functions. They are frequently found in highly variable ("hyperplastic") regions of the genome, and they appear to easily move in and out of viral genomes.

Some of the core and noncore genes are for proteins that are not directly tied to viral replication but rather are involved in host metabolism. These are the "auxiliary metabolic genes." Although they come from the viral genomes, these auxiliary genes code for functions already carried out by the host directed by host genes. These functions boost aspects of host metabolism that directly aid viral reproduction. By bringing along its own version of the genes, the virus ensures that the function continues even if the host shuts down due to the viral infection.

One of the best examples is the set of genes for photosynthesis carried by viruses (cyanophages or cyanoviruses) that attack the cyanobacterium *Prochlorococcus*. The initial work showed that some of these viruses carry the genes *psbA* and *psbD* that encode key components of photosystem II, while later work found evidence of some cyanophages carrying genes for photosystem I as well (Krupovic et al., 2011). Cyanophage replication is higher in the light than in the dark or when photosystem II is inhibited by a herbicide (Figure 10.16). Over time as infection proceeds, expression of the viral genes for the photosystem

components increases while expression of the host genes decreases. The net effect is the continuation of photosynthesis by the host to support the synthesis of components needed for viral replication, even as the host is being killed off by the infection.

Infections by cyanophages would seem to be only bad news for cyanobacteria, but over long timescales, the infections may benefit the species. Some of these viral genes may provide the raw material needed by evolution to invent new genes, or new variants of old genes that ultimately benefit the host (Krupovic et al., 2011). There is some evidence that this has happened for a light-harvesting pigment and for some photosystem I components. The amount of DNA and RNA in viruses is vast on a global scale, and this viral genetic material may evolve more rapidly than host genomes. The net result is that viral genomes potentially provide a rich source of genetic material for evolution of not only cyanobacteria and other microbes but also of all organisms.

Viruses of other organisms

Viruses infect every organism in the biosphere, with potentially large impacts on the biology and ecology of those organisms. As mentioned in Chapter 1, viruses kill off plants and animals in natural environments, but these are not as well studied. Better characterized are the many connections between viruses of wild and domesticated animals and humans (Figure 10.17), vividly portrayed by David Quammen in his book *Spillover* (Quammen, 2012). Many influenza viruses start in pigs and birds before

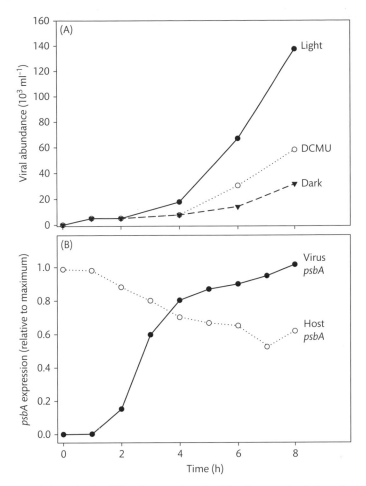

Figure 10.16 Effect of light on viral production (A) and expression of *psbA*, a key gene in photosystem II (B). "DCMU" is the herbicide 3-(3,4-dichlorophenyl)-1,1-dimethylurea, that blocks photosystem II. Expression of *psbA* and other genes helps to ensure that photosynthesis continues and viral replication is uninterrupted as the cyanobacterial host dies. Data from Lindell et al. (2005).

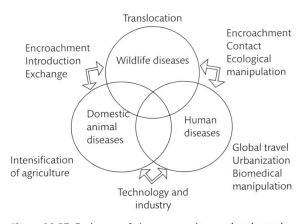

Figure 10.17 Exchange of viruses, parasites, and pathogenic bacteria among wildlife, domestic animals, and humans, and anthropogenic impacts that lead to new diseases. Based on Hassell et al. (2017).

evolving to infect humans, and other viruses now capable of infecting humans apparently first used wild primates or other mammals as hosts. These viruses include the Ebola and Marburg viruses, as well as HIV (Lloyd-Smith, 2017).

Here, we focus on viruses attacking other microbes. Although much less numerous than phages, non-phage viruses are important in the biology and ecology of other microbes. They also provide good examples of the fascinating diversity in the viral world and the challenges to studying viruses in natural environments.

Mycoviruses: Viruses in fungi
Given the high abundance of fungi in soils, one may think that viruses attacking fungi would be as abundant

as those attacking bacteria. In fact, very few "free" viruses (those not in hosts) are "mycoviruses," viruses that infect fungi, and the reasons why illustrate some of the oddities of these viruses. First, it would be difficult to detect mycoviruses because of the nature of their genomes. The genetic material for mycoviruses includes just about every possible version of RNA and DNA, except for the dsDNA found in most bacteriophages (and cellular life). Roughly 70% of mycoviruses have dsRNA, while most of the other 30% have (+)ssRNA genomes (Ghabrial et al., 2015). Some also have (−)ssRNA or even circular ssDNA, but no mycovirus so far has been shown to have dsDNA. The ssRNA viruses might be difficult to detect using conventional methods because the commonly used stains are brightest for dsDNA.

Regardless of stains and detection, free mycoviruses are not numerous because they are transmitted from one fungus to another, without being released as a free particle into the external environment. Mycoviruses can be transmitted vertically via fungal spores, from mother to daughter cell during asexual reproduction, and horizontally via hyphal anastomosis. The horizontal mechanism involves the fusion of two fungal hyphae that results in the exchange of genetic and cytoplasmic material (Figure 10.18). Fungi cannot be infected by free mycoviral particles in controlled experiments, although some yeasts can be infected by this external mechanism. The chitin cell wall of fungi is one defense against infection by free mycoviruses, but that cannot be the whole explanation. The peptidoglycan found in the cell wall of bacteria is similar to chitin, yet it is not effective against stopping infection by bacteriophages.

Still another odd aspect of mycoviruses is the impact on their fungal hosts. Often, there is none (Ghabrial et al., 2015). The rather benign effect of a mycovirus infection results from the apparently innocuous mode of transmission of a mycovirus to a new host; it does not have to lyse its host in order to complete its replication cycle. Mycovirus infection can affect the physiology of fungi, however, even if it is not lethal. Some fungi infected with mycoviruses are less virulent plant pathogens than uninfected strains. An example is the fungus, *Cryphonectria parasitica*, that causes chestnut blight which destroyed the American chestnut tree at the turn of the twentieth century. A mycovirus has been used with some success in Europe as a biocontrol to lessen the virulence of the blight fungus, although it has not worked as well in the United States.

As with nearly all of the microbes encountered in this book, uncultivated mycoviruses infecting uncultivated fungi are probably even more diverse and may yield more surprises than the viruses we know that infect cultivated organisms. RNA mycoviruses have been examined with a metagenomic approach in which the RNA is first reverse-transcribed to DNA and sequenced. Using that approach, one study of soybean leaves found 25 partial genomes that represented at least 22 mycoviruses, only one of which had been known before (Marzano and Domier, 2016). Metagenomic approaches continue to uncover new aspects of the incredible diversity of viruses.

Viruses infecting algae and amoebae

In contrast with the non-lethal effect of mycoviruses on fungi, viruses are important in killing off other eukaryotic

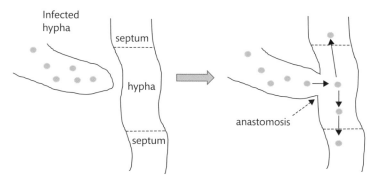

Figure 10.18 One mechanism, anastomosis, for the transmission of mycoviruses from one fungus to another. The septum is a cross wall with pores large enough for ribosomes, mitochondria, and viruses to pass through. This and other mechanisms for mycoviral transmission do not involve release of the viruses into the external environment.

microbes and controlling their population size. The viruses attacking algae have been extensively examined. Just as with bacteria, the fraction of algal mortality attributable to grazing versus viral lysis varies with location and over time (Mojica et al., 2016). Viral lysis is thought to terminate some algal blooms, even blooms of organisms like diatoms and coccolithophores with thick, inorganic cell walls. In the North Sea, as many as 50% of all coccolithophores were infected by viruses at the end of one bloom. Later work demonstrated that a coccolithophore virus has genes for glycosphingolipid synthesis which are turned on when the virus infects the alga (Vardi et al., 2009). The isolated glycosphingolipid alone is sufficient to kill the alga by setting off a series of biochemical events similar to programmed cell death. There is much interest in viruses that kill off and thus potentially control harmful algae in coastal waters.

Many viruses attacking eukaryotic algae are in the *Phycodnaviridae* family with large (100–560 kb) dsDNA, but many eukaryotic viruses, such as the mycoviruses already discussed, have RNA genomes (Koonin et al., 2015). RNA viruses may be more adept than DNA viruses at getting past the membrane barrier of the eukaryotic nucleus. The few viral metagenomic studies completed so far have found that RNA viruses are less diverse than DNA viruses and have a high fraction of genes similar to those in known RNA viruses (Kristensen et al., 2010). Many (−)ssRNA and dsRNA viruses are known to infect

plants and animals, but RNA viruses in marine waters at least seem to have positive-sense RNA genomes and are in the order *Picornavirales*. This order accounted for nearly all of the RNA viruses in one study of coastal waters of British Columbia (Culley et al., 2006). Although other viruses in *Picornavirales* are known to infect animals and higher plants, the marine picorna-like viruses appear to attack a diverse group of protists, including members of algal class Raphidophyceae, *Schizochytrium* (fungus-like), and diatoms (Kristensen et al., 2010).

Some fascinating eukaryotic viruses are very large and infect the amoeba *Acanthamoeba polyphaga* (Figure 10.19) (Colson et al., 2017). This family of viruses, one commonly known as "mimivirus" ("mimicking microbe" virus), blurs the boundaries between viruses and cellular life. The particle and genome sizes of mimiviruses and other giant viruses are as large as those for oligotrophic bacteria. These viruses are between 0.5 and 1 μm, and one type, APMV, is covered by 120–140 nm-long fibrils that may enable attachment to amoebae, bacteria, arthropods, or fungi. The genome size for *Mimiviridae* is about 1.2 Mb in the form of dsDNA, and is protected by a spherical lipid bilayer and protein capsid. Most unusual for viruses is the presence of mRNA, essential of course for cellular life, but not seen in other viruses. New mimiviruses have been isolated by enriching for amoebae from water, soil, insects, and even humans, and sequences related to known giant viruses

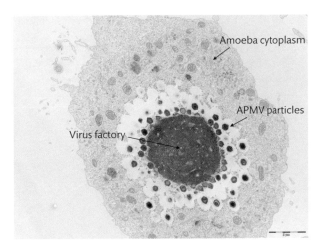

Figure 10.19 Mimiviruses inside an amoeba. The "APMV particles" are the mimiviruses. "Virus factory" is where the membrane, DNA, capsid, and other viral components are assembled. Courtesy of IHU Méditerranée Infection and used with permission. See also Colson et al. (2017).

have been found in metagenomic surveys of natural environments, indicating they must be abundant. (Sequences from rare viruses would be equally rare and not sampled using current sequencing technology.) Although their ecological impact is largely unknown, the existence of these giant viruses has shaken the field of virology by challenging our definitions of viruses and of cellular life.

The story of giant viruses illustrates again that there is more to viruses than simply killing off cellular life. Viruses are likely intimate players in the evolution of microbes and other organisms, a reason why Baltimore is misguided in saying we wouldn't miss viruses if they weren't here. Of course, we should work to eradicate viruses that cause disease in humans in order to minimize the "bad news" disliked by Medawar. But the vast majority of viruses do not cause disease in humans or in organisms we value, and many have positive impacts on cellular life (Roossinck, 2011). Viruses play essential roles in nature and are essential to life as we know it.

Summary

1. Viruses consist of nucleic acids surrounded by a protein coat and for some, membrane material from a previous host. The nucleic acids occur in every possible variety of DNA and RNA and vary among different types of viruses.

2. Viruses reproduce by taking over the biochemical machinery of the infected host. Virulent viruses have only a lytic phase, whereas temperate viruses have a lysogenic phase as well as the lytic phase.

3. Microscopic methods indicate that viruses are very abundant, about ten-fold more abundant than bacteria, their most probable hosts. In contrast, the plaque assay recovers very few viruses from natural environments, which means that few viruses can be isolated and grown in the laboratory.

4. Viral lysis accounts for about 50% of bacterial mortality, with the other 50% due to grazing, although these percentages vary greatly over time and among environments. Viruses can also be important in stopping algal blooms and in the mortality of other microbes, except for fungi which are not killed off by their mycoviruses.

5. Viruses mediate the exchange of genetic material among hosts, affecting the evolution of microbes and other organisms. In contrast with grazing, viral lysis leads to the release of labile organic material via the "viral shunt."

6. Metagenomic approaches are essential for exploring the diversity of viruses in natural environments and have revealed an incredibly large number of unknown genes and viruses.

CHAPTER 11

Processes in anoxic environments

Nearly all of the organisms discussed in detail so far in this book live in an oxygen-rich world. The production of oxygen by light-driven primary producers is enough for use by heterotrophic organisms, not only in sunlit environments, but also in many habitats without direct exposure to the sun. Oxygen can be high in these dark habitats if the supply by physical processes exceeds the use of oxygen by aerobic oxidation of organic material; the supply of oxygen has to exceed the supply of organic material. In those cases, there is enough oxygen to degrade organic material, to drive other biogeochemical cycles, and to support growth of microbes and other organisms in many dark habitats. But there are times and places when the supply is not enough and oxygen runs out. What happens then? This chapter provides some answers.

Some oxygen-deficient habitats are far removed from oxic systems, while others are close by. Most of the oceans, for example, have lots of oxygen, but as soon as water meets the mud at the bottom, things change. Intense aerobic heterotrophy in superficial sediments prevents oxygen from penetrating far, creating anoxic mud only millimeters away from oxic waters. The same mechanism explains why waterlogged soils turn anoxic. Likewise, aerobic heterotrophy can use up the oxygen in the middle of organic-rich particles, resulting in anoxic microhabitats in otherwise oxic soil and water. Other anoxic systems are distant from sunlit oxic ones. The subsurface environment below soils and ocean sediments is largely devoid of oxygen. A few marine basins, like the Black Sea and the bottom waters of many lakes, are anoxic because mixing in of oxygen-rich waters is restricted and because aerobic respiration in surface layers uses up the oxygen before it can diffuse deeper in the water column. In all anoxic systems, bacteria and archaea dominate, and only a few anaerobic eukaryotic microbes can flourish.

While Earth's surface is now oxic, that has not always been the case. The entire planet was anoxic for the first half of its existence (Figure 11.1), with oxygen becoming abundant in the atmosphere only about 2.2 billion years ago after the evolution of oxygenic photosynthesis in cyanobacteria. Atmospheric oxygen increased to modern day levels during the Carboniferous period, about 359 to 299 million years ago, when massive forests on land were buried without being decomposed, eventually turning into coal. Only after atmospheric oxygen became sufficient did the evolution of larger, multi-cellular eukaryotes become possible. Although anoxic environments have receded and anaerobic organisms no longer rule the biosphere, both are very important in many biogeochemical cycles and ecological interactions. The anoxic world is populated by many diverse microbes carrying out many exotic processes not seen in the oxic world.

This chapter will discuss many of these processes with a focus on the carbon and sulfur cycles. Chapter 12 will have more about anaerobic processes in the nitrogen cycle.

Introduction to anaerobic respiration

In terms of the carbon cycle, an important process occurring in anoxic environments is the mineralization

Processes in Microbial Ecology. Second Edition. David L. Kirchman. Oxford University Press (2018). © David L. Kirchman 2018.
DOI 10.1093/oso/9780198789406.001.0001

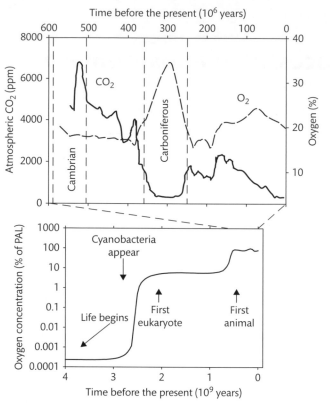

Figure 11.1 Atmospheric carbon dioxide and oxygen concentrations over geological time. Oxygen concentrations are given as either a percentage of total atmospheric gases (top panel) or a percentage of present atmospheric levels (PAL, in the bottom panel). The top graph is the last 600×10^6 years, while the bottom graph covers the last 4.0×10^9 years. Two important geological periods are also given in the top panel. The Cambrian saw a huge explosion of metazoan diversity, while lignin-rich plants were abundant during the Carboniferous period, leading eventually to the formation of extensive coal beds. Data from Berner (1999), Berner and Kothavala (2001), Donoghue and Antcliffe (2010), and Kump (2008).

of organic material by anaerobic respiring microbes. To understand anaerobic respiration, let us go back to aerobic mineralization and break down the familiar equation describing it. As seen before (Chapter 7), the equation for aerobic oxidation of organic material is

$$CH_2O + O_2 \rightarrow CO_2 + H_2O \qquad (11.1)$$

where CH_2O again symbolizes generic organic material, not a specific compound. Equation 11.1 describes a redox reaction that can be split apart into two half-reactions. One half reaction generates electrons (e^-):

$$CH_2O + H_2O \rightarrow CO_2 + 4e^- + 4H^+ \qquad (11.2)$$

while the other half-reaction describes how oxygen takes on the electrons:

$$O_2 + 4e^- + 4H^+ \rightarrow 2H_2O \qquad (11.3).$$

Combining Equations 11.2 and 11.3 yields Equation 11.1. Organic material is the electron donor, while oxygen is the electron acceptor.

We can write a more general form of Equation 11.1:

$$CH_2O + A_2 + H_2O \rightarrow CO_2 + H_2A \qquad (11.4)$$

to illustrate that organic material can be oxidized to carbon dioxide with a generalized electron acceptor, A_2. ("A_2" nicely fits O_2, but nitrate (NO_3^-) and many other electron acceptors are not diatomic.) Anaerobic respiration uses various electron acceptors, symbolized by A_2 in Equation 11.4. Of the many elements and compounds that can take the place of A_2, all are in an oxidized state,

Box 11.1 Balancing equations

A chemical equation balanced in terms of electrons and elements is a succinct and powerful description of a biogeochemical process potentially occurring in an environment. To balance a chemical equation, the starting point is to make sure the number of electrons from the electron donor matches the electrons being received by the electron acceptor. These are set by the valence of the elements being oxidized and reduced. The main elements other than hydrogen and oxygen should be balanced and equal in number on both sides of the equation. To balance hydrogen and oxygen atoms, H^+ or OH^- (but not O_2) can be added to either side as needed, because the reaction is in an aqueous solution, even in soils, where H^+ and OH^- are plentiful. If done correctly, at this point everything should be in balance: electrons, elements, and charge. Among many resources, *Brock Biology of Microorganisms* (Madigan et al., 2012) gives a primer on how to balance chemical equations and to calculate energy yields.

meaning they can take on more electrons. For example, NO_3^- can be an electron acceptor because of the oxidized state of its nitrogen (+5), whereas ammonium (NH_4^+) cannot be because its nitrogen is highly reduced (−3) and cannot take on more electrons (see Table 3.2 in Chapter 3).

Microbes use electron acceptors in order of redox potential

Realizing that several electron acceptors are possible, the next question becomes, which ones are preferred by microbes and which are most important in oxidizing organic material? Clues to the answers come from looking at how three common electron acceptors, O_2, NO_3^-, and sulfate (SO_4^{2-}), vary with depth in a typical sediment profile (Figure 11.2). Invariably, going from the top to the bottom of the profile, oxygen disappears before sulfate begins to decline. Nitrate also disappears quickly soon after oxygen but before sulfate. Figure 11.2 gives the variation in space (depth in the sediment profile), but it also illustrates how these compounds would vary through time, starting at the surface. If enough organic material

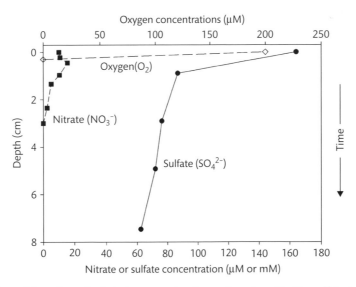

Figure 11.2 Concentrations of three important electron acceptors in a sediment profile. The sulfate concentrations are in mM but multiplied by 10 to fit on this graph (the highest concentration was about 17 mM in this example), whereas nitrate concentrations are in µM. Going down a profile is equivalent to the passing of time. In closed incubations, these three electron acceptors are used up over time in the following order: oxygen, nitrate, and sulfate. Data from Sørensen et al. (1979).

were placed in a bottle along with possible electron acceptors, O_2 would disappear first, followed by NO_3^-, and then SO_4^{2-}. Why this order?

The order can be explained by the tendency of these compounds to accept electrons. This tendency is measured relative to the reduction of H^+ to H_2, which is set at 0 volts (V). Possible electron acceptors are put in an "electron tower" of reduction half-reactions (Figure 11.3), with oxygen at the top (+0.82 V) and CO_2 at the bottom (–0.24 V). A more positive half-reaction indicates a greater tendency for the compound to accept electrons. So, oxygen is the strongest electron acceptor, while CO_2 is the weakest. The strength of an electron acceptor is an important characteristic in explaining the contribution of various elements

and compounds to anaerobic respiration and to the mineralization of organic material in anoxic environments.

The electron tower explains the order of electron acceptors used up over time and down a depth profile, but it is insufficient for exploring the benefit to an organism of using one acceptor over another and for predicting which is most important in oxidizing organic material in the absence of oxygen. To explore these issues, it is useful to calculate a theoretical energy yield for an electron acceptor oxidizing an organic compound. This energy yield is the Gibbs free energy change ($\Delta G^{o'}$), where the superscripts indicate that standard biochemical conditions are assumed: pH = 7, the temperature is 25 °C, and each of the compounds other than H^+ in the reaction occurs in equal molar amounts. To compare the electron acceptors, we assume for now that the same electron donor, here an organic compound, is oxidized for all electron acceptors; in Table 11.1, the "compound" is in fact a hypothetical one with the main elements (C, N, and P) occurring in Redfield ratios (Chapter 2). As we will soon see, sulfate reducers and carbon dioxide reducers do not use the same electron donors. But the calculations and the theoretical energy yields are still useful in thinking about these various electron acceptors.

The order of electron acceptors in Table 11.1 is the same as seen in the electron tower (Figure 11.3). Because the electron donor is the same in Table 11.1, the differences in energy yield are due to the electron acceptors. But a couple of new points are illustrated by the energy yield calculations. Note the small difference in energy yield between oxygen and the next electron acceptors, especially nitrate, in Table 11.1. The implication is that these other electron acceptors should be nearly as commonly used by microbes as oxygen. In contrast, using sulfate or carbon dioxide

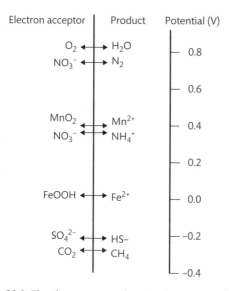

Figure 11.3 The electron tower showing the potential (volts) for some half-reactions involving electron acceptors commonly used by microbes. Data from Canfield et al. (2005) and Thauer et al. (1977).

Table 11.1 Theoretical yield of energy from organic material oxidation using various electron acceptors. C_o is $(CH_2O)_{106}$ $(NH_3)_{16}(H_3PO_4)$–an idealized organic material with Redfield ratios for C, N and P. The oxidized form of manganese used here is pyrolusite, and the oxidized iron is goethite. Data from Froelich et al. (1979).

Electron acceptor	Reaction	Energy yield (kJ mol⁻¹)
Oxygen	$C_o + 138O_2 \rightarrow 106CO_2 + 16HNO_3 + H_3PO_4 + 122H_2O$	–3190
Nitrate	$C_o + 94.4HNO_3 \rightarrow 106CO_2 + 55.2N_2 + H_3PO_4 + 177.2H_2O$	–3030
Manganese	$C_o + 236MnO_2 + 472H^+ \rightarrow 106CO_2 + 236Mn^{2+} + 8N_2 + H_3PO_4 + 366H_2O$	–2920
Iron	$C_o + 424FeOOH + 848H^+ \rightarrow 106CO_2 + 424Fe^{2+} + 16NH_3 + H_3PO_4 + 742H_2O$	–1330
Sulfate	$C_o + 53SO_4^{2-} \rightarrow 106CO_2 + 53S^{2-} + 16NH_3 + H_3PO_4 + 106H_2O$	–380
CO_2	$C_o \rightarrow 53CO_2 + 53CH_4 + 16NH_3 + H_3PO_4$	–350

yields very little energy, nearly ten-fold less, during the oxidation of the same hypothetical organic carbon, implying both sulfate and carbon dioxide are less desirable as electron acceptors and unimportant in organic material oxidation. We will see in a moment that this is not the case.

The energy yield also explains why eukaryotic microbes do not use the electron acceptors at the bottom of the list, such as sulfate and carbon dioxide. The energy yield with these electron acceptors is too small to support the high energy requirements of the eukaryotic lifestyle. Some of the other electron acceptors are not as easily ruled out. There is a seemingly small drop-off in the theoretical energy yield in switching from nitrate to manganese. In fact, the only electron acceptor used by eukaryotic microbes other than oxygen is nitrate (Risgaard-Petersen et al., 2006; Kamp et al., 2016).

Oxidation of organic carbon by different electron acceptors

So far, oxygen and other electron acceptors have been evaluated on theoretical grounds using basic thermodynamics under standard biochemical conditions. But standard conditions are not necessarily realistic. How do the predictions compare with the real world—what is the dominant electron acceptor under real-world conditions?

Globally, the answer is oxygen (Figure 11.4). This should be no surprise after seeing its high energy yield in oxidizing organic material, and after remembering that production of organic material and oxygen production are intimately coupled. Next on the energy yield list is nitrate. But this electron acceptor is responsible for relatively little organic material oxidation, except for polluted waters and some water-saturated soils with high nitrate concentrations. Nitrate reduction may account for as much as 50% of organic carbon mineralization in low-oxygen basins in the Pacific Ocean (Liu and Kaplan, 1984). Still, nitrate reduction accounts for little organic carbon oxidation on a global scale. So, why isn't nitrate more important?

Before answering that question, here is yet another puzzle. Even though sulfate and carbon dioxide are near the bottom of the electron acceptor tower, both are important in anaerobic organic material oxidation. Sulfate reduction is crucial in marine environments, while carbon dioxide reduction fills that role in freshwater environments, such as wetlands and rice paddies. So, energetic yield

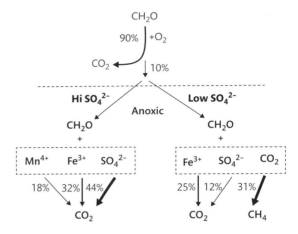

Figure 11.4 Relative importance of electron acceptors in oxidizing organic material with or without oxygen and with low or high sulfate concentrations. The fraction oxidized by anaerobic processes (10%) was set to the fraction of organic material buried in anoxic environments. The total was then re-set to 100% to calculate averages for the anoxic environments. The contributions by some electron acceptors are not shown because they are low. These include CO_2 in high-sulfate environments, Mn^{4+} in low-sulfate environments, and NO_3^- in both types of environment. Data for the high-sulfate environments are from studies cited in Canfield et al. (2005), and the low-sulfate environment data are from Keller and Bridgham (2007), Yavitt and Lang (1990), Roden and Wetzel (1996), and Thomsen et al. (2004).

only partially explains why some electron acceptors are more important than others. The two other factors are concentration and chemical form of the electron acceptors. These answer the question about nitrate and explain the puzzle about sulfate and carbon dioxide.

Limitations by concentration and supply

The thermodynamic calculations for energy yields given in Table 11.1 assume equal concentrations of everything. However, that is far from being true in real environments. One reason why oxygen is so important is that it is often readily available and concentrations are high. In fact, the production of oxygen is as high as the production of organic material because the two processes are mechanistically linked by oxygenic photosynthesis. The dominant form of primary production by far is oxygenic photosynthesis, which produces enough oxygen to oxidize all of the organic material made by photosynthesis.

In contrast with oxygen, nitrate production is not intimately linked to organic carbon production. Nitrate formation starts with nitrogen fixation ($N_2 \rightarrow NH_3$), a slow, energy-intensive process, carried out by only a few organisms (Chapter 12), unlike the widespread capacity of oxygen formation by oxygenic photosynthesis. In contrast with nitrogen fixation, nitrate reduction is a rapid process that can lead to nitrogen gases (N_2 and N_2O) and the loss of fixed nitrogen from the system. Another sink for nitrate is its use as a nitrogen source for biomass synthesis by higher plants, phototrophic microbes, and heterotrophic bacteria. For these reasons, concentrations and supply rates of nitrate are low, explaining why nitrate respiration does not consume more organic material even with its high theoretical energy yield.

Concentrations and supply explain much about the contribution of the other electron acceptors to organic material oxidation. The concentration of sulfate is the reason why organic material oxidation by sulfate reduction is so high in marine environments. Sulfate is not lost from these environments because the end product of sulfate reduction (H_2S) is usually easily converted back to sulfate. Sulfate concentrations are low in freshwaters and soils, explaining why sulfate reduction is usually not important in those environments. Carbon dioxide reduction is never limited by carbon dioxide concentrations. Likewise, iron and manganese are often abundant, explaining why these two elements are often important in organic material oxidation. But use of oxidized iron and manganese as electron acceptors is complicated by yet another property: their chemical form.

Effect of physical state and chemical form

The physical state of the acceptors used by microbes ranges from a gas to a solid. Oxygen once again has a form most conducive to use by microbes. It has the lowest molecular weight of all compounds in the electron tower, resulting in its diffusion rate being highest of all electron acceptors. Because it is a gas, it can be transported to microbes without water, an important feature in soils. Finally, because it is uncharged and small, oxygen easily enters into cells without special transport mechanisms. The only other electron acceptor with any of these traits is carbon dioxide, but its physical state and chemical form (low molecular weight, gas, and uncharged) and high concentrations are not enough to offset its low energy yield.

Concentration is the main reason why nitrate reduction contributes so little to organic carbon oxidation, but its chemical form doesn't help. Its charge and thus non-gaseous state means that it is made available to microbes only via water. The charge also means that specialized transport mechanisms and energy are required to bring it across membranes and into cells.

Chemical state has a big impact on the use of electron acceptors such as ferric iron (Fe(III)) and oxidized manganese (Mn(IV)). Most importantly, because Fe(III) is insoluble at the near-neutral pH of most environments, it occurs as particulate oxides that are much too large to transport across membranes. Consequently, iron-reducing bacteria may have to be in physical contact with iron oxides and somehow transport electrons from organic material oxidation to the iron oxide (Figure 11.5). This contact may be via

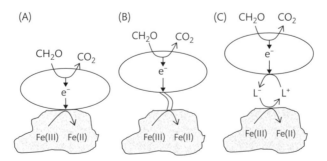

Figure 11.5 Three hypothesized strategies for using insoluble iron oxides by iron-reducing bacteria. The first (A) is to be in direct physical contact with the iron oxide. In the second strategy (B), the electrons are transferred via a conductive pilus or nanowire. The third strategy (C) involves a soluble ligand (L^+) that is reduced by the bacterium. The now reduced ligand (L^-) reduces Fe(III) to Fe(II). Based on Nealson and Rowe (2016).

"nanowires" (Shi et al., 2016), as mentioned in Chapter 2. Alternatively, direct physical contact may not be necessary if electrons can be shuttled from the bacterium to the insoluble iron oxide via soluble ligands.

Another complication is the type of crystal form iron takes on. The crystallinity of oxidized iron, which ranges from amorphous oxides to highly crystallized ones, affects the access of Fe(III) to iron-reducing bacteria. Iron in amorphous oxides is more easily reduced and thus supports more organic material oxidation than highly crystallized iron. So, even though concentrations of iron are high in soils and sediments and its energy yield as an electron acceptor is also high, the chemical form of iron can limit its contribution to the oxidation of organic material in anoxic environments.

The anaerobic food chain

So far, we have assumed that anaerobic respiring organisms can all use the same organic compounds as electron donors. In fact, this is far from being the case. It is true that the suite of organic compounds used by nitrate reducers is about the same as oxygen reducers (aerobic respiration), and geochemists often treat the two processes as being nearly equivalent. With these electron acceptors, in theory, any labile organic material could be degraded, oxidized, and mineralized by a single organism. In contrast, most of the anaerobic respiring bacteria and archaea cannot use many organic compounds as electron donors. Consequently, an entire consortium of organisms working sequentially is needed to mineralize organic material in anoxic habitats. The consortium is called the "anaerobic food chain." Although members of this consortium are not eating one another, "food chain" does convey the correct idea of organic carbon being passed from one organism to another (Figure 11.6). The anaerobic food chain model assumes that sulfate and carbon dioxide are dominant terminal electron acceptors.

Before discussing specific parts of the anaerobic food chain, let us start with an overview by following plant detritus as it is degraded and eventually oxidized back to carbon dioxide. The detritus is first broken up into smaller fragments by larger, eukaryotic organisms (Chapter 7), creating more surface area for bacteria and fungi to attack the macromolecules making up the detritus. These microbes have hydrolases (cellulase, in the case of cellulose) that cleave the detrital macromolecules to monomers (glucose, in the case of cellulose). The monomers and other by-products from macromolecule hydrolysis are not used directly by sulfate or carbon dioxide reducers. Instead, these by-products are used by bacteria carrying out fermentation. The fermenting bacteria in turn produce several compounds, most importantly acetate

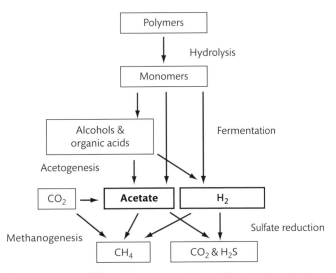

Figure 11.6 Anaerobic food chain. Acetate and H_2 are key compounds supplying electrons that eventually end with the terminal electron acceptors, sulfate and carbon dioxide.

and hydrogen gas (H_2). These two compounds and a few others are then used by sulfate or carbon dioxide reducers, thus completing the anaerobic food chain.

The anaerobic food chain model is a simple depiction of a complicated suite of possible electron acceptors and donors, and of microbes and microbial physiologies. How terminal electron acceptors, such as iron and manganese, fit into this model is not clear. More problematic, one study found that about 80% of sulfate and iron reduction was supported by other electron donors besides acetate and H_2 (Finke et al., 2007). These other electron donors could include alcohols, sugars, and amino acids, and several other compounds—all now assumed to be used by microbes in the initial steps of the anaerobic food chain. Still, while the model is not perfect, it does serve to illustrate connections among several microbial metabolic pathways important in the anaerobic degradation of organic material.

Fermentation

This form of catabolism is an important intermediate step between biopolymer hydrolysis and oxidation by the terminal electron acceptors. Fermentation refers to several anaerobic processes used by organisms to generate energy (ATP) without external electron acceptors. Using simple monomers like sugars and amino acids, fermenting organisms produce organic compounds like organic acids or alcohols and sometimes CO_2, while generating energy.

Fermentation pathways have been extensively examined for many years, but not much is known about fermentation in natural environments. It is known that fermentation is common among microbes (Box 11.2) and even among eukaryotes, including the muscle cells of mammals. When the supply of oxygen is insufficient, our muscles carry out lactic acid fermentation:

$$\text{Glucose} \rightarrow 2\,\text{lactate} + 2H^+ + 2\,\text{ATP} \qquad (11.5)$$

which yields 196 kJ mol^{-1}, much less than the roughly 3000 kJ mol^{-1} released by aerobic respiration. When the supply of oxygen is insufficient (such as when we are out of shape), muscle cells are forced to do lactic acid fermentation because of the lack of an external electron acceptor. In this form of fermentation, no carbon dioxide is released because there is no net oxidation of the glu-

---

> **Box 11.2 Trash as treasure**
>
> Many types of fermentation are carried out by microbes. These pathways take their name from their end product, excreted as waste. One example is lactic acid fermentation, used for centuries to make yogurt, cheeses, and other food. Another example is ethanol fermentation, a key process in making wine and beer. Many commercially valuable products are made by fermentation, including enzymes, vitamins, and antibiotics. Compounds, such as insulin, are produced by fermenting microbes with cloned genes from other organisms. Fermentation can produce hydrocarbons, including butane and oil, suitable for use as biofuels.

---

cose carbon; there is no place for electrons from glucose oxidation to go. Our muscle cells cannot go for long without oxygen, but many microbes can grow using energy only from fermentation.

Which bacteria and eukaryotes carry out fermentation in nature is still largely unknown. Stable isotope experiments have suggested bacteria related to *Acidobacterium capsulatum* in the *Acidobacteria* phylum, which is very abundant in soils (Chapter 4), are important in fermenting sugars in fens and peatlands (Hamberger et al., 2008). Most fermentation in natural environments is generally assumed to be carried out by bacteria (Valdemarsen and Kristensen, 2010), even though fermentation by yeasts is very important in producing many practical by-products from alcohol to leavened bread, and some parasitic protists, such as *Giardia* and *Entamoeba*, are known to carry out fermentation when oxygen becomes limiting (Müller et al., 2012). Less appreciated is the possible role of algae in fermentation. When buried in dark, anoxic environments, some algae may switch to fermentation or nitrate respiration (Bourke et al., 2017).

A key feature of fermentation is the release of organic compounds. These compounds are not as energy-rich as the starting material but they are still reduced enough to yield energy when oxidized by sulfate and carbon dioxide reducers. These organisms use many of the compounds released by various fermentation pathways, but the main fluxes of carbon and energy are through acetate and hydrogen gas. This conclusion was reached by studies examining concentrations and fluxes of short chain organic acids like acetate, lactate, and propionate (Parkes et al., 1989). Fluxes can be examined by measuring concentrations of these compounds coupled with uptake rates estimated from ^{14}C- or ^{13}C-labeled compounds in nature. Stable isotope probing (SIP) experiments using ^{13}C-acetate have been used to identify bacteria in the gammaproteobacterial family *Oceanospirillaceae* and others as acetate users in sediments (Vandieken and Thamdrup, 2013). A complementary approach is to follow the build-up of compounds after sulfate reduction or methanogenesis has been inhibited by adding either molybdate or 2-bromochloromethane. These inhibition experiments as well as the flux measurements have demonstrated that acetate is usually more important than lactate and other short chain fatty acids in supporting sulfate reduction and methanogenesis.

Why are acetate and H_2 key compounds in the anaerobic food chain? Both can be produced directly by fermentation pathways, but they are also produced by another group of microbes in another step in the anaerobic food chain.

Acetogenesis, interspecies hydrogen transfer, and syntrophy

The next step in the anaerobic food chain is the production of acetate and H_2 by acetogenic bacteria, using another metabolic pathway, acetogenesis. There are about 20 genera of acetogenic, mainly Gram-positive bacteria, with most known strains in the *Acetobacterium* and *Clostridium* genera (Drake et al., 2008). These organisms have been isolated from a wide variety of environments, including soils, animal guts, and sediments.

Acetogenic microbes can use several organic compounds, such as ethanol, a common end product of fermentation. The reaction describing the use of ethanol by acetogens is

$$\text{Ethanol} + H_2O \rightarrow \text{acetate} + H^+ + H_2 \quad (11.6).$$
$$\Delta G^{o'} = +9.6 \text{ kJ mol}^{-1}$$

The reaction has a serious problem as now written; the change in Gibbs free energy ($\Delta G^{o'}$) is positive, implying that the reaction is thermodynamically impossible. But, experimental studies have demonstrated that it occurs and that organisms grow with energy from it. How is this

possible? This is one example of many where biology seems to break the laws of thermodynamics. What actually happens is less dramatic. No thermodynamic laws are broken because biology does not necessarily operate under standard biochemical conditions.

Note that the energy yield given for Equation 11.6 assumes that the reaction occurs under standard biochemical conditions—most importantly, equal concentrations of the reactants and by-products. In fact, the reaction does go forward when H_2 is removed and its concentration drops far below the hydrogen to ethanol ratio implied by Equation 11.6. According to theoretical calculations, acetate formation from ethanol becomes thermodynamically feasible when the partial pressure of H_2 drops below one atmosphere, and becomes energetically profitable for growth when it is less than about 0.01 atmosphere (Figure 11.7). In theory, removing the other end product, acetate, could also "pull" the reaction to the right and make it thermodynamically possible, but that is easier to accomplish with a gas (here hydrogen) which can diffuse away more quickly than a charged, larger compound like acetate. In any case, uptake of both acetate and H_2 would pull the reaction to the right.

In addition to diffusion, H_2 concentrations are reduced by its use by other organisms. This connection between a H_2 producer (the acetogenic bacterium) and a H_2 user (sulfate or carbon dioxide reducer) is referred to as "interspecies hydrogen transfer." It is greatly facilitated when the two organisms are physically close together in a mutually beneficial arrangement called "syntrophy" (Morris et al., 2013). A famous example of syntrophy is "*Methanobacillus omelianskii*" described by H. A. Barker in 1940. This seemed to be one organism that used ethanol and carbon dioxide to produce acetate and methane. The reaction is thermodynamically favorable ($\Delta G^{o'} = -116.4 \text{ kJ} / \text{reaction}$) and thus seemed possible for one organism to carry out. However, later it was shown that this reaction was actually carried out by two organisms: an acetogenic bacterium (*Acetobacterium woodii*) that produces H_2 and acetate from ethanol, and a methanogen (*Methanobacterium bryantii*) that uses H_2 and carbon dioxide (but not ethanol) and produces methane. That these microbes were isolated and maintained together for years is indicative of the tight physical relationship between the two. Single cell genomics, rRNA tag sequencing, and other cultivation-independent methods have been used to examine the

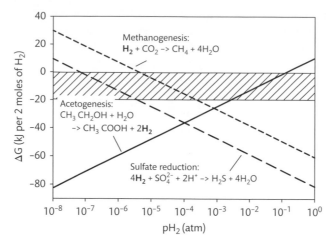

Figure 11.7 Energetic yield for three reactions as a function of hydrogen gas (H_2) partial pressure (pH_2). The top line of the hatched area is at a Gibbs free energy equal to zero, below which reactions are thermodynamically favorable, whereas the bottom line indicates the minimal energy thought to support microbial metabolism. Data from Canfield et al. (2005).

syntrophic acetate oxidation to H_2 and its use by methanogens (Gies et al., 2014).

Acetogenic bacteria do not appear to be very diverse. The culturable acetogenic bacteria include the thermophile *Thermacetogenium phaeum* and mesophiles *Clostridium ultunense* and *Synthrophaceticus schinkii* (Mosbaek et al., 2016). Metagenomics and stable isotope probing have been used to examine the microbes taking up ^{13}C-acetate and also those involved in acetogenesis. This work turned up ^{13}C-labeled peptides in methanogenic archaea and also five species of *Clostridia*, a class of bacteria in the *Firmicutes* phylum. Metagenomic data indicate that the bacteria have the FTFHS gene for formyltetrahydrofolate synthetase, a key enzyme for reductive acetogenesis. Metagenomic analyses of marine sediments suggest that archaea in the *Bathyarchaeota* may be capable of acetogenesis (He et al., 2016).

The sulfur cycle and sulfate reduction

The next step in the anaerobic food chain in environments with high sulfate concentrations is sulfate reduction, which oxidizes acetate, H_2, and other by-products of fermentation and acetogenesis. This process and the rest of the sulfur cycle are worthy of more discussion because sulfate-reducing bacteria oxidize a large amount

of organic material in the biosphere, and because these microbes and others involved in the sulfur cycle are abundant in many natural ecosystems, not just marine ones. Sulfur biogeochemistry also plays a big part in figuring out the history of early life on the planet. The sulfur cycle consists of several biogeochemical reactions mediated by several types of bacteria and archaea (Figure 11.8).

The focus here is on those organisms using sulfur compounds to generate energy (catabolism), in contrast with its use by organisms for biosynthesis of protein and other macromolecules (anabolism). Because it is the most available form of sulfur in many environments, sulfate is the main sulfur source for many organisms, ranging from microbes to higher plants. Sulfate needs to be reduced before its sulfur can be assimilated and used in biosynthetic pathways. Assimilatory sulfate reduction is quite different from dissimilatory sulfate reduction (Table 11.2), the main topic of this section. There is a similar difference between dissimilatory nitrate reduction, previously mentioned as just "nitrate reduction," and assimilatory nitrate reduction.

The capacity to carry out dissimilatory sulfate reduction is not as common as assimilatory sulfate reduction, dissimilatory nitrate reduction, and oxygen reduction (aerobic respiration). Most of the known sulfate reducers are in the class *Deltaproteobacteria*, although there

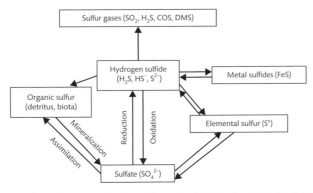

Figure 11.8 Some of the compounds and processes in the sulfur cycle. DMS is dimethylsulfide (Chapter 6), and COS is carbonyl sulfide.

Table 11.2 Comparison of assimilatory and dissimilatory sulfate reduction.

Characteristic	Assimilatory	Dissimilatory
Purpose	Biosynthesis	Energy production
Fate of reduced sulfur	Assimilated into organic compounds	Excreted
Requires energy?	Yes	No
Membrane-associated	No	Yes
Key enzyme (gene)	ATP sulfurylase	Dissimilatory sulfite reductase (*dsr*)
Organisms	Widespread	*Deltaproteobacteria* and a few others

are some Gram-positive sulfate reducers in the genus *Desulfotomaculum* in the *Firmicutes* phylum (Rabus et al., 2015). Single cell and metagenomic approaches have uncovered evidence (*dsr* genes for dissimilatory sulfite reductase) that some members of the *Chloroflexi* phylum are sulfate reducers (Wasmund et al., 2017).

Sulfate reduction in the *Archaea* is restricted, as far as is known, to the genus *Archaeoglobus*. This is an interesting organism because it grows at high temperatures, reaching maximum growth at over 90 °C. Probably other, uncultivated sulfate reducers grow at even higher temperatures, because sulfate reduction by natural communities has been observed at temperatures hotter than 90 °C. Another interesting aspect of this archaeon is that it is most closely related to methanogenic archaea, suggesting *Archaeoglobus* lost its methanogenic pathway and gained sulfate reduction genes later from bacteria by horizontal gene transfer. Support for this hypothesis comes from comparing phylogenetic trees of genes for 16S rRNA and dissimilatory sulfite reductase (*dsr*).

Electron donors for sulfate reduction

An individual sulfate-reducing bacterium may not be able to use many organic compounds, but the entire collection of sulfate reducers can use a great variety of compounds as electron donors. In fact, the list is quite long: H_2, hydrocarbons, organic acids, alcohols, amino acids, sugars, and aromatic compounds, to name just the broad classes. Some of these electron donors are given in Figure 11.9. Still, acetate is the most important organic compound for natural communities of sulfate reducers.

This finding was surprising to microbiologists because many sulfate reducers were isolated and grown in the laboratory with lactate. In spite of acetate and lactate being quite similar, both being organic acids that differ by only one carbon, several sulfate reducers able to use lactate cannot grow on acetate. This apparent preference for lactate over acetate may be a bias introduced through cultivation of organisms and growing them in the laboratory. Another aspect of sulfate reduction, incomplete oxidation of organic compounds, may be more important in the laboratory than in natural environments.

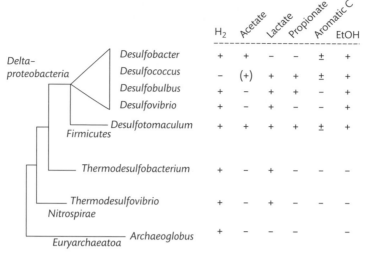

		H$_2$	Acetate	Lactate	Propionate	Aromatic C	EtOH
Delta-proteobacteria	Desulfobacter	+	+	−	−	±	+
	Desulfococcus	−	(+)	+	+	±	+
	Desulfobulbus	+	−	+	+	−	+
	Desulfovibrio	+	−	+	−	−	+
	Desulfotomaculum	+	+	+	+	±	+
Firmicutes							
	Thermodesulfobacterium	+	−	+	−	−	−
	Thermodesulfovibrio	+	−	+	−	−	−
Nitrospirae							
Euryarchaeatoa	Archaeoglobus	+	−	−	−	−	

Figure 11.9 Selected electron donors for some sulfate reducers. "EtOH" is ethanol. Data from Canfield et al. (2005) and Rabus et al. (2015).

Box 11.3 Biocorrosion by sulfate-reducing bacteria

In addition to their roles in natural environments, sulfate-reducing bacteria (often abbreviated as SRB) are big contributors to microbially influenced corrosion of ferrous metals. This problem costs hundreds of millions of dollars each year in the USA alone. The problem starts when aerobic bacteria colonize metal surfaces and create anoxic micro-environments for sulfate-reducing bacteria. This results in an uneven distribution of microbes and biofilm along the metal surface, a key feature of biocorrosion. Fueled by H$_2$ gas and other electron donors, sulfate-reducing bacteria accelerate corrosion by producing sulfides which combine with Fe^{2+} from ferrous metals to form iron sulfide (FeS), and eventually iron oxides, better known as rust (see figure). As these bacteria grow and excrete extracellular polymers, other microbes join the biofilm, exacerbating the corrosion problem.

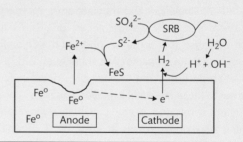

Oxidation of reduced sulfur compounds and the rest of the sulfur cycle

Sulfate reduction produces hydrogen sulfide and several other reduced sulfur compounds. Concentrations of these compounds are high in sulfate-rich environments, but they do not build up indefinitely, because both biotic and abiotic processes oxidize the reduced sulfur compounds back eventually to sulfate. Abiotic reactions of sulfide and amorphous oxides of iron and manganese are very fast. The half-life of sulfide in the presence of colloidal manganese oxides can be as short as 50 seconds, for example. However, calculations indicate that sulfide should persist in the presence of oxygen much longer than it actually does if abiotic oxidation were the sole process operating, suggesting that biotic oxidation dominates. The sulfur cycle is complicated because both biotic and abiotic reactions are important and because sulfur can take on many oxidation states (Figure 11.10).

There are two types of reduced sulfur oxidation metabolisms (Table 11.3). One depends on light and is a form of phototrophy, including both photoautotrophy and photoheterotrophy; the other does not depend on light ("non-phototrophic sulfur oxidation") and is a form of chemolithoautotrophy. The two types are carried out by very different organisms and are quite different from the phototrophic and heterotrophic metabolisms discussed so far in this book.

Non-phototrophic sulfur oxidation

A wide variety of organisms, often called colorless sulfur bacteria (they lack pigments), obtain energy from oxidizing sulfide and other reduced sulfur compounds in the dark. Sulfide oxidation is carried out by bacteria in the *Alpha-*, *Beta-*, *Gamma-*, *Delta-*, and *Epsilonproteobacteria*, and by archaea in the *Sulfolobales* family. These organisms also oxidize other reduced sulfur compounds, such as elemental sulfur and thiosulfate. A common reaction is

$$H_2S + O_2 \rightarrow SO_4^{2-} + 2H^+ \qquad \Delta G^{o'} = -796 \text{ kJ mol}^{-1} \quad (11.7).$$

The reaction can stop at elemental sulfur, which is deposited within the cell where it serves as an energy store. *Thiothrix nivea* oxidizes it further to sulfate, while *Beggiatoa alba* uses elemental sulfur as an electron acceptor and reduces it back to hydrogen sulfide.

Sulfide oxidation is an example of chemolithotrophy, meaning that these microbes gain energy from the oxidation of inorganic compounds, in this case hydrogen sulfide. In essence, hydrogen sulfide takes the place of organic compounds (CH_2O) in Equation 11.2 and is the electron donor:

$$H_2S + 4OH^- \rightarrow SO_4^{2-} + 6H^+ + 8e^- \quad (11.8).$$

The electron acceptor is often oxygen, as in Equation 11.3. Putting Equation 11.3 and 11.8 together yields Equation 11.7 and energy for the sulfur-oxidizing microbe.

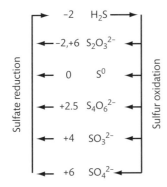

Figure 11.10 Inorganic sulfur compounds in the sulfur cycle, including the major transformations mediated by microbes. The numbers refer to the oxidation state of sulfur in each compound. Thiosulfate ($S_2O_3^{2-}$) can be pictured as a sulfate molecule with one of the oxygen atoms replaced with sulfide (S^{2-}), resulting in the outer sulfur having an oxidation state of −2 and the inner one +6. In tetrathionate ($S_4O_6^{2-}$), two of the S atoms are at +5 while two others are at 0, giving an average of +2.5.

Table 11.3 Major characteristics of the two main types of sulfur (S) oxidizing bacteria and archaea in nature.

Characteristic	Non-phototrophic S oxidizers	Phototrophic S oxidizers
Pigments	None	Bacteriochlorophyll *a* and others
Role of light	None	Energy source
Role of reduced S	Source of energy and reducing power	Source of reducing power and energy
Role of oxygen	Electron acceptor for S oxidation	Represses photosynthesis, used as electron acceptor for oxidation of reduced S (chemolithotrophy) or organic carbon (heterotrophy), or kills cells*
Carbon source	CO_2	CO_2 (when not growing heterotrophically)

* Phototrophic sulfur-oxidizing bacteria are anaerobes that vary in their response to oxygen, depending on the species. Some are strict anaerobes that are killed by oxygen.

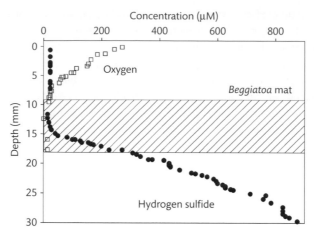

Figure 11.11 Oxygen and sulfide concentrations above and below a mat of the colorless sulfur bacterium, *Beggiatoa* (hatched area). Data from Kamp et al. (2006).

Optimal conditions for sulfide oxidation are at the interface between the oxic world, where oxygen concentrations are high, and the anoxic world which produces sulfides (H_2S) by sulfate reduction (Figure 11.11). At this interface, oxygen and hydrogen sulfide may overlap but only for millimeters. One gammaproteobacterial genus, *Beggiotoa*, is well known to reside at this interface and to glide away from high concentrations of either oxygen or sulfide; it seeks the interface, not the extremes of either compound. It also demonstrates negative taxis against light as part of its strategy to avoid high oxygen concentrations produced by oxygenic photosynthesis. *Beggiotoa* may migrate several millimeters over a day as oxygen concentrations vary due to photosynthesis and aerobic respiration. This bacterium and other sulfide oxidizers are microaerophilic, meaning they prefer low oxygen concentrations, about 5–10% of atmospheric levels.

Other sulfide oxidizers do not remain at the oxic–anoxic interface and migrate between the oxic and anoxic layers. These organisms use nitrate as an electron acceptor for sulfide oxidation, yielding 785 kJ per reaction, only slightly less than with oxygen as the electron acceptor. The nitrogenous end product is ammonium, although some bacteria, such as *Thiobacillus denitrificans*, produce nitrogen gas. Some well-studied examples of nitrate-reducing, sulfide-oxidizing bacteria are *Thioploca* and *Thiomargarita*. These bacteria are interesting because their cells are huge (750 μm), taken up mostly by a nitrate-filled vacuole (Schulz and Jørgensen, 2001). The

bacteria fill the vacuole with nitrate at the sediment–water interface where concentrations are high, and then migrate deeper into sediments where hydrogen sulfide is available.

Still another group of bacteria, "cable bacteria," don't bother to move up and down to get enough sulfide and electron acceptor. Bacteria in the deltaproteobacterial family *Desulfobulbaceae* form long chains of cells with one end sticking down into sulfide-rich sediments and the other end sticking up into sediment layers with sufficient oxygen or sometimes nitrate (Lovley, 2017). It appears that electrons from sulfide oxidation travel up the chain to cells with sufficient electron acceptors to take on the electrons. In support of the hypothesized mechanism, experiments showed that the reaction was stopped by slicing horizontally through sediments with a thin tungsten wire. How electrons travel along the chain is not clear, although it seems the mechanism involves a ridge running the length of the chain. Chains of cable bacteria can stretch for thousands of cells over several centimeters through sediments.

Sulfide oxidation by anoxygenic photosynthesis
The other biotic mechanism for oxidizing reduced sulfur is carried out by anaerobic anoxygenic photosynthetic (AnAP) bacteria; no archaeon or eukaryote is known to have this form of metabolism. The AnAP bacteria and their contribution to primary production were discussed

in Chapter 6. Although both oxidize sulfide, AnAP bacteria and colorless sulfide oxidizers are quite different in physiology and phylogeny. Unlike many colorless sulfide oxidizers, AnAP bacteria oxidize sulfide without oxygen; they are anaerobic and carry out anoxygenic photosynthesis only in the absence of oxygen. ATP synthesis in AnAP bacteria is driven by light when growing photosynthetically, supplemented by ATP from sulfide oxidation. The phototrophic sulfur oxidizers also use the reduced sulfur as an electron source for synthesis of the NADH needed for carbon dioxide reduction. Because the reduced sulfur replaces the water used by oxygenic phototrophic organisms, these AnAP bacteria do not produce oxygen (they are anoxygenic), as mentioned in Chapter 6. Similar to colorless sulfide oxidizers, AnAP bacteria are found at interfaces where light and hydrogen sulfide are both present. They are common in waterlogged soils, salt marshes, and stagnant pools where their unusual pigments can color the water brilliant purples and reds. Nearly all AnAP bacteria have bacteriochlorophyll *a* and often several other types of bacteriochlorophyll and carotenoids (see Table 6.7 in Chapter 6).

The five main groups of AnAP bacteria differ in their potential for heterotrophy, tolerance of oxygen, use of different electron donors, and the capacity to grow under low light. For example, the purple sulfur bacteria are mainly obligate anaerobes (they are killed off by oxygen) and rely on photolithoautotrophy. The purple nonsulfur bacteria use organic compounds and a great variety of other electron donors in place of reduced sulfur for photosynthesis; photoorganotrophy is the group's preferred mode of metabolism. The green sulfur bacteria, such as *Chlorobaculum* (formerly *Chlorobium*) *tepidum*, are obligate anaerobic phototrophs and can grow under low light at rates observed for purple sulfur bacteria grown with much higher light intensities. For these AnAP bacteria, taxonomy and functioning are connected, both responding to the patchiness and gradients in light, substrate supply, and oxygen availability.

The carbon source for sulfur oxidizers

Sulfide oxidizers, other chemolithotrophs, and AnAP bacteria when not growing heterotrophically use carbon dioxide as their carbon source, making them autotrophs. The full name of the metabolism carried out by the colorless sulfide-oxidizing bacteria is chemolithoautotrophy, whereas it is photolithoautotrophy for the AnAP bacteria. For colorless sulfide oxidizers, the carbon dioxide fixation pathway is the same as for higher plants, eukaryotic algae, and cyanobacteria: the Calvin–Benson–Bassham (CBB) cycle (Hanson et al., 2012b). Bacteria capable of oxidizing elemental sulfur also use the reverse trichloroacetic acid cycle (rTCA) or the 3-hydroxypropionate/4-hydroxybutyrate (3-HPP) pathway. Depending on the species, AnAP bacteria fix carbon dioxide by the rTCA cycle or by the 3-HPP pathway, in addition to the CBB cycle.

Methane and methanogenesis

Carbon dioxide reduction is another branch of the anaerobic food chain that is common in freshwaters and waterlogged soils where concentrations of sulfate and all other electron acceptors are low. What gives this process added importance and global significance is the end product of carbon dioxide reduction, methane. Although its concentration is 100-fold less than that of carbon dioxide in the atmosphere, methane is over 20-fold more effective than carbon dioxide in trapping heat, as mentioned in Chapter 1. Both gases have been increasing, albeit at different rates, since the nineteenth century (Figure 11.12).

Unlike carbon dioxide, the anthropogenic inputs of methane now exceed natural ones—human activity accounts for two-thirds of methane emissions (Nisbet et al., 2014). Again unlike carbon dioxide, prominent anthropogenic inputs of methane are agricultural, with emissions via belches and flatulence by cows, other ruminants, and termites high on the list. Rice paddies and other anoxic habitats on land are also major sources of methane, and some methane escapes into the atmosphere during mining or transport of natural gas and other fossil fuels. Natural gas is mostly methane, of which some is directly from methanogens, while the rest is from geothermal reactions working on preserved organic material.

Methanogenesis is carried out by strict anaerobes in the *Euryarchaeota* phylum of archaea (Whitman et al., 2001). There are five well-defined orders of methanogens: *Methanobacteriales*, *Methanomicrobiales*, *Methanococcales*, *Methanosarcinales*, and *Methanopyrales*, a deeply branching order of hyperthermophiles. Its ecophysiology and

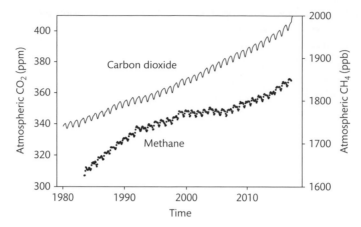

Figure 11.12 Methane and carbon dioxide concentrations in the atmosphere (global means). After pausing between 2000 and 2006 for unclear reasons, methane concentrations are again on the rise (Prather and Holmes, 2017). The data sets were used with permission from Edward J. Dlugokencky at the NOAA Earth System Research Laboratory.

position in phylogenetic trees suggest that members of *Methanopyrales* arose early in the history of life. As mentioned in Chapter 1, one hypothesis is that the first cell on the planet used carbon dioxide as an electron acceptor and H_2 as the electron donor. The methanogens differ in their cell wall and how they produce methane, among other characteristics.

The anaerobic food chain model emphasizes the role of acetate, carbon dioxide, and H_2 in producing methane, and these compounds are arguably the most important. However, several other compounds are potentially involved as well (Table 11.4). These include several single carbon compounds (C_1), such as methanol and even carbon monoxide (CO). When carbon dioxide is used, the electron donor is H_2. The other compounds used in methanogenesis undergo a disproportionation reaction (Box 11.4), so a reductant is not needed. The following equation with formate ($HCOO^-$) is one example of this type of reaction:

$$4HCOO^- + 4H^+ \rightarrow CH_4 + 3CO_2 + 2H_2O \qquad (11.9).$$

This reaction yields slightly less energy than the reduction of CO_2 with H_2, but another prominent disproportionation reaction, that with acetate to produce methane, yields much less energy (Table 11.4). Less understood is methane production by methanogens using more complicated starting material, such as aromatic organic material (Mayumi et al., 2016).

Table 11.4 Reactions and energetic yields for methanogenesis. Taken from Garcia et al. (2000).

Reaction	Energetic yield ($\Delta G^{o\prime}$, kJ mol^{-1} CH$_4$)
$H_2 + CO_2 \rightarrow CH_4 + 2H_2O$	−135.6
$4\,Formate \rightarrow CH_4 + 3CO_2 + 2H_2O$	−130.1
$2\,Ethanol + CO_2 \rightarrow CH_4 + 2\,Acetate$	−116.3
$4\,Methanol \rightarrow 3CH_4 + CO_2 + 2H_2O$	−104.9
$4\,Methylamine + 2H_2O \rightarrow 3CH_4 + CO_2 + 4NH_4^+$	−75.0
$4\,Trimethylamine + 6H_2O \rightarrow 9CH_4 + 3CO_2 + 4NH_4^+$	−74.3
$2\,Dimethylsulfide + 2H_2O \rightarrow 3CH_4 + CO_2 + H_2S$	−73.8
$Acetate \rightarrow CH_4 + CO_2$	−31.0

Thermodynamics seems to explain why methanogens are not abundant and why methanogenesis does not operate where other electron acceptors such as oxygen, ferric iron, and sulfate are present. Sulfate reduction and all other forms of respiration are energetically more favorable than methanogenesis. However, thermodynamics does not explain how this energetic advantage is manifested at the physiological level of the microbes. The energetic advantage shows up in uptake kinetics for two key compounds used by both methanogens and sulfate reducers. When sufficient sulfate is available, sulfate reducers outcompete methanogens because the half-saturation constants (K_m) for acetate and H_2 use by sulfate reducers are much lower than those for

Box 11.4 Disproportionation

Methanogenesis using acetate is a disproportionation reaction. This type of reaction (also called dismutation) can be described as $2A \rightarrow A' + A''$, where A, A', and A'' are different chemicals but contain the same main element, such as carbon in Equation 11.9. Fermentation can be considered as a type of disproportionation. Other examples of disproportionation are those involving sulfur compounds of intermediate oxidation state. One reaction is

$$4S° + 4H_2O \rightarrow 3H_2S + SO_4^{2-} + 2H^+$$

A variety of obligate anaerobic bacteria carry out disproportionation reactions of sulfur compounds. These reactions have only relatively recently been discovered, long after fermentation was well understood.

methanogens (Muyzer and Stams, 2008). Consequently, uptake by sulfate reducers and other microbes decreases acetate and hydrogen concentrations to levels too low for methanogenesis to operate.

The most important biotic source of methane is methanogenesis by strict anaerobic archaea that are active only in anoxic habitats. So any hint of another methane-producing process carried out by other organisms, or of methane in an oxic habitat, attracts lots of attention. Anything about this potent greenhouse gas is newsworthy. Some oxic habitats can have measurable methane because they are near anoxic habitats with high methanogenic activity, and anoxic micro-habitats in an otherwise oxic habitat can harbor active methanogens. More unusual are reports of methane production by organisms other than archaea, such as bacteria, fungi, algae, and even higher plants by a variety of mechanisms, some better understood than others (Lenhart et al., 2016; Nisbet et al., 2009). One mechanism involves bacteria using methylphosphonate ($CH_3HPO_3^-$) in oxic habitats (Yao et al., 2016); this unusual organic compound with its C–P bond is broken down by microbes to obtain the P, releasing methane as a waste product. Other work has suggested that the C–P lyase pathway acting on other C–P compounds, such as polysaccharide esters of phosphonic acids, produces methane (Repeta et al., 2016).

However, the amount of methane from this and all of the other aerobic pathways is small compared with that from methanogenesis by archaea.

Methanotrophy

Methane is produced by methanogens as a by-product of energy-generating reactions, yet ironically the gas is used by many "methanotrophic" microbes as an energy source. Methanotrophy is an important part of the carbon cycle in methane-rich environments and contributes to setting methane fluxes out of those environments. Atmospheric methane concentrations would be even higher if not for methanotrophy. Environments such as rice paddies with high rates of methanogenesis can have equally high rates of methanotrophy, although only about 20% of the methane produced in rice paddies is oxidized by methanotrophs before reaching the atmosphere (Conrad, 2009). Globally, soils are the dominant biological sink for atmospheric methane. Once in the atmosphere, methane can be oxidized by ·OH radicals, resulting in a residence time of about eight years.

Aerobic methane degradation
As with sulfide, another by-product of anaerobic metabolism, methane is oxidized by aerobic methanotrophs situated at the interface between oxic and anoxic habitats. Methane oxidation with oxygen as the electron acceptor is described by

$$CH_4 + 2O_2 \rightarrow HCO_3^- + H^+ + H_2O$$
$$\Delta G^{o'} = -814 \text{ kJ mol}^{-1} \quad (11.10).$$

The known aerobic methanotrophs mostly are *Alpha-* and *Gammaproteobacteria* and differ in key steps in the methane oxidation pathway (Figure 11.13). All methanotrophs have internal membranes presumed to be involved in methane oxidation, although the arrangement of these membranes differs among the types of methanotrophs. All known methanotrophs also have the same first step in the pathway, the oxidation of methane to methanol by particulate methane mono-oxygenase (pMMO); "particulate" here refers to membranes. Only one methanotroph (*Methylocella*) does not have pMMO, while many also have a soluble methane mono-oxygenase.

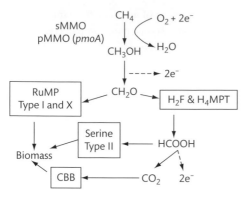

Figure 11.13 Pathway for aerobic methane oxidation. Two key enzymes are pMMO = particulate methane mono-oxygenase (and the gene *pmo*A for one of the pMMO subunits, a common target for studying uncultivated aerobic methanotrophs) and sMMO = soluble methane mono-oxygenase. RuMP = ribulose monophosphate, H_4F = tetrahydrofolate, H_4MPT = tetrahydromethanopterin, and CBB = Calvin–Benson–Bassham cycle. Type 1 and Type X methanotrophs use the RuMP pathway, while Type II uses the serine pathway. The steps with electrons (e⁻) indicate connections to the electron transfer system and ATP production. Modified from Chistoserdova et al. (2005).

Methanotrophs generally use only methane, although many can also use methanol (CH_3OH), and at least one has been shown to use acetate (Conrad, 2009). While some methanotrophs can oxidize other compounds, they cannot grow on these. A soil bacterium in the *Verrucomicrobia* phylum is able to simultaneously use H_2 and methane (Carere et al., 2017). Methanotrophs are often included in a broader group of bacteria, "methylotrophs," defined by the capacity to oxidize and grow on compounds with only one carbon atom (C_1) such as formate and carbon monoxide, in addition to methane and methanol.

Anaerobic methane oxidation

Long after the discovery of aerobic methanotrophy, microbes were found to be carrying out the anaerobic oxidation of methane (AOM). AOM has generated much interest and research over the last few years. Just as any methane-producing process attracts attention, so too do processes that consume this potent greenhouse gas.

Microbiologists and microbial physiologists are intrigued by the intricate, anaerobic biochemical pathways that turn methane back into CO_2, and microbial ecologists are fascinated by the novel interactions between archaea and bacteria involved in AOM. There is even the possibility of harnessing AOM to produce electricity in microbial fuel cells (McAnulty et al., 2017).

The discovery of AOM was delayed in part because microbiologists did not think it was possible. But geochemists long had evidence that methane was consumed in anoxic sediments by the following reaction (Figure 11.14):

$$CH_4 + SO_4^{2-} \rightarrow HS^- + HCO_3^- + 2H_2O$$
$$\Delta G^{\circ'} = -16.7 \text{ kJ mol}^{-1} \qquad (11.11).$$

Microbiologists were not convinced that microbes mediated this reaction because they were unable to isolate an organism capable of carrying it out. Early work did show that isolates of methanogens can oxidize methane anaerobically (Zehnder and Brock, 1979), but the rates were not high enough to explain the geochemical evidence. Nevertheless, Zehnder and Brock proposed that in nature methane is oxidized by methanogens carrying out "reverse methanogenesis" that produce H_2 or acetate which is subsequently used by a sulfate reducer.

An important piece of the puzzle was more geochemical evidence indicating that the lipids of archaea were highly depleted in ^{13}C in sediments near a methane seep (Hinrichs et al., 1999). The ^{13}C data made sense only if the lipid carbon came from methane; both methanogenic and geothermal production of methane favors ^{12}C over the heavier ^{13}C, resulting in very negative $\delta^{13}C$ values for methane. Analysis of the 16S rRNA genes turned up a new cluster of archaea, ANME-1 (anaerobic methanotrophic *Euryarchaeota*), related to methanogens. Soon after the geochemical study, microbial ecologists applied fluorescence in situ hybridization (FISH) to samples from methane seeps and found aggregates of archaeal cells surrounded by sulfate-reducing bacteria (Boetius et al., 2000). More 16S rRNA gene sequencing revealed that the archaeal cells were in another cluster, ANME-2, related to ANME-1. Data showing that ANME-2 cells carry out methane oxidation came from FISH-type experiments

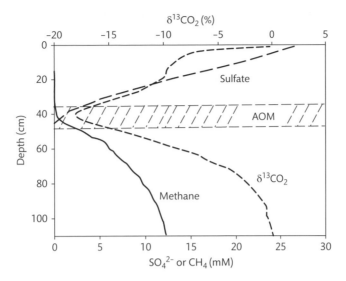

Figure 11.14 Geochemical evidence for the anaerobic oxidation of methane (AOM). The minimum in sulfate concentrations coinciding with the disappearance of methane without any oxygen can only be explained by AOM. It would also explain the low amount of ^{13}C in the CO_2 pool (low $\delta^{13}CO_2$). Data from Reeburgh (1980).

using secondary ion mass spectrometry (SIMS), which demonstrated that organic carbon in the archaeal cells was highly depleted in ^{13}C (Orphan et al., 2001).

The initial studies focused on AOM coupled to sulfate reduction in which an archaeon oxidized the methane via reverse methanogenesis, producing electrons that were somehow transferred via "interspecies electron carriers" to a sulfate-reducing bacterial partner (Figure 11.15). More recent work has addressed the interspecies electron carriers and the need for a bacterial partner. Some obvious candidates for the carriers were the two originally suggested by Zehnder and Brock: H_2 and acetate, both well-known to be important in anaerobic metabolism. But these and others were shown not to be involved in AOM. There is some evidence for polysulfides. Even more intriguing is the direct transfer of electrons from a ANME cell to the sulfate-reducing bacterium via multiheme cytochrome c proteins (McGlynn et al., 2015) or electrically conductive pili, an example of direct interspecies electron transfer (Lovley, 2017).

There is evidence, however, of archaea carrying out AOM without any syntrophic bacterial partner. For example, ANME-2d alone can anaerobically oxidize methane using nitrate as the electron acceptor (Haroon

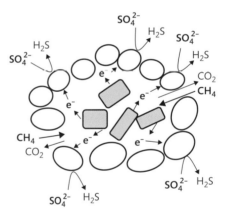

Figure 11.15 Anaerobic methane oxidation by archaea (shaded rectangles) surrounded by sulfate-reducing bacteria (open ovals). The cells are much closer together than actually depicted here. The main text describes possible mechanisms for transferring electrons (e⁻) from the ANME archaea to the sulfate reducers.

et al., 2013). Still, ANME-2d do form consortia with another type of bacterium, one that anaerobically oxidizes nitrite to N_2 gas (Haroon et al., 2013). This helps ANME-2d by removing an end product of AOM (nitrite), thus making the reaction more thermodynamically favorable.

Table 11.5 Possible electron acceptors, energy yield, and ANME clades involved in the anaerobic oxidation of methane. The potential energetic yield is the change in the Gibbs free energy under standard conditions (ΔG°). Data from Timmers et al. (2017).

Electron acceptor	Reaction	ΔG° (kJ mol^{-1})	ANME clades
Sulfate	$CH_4 + SO_4^{2-} \rightarrow HCO_3^- + HS^- + H_2O$	−16.3	ANME-1; ANME-2a,-2b, and -2c
Nitrate	$CH_4 + 4NO_3^- \rightarrow HCO_3^- + 4 NO_2^- + H_2O + H^+$	−517.2	ANME-2d
Iron oxide	$CH_4 + 8Fe(OH)_3 + 16H^+ \rightarrow CO_2 + 8Fe^{2+} + 22H_2O$	−571.2	ANME-2d; ANME-3
Manganese oxide	$CH_4 + 4MnO_2 + 8H^+ \rightarrow CO_2 + 4Mn_2^+ + 6H_2O$	−763.2	ANME-3
Chromium	$CH_4 + 4/3Cr_2O_7^{2-} + 32/3H^+ \rightarrow 8/3Cr_3^+ + CO_2 + 22/3H_2O$	−841.4	ANME-2d

More has been learned about the diversity of microbes carrying out AOM from additional work with 16S rRNA, other genes, and metagenomic approaches; so far, no ANME strain has been isolated and maintained in pure culture. Studies of a key functional gene, the alpha subunit of methyl-coenzyme M reductase (*mcrA*), have told us much about ANME diversity and relationships to known methanogens. Many insights into possible AOM metabolism have also come from metagenomes of these organisms (Timmers et al., 2017). In addition to ANME-1, the ANME-2 clade now appears to have at least four subclades (ANME-2a, -2b, -2c, and -2d), and there is ANME-3, related to the methanogen *Methanococcoides*. These ANME clades are not closely related, having 16S rRNA genes that are only 75–92% similar to each other. They appear to occur in different environments, possibly use different electron acceptors, and interact or not with bacterial partners by different mechanisms. For example, ANME-2d is found in freshwaters and soils, while the other clades have been found in marine sediments. AOM may be substantial in anoxic soils (Gauthier et al., 2015).

AOM may not seem likely in soils or other environments with low sulfate concentrations, but methane oxidation is now known to occur in non-marine anoxic environments and to involve other electron acceptors that potentially yield even more energy than sulfate (Table 11.5). Some AOM activity may be carried out by bacteria coupled to the reduction of nitrite (NO_2^-) without help from archaea (Ettwig et al., 2010). These bacteria, belonging to a poorly characterized phylum known only as "NC10," reduce nitrite to N_2 gas and generate molecular oxygen which then is used by the bacterium's pMMO in aerobic methane oxidation. A partnership between ANME-2d and an anaerobic ammonium-oxidizing (anammox) bacterium may also produce N_2 gas during AOM (Haroon et al., 2013). Whatever the mechanism, methane could fuel the loss of nitrogen from the environment.

Microbial ecologists have progressed greatly from the days of not believing that AOM existed, yet much more needs to be done before we completely understand this important process.

Anaerobic microbial eukaryotes

While bacteria and archaea dominate anoxic environments, some eukaryotes are present and are active. Some metazoans can survive in the absence of oxygen, occasionally for extended periods of time, but none can reproduce and live indefinitely without oxygen. Metazoans are also affected by by-products from anaerobic metabolism, most notably hydrogen sulfide (H_2S), which is toxic to many organisms. Much more successful in anoxic environments are protists and fungi. Yeasts are well known for their fermentation pathways and end products, and there are studies of their distribution in vineyards, fruits, and soils, but little is known about their role in anoxic environments. Anaerobic protists include species of flagellates (including dinoflagellates) and ciliates. Like metazoans, some protists may switch metabolisms to survive anoxia, or form resting stages (cysts) and wait until oxygen returns.

The ecological roles of eukaryotic microbes in anoxic environments are similar to those seen in oxic environments, but overall the contribution of these microbes is much reduced. Fungi are not as abundant nor do they degrade as much organic material in anoxic as in oxic environments. Mineralization via fermentation by diatoms may occasionally be substantial (Bourke et al., 2017), but bacteria and archaea usually are more important. Anaerobic protists can graze on other microbes in anoxic environments, but the few studies examining this process have found low rates (Oikonomou et al., 2014). A few microbial eukaryotes can carry out one form of anaerobic respiration, dissimilatory nitrate reduction, and contribute to the N cycle (Kamp et al., 2015). The first eukaryote shown to carry out dissimilatory nitrate reduction was the freshwater ciliate *Loxodes*, which reduces nitrate to nitrite. Since then, the list of nitrate-respiring eukaryotes has grown to include diatoms, other protists (foraminifera), and fungi.

Overall, however, the abundance and diversity of eukaryotes in anoxic environments is low. Except for the assimilation of sulfate and organic sulfur for biomass synthesis, eukaryotes are not directly involved in the sulfur cycle. With regard to the carbon cycle, eukaryotes may produce some methane but only in oxic environments, and they do not consume this important greenhouse gas.

While their diversity and abundance may be low, some anaerobic protists are important and interesting for other reasons. *Giardia* is an anaerobic flagellate that lives in the small intestine of humans and other animals, causing diarrhea when it attaches to epithelial cells of its host. It survives outside its host as a cyst and is transmitted by ingestion of fecal-contaminated water. Aside from public health concerns, *Giardia* and related organisms, including those in the trichomonad order, are interesting because they provide clues about the evolution of primitive eukaryotes. *Giardia* and other anaerobic protozoa occupy deep branches in the Tree of Life, suggesting that they were among the first eukaryotes to appear on the planet (Figure 11.16).

Further support for that hypothesis came from the observation that these protozoa lack mitochondria. At first, the microbes appeared to be missing links between prokaryotes and fully equipped eukaryotes, but subsequent work demonstrated that amitochondriate protozoa lost their mitochondria during evolution while

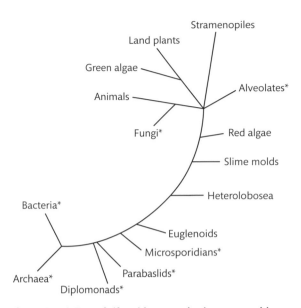

Figure 11.16 Tree of Life, with an emphasis on anaerobic organisms. All taxa with a * have representatives that can grow in anoxic habitats. The protistan groups without mitochondria are Diplomonads (which includes *Giardia*), Parabasalids, and Microsporidians. This tree was modified from one in Dacks and Doolittle (2001).

retaining some mitochondria-like proteins (Smith and Keeling, 2015). In some protozoa, the mitochondrion evolved into a hydrogenosome where ATP is generated from the oxidation of pyruvate, producing H_2 as a by-product. Other anaerobic protozoa, including *Giardia*, have a mitochondrion-like organelle, called a mitosome, whose function remains unknown. Mitosomes are much smaller than mitochondria and are not involved in ATP generation. They may synthesize Fe–S proteins. Some anaerobic protozoa have symbiotic bacteria or methanogenic archaea. One cellulose-degrading protozoan is lined with a wriggling fringe of motile spirochetes, a type of bacterium, that has been hypothesized to be the predecessor of cilia in eukaryotes (Wier et al., 2010).

While protists and other microbial eukaryotes may not contribute much to biogeochemical processes in anoxic habitats, they are important in thinking about early life on the planet and provide more fascinating examples of diversity in the microbial world.

Summary

1. Thermodynamics explains why oxygen and nitrate are preferred electron acceptors, whereas sulfate and carbon dioxide are major terminal electron acceptors in anoxic environments because of their high concentrations.

2. In the absence of oxygen, organic material is mineralized by a complex consortium of microbes in the anaerobic food chain. Two key compounds include acetate and H_2 produced by fermenting bacteria and acetogens. These compounds are then taken up by bacteria using sulfate or carbon dioxide as terminal electron acceptors.

3. Hydrogen sulfide and other reduced sulfur compounds produced by sulfate reduction are oxidized in the dark by colorless sulfur-oxidizing bacteria and in the light also by anaerobic anoxygenic phototrophic bacteria.

4. Methane, an important greenhouse gas, is produced only by strictly anaerobic archaea from the reduction of carbon dioxide coupled to the oxidation of H_2, or from the disproportionation of acetate, methanol, methylamines, and a few other compounds. Methanogens are outcompeted by sulfate reducers who use many of these same compounds, especially acetate and H_2.

5. Methane is degraded aerobically by specific methanotrophic bacteria or anaerobically by archaea in the ANME clades, carrying out reverse methanogenesis, sometimes in tight consortia with sulfate-reducing bacteria.

6. Although a few protists are capable of generating ATP from dissimilatory nitrate reduction, most anaerobic protists gain energy from fermentation. In many of these microbes, the mitochondrion has evolved into hydrogenosomes or mitosomes, providing fascinating examples of evolution in progress.

CHAPTER 12

The nitrogen cycle

We have already seen some important parts of the nitrogen cycle, such as N_2 fixation and ammonium assimilation and regeneration (Figure 12.1). Fixation of N_2 gas by a few bacteria and archaea produces a nitrogenous form, ammonium (NH_4^+), usable as a nitrogen source for biosynthesis by many prokaryotic and eukaryotic microbes and higher plants. Ammonium is an important by-product of organic material mineralization (Chapter 7). Chapter 11 mentioned briefly another important part of the cycle, the use of nitrate as an electron acceptor in a type of anaerobic respiration. But we have not discussed yet how nitrate is formed. This is one of several parts of the nitrogen cycle discussed in detail in this chapter.

Nitrogen is the only element with its own chapter in this book. It deserves special treatment for several reasons. Because microbes require so much nitrogen (Chapter 2), the supply of fixed nitrogen compounds often limits the growth and biomass of all organisms in terrestrial

and aquatic ecosystems. Nitrogen is involved in several important redox reactions because it can take on many oxidation states (Figure 12.2), ranging from −3 in ammonium $\left(NH_4^+\right)$ to +5 in nitrate $\left(NO_3^-\right)$. Consequently, many nitrogenous compounds are involved in catabolic, energy-generating reactions, either as electron donors or acceptors. Nitrogen is involved in so many biogeochemical cycles and microbial physiologies that it needs its own chapter.

Another reason for paying special attention to nitrogen is that one nitrogenous compound, nitrous oxide (N_2O), is a potent greenhouse gas, being about 270-fold more effective in trapping heat than carbon dioxide, as mentioned in Chapter 1. This nitrogen gas is third behind carbon dioxide and methane (not counting water vapor) in contributing to the overall greenhouse effect. Nitrous oxide is also now the most potent destroyer of ozone (Ravishankara et al., 2009). Concentrations of N_2O have increased in the atmosphere, along with carbon dioxide and methane, over the past 100 years. Similar to methane, increases in atmospheric N_2O are mainly due to increases in agriculture.

Humans have several other impacts on the nitrogen cycle (Fowler et al., 2015), starting with nitrogen fixation. The Haber–Bosch process fixes N_2 to make ammonium for fertilizer and other industrial applications. The rate of this anthropogenic N_2 fixation, together with nitrogen fixation by human-managed legumes (Chapter 14), is about equal to or may even exceed natural N_2 fixation rates in the biosphere. On land, its impact is even greater: anthropogenic sources of fixed nitrogen are two-fold larger than natural ones in terrestrial systems. Humans

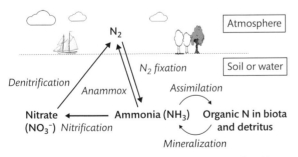

Figure 12.1 A simple depiction of the nitrogen cycle with only some of the relevant processes and nitrogenous compounds. "Anammox" is anaerobic ammonium oxidation, which produces N_2 from ammonium and nitrite.

Processes in Microbial Ecology. Second Edition. David L. Kirchman. Oxford University Press (2018). © David L. Kirchman 2018.
DOI 10.1093/oso/9780198789406.001.0001

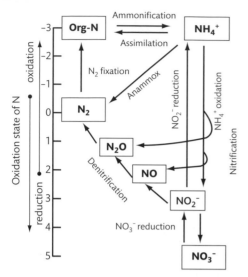

Figure 12.2 Some compounds and reactions in the N cycle. "Org-N" refers to organic nitrogen. Adapted from Capone (2000).

add still more nitrogen in the form of nitric oxide (NO) to the biosphere from wildfires and the burning of fossil fuels. Nitric oxide is a major component of acid rain. Over 80% of all NO emissions are thought to be from human activities, and in some regions, the anthropogenic sources exceed natural inputs by ten-fold.

Nitrogen fixation

The capacity for N_2 fixation (diazotrophy) is widespread among bacteria and archaea, but it is a rather specialized process, carried out by relatively few microbes. It is not done by any eukaryote (not counting the Haber–Bosch process), although some eukaryotic microbes and higher plants form symbiotic relationships with diazotrophs (Chapter 14). Nitrogen fixation is found in prokaryotes carrying out every energy-generating form of metabolism: aerobic and anaerobic heterotrophy (chemoorganotrophy), oxygenic phototrophy (cyanobacteria only), anaerobic anoxygenic phototrophy, and chemolithotrophy. The one exception is aerobic anoxygenic phototrophy (AAP); no known AAP bacterium is capable of N_2 fixation.

It is remarkable that diazotrophs can accomplish at room temperature under normal atmospheric pressure what the Haber–Bosch process can do only at high temperature (300–550 °C) and pressure (15–26 MPa).

Nitrogenase, the N_2-fixing enzyme

Nitrogen fixation is the reduction of nitrogen gas to ammonia and can be described by

$$N_2 + 8H^+ + 8e^- + 16ATP \rightarrow 2NH_3 + H_2 + 16ADP \qquad (12.1)$$

where the reducing power ($8e^-$) is supplied by NAD(P)H. Although Equation 12.1 indicates that N_2 fixation needs 16 ATPs, energetic costs may be even higher under natural conditions, up to 30 ATPs for some microbes. The huge energetic cost of N_2 fixation helps to explain the limited distribution of the process among organisms and environments. It also helps to explain why every prokaryote does not routinely carry out N_2 fixation, even in nitrogen-limited environments. Finally, it explains why microbes turn off N_2 fixation when fixed nitrogen, especially ammonium, is available. Energetic costs are high due to the difficulty of breaking the triple bond of nitrogen gas (N≡N) to form ammonia.

The key enzyme carrying out N_2 fixation is nitrogenase, a huge enzyme, as large as 300 kDa, which can make up as much as 30% of cellular protein in diazotrophs. Nitrogenase is actually a complex of two proteins, dinitrogenase and dinitrogenase reductase (Figure 12.3). One contains iron (Fe) and either molybdenum (Mo) or vanadium (V), while the other contains only Fe. Dinitrogenase reductase, the Fe protein coded for by the *nifH* gene, has two

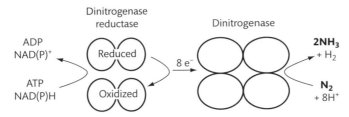

Figure 12.3 N$_2$ fixation by the nitrogenase complex. Dinitrogenase reductase, the Fe protein, has two subunits encoded by the *nifH* gene. Dinitrogenase is a Mo–Fe protein with four subunits, two encoded by *nifD* and another two encoded by *nifK*. Dinitrogenase reductase is about 60,000 Da, while dinitrogenase is about 240,000 Da.

Table 12.1 Some types of N$_2$ fixing prokaryotes and strategies to protect nitrogenase from oxygen poisoning. The O$_2$ protection strategy for *Rhizobium* is labeled as "physical" because the host plant for this symbiotic diazotroph physically limits the flow of oxygen to the bacterium. Several nitrogen fixers use more than one strategy. *Oscillatoria* partially avoids oxygen by living in a low-oxygen environment ("avoidance") as well as separating oxygen-producing photosynthesis from N$_2$ fixation over time ("time separation"). Some cells in filaments of the non-heterocystous cyanobacterium *Trichodesmium* carry out photosynthesis, while others fix nitrogen ("spatial separation"). The soil bacterium *Azotobacter* is famous for lowering oxygen to tolerable levels with high respiration.

O$_2$ protection	Genus	Phylum or class	Environment
Heterocyst	*Anabaena*	*Cyanobacteria*	Freshwaters
Heterocyst	*Nostoc*	*Cyanobacteria*	Microbial mats
Heterocyst	*Nodularia*	*Cyanobacteria*	Microbial mats
Heterocyst	*Richelia*	*Cyanobacteria*	Endosymbiosis
Physical	*Rhizobium*	*Alphaproteobacteria*	Soil endosymbiosis
Time and spatial separation	*Trichodesmium*	*Cyanobacteria*	Marine waters
Time separation	*Oscillatoria*	*Cyanobacteria*	Microbial mats
Respiration	*Azorhizobium*	*Alphaproteobacteria*	Soils
Respiration	*Azotobacter*	*Gammaproteobacteria*	Soils
Respiration	*Azospirillum*	*Alphaproteobacteria*	Soils
Avoidance	*Methanosarcina*	*Euryarchaeata*	Various
Avoidance	*Clostridium*	*Firmicutes*	Various
Avoidance	*Chlorobium*	*Chlorobi*	Freshwaters

identical subunits with about four Fe atoms. The subunits of the Mo–Fe protein are encoded by the *nifD* and the *nifK* genes. Nitrogenase is thought to have appeared early in evolution. The diversity of diazotrophs is one argument for it being an ancient process, potentially evolving soon after the first cell came into existence. There is some speculation that nitrogenase had another main function, such as reduction of cyanide or carbon monoxide, when it first appeared in the biosphere (Lee et al., 2010).

Solving the oxygen problem

Nitrogenase is irreversibly damaged by oxygen, creating a problem for the many diazotrophs that live in oxic environments, particularly those that evolve oxygen during photosynthesis. Reviewing the strategies for solving this oxygen problem serves to introduce some of the best-known nitrogen fixers and to illustrate their diversity (Table 12.1). The oxygen protection strategy taken by filamentous cyanobacteria, such as *Anabaena*, was mentioned in Chapter 6. These microbes house nitrogenase in a specialized cell, the heterocyst. Its thick cell walls physically limit oxygen diffusion and help to keep oxygen concentrations around the nitrogenase low. Unique among the cells of a cyanobacterial filament, heterocysts lack the oxygen-producing part of photosynthesis, photosystem II (PS II). The other, vegetative cells of the filament feed sugars and organic acids to heterocysts in exchange for fixed nitrogen in the form of glutamate. Organic carbon-fueled respiration also helps to keep oxygen concentrations low in heterocysts. Analogous to the heterocyst strategy, the soil actinomycete, *Frankia*, produces a vesicle to help protect its nitrogenase. Other microbes fix nitrogen when coated in a polysaccharide-rich slime that minimizes oxygen diffusion. Free-living soil bacteria in the genus *Azobacteria* use a couple of oxygen protection strategies. One is to maintain high respiration rates to consume oxygen. This respiration may even be uncoupled from ATP synthesis. Some species in this genus produce proteins that bind to nitrogenase and protect it from oxygen.

N_2 fixation in nature

On a global scale, rates of N_2 fixation are roughly equal in the oceans and on land (Fowler et al., 2015). Ignoring anthropogenic sources, diazotrophs fix about 120×10^{12} g N y^{-1} in both systems (Figure 12.4), and diazotrophy is much larger than other natural sources of reactive nitrogen, such as fixation by lightning in the atmosphere.

Different types of diazotrophs are in soils, freshwaters, and marine systems. In soils, most studies have focused on agriculturally important symbiotic diazotrophs, such as the heterotrophic bacterium *Rhizobium* and its host, the legumes, which will be discussed in more detail in Chapter 14. Plants such as alder and buckthorn in natural environments also harbor symbiotic diazotrophs. Diazotrophic cyanobacteria associated with mosses account for as much as half of total N input in boreal forests

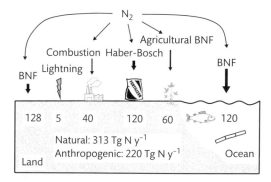

Figure 12.4 Global N_2 fixation by natural and anthropogenic processes. "BNF" is biological N_2 fixation mediated by microbes. Agricultural BNF refers to fixation by symbiotic bacteria in legumes. Based on Fowler et al. (2015).

(Rousk et al., 2013b). Terrestrial systems also have many free-living diazotrophs. Filamentous cyanobacteria with heterocysts such as *Anabaena* are the main N_2 fixers in freshwaters.

It had long been thought that *Trichodesmium* was the main diazotroph in the oceans, until small coccoid cyanobacteria capable of N_2 fixation were discovered by a combination of ^{15}N rate measurements and cultivation-independent studies of *nifH* (Zehr et al., 2016). Nitrogen fixation by these coccoid cyanobacteria and perhaps other small, heterotrophic bacteria can be substantial and exceed those by *Trichodesmium*, as shown by ^{15}N size fractionation experiments. The newly discovered cyanobacterial diazotrophs are much smaller, with a cell diameter of 10 μm or smaller, than *Trichodesmium* with its millimeter-length filaments. This difference in size has many implications for control of N_2 fixation and the interactions between diazotrophs and other organisms, including grazers.

Microbial ecologists had a hard enough time understanding how *Trichodesmium* fixes N_2 without heterocysts, so it was even harder to figure out how a small, unicellular cyanobacterium does it. Clues about the mechanism came from the genomic sequence of one abundant type (Zehr et al., 2016), termed UCYN-A (uncultivated cyanobacterium A), obtained by a single-cell genomic approach using flow cytometry (Chapter 5). The genomic data had many surprises: the UCYN-A genome is missing many genes, including those for the oxygen-evolving part of photosynthesis (photosystem II), the CO_2-fixing enzyme Rubisco, and parts of heterotrophic pathways for oxidizing organic carbon. The lack of photosystem II genes explained how it could fix N_2 during the day (solving another enigma at the time), but the absence of the other genes raised questions about how it obtained its energy and carbon. One possibility is that UCYN-A is a symbiont living with a host.

To identify the host, studies using flow cytometry focused on photosynthetic "picoeukaryotes," microbes smaller than 3 μm which could be isolated based on their pigments and size (Thompson et al., 2012). The isolated photosynthetic picoeukaryotes were tested for UCYN-A by looking for its *nifH* by PCR. This single-cell work found that the host picoeukaryote is a prymnesiophyte (also referred to as haptophytes), an algal class that includes calcifying species such as *Emiliania huxleyi*. We now know

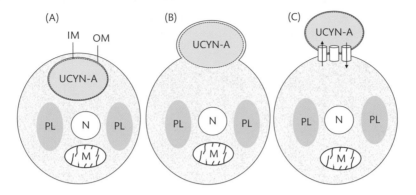

Figure 12.5 Possible relationships between the symbiotic diazotroph UCYN-A and its prymnesiophyte host. UCYN-A may be an endosymbiont fully embedded in the host (A), only partially enclosed by the host (B), or attached to the outside of the host with exchange facilitated by shared transporters (C). The inner membrane (IM) is represented by a dashed line, the outer membrane (OM) by a solid line. PL=plastids; N=nucleus; M=mitochondrion; Based on Zehr et al. (2016).

of several distinct pairs of hosts and UCYN-A strains, and more probably await discovery (Zehr et al., 2016). What is not known is the physical relationship between UCYN-A and its host (Figure 12.5). There is some evidence that UCYN-A is an endosymbiont like other diazotrophs known to live within diatoms, such as the relationship between the diatom *Epithemia turgida* and its diazotroph symbiont which is somewhat similar to UCYN-A. However, other evidence suggests that UCYN-A has a looser physical connection to its prymnesiophyte host and may even be attached to the outer membrane.

Box 12.2 Measuring rates in the N cycle

Nitrogen fixation and other N cycle rates are difficult to measure because of the lack of a convenient radioactive isotope of N and the difficulties of working with ^{15}N, a stable nitrogen isotope. There are alternatives to ^{15}N. Nitrogen fixation has been estimated by following the reduction of acetylene (C_2H_2) to ethylene (C_2H_4), which is easily measured by gas chromatography. Nitrogenase works on acetylene because it has a triple bond like that in N_2. Another part of the nitrogen cycle, nitrification (to be discussed further), has been studied using inhibitors, such as nitrapyrin and allylthiourea (inhibitors of ammonia oxidation), or chlorate (inhibitor of nitrite oxidation). These alternatives have proved useful in many studies, but their limitations prevent them from replacing ^{15}N for obtaining accurate estimates of N cycle rates.

Limitation of N_2 fixation

Microbial ecologists have wondered why rates of N_2 fixation are not higher and diazotrophs more abundant in nitrogen-limited systems, such as the oligotrophic oceans and boreal forests. There are several answers to these questions, reflecting the several factors governing N_2 fixation rates in nature (Figure 12.6). The relative importance of each factor varies with the diazotroph and the environment. Because of energetic costs, light limitation would lead to low N_2 fixation by phototrophs and a low supply of organic material would have the same effect on heterotrophic diazotrophs. The latter explains why heterotrophic bacteria are not more important in fixing N_2 in the oceans, whereas they account for much of the N_2 fixation in soils supplied with organic carbon from higher plants. Energetic cost also accounts for why N_2 fixation is shut down by high concentrations of ammonium and nitrate. Microbes can save energy by using these inorganic sources of nitrogen rather than fixing it.

Several inorganic nutrients in addition to ammonium and nitrate also affect N_2 fixation. The very low

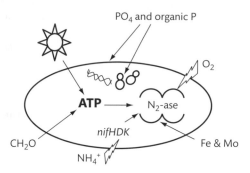

Figure 12.6 Factors affecting N_2 fixation by diazotrophs. The positive factors are indicated by simple arrows, while the two negative factors are depicted by lightning bolts. Because this process is so energetically expensive, fixation by heterotrophs and phototrophs may be limited by organic material (CH_2O) or light. Nitrogenase ("N_2-ase") requires iron (Fe) and molybdenum (Mo), whose concentrations may be too low. Diazotrophs need phosphorus supplied as phosphate ("PO_4^-") or in organic compounds for making DNA and ribosomes. Expression of nitrogenase genes (*nifHDK*) is negatively regulated by ammonium.

concentration of iron is a big reason why N_2 fixation is not higher in open-ocean regions. An alternative hypothesis is that diazotrophs in the open oceans are limited by low phosphate concentrations (Sañudo-Wilhelmy et al., 2001). A high input of phosphate can switch a lake from phosphorus limitation to nitrogen limitation, stimulating cyanobacterial diazotrophs; nasty cyanobacterial blooms have been caused by phosphorus in detergents and other phosphorus-rich contaminants polluting reservoirs and lakes. Concentrations of the other trace element used by nitrogenase, molybdenum, are also low but are apparently sufficient for N_2 fixation in most aquatic and terrestrial habitats. The exceptions may include some freshwaters and highly weathered acidic soils, such as those found in the tropics (Glass et al., 2010; Barron et al., 2009).

Temperature affects N_2 fixation and diazotroph abundance, as it does all microbial processes and all microbes. Temperature may explain why cyanobacterial diazotrophs in the oceans, such as *Trichodesmium*, occur as filaments without heterocysts, or are unicellular. One hypothesis is that solubility and diffusion of oxygen into regular cells is sufficiently limited by temperature and salinity such that heterocysts are not needed to protect nitrogenase from oxygen (Stal, 2009). Still, why heterocystous cyanobacteria are not more common in marine habitats remains a mystery.

Ammonium assimilation, regeneration, and fluxes

Once N_2 is fixed into ammonium by diazotrophs, a complex suite of reactions and organisms take over. Ammonium is an important nitrogen source for heterotrophic bacteria, eukaryotic algae, cyanobacteria, and higher plants. It is preferred over nitrate because the oxidation state of nitrogen in ammonium (−3) is the same as that in amino acids and other nitrogenous biochemicals in cells. Because of this preference, rates of ammonium uptake are often faster than nitrate uptake rates in both soils and aquatic habitats, even when nitrate concentrations are higher; nitrate uptake by microbes exceeds ammonium uptake only when ammonium concentrations are very low. Fluxes through the ammonium pool are high even though concentrations are low because of rapid production and uptake and, in soils, rapid exchange with clay and other negatively charged soil constituents.

In order for ammonium to be used for biosynthesis of all N-containing compounds in cells, it is first assimilated into the amino acid glutamate by one of two pathways found in all microbes, both prokaryotic and eukaryotic. The first pathway, designed for high ammonium concentrations, relies on the enzyme glutamate dehydrogenase (GDH) and uses only one NAD(P)H. The second is a high-affinity system for low ammonium concentrations. The first step of the second pathway consists of the synthesis of glutamine from glutamate, catalyzed by glutamine synthetase (GS):

$$glutamate + NH_4^+ + ATP \rightarrow glutamine + ADP \quad (12.2)$$

followed by the second step that yields the net production of one glutamate:

$$glutamine + \alpha - oxoglutarate + NADPH \rightarrow \\ 2glutamates + NADP^+ \quad (12.3).$$

The enzyme for the second step is glutamate synthase, also called glutamine-α-oxoglutarate transferase (GOGAT). The entire two-step pathway is called the GS-GOGAT pathway. With either pathway, the resulting glutamate supplies the nitrogen needed for all other amino acids and nitrogenous compounds within a cell.

Diazotrophs use the GS-GOGAT pathway to synthesize nitrogenous compounds, and then this fixed nitrogen enters the microbial food web by all of the mechanisms discussed in previous chapters: excretion, grazing, and viral lysis (Figure 12.7). Exchanges of nitrogen, depicted in Figure 12.7, are dominated by organisms other than the diazotrophs; rates of N_2 fixation are small compared with other internal nitrogen fluxes. In both aquatic and terrestrial environments, the internal nitrogen fluxes are roughly ten-fold greater than N_2 fixation.

When the organic nitrogenous compounds enter the detritus pool, they are mineralized eventually to ammonium by bacteria, fungi, and to a lesser extent other organisms, as discussed in Chapter 7. This production of ammonium is also referred to as "regeneration" or "ammonification." Mineralization of detrital protein to ammonium occurs after protein has been hydrolyzed to amino acids. Ammonium results from the deamination of an amino acid, $R-C(NH_2)COOH$, where "R" represents the various side chains of amino acids:

$$R-CHCOOH+NAD^+ +H_2O \rightarrow \quad R-CHCOOH$$
$$| \qquad\qquad\qquad\qquad\qquad\qquad \|$$
$$NH_2 \qquad\qquad\qquad\qquad\qquad\qquad O$$
$$+NH_4^+ +NADH$$

(12.4).

The by-products of deamination include an alpha-keto acid as well as ammonium. The reactions producing ammonium from other detrital organic nitrogenous compounds are more complicated and not as well understood.

Ammonium is the nitrogenous excretory product of many organisms, and it is an end product of the degradation of other excreted nitrogenous wastes. Many aquatic organisms excrete ammonium, but terrestrial animals and some aquatic ones excrete urea ($CO(NH_2)_2$) (Wright, 1995). Birds, some reptiles and amphibians, and most insects excrete uric acid ($C_5H_4N_4O_3$). Urea is degraded by urease to ammonium, whereas uric acid is first degraded by uricase (also called urate oxidase) to allantoin, and then eventually to ammonium.

In the absence of oxygen, ammonium is produced by anaerobic microbes using other electron acceptors by the mechanism just described (Equation 12.4). It is also produced by fermenting bacteria acting on the amino acids and nucleotide bases released by hydrolysis of nitrogenous macromolecules. Some of the best-known bacteria carrying out this type of fermentation are in the genus *Clostridium*. One example is fermentation of glycine,

$$4H_2N-CH_2-COOH+2H_2O \rightarrow 4NH_3 +2CO_2$$
$$+3CH_3-COOH$$

(12.5).

Ammonia oxidation and nitrification

We have seen that mineralization of organic material by both aerobic and anaerobic processes yields ammonium, yet concentrations of ammonium are often quite low in

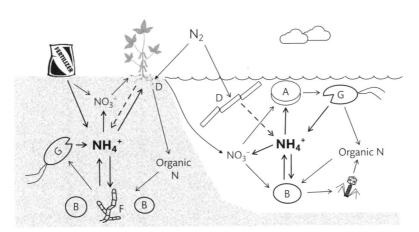

Figure 12.7 Internal cycling of nitrogen via ammonium (NH_4^+). Ammonium is released from diazotrophs ("D"; in water, represented here as a filamentous microbe and in soil by symbiotic bacteria) by several processes (dashed line) and is taken up by higher plants, algae ("A"), or bacteria ("B"). Grazers ("G"), bacteria, and fungi ("F") are sources of ammonium. In addition to N_2 fixation, external sources of nitrogen include fertilizer on land and runoff of NO_3^- from land into aquatic systems. Fluxes involving ammonium are in bold.

many natural environments. The more common inorganic form of fixed nitrogen is nitrate, one of the largest pools of nitrogen in the biosphere. In the deep ocean, for example, nitrate concentrations reach 40 μM, whereas all other forms of nitrogen, except N_2, are low, less than 10 nM. Nitrate concentrations vary greatly in soils, depending on water content and fertilization, but usually there is more nitrate than ammonium. Where does this nitrate come from?

The answer is "nitrification", the oxidation of ammonia to nitrite and then to nitrate (Figure 12.8). (Here we use "ammonia" (NH_3) because it is the actual substrate for this process even though concentrations of ammonium (NH_4^+) are higher in most environments. The switch from ammonium to ammonia is set by pH and the pKa of the reaction (Chapter 3).)

Bacteria and archaea oxidize ammonia and nitrite to gain energy, making them a type of chemolithotroph. Most of them are also chemoautotrophs and get their carbon from carbon dioxide. Nitrification has many practical ramifications because the end product, nitrate, is the starting compound for the process producing nitrogenous gases. Loss of nitrogen as a gas is good in a waste water treatment plant trying to minimize the release of excess nutrients, but bad in an agricultural field that has been fertilized to increase crop production. Nitrogen loss from fields is also potentially higher with nitrate, which is more mobile than ammonium in soils, because the negatively charged nitrate does not bind as readily as the positively charged ammonium to soil components.

There are two nitrification pathways. We will first discuss the most well-known pathway, the one with two steps, in which nitrification is carried out by two different groups of bacteria and archaea. The bacteria involved in this pathway were discovered by the Russian microbial ecologist Sergei Winogradsky over a hundred years ago (Box 12.3). Decades after Winogradsky's work, the second pathway in which one organism alone carries out nitrification was discovered.

The first step in nitrification is usually considered to be the rate-limiting one that sets the overall pace of the process. This first step is the oxidation of ammonia to nitrite (NO_2^-),

$$NH_3 + 1.5\,O_2 \rightarrow NO_2^- + H_2O + H^+$$
$$\Delta G^{\circ\prime} = -272\ \text{kJ mol}^{-1} \qquad (12.6)$$

Box 12.3 Winogradsky columns

Sergei Winogradsky (1856–1953) made several seminal discoveries in microbiology and microbial ecology, but probably only science historians would recognize his name if not for the eponymous "Winogradsky columns." These are made by mixing together mud, a bit of water, shredded newspaper, and perhaps an egg yolk, with or without sea salt. Pour the mixture into a glass or clear plastic bottle, preferably tall and narrow, and incubate undisturbed with some sunlight for several weeks. Over time, layers colored different hues of green and red will appear as the aerobic photoautotrophs grow at the surface and anaerobic anoxygenic phototrophs proliferate deeper down in the column.

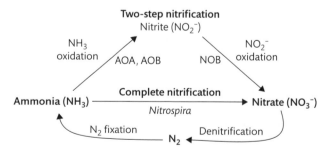

Figure 12.8 Two nitrification pathways and some other parts of the N cycle. The two-step pathway has been examined the most. The first step is ammonia (NH_3) oxidation by ammonia-oxidizing archaea (AOA) or bacteria (AOB). The second step is nitrite (NO_2^-) oxidation by bacteria (NOB). Complete nitrification is done by single organisms in the *Nitrospira* genus carrying out both ammonia and nitrite oxidation.

while the second step, the oxidation of nitrite to nitrate, completes the process:

$$NO_2^- + 0.5O_2 \rightarrow NO_3^- \qquad \Delta G^{\circ'} = -76 \text{ kJ mol}^{-1} \quad (12.7).$$

As is typical of chemolithotrophy, neither of these reactions yields much energy. Note also that nitrification is an aerobic process, and the microbes being discussed here depend on oxygen. Later we consider the anaerobic oxidation of ammonia, which shares very little in common with aerobic ammonia oxidation except that both processes oxidize ammonia.

In addition to the chemolithoautotrophic process, there is some production of nitrate during the aerobic oxidization of organic material by bacteria and fungi, by a process called "heterotrophic nitrification" (Zhang et al., 2015). However, rates by this type of nitrification are 10^3- to 10^4-fold slower than rates for chemolithotrophic nitrification. The mechanisms of heterotrophic nitrification are not well understood, but it is thought that heterotrophic nitrifiers do not gain energy from nitrogen oxidation, in contrast with chemolithotrophic nitrification. Heterotrophic nitrification has been examined mostly in soils, but it is thought to occur in aquatic habitats as well.

Aerobic ammonia oxidation by bacteria

Bacteria capable of ammonia oxidation make up a tight cluster of closely related organisms mostly in the *Betaproteobacteria*, along with a few in the *Gammaproteobacteria*. Classically, bacteria in the betaproteobacterial genera *Nitrosomonas* and *Nitrosospira* were considered to be the main microbes oxidizing ammonia in oxic environments. The classic picture is still largely correct, with some important additions. The cultivated ammonia oxidizers are strict chemolithoautotrophs, relying solely on ammonia as an energy source, although some can use urea to produce ammonia; urea degradation does not itself yield energy. As with all microbes and microbial processes, however, many ammonia oxidizers cannot be cultivated and grown in the laboratory. To examine these uncultivated ammonia oxidizers, microbial ecologists use PCR-based approaches to examine the gene for a subunit of a key enzyme, ammonia monooxygenase (*amoA*), catalyzing ammonia oxidation.

The *amoA* gene has proven to be a very powerful tool for exploring ammonia oxidation in nature. It can be used to identify these chemolithotrophic microbes because the phylogeny implied by *amoA* appears to match the 16S rRNA gene phylogeny of cultivated ammonia oxidizers (Figure 12.9). Using the *amoA* gene, microbial ecologists have mapped out the biogeography of ammonia oxidizers and have estimated their abundance in various habitats. These PCR-based surveys have found different clades of *amoA* genes in different habitats. Not surprisingly, marine ammonia oxidizers differ from those in soil and both differ from those in freshwaters. Betaproteobacterial ammonia oxidizers greatly outnumber gammaproteobacterial ammonia oxidizers in the habitats examined so far.

Ammonia oxidation by archaea

Chapter 5 briefly mentioned how metagenomic research opened up a new chapter in the study of ammonia oxidation and of archaea in natural environments. A metagenomic survey of the Sargasso Sea found an *amoA* gene linked to a phylogenetic marker from the *Thaumarchaeota* phylum, called *Crenarchaeota* at the time (Venter et al., 2004). Ammonia oxidation by archaea in pure culture had not been observed before. Although the archaeal *amoA* gene is similar enough to the bacterial to be recognized as an *amoA* gene, there are sufficient differences that archaeal and bacterial versions of this gene can be distinguished by standard PCR methods. The Sargasso Sea finding was soon applied to soil communities and many more. Ammonia-oxidizing archaea are the only group of archaea that are really abundant in oxic terrestrial and aquatic environments (Offre et al., 2013).

The abundance of ammonia-oxidizing archaea (AOA) and ammonia-oxidizing bacteria (AOB) has been assessed with quantitative PCR (qPCR), also called real-time PCR. qPCR studies have found that AOA abundance often exceeds AOB abundance in many environments, ranging from soils to the water column of the oceans (Offre et al., 2013). The implication is that AOA carry out more ammonia oxidation than do AOB. But it is difficult to determine conclusively whether ammonia oxidation is mostly by AOA or by AOB, because actual rate measurements by current methods cannot distinguish oxidation by archaea from that by bacteria.

One approach used in soils to explore this issue has been stable isotope probing (SIP), encountered several

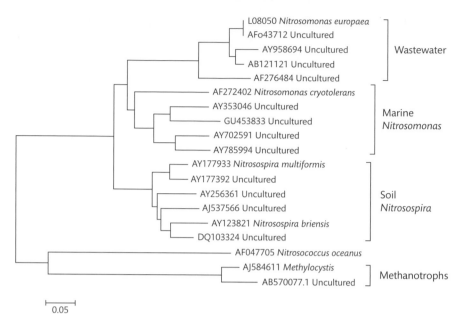

Figure 12.9 A phylogenetic tree of *amoA*, a key gene for ammonia oxidizers. The label with letters and numbers, such as LO8050, is a unique identifying number (the "accession number") for the sequence in GenBank. The phylogeny suggested by *amoA* sequences is very similar to that defined by 16S rRNA gene sequences. The methanotrophs are represented by genes for particulate methane mono-oxygenase (*pmoA*), which is distantly related to *amoA*. Ammonia oxidizers can oxidize methane and methane oxidizers can oxidize ammonia, albeit at ten-fold lower rates than the microbes specialized for the substrate. Tree provided by Glenn Christman and used with permission.

times already in this book. Being chemoautotrophs, the active ammonia oxidizers incorporate $^{13}CO_2$ into DNA, which can be separated by density gradient centrifugation from unlabeled, "light" DNA in organisms not active in assimilating the added $^{13}CO_2$. Jia and Conrad (2009) found that the ^{13}C-rich DNA contained AOB *amoA* genes but not those from AOA in agricultural soils, suggesting that bacteria were more active than archaea in incorporating CO_2 and presumably oxidizing ammonia. They also found that treatments that inhibited nitrification (addition of acetylene) or stimulated it (ammonium additions) affected the abundance of AOB *amoA* genes, but not those from AOA. The two lines of evidence argue strongly that bacteria dominate ammonia oxidation in these soils, even though AOA were much more abundant. However, the situation is different in acidic soils where AOB *amoA* genes are hard to detect (Prosser and Nicol, 2012). SIP assays found more $^{13}CO_2$ incorporation by archaea than bacteria in acidic soils (Zhang et al., 2017). So, sometimes the high abundance of AOA translates into high rates of ammonia oxidation, but sometimes it does not.

The abundance of AOA, as a fraction of total prokaryotic abundance, is quite high in the deep ocean. In this vast habitat, fluorescence in situ hybridization (FISH) studies has shown that archaea in the *Thaumarchaeota* phylum are abundant and account for as much as half of all microbes, as discussed in Chapter 4. Subsequent work with qPCR has confirmed the FISH results (Figure 12.10A). Two lines of evidence point to most, if not all, of the *Thaumarchaeota* being ammonia oxidizers.

First, qPCR studies found that the abundance of *Thaumarchaeota amoA* genes is high in deep waters, relative to archaeal 16S rRNA genes (Figure 12.10B). The ratio of *Thaumarchaeota amoA* genes to *Thaumarchaeota* 16S rRNA genes is about one or higher, suggesting that most of the *Thaumarchaeota* have *amoA* genes and carry out ammonia oxidation. A ratio of one is expected based on the number of *amoA* and 16S rRNA genes found in cultivated bacterial ammonia oxidizers, and the few *Thaumarchaeota* ammonia oxidizers isolated to date. Studies of other environments, including sediments, soils, and shallow waters, also found similar

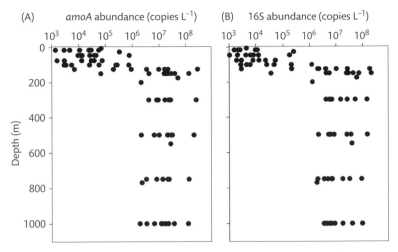

Figure 12.10 Abundance of the ammonia-oxidization gene, *amoA* (A), and 16S rRNA gene (B) for *Thaumarchaeota* in the North Pacific Ocean. Data from Church et al. (2010).

ratios, suggesting that most *Thaumarchaeota* are ammonia oxidizers.

The second piece of evidence came from studies of the natural abundance of ^{14}C in archaeal lipids (Hansman et al., 2009). To get these data, microbes were collected by filtration from over 200,000 liters of water at a unique shore-based facility on the Big Island of Hawai'i. It was known that organic material in the deep waters is relatively young and thus has high amounts of ^{14}C, produced by cosmic bombardment of nitrogen in the upper atmosphere, whereas inorganic carbon in deep waters is much older and has much lower ^{14}C levels. The archaeal lipids also had low ^{14}C levels, suggesting that the carbon came from inorganic pools. Because photosynthesis is not possible at these depths, CO_2 must have been fixed by a chemoautotrophic process, most likely ammonia oxidation. Together, these and other data indicate that most *Thaumarchaeota* are chemoautotrophic ammonia oxidizers.

Controls of aerobic ammonia oxidation

The two most important factors limiting rates of ammonia oxidation in natural environments are energetic yield and ammonium concentration. As with other chemolithotrophs, the energetic yield of ammonia oxidation is low, explaining why cell yields of these organisms (Figure 12.11), and thus abundances, are low relative to the rest of

the community fueled by other forms of catabolism. The low energetic yield also explains why ammonia oxidizers cannot compete with algae, heterotrophic bacteria, and higher plants for ammonium, although there may be exceptions (Inselsbacher et al., 2010). The second factor limiting ammonia oxidation is the low ammonium concentrations found in most oxic environments. AOA appear to outcompete AOB when concentrations are low, because AOA have a higher affinity (lower K_m) for ammonium than do AOB (Martens-Habbena et al., 2009). This would account for the relatively high numbers of AOA in the deep ocean. In contrast, AOB may have a higher V_{max} for ammonia oxidation because AOB respond more than AOA to ammonium additions in soils (Carey et al., 2016a).

Three other factors affecting aerobic ammonia oxidation are oxygen, light, and pH. Lack of oxygen prevents nitrification from occurring in sediments below the oxic surface layer and in waterlogged soils. However, atmospheric levels of oxygen inhibit ammonia oxidation by archaea in pure cultures (Qin et al., 2017), and the ammonia oxidation pathway in AOA may require less oxygen than the AOB pathway (Levy-Booth et al., 2014). Inhibition by light is thought to be one reason why ammonia oxidation is not high in the upper surface layer of aquatic habitats. Laboratory experiments have demonstrated that both AOA and AOB strains are inhibited by continuous light at high intensities (500 µE m^{-2} s^{-1}), but AOA are also sensitive at low intensities (<60 µE m^{-2} s^{-1}), whereas

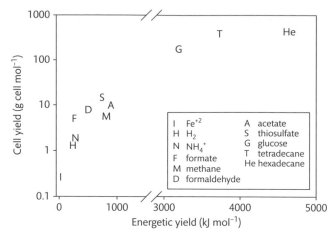

Figure 12.11 Cell yield as a function of the energetic yield of the growth substrate (Gibbs change in free energy). The energetic yield of ammonia oxidation and other chemolithotrophic reactions is low. Data from Bongers (1970), Candy et al. (2009), Farmer and Jones (1976), Goldberg et al. (1976), Jetten (2001), Kelly (1999), and Winkelmann et al. (2009).

AOB are unaffected. The importance of acidity in selecting for ammonia-oxidizing archaea over bacteria in soils has already been pointed out. In marine habitats, however, both archaeal and bacterial ammonia oxidizers may be affected by ocean acidification (Beman et al., 2011).

The second step in nitrification: nitrite oxidation

Ammonia oxidation produces nitrite which is further oxidized to nitrate (Figure 12.8). In the two-step pathway, nitrite oxidation is carried out by a separate group of microbes. The ecology of these microbes has not been studied extensively (Daims et al., 2016).

The known nitrite-oxidizing bacteria (NOB) are phylogenetically different from the ammonia-oxidizing bacteria and belong to seven genera in four phyla which differ somewhat in physiology and distribution among environments (Daims et al., 2016). A few of these bacteria are worth commenting on here. Members of the *Nitrospinae* phylum, with its prominent genus *Nitrospina*, are the dominant NOB in the oceans, especially in oxygen minimum zones (OMZs), where water layers with low oxygen concentrations are sandwiched between the surface layer and the deep ocean with high oxygen concentrations. The phylum *Nitrospirae* with its prominent genus *Nitrospira* is the most diverse NOB lineage and can be abundant in soils. The alphaproteobacterial genus *Nitrobacter* has

aquatic and soil species that are facultative nitrite oxidizers. In addition to growing chemolithoautotrophically, *Nitrobacter* can grow heterotrophically on simple organic compounds.

Physiological and molecular studies with these cultivated bacteria have found that a key enzyme in nitrite oxidation is nitrite oxidoreductase (Nxr), and two of its genes, *nxrA* and *nxrB*, have been used to follow nitrite-oxidizing bacteria in the environment (Daims et al., 2016). Pure culture work also has made possible development of FISH probes targeting 16S rRNA to enumerate the main NOB genera in complex, natural communities (Fussel et al., 2012; Schramm et al., 1998). Studies using these tools have found that NOB can comprise up to 10% of microbial communities in habitats where nitrification is high.

In spite of being discovered over 100 years ago, there are still many unanswered questions about nitrite-oxidizing bacteria (Daims et al., 2016). We don't know, for example, whether archaea are involved or whether only bacteria oxidize nitrite. Another unknown is whether nitrite oxidation is the main energy-harvesting mechanism for these bacteria. There is some evidence of NOB using hydrogen gas and formate as electron donors, and these bacteria may use electron acceptors other than oxygen, such as iodate, Mn(IV), or Fe(III). Other studies, however, indicate the importance of oxygen even at low concentrations. Bristow et al. (2016), for example, found that aerobic nitrite oxidation as well as ammonia oxidation in some marine regions continued even when oxygen concentrations drop to 5–30 nM or about 0.01% of the concentration in equilibrium with the atmosphere.

Complete nitrification by one organism

The two groups of bacteria carrying out the two-step nitrification pathway have been known since Winogradsky's discovery in 1890, so it was a great surprise to discover in 2015 that a single microbe can oxidize ammonia to nitrate (Figure 12.12). It should not have been a surprise because calculations showed that complete oxidation of ammonia to nitrate ("comammox") is thermodynamically favorable and would yield sufficient energy to support microbial growth (Costa et al., 2006). As we have seen several times already, if a reaction yields energy, usually a microbe will be there to take advantage of it.

Two studies simultaneously reported the comammox discovery (Daims et al., 2015; van Kessel et al., 2015). They both used similar cultivation-independent approaches on enrichment cultures obtained from very different samples: a Russian oil well and a Dutch water recirculation system for aquaculture. The ammonium-fed enrichment cultures oxidized ammonia to nitrate under oxic and suboxic conditions, but the typical ammonia oxidizers could not be detected by standard PCR for *amoA* or 16S rRNA genes from ammonia-oxidizing archaea. In both studies, FISH demonstrated the presence of species related to *Nitrospira*, and one study used FISH combined with microautoradiography to find this *Nitrospira* species to be the only chemoautotrophic microbe in the culture. Metagenomic work on these enrichment cultures recovered enough genomic information to confirm that the *Nitrospira* species have the expected genes for nitrite

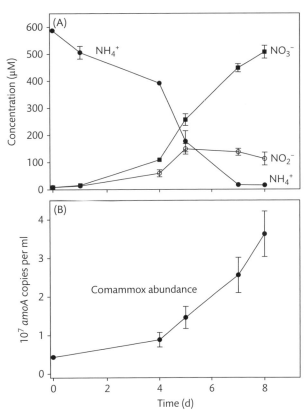

Figure 12.12 Complete oxidation of ammonia to nitrate by an enrichment culture with *Nitrospira*, the comammox bacterium. Ammonium (NH$_4^+$) was completely oxidized to nitrate (NO$_3^-$) with production of some nitrite (NO$_2^-$) (A). Abundance of *Nitrospira* was followed with the ammonia monooxygenase gene (*amoA*) (B). Data from Daims et al. (2015).

oxidoreductase, but they also have genes for ammonia oxidation, including a version of *amoA* that was different enough not to be detected by the initial PCR assays. So, the physiological and genomic data indicate that one organism is capable of oxidizing ammonia all the way to nitrate.

Armed with the unique *amoA* sequence, investigators could then search for evidence of comammox in the public metagenomic data sets (van Kessel et al., 2015). They found comammox *amoA* in many soils, freshwaters, sediments, and engineered habitats like waste water treatment plants, but the comammox *amoA* seems to be missing from one of the biggest habitats, the water column of the oceans. Comammox bacteria may really not survive in the oceans for unknown reasons, or perhaps they have not been detected because the marine version of the comammox *amoA* is too different to be found by current methods. The need for further work is evident.

Dissimilatory nitrate reduction and denitrification

The end product of nitrification, nitrate, is the starting point for denitrification, a form of anaerobic respiration used by organisms to oxidize organic material (Chapter 11). The first step in that process is dissimilatory reduction of nitrate to nitrite, which then in turn is reduced eventually to nitrogen gas (N_2). One equation describing denitrification is

$$5\,glucose + 24NO_3^- + 24H^+ \rightarrow 30CO_2 + 12N_2 + 42H_2O$$
$$\Delta G^{o\prime} = -2657 \text{ kJ mol}^{-1} \tag{12.8}$$

So, nitrate is the electron acceptor, and here, glucose is the electron donor, but in fact many other organic compounds can serve as carbon and electron (energy) sources for this heterotrophic reaction. In addition to N_2, dissimilatory nitrate reduction can stop at nitrite, which is excreted, resulting in a smaller energetic yield compared with reduction to N_2: $\Delta G^{o\prime} = -2657$ kJ mol^{-1} glucose for N_2 gas versus $\Delta G^{o\prime} = -1926$ kJ mol^{-1} glucose for nitrite (Buckel, 1999). Two other possible nitrogenous end products, ammonium and nitrous oxide, will be further discussed. Only when the nitrogenous end product is a gas, either N_2 or nitrous oxide, is the pathway called denitrification.

The denitrification pathway producing N_2 requires four enzyme complexes, starting with nitrate reductase encoded by the *nar* genes (Figure 12.13). All of these

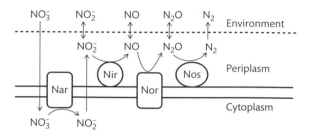

Figure 12.13 Pathway for denitrification. The enzymes are nitrate reductase (Nar), nitrite reductase (Nir), nitric oxide reductase (Nor), and nitrous oxide reductase (Nos), encoded by *nar*, *nir*, *nor*, and *nos*, respectively. Based on Ye et al. (1994) and Zumft (1997).

redox-mediating enzymes require iron. In addition, nitrate reductase has a molybdenum co-factor, and nitrous oxide reductase (Nos) and one of the two nitrite reductases (NirK) contain copper. Studies using cultivation-independent methods have examined the genes *nirS*, *nirK*, and *nosZ* to explore questions about the potential for denitrification and the diversity of denitrifiers in natural environments. Not surprisingly, the microbes carrying out denitrification in nature are quite different from those studied in laboratory cultures.

Dissimilatory nitrate reduction and denitrification can be carried out by many types of prokaryotes and some eukaryotic microbes. Some fungi may contribute substantially to the process in some soils, and even some protists, such as marine foraminifera and diatoms, can do it, as mentioned in Chapter 11 (Kamp et al., 2015). Many of these eukaryotic microbes as well as prokaryotes are facultative anaerobes with the capacity to switch to oxygen when concentrations are favorable.

The difference in energy yield implies that denitrification would shut down if any oxygen were available. However, some denitrification can occur with measurable oxygen concentrations (Figure 12.14). It seems that the small advantage in energy yield in using oxygen over nitrate (Chapter 11) is not enough to prevent dissimilatory nitrate reduction when the oxygen supply is inadequate. Rates of denitrification were inhibited by only 50% in suboxic waters with about 300 nM oxygen in the oxygen minimum zone off northern Chile (Dalsgaard et al., 2014); oxygen <1000 nM is quite low, <1% of values for water in equilibrium with the atmosphere. Metatranscriptomic data from these waters indicated that transcripts for nitrous oxide reductase (*nosZ*), nitrite

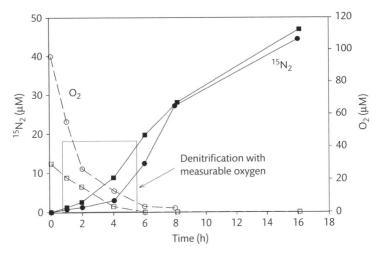

Figure 12.14 An example of denitrification in the presence of oxygen. The data are from two layers 0–2 cm (circles) and 2–4 cm (squares) below the surface of a sand flat in the German Wadden Sea. Denitrification was measured by following $^{15}NO_3^-$ into N_2 (solid symbols) while oxygen (open symbols) was measured by micro-electrodes. Data from Gao et al. (2010).

reductase (*nirS*), and nitric oxide reductase (*norB*) genes decreased when oxygen concentrations were higher than 200 nM. However, 200 nM of oxygen did not affect transcription of nitrate reductase (*narG*) and nitrite reductase (*nirK*) genes. The continued operation of denitrification when oxygen is present means more N may be lost from N-limited environments if oxygen concentrations were to decline, as seen for the oceans and elsewhere (Schmidtko et al., 2017).

Other environmental factors affecting denitrification include organic carbon and nitrate concentrations, as discussed in Chapter 11. The need for nitrate means that denitrification depends on nitrification. These two processes are often said to be "coupled," here meaning that the end product of one reaction, nitrate from nitrification, is used by another, denitrification. But these processes have to be separated in time or more commonly in space because nitrification is an aerobic process while denitrification occurs mainly in anoxic environments. A classic example occurs in sediments where nitrification in the top, oxic layer supplies nitrate used by denitrifying microbes in lower, anoxic sediment layers. However, there are exceptions to the general rule that denitrification occurs only in anoxic environments. Oxic (unsaturated) soils may have some denitrification activity because of anoxic microhabitats.

In addition to heterotrophic denitrification, some chemolithotrophs reduce nitrate while oxidizing hydrogen sulfide, a process called "chemoautotrophic denitrification" or "lithotrophic denitrification." But heterotrophic organisms that use nitrate as a terminal electron acceptor to oxidize organic material are much more important in denitrification and releasing nitrogenous gases.

Nitrate reduction can produce ammonium by a process called "dissimilatory nitrate reduction to ammonium" (DNRA), which is carried out by bacteria and fungi in aquatic and terrestrial environments (Rütting et al., 2011; Thamdrup, 2012). One equation describing DNRA is

$$glucose + 3NO_3^- + 6H^+ \rightarrow 6CO_2 + 3NH_4^+ + 3H_2O$$
$$\Delta G^{o'} = -1767 \text{ kJ mol}^{-1} \tag{12.9}.$$

DNRA yields about 50% less energy than reducing nitrate to N_2, raising the question why any microbe would do it. Note that the C:N ratio for the organic carbon and nitrate required by DNRA is 2:1 (Equation 12.9) whereas it is 1.25:1 for bacteria denitrifying nitrate to N_2 gas (Equation 12.8). That is, DNRA requires less nitrate than denitrification to oxidize one mole of organic carbon, leading to the hypothesis that DNRA is advantageous in organic-rich but nitrate-poor habitats. Evidence from studies in natural environments supports this hypothesis (Thamdrup, 2012). In some environments, more nitrate goes through DNRA than through denitrification (Dong et al., 2011).

Anaerobic ammonia oxidation

In 1965, the chemical oceanographer F. A. Richards pointed out the lack of ammonium accumulation in oxygen minimum zones of the oceans and speculated that ammonium was being oxidized with nitrate as the electron acceptor (Strous and Jetten, 2004). Over ten years later, the Austrian microbiologist and alleged KGB spy, E. Broda, suggested the reaction

$$NH_4^+ + NO_2^- \rightarrow N_2 + 2H_2O \qquad (12.10)$$

but it took another 20 years before experimental evidence was found in a wastewater reactor. The stoichiometry of the reactants and products confirmed that nitrite rather than nitrate was the electron acceptor in anaerobic ammonia oxidization ("anammox"). After the initial work with wastewater, anammox was found in natural anoxic aquatic and terrestrial environments (Oshiki et al., 2016). While aerobic ammonia oxidation and anammox both act on ammonia, they are very different processes involving very different organisms (Table 12.2).

Unlike aerobic ammonia oxidation, anammox is carried out by a very limited, monophyletic group of bacteria belonging to the phylum *Planctomycetes* (Oshiki et al., 2016). No anammox has been grown in pure culture to date, but much has been learned from enrichments dominated by these bacteria. A metagenomic approach applied to an enrichment culture was used to deduce the genome of the wastewater anammox bacterium, *Candidatus* Kuenenia stuttgartiensis (Strous et al., 2006). Among several unusual features, these bacteria have an intracellular compartment, the "anammoxosome," where ammonia oxidation takes place. The membranes of anammox bacteria contain an unusual lipid, ladderane, thought to be important in compartmentalizing a potent intermediate, hydrazine (N_2H_4), formed during

ammonia oxidation. Hydrazine is used in rocket fuel and is highly unstable. Based on transcripts of hydrazine synthase (*hzsA*, *hzsB*, and *hzsC*) and hydrazine oxidoreductase (*hzoA* and *hzoB*), a metatranscriptomic study suggested that anammox bacteria accounted for nearly 20% of total microbial gene expression in a suboxic aquifer (Jewell et al., 2016).

Denitrification versus anaerobic ammonia oxidation

After anammox was discovered, an obvious question was, how much N_2 does it produce compared with traditional, heterotrophic denitrification? The answer varies from nothing to >75% of N_2 production, depending on the environment. Reviews of the topic often say the average is about 50%. A more nuanced view gives some insights into factors controlling both processes.

It is possible to predict the fraction of N_2 production attributable to anammox and denitrification by looking at the stoichiometry of the anaerobic degradation of organic material with a Redfield composition (Chapter 2). When F. A. Richards noticed back in 1965 the lack of any ammonium in anoxic waters, he proposed that degradation of Redfield organic material using nitrate as the electron acceptor could be described by

$$(CH_2O)_{106}(NH_3)_{16}H_3PO_4 + 94.4NO_3^- + 94.5H^+ \rightarrow$$
$$106CO_2 + 55.2N_2 + 177.2H_2O + H_3PO_4 \qquad (12.11).$$

The problem is that denitrification alone could not produce all of the N_2 implied by Equation 12.11. Denitrification can only do this:

$$(CH_2O)_{106}(NH_3)_{16}H_3PO_4 + 94.4NO_3^- + 94.5H^+ \rightarrow$$
$$106CO_2 + 16NH_4^+ + 16NO_2^- + 39.2N_2 + \qquad (12.12).$$
$$145H_2O + H_3PO_4$$

Table 12.2 Comparison of aerobic and anaerobic ammonia oxidation. Data from Jetten (2001).

Property	Aerobic	Anaerobic
Energy yield (kJ mol⁻¹)	272	357
Oxidation rate (nmol mg⁻¹ min⁻¹)	400	60
Generation time (days)	1	10
Organisms	*Betaproteobacteria, Thaumarchaeota*, others	*Planctomycetes*
Carbon source	CO_2	CO_2
End product	Nitrite	N_2 gas
Ecosystem role	Fuels denitrification	Removes fixed nitrogen

Denitrification produces NH_4^+ during mineralization of the organic material with its 16 N, but it cannot oxidize it further. It produces NO_2^- and N_2 from the reduction of NO_3^-, the electron acceptor in this reaction. NO_2^-, not more N_2, is produced because there are not enough electrons (reducing power) in the organic material to reduce all 94.4 moles of NO_3^- all the way to N_2 gas. In environments where neither NH_4^+ nor NO_2^- are produced during anaerobic degradation, anammox is needed to remove both and to explain their absence. So, to get Equation 12.11, Equation 12.12 must be coupled with

$$16NH_4^+ + 16NO_2^- \rightarrow 16N_2 + 32H_2O \quad (12.13).$$

(Equation 12.13 is simply Equation 12.10 multiplied by 16, the number of NH_4^+ moles originally in the starting Redfield organic material and released during its degradation.) Looking at the 39.2 N_2 produced by denitrification (Equation 12.12) and the 16 N_2 from anammox (Equation 12.13), we can predict that anammox should account for about 29% of total N_2 production.

Actual percentages in aquatic sediments, where anammox has been examined the most, vary greatly around the 29% prediction, probably because of variation in organic material input (Thamdrup, 2012; Babbin et al., 2014). Organic material input is perhaps the best explanation for why the contribution of anammox to total N_2 production in marine sediments increases with increasing water depth (Figure 12.15). In shallow sediments, the anammox contribution is small because it is limited by nitrite, whose concentrations are driven down by the high activity of denitrifiers, in turn driven up by high inputs of organic material. By contrast, in deep water column sediments, the fraction of N_2 production attributed by anammox is close to the 29% predicted by stoichiometry and even exceeds it, probably because denitrification is limited by the low supply of organic material to deep sediments. Anammox rates are relatively high in Mn-rich sediments, where denitrifying bacterial activity is reduced by competition with Mn-reducing bacteria for organic material.

Anammox bacterial abundance and rates also vary greatly in soils for unclear reasons. Organic material availability may explain the increase in anammox 16S rRNA genes with depth in soils (Humbert et al., 2012). Anammox accounted for 37 to 68% of N loss in the water-saturated zone of an aquifer, but the process was unmeasurable in the overlying soil surface (Wang et al., 2017). Although anammox can explain up to nearly 40% of N loss from paddy soils, it contributed <15% in temperate forest soils (Xi et al., 2016). More data from terrestrial systems are needed before any generalizations can be proposed.

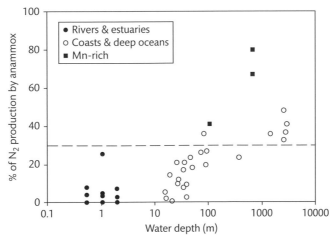

Figure 12.15 Contribution of anaerobic ammonia oxidation (anammox) to N_2 production in sediments as a function of water depth. The horizontal dashed line is at 29%, the percentage of N_2 production contributed by anammox predicted from degradation of Redfield organic material. Data from studies reviewed by Thamdrup (2012).

The contribution of denitrification versus anammox in N_2 production is important for thinking about the loss of crucial fixed nitrogen from the biosphere. It is also important in thinking about the fluxes of another nitrogenous gas, nitrous oxide. Anammox bacteria have no direct role in these fluxes, in contrast with denitrifying microbes, discussed next.

Sources and sinks of nitrous oxide

Nitrogen gas is the main gas produced by denitrification and the only one produced by anammox, but another gas potentially produced by denitrification, nitrous oxide (N_2O), deserves a closer look. As mentioned before, N_2O is a potent greenhouse gas, it depletes atmospheric ozone, and it is a route by which nitrogen exits an ecosystem. Terrestrial ecosystems account for about twice as much to natural N_2O emissions as the oceans (Fowler et al., 2015).

Production by bacteria and archaea
Several organisms and processes are potential sources of N_2O. Here we focus on the main sources: denitrifiers and nitrifying microbes. Denitrifiers can produce the gas if nitrate concentrations are too low or if oxygen is too high; the last step in denitrification, reduction of N_2O to N_2, is the most sensitive to oxygen. Heterotrophic denitrification by bacteria and fungi can dominate N_2O

production in anoxic environments such as saturated soils, and sometimes even in moist soils with anoxic microhabitats (Müller et al., 2014). But ammonia oxidizing microbes are the most important organisms in producing N_2O in aerated soils and oxic oceans (Freing et al., 2012; Hu et al., 2015). The process is sometimes referred to as "nitrifier denitrification," because nitrifiers are causing the loss of fixed N in the form of N_2O.

Pure culture experiments had demonstrated that both ammonia-oxidizing bacteria and ammonia-oxidizing archaea produce N_2O. The question then became, what prokaryotic group is more important in natural environments? The contribution by archaea, bacteria, and fungi to N_2O production has been explored by inhibiting ammonia oxidizers with the addition of acetylene, or all bacteria with antibiotics. Other insights have been gained from the stable isotope content of N_2O, which varies with the production mechanism and organism. Pure culture studies have shown that the N_2O production pathway in ammonia-oxidizing bacteria may differ from that in ammonia-oxidizing archaea (Figure 12.16).

In contrast with the relatively straightforward pathway in ammonia-oxidizing bacteria, the pathway in ammonia-oxidizing archaea involves the "hybrid formation" of N_2O from nitric oxide (NO) and hydroxylamine (NH_2OH). This hybrid formation is thought to be spontaneous and not mediated by enzymes (Kozlowski et al., 2016); ammonia-oxidizing archaea lack the gene for nitric oxide reductase (Nor) necessary for reducing nitrite to N_2O.

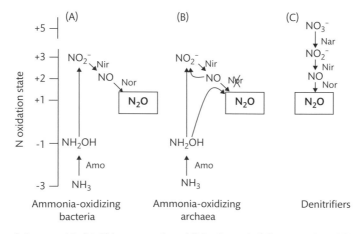

Figure 12.16 Production of nitrous oxide (N_2O) by ammonia-oxidizing bacteria (A), ammonia-oxidizing archaea (B), and denitrifiers (C). The enzymatic reduction of NO_2^- to N_2O is not seen in ammonia-oxidizing archaea because they lack nitric acid reductase (Nor). Other key enzymes include ammonia monooxygenase (Amo), nitrite reductase (Nir), and nitrate reductase (Nar). Based on Kozlowski et al. (2016).

These differences between bacteria and archaea affect the stable isotope content of N_2O ($\delta^{15}N$ and $\delta^{18}O$) and the fate of $^{15}NH_4^+$ and $^{15}NO_2^-$ added to ammonia-oxidizing microbial cultures and natural samples.

The ^{15}N content of N_2O and other data have been used to address the question of whether ammonia-oxidizing bacteria or archaea are most important in producing N_2O. The answer is, ammonia-oxidizing archaea dominate N_2O production in the oceans (Ji and Ward, 2017), whereas both archaea and bacteria contribute in soils, depending on the soil type and the relative abundance of the two prokaryotic domains (Breuillin-Sessoms et al., 2017).

The mechanism of N_2O production helps us to understand effects of oxygen concentrations, an important factor governing rates and yields of this greenhouse gas. N_2O production by ammonia-oxidizing bacteria in pure culture increases substantially when oxygen drops, consistent with studies of complex communities in the oceans (Qin et al., 2017). The same response is seen with most ammonia-oxidizing archaea in pure culture. More moisture and thus less oxygen in soils can lead to more N_2O production. Whether ammonia-oxidizing bacteria or archaea dominate N_2O production could be important for predicting how production responds to decreasing oxygen concentrations, potentially caused by climate change.

Consumption of N_2O

In contrast with N_2O production, consumption of N_2O by biological processes is straightforward. The gas is consumed when it is reduced by heterotrophic denitrifiers and by other, recently discovered N_2O-reducing organisms that are not denitrifiers (Hallin et al., 2018). Heterotrophic denitrifier microbes use N_2O as an electron acceptor for oxidizing organic material and reduce N_2O at the last step in denitrification, producing N_2. Although less energy is gained compared with using nitrate, denitrifiers turn to N_2O when nitrate concentrations are insufficient. Assuming that abundance is indicative of importance, bacteria seem to dominate N_2O reduction, although archaea have the critical enzyme, nitrous oxide reductase (Nos). Dissimilatory nitrate-reducing fungi do not (Maeda et al., 2015), perhaps one reason why their rates of N_2O production are so high. In fact, about 40% of genomes with

a key denitrification gene, *nir*, for nitrite reductase, do not have the gene, *nosZ*, for N_2O consumption (Graf et al., 2014).

Metagenomic studies have found that *nir* can be as much as ten-fold more abundant than *nosZ*, implying that many organisms have incomplete denitrification pathways which produce but do not consume N_2O, consistent with the observation that natural environments are usually net sources of N_2O. Abundance and sequence analyses of *nirS*, *nirK*, and *nosZ* give insights into the production and consumption of N_2O not possible by standard biogeochemical analyses.

N budgets: global balance, local imbalances

This chapter began with two observations that raise questions about whether the global N budget is in steady state or is changing. In spite of N_2O consumption by microbes and atmospheric chemical reactions, concentrations of this important greenhouse gas are increasing in the atmosphere. On the other hand, rates of anthropogenic N_2 fixation are also increasing, introducing more and more N to the biosphere. These observations raise the questions: do the rates of natural and anthropogenic N_2 fixation now exceed the return of N to the atmosphere via production of N_2 and N_2O? Is the N budget balanced with inputs equaling outputs?

Some biogeochemists argue that the N cycle is out of balance, while others say it is in balance, or that we don't have enough data to say one way or the other (Devol, 2015; Gruber and Galloway, 2008; Fennel, 2017). The increase in atmospheric N_2O does indicate an imbalance but the increase is small compared with N_2 fixation (<10% of the Haber–Bosch process alone). The higher rates of anthropogenic N_2 fixation may be partially offset by lower natural N_2 fixation, because the fixed N from anthropogenic sources would repress N_2 fixation by microbial diazotrophs. In any case, given the size of N reservoirs relative to the fluxes, it would be difficult to detect any change in nitrate or organic N concentrations in soils and the oceans, although the effect of the Haber–Bosch process should be measurable in the $\delta^{15}N$ of fixed N (Yang and Gruber, 2016). More work is needed to explore how anthropogenic N_2 fixation and other processes affect the N cycle on a global scale.

Regardless of the status of the global N cycle, the increase in fertilizer production and use is certainly causing local imbalances in many environments. Fertilizer nitrogen runoff into estuaries, coastal oceanic waters, and lakes leads to harmful algal blooms and to "dead zones" or anoxic and hypoxic waters with several deleterious side effects. The cost of nitrogen pollution has been estimated to be an incredible USD $800 billion per year (Fowler et al., 2015). We need fertilizer to feed the world's burgeoning population, but we also need better management strategies to maximize the effectiveness of fertilizer use, while minimizing its loss to natural environments. Although these topics are beyond microbial ecology, microbial ecologists are needed to understand and predict the consequences of perturbing the nitrogen cycle.

Summary

1. N_2 fixation is carried out by a select but diverse group of prokaryotes. Many of these prokaryotes live symbiotically with eukaryotic microbes, mosses, and higher plants.

2. Ammonia oxidation, the rate-limiting step of nitrification (the "making" of nitrate), is carried out mainly by *Betaproteobacteria* and *Thaumarchaeota*. Ammonia-oxidizing *Thaumarchaeota* are the most abundant type of archaea in oxic environments.

3. The end product of aerobic ammonia oxidation, nitrite, is used by a few nitrite-oxidizing bacteria. Taxa within at least one genus of bacteria, *Nitrospira*, can carry out the complete oxidation of ammonia to nitrate ("comammox").

4. Denitrification, a type of anaerobic respiration, releases the gases nitrous oxide (N_2O) or N_2 and starts with dissimilatory nitrate reduction, which is carried out by a wide variety of bacteria, archaea, and even some eukaryotic microbes. Another possible by-product of dissimilatory nitrate reduction is ammonium.

5. In contrast with aerobic ammonia oxidation, anaerobic ammonia oxidation (anammox) produces N_2 and is carried out only by bacteria in the phylum *Planctomycetes*. Anammox can release more N_2 than denitrification in some environments, but overall anammox is predicted to account for 29% of total N_2 release.

6. The greenhouse gas nitrous oxide (N_2O) is produced by ammonia-oxidizing bacteria and archaea and in some environments by denitrifying bacteria and fungi. N_2O is consumed by denitrifiers and ammonia-oxidizing bacteria and archaea.

7. It is unclear whether the huge increase in anthropogenic inputs of fixed N is upsetting the global N budget. What is quite clear is that these inputs are causing local imbalances and are leading to many environmental problems.

CHAPTER 13

Introduction to geomicrobiology

Previous chapters have mentioned a few examples of the impact of microbes on the physical world of the Earth. Formed by the marriage between geology and microbiology, the field of geomicrobiology is even more focused on those impacts. This chapter will highlight processes important in thinking about how microbes shape Earth materials, the inanimate things making up our planet. Many of these processes involve microbe–rock and microbe–mineral interactions: processes involving microbes and harder materials than what is most common in the soils and water discussed so far. Geomicrobiology is in the newspapers these days because of discoveries in exotic habitats like caves and gold mines, some kilometers below the Earth's surface (Onstott, 2017). But the topics to be discussed in this chapter are relevant for many habitats in the biosphere.

Geomicrobiologists face the challenge of meshing vast scales of time and space over which microbes interact with geology. We have already seen that environmental conditions of microhabitats have ramifications for global phenomena. The net production or consumption of greenhouse gases such as methane and nitrous oxide, for example, depends on oxygen concentrations in the micron-sized space surrounding bacteria, archaea, and fungi. In geomicrobiology, the range of timescales is especially evident. Microbes important in geomicrobiological processes often grow slowly, and their impact seems minute at any particular time. But these impacts have huge consequences when the process continues for millennia or longer. Lots of small things occurring over a long time add up to big consequences.

Cell surface charge and metal sorption

Sorption of material to cell surfaces is important in all of microbial ecology, but it arguably is most pertinent in geomicrobiology. Metal–microbe and mineral–microbe interactions take up a large part of this chapter. Sorption of a metal such as iron may be crucial for the microbe's survival and growth, while sorption of other metals, such as copper and cadmium, which are toxic in high concentrations, is crucial for the fate of heavy metal contaminants in an environment. An important factor governing all of these interactions is the surface charge of a microbial cell.

The net charge of a microbial cell is usually negative, although the precise value depends on the microbe, growth conditions, and the environment. Because of this net negative charge, the cell surface attracts positively charged atoms and compounds, giving rise to a gradient of charge between the cell and the surrounding environment. The classic model for this gradient, referred to as the "electric double layer," consists of an inner layer of positively charged ions (the Stern layer) tightly held next to the negative cell surface, followed by a diffuse layer of counter-ions (the Gouy layer) (Figure 13.1).

One term used to describe the electrical charges around a cell or any particle is "isoelectric point," found experimentally by monitoring the movement of cells in an electric field as a function of pH. Increasing the pH leads to cells becoming more negatively charged, causing them to move towards the cathode, the positive end of the electrical field; when pH decreases, they move towards the anode, the negative end. The pH at which

Processes in Microbial Ecology. Second Edition. David L. Kirchman. Oxford University Press (2018). © David L. Kirchman 2018.
DOI 10.1093/oso/9780198789406.001.0001

the cell remains stationary is the isoelectric point. The isoelectric point for most microbes is between pH 2 and 4. That most environments have pH > 4 explains why most microbes are overall negatively charged. An analogous concept is the point of zero charge (pzc) for mineral surfaces.

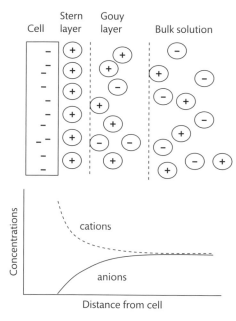

Figure 13.1 A model of a cell surface surrounded by an aqueous solution of various ions. The main feature of the model is the double layer of ions that build up on microbial cell surfaces.

Microbes provide extensive reactive, mostly negatively charged surfaces onto which metals and other positively charged atoms and compounds potentially sorb. The cumulative surface area of microbes is larger than that of other particles in aquatic habitats, and is substantial even in soils and sediments. The microbial surface is more reactive than many other particles, evident in the high cation-exchange capacity of microbial cell surfaces. Consequently, microbe–metal interactions are important in thinking about the environmental fate of metals, including toxic ones, in addition to understanding the impact of metals on microbes. Sorption of metals is a form of "bioremediation" or the cleaning up by organisms of toxic metals or organic contaminants (Figure 13.2). The topic of metal sorption has been examined by hundreds of studies, mostly focusing on bacteria, although fungi can be important in soils as pH declines. Even

viruses may have a role, as shown for mineral precipitation in microbial mats (Pacton et al., 2014).

Metals can passively sorb onto microbes because of electrostatic attraction. Being positively charged, metals sorb onto negatively charged carboxyl and phosphoryl groups of cell walls, membranes, and extracellular material. The precise identity and number of sorption sites vary with the microbe and growth conditions (Ledin, 2000). In addition to cell surface properties, the amount of sorption also varies with the metal, dissolved metal concentrations, and pH. Several models exist for describing how sorption varies as a function of dissolved metal concentration. One simple model posits that the cell surface has a finite number of reactive sites where sorption can occur. Once those sites are filled, sorption stops. These simple assumptions lead to the Langmuir equation,

$$q_e = q_{max} \cdot K \cdot C_e / (1 + K \cdot C_e) \qquad (13.1)$$

where q_e is the sorption at a particular equilibrium dissolved concentration (C_e), q_{max} is the maximum sorption, and K is the Langmuir equilibrium parameter. This variation in sorption as a function of concentration is reminiscent of the uptake–concentration relationship modeled by the Michaelis–Menten equation (Chapter 6). Other models for sorption include Freundich and Brunauer–Emmett–Teller (BET) isotherms.

Metal sorption is thought to be passive, but the word may be misleading in thinking about metal–microbe interactions. Microbes can modify functional groups at cell surfaces and thus indirectly control sorption to some extent, with several possible rewards. The binding of some metals can help to stabilize cell walls and membranes by neutralizing otherwise destabilizing interactions between anionic wall and membrane components. Sorption to cell surfaces helps in securing some metals, such as iron and copper, needed for biosynthesis, while it may lessen toxic effects of other metals present in high concentrations. Toxic effects may be particularly minimized by sorption to extracellular polymers. Bacteria with capsules (Chapter 2) are able to withstand high metal concentrations better than capsule-less mutants.

Biomineralization by microbes

The principles of sorption just reviewed are useful for thinking about microbes and the formation of iron-rich

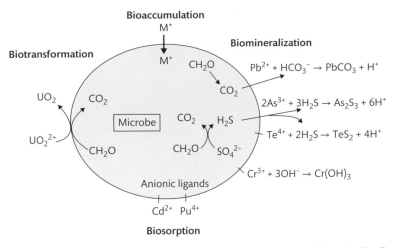

Figure 13.2 Some mechanisms by which microbes contribute to bioremediation. Inspired by a similar figure by Konhauser (2007).

minerals, an important example of "biomineralization." Chapter 7 used the term "mineralization" in discussing the degradation of organic material to inorganic compounds such as carbon dioxide, ammonium, and phosphate. The term takes on a different meaning in geology and geomicrobiology. It is used to describe the precipitation of minerals from dissolved constituents, such as the formation of iron oxide minerals from dissolved Fe^{2+} or colloidal Fe(III) oxides. When microbes or other organisms are involved in the process, it is called "biomineralization."

The extent to which microbes control biomineralization varies with the organism and reaction. In cases of "biologically induced" biomineralization, the microbial role may be indirect, affecting mineralization only through the release of metabolic waste products (Kendall et al., 2012). The microbe may gain no direct advantage from mineral formation, as seems to be the case in the formation of amorphous iron oxide minerals around some bacteria. At the other extreme, "biologically controlled" biomineralization, microbes regulate the entire process of mineral formation, including nucleation, mineral phase, and location in or around the cell. Two examples were mentioned briefly in Chapter 6: diatoms form the mineral opal ($SiO_2 \cdot nH_2O$) from dissolved silicate to make exquisitely designed cell walls of glass, and coccolithophorids guide the precipitation of calcium carbonate during the construction of their coccolith cell walls.

More examples of both types of biomineralization will be discussed further.

Iron minerals and microbes
Biomineralization of iron minerals is important because iron is a required micro-nutrient for nearly all organisms (Chapter 2) and because it is a crucial process shaping Earth materials. Also, iron minerals can hold clues about the evolution of early life and interactions with geology during the Precambrian. Several iron-rich minerals are common, with many interactions with microbes (Table 13.1).

Some iron minerals are formed by biologically induced biomineralization, due to the electrostatic interactions between the negatively charged microbes or their extracellular polymers and ferrous iron (Fe^{2+}). The microbial surfaces can act as passive nucleation sites of ferric hydroxide ($FeO(OH) \cdot nH_2O$) from dissolved Fe^{2+} in the presence of oxygen. The precipitated ferric hydroxide in turn promotes more precipitation until the entire bacterium is encased in a mineral matrix. This is a good example of biologically induced biomineralization. The bacterium seems intimately involved in the mineral formation, yet its role is passive; the cell surface, extracellular polymers, and its metabolism are not necessarily designed to promote precipitation of iron minerals. Other microbes, referred to as iron-depositing bacteria or simply iron bacteria, do synthesize cell surface ligands to

Table 13.1 Some iron-rich minerals and possible connections to microbes. "Oxides" includes hydroxides.

Type	Mineral	Composition	Connection with microbes
Carbonate	Siderite	$FeCO_3$	Fe(II) produced by iron reducers
Oxide	Amorphous ferric hydroxide	Fe_2O_3	No crystallinity, most easily used by iron reducers
Oxide	Ferrihydrite	$Fe_2O_3 0.5(H_2O)$	Some crystallinity
Oxide	Hematite	Fe_2O_3	Stable form of iron oxide
Oxide	Goethite	$FeO(OH)$	Stable form of iron oxide
Oxide	Magnetite	Fe_3O_4	Produced by magnetotactic bacteria
Sulfide	Greigite	Fe_3S_4	Produced by magnetotactic bacteria
Sulfide	Pyrite	FeS_2	Produced by sulfate reducers

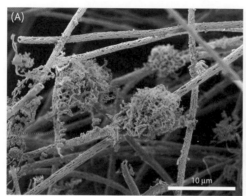 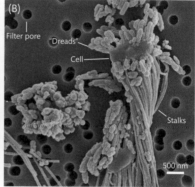

Figure 13.3 Examples of iron-oxidizing bacteria, visualized by scanning electron microscopy. Panel A: Marine iron-oxidizing bacteria with associated iron oxides. Fibril Fe-rich nests with cells inside appear to be secondary colonizers of the sheaths. Courtesy of Clara Chan from Chan et al. (2016) and used with permission of the authors. Panel B: Strain R-1, a neutrophilic microaerophilic iron-oxidizing bacterium. "Dreads" are iron-rich particles produced by the bacterium. Courtesy of Shingo Kato from Kato et al. (2015) and used with permission of the authors.

promote Fe^{2+} oxidation, although these bacteria may not necessarily harvest energy from the process, in contrast with the true iron oxidizers, to be further discussed.

One genus of iron-depositing bacteria is *Leptothrix*, commonly found in iron-rich, low-oxygen freshwater habitats (Emerson et al., 2010; Kunoh et al., 2017). Some members of this genus appear to be true chemolithotrophic iron oxidizers, while others may be heterotrophs. In either case, this bacterium makes tubular sheaths that become encrusted in iron minerals. Many of these sheaths do not contain living cells, evidence that the bacterium escaped from being entombed in an iron casket. Other iron-oxidizers also make sheaths and other structures, such as the iron-rich "nests" (Figure 13.3), apparently as part of strategies to minimize negative effects of the precipitated iron oxides. The precipitated iron, however, may convey advantages

to a cell. It may sorb needed trace metals and nutrients and protect the cell from dissection and attack by viruses and protist grazers.

Iron oxide minerals formed by iron-oxidizing bacteria
In addition to forming by passive sorption, iron oxide minerals can also result from the oxidation of ferrous iron (Fe^{2+}) by several types of chemolithoautotrophic bacteria and archaea. This microbe-mediated oxidation explains formation of iron oxides under low pH and low oxygen concentrations, where purely abiotic oxidation is too slow. The oxidation of Fe^{2+} is described by

$$4Fe^{2+} + O_2 + 4H^+ \rightarrow 4Fe^{3+} + 2H_2O \qquad (13.2).$$

The energy potentially gained from this reaction is only 29 kJ mol^{-1}, making it the lowest energy-yielding process of all chemolithoautotrophic metabolisms (Emerson et al., 2010) (see also Figure 12.11 in Chapter 12). The energy yield doubles if the ferric iron by-product precipitates to form ferrihydrite, which occurs spontaneously at near-neutral pH under oxic conditions. Even more energy is to be had (ΔG° = −90 kJ mol^{-1}) under low partial pressures of oxygen. While iron-oxidizing microbes gain more energy when the pH is near neutral, they must compete with abiotic oxidation of Fe^{2+} when oxygen concentrations are high (Figure 13.4). Acidophilic prokaryotes dominate iron oxidation when pH is low or in anoxic environments, but neutrophilic prokaryotes can still account for as much as half of iron oxidation when the pH is about 7 in suboxic environments (Neubauer et al., 2002). Even though iron-oxidizing bacteria were described back in 1837, much remains unknown about their physiology and ecology.

Several types of bacteria can oxidize iron by different metabolic pathways (Table 13.2). The best-known example is the chemolithotrophic oxidation by the betaproteobacterium *Gallionella*, first described back in the nineteenth century (Emerson et al., 2013). This genus is typically found in freshwaters, and strains have been isolated from a groundwater iron seep. Cultivation-dependent studies have found other, closely related *Betaproteobacteria* (*Sideroxydans*, *Ferriphaselus*, and *Ferritrophicum*) in saturated soils and in the rhizosphere

of wetland plants. Many marine iron-oxidizing bacteria are in the *Zetaproteobacteria* (Scott et al., 2015). Iron-oxidizing chemolithotrophs are microaerophilic and do best when low oxygen concentrations limit abiotic iron oxidation (Emerson et al., 2010). Perhaps to maintain their position at the interface between the oxic and anoxic layers with high Fe^{2+}, some *Gallionella* species and *Zetaproteobacteria* iron-oxidizers have a bean-like cell at the top of a stalk.

The electron acceptor for the *Gallionella*-related strains and iron-oxidizing *Zetaproteobacteria* is oxygen. Other iron-oxidizing bacteria may use nitrate as an electron acceptor rather than oxygen, according to the equation

$$10Fe^{2+} + 2NO_3^- + 6H^+ \rightarrow 10Fe^{3+} + N_2 + 6OH^- \quad (13.3).$$

However, there is some doubt that bacteria are directly involved in this iron oxidation reaction (Kappler et al., 2015). The microbial cultures capable of nitrate-dependent Fe^{2+} oxidation appear to require organic carbon, raising the possibility that Fe^{2+} is oxidized abiotically by nitrite formed during heterotrophic nitrate reduction. Regardless, the overall process exemplified by Equation 13.3 is important in explaining iron geochemistry in anoxic environments, because it is not clear how Fe^{2+} could be oxidized abiotically in those environments.

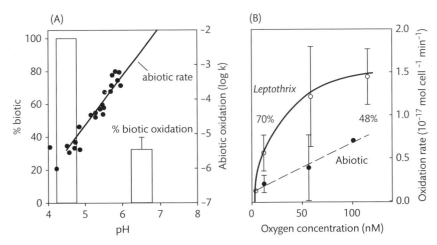

Figure 13.4 Biotic and abiotic oxidation of Fe(II). Biotic rates were measured in a wetland plant rhizosphere as a function of pH, with oxygen levels between 0.5 and 3% of saturation. The abiotic oxidation rate is given as the log of the rate constant with units of min^{-1} (A). Biotic oxidation by *Leptothrix cholodnii* along with abiotic oxidation at pH 7.0 as a function of oxygen concentrations (B). The percentages are the relative biotic oxidation at two oxygen concentrations. Data from Neubauer et al. (2002), Singer and Stumm (1970), and Vollrath et al. (2012).

Table 13.2 Some types of iron-oxidizing bacteria. The iron oxidizers using nitrate as an electron acceptor are chemolithotrophic denitrifiers and produce N_2. The chemoorganotrophs oxidize iron but gain energy from the oxidation of organic material rather than from the iron. The photolithotrophic iron oxidizers use the electrons from ferric oxidation to reduce CO_2 and synthesize organic material autotrophically. Based on Emerson et al. (2010) and Hedrich et al. (2011).

Metabolism	Electron acceptor	Phylum or class	Example
Neutrophilic lithotrophy	O_2	*Betaproteobacteria*	*Gallionella*
	O_2	*Zetaproteobacteria*	*Mariprofundus ferrooxydans*
	NO_3^-	*Alphaproteobacteria*	*Paracoccus ferrooxidans*
	NO_3^-	*Betaproteobacteria*	*Thiobacillus denitrificans*
	NO_3^-	*Gammaproteobacteria*	*Thermomonas*
Acidophilic lithotrophy	O_2	*Betaproteobacteria*	*Ferrovum myxofaciens*
	O_2	*Actinobacteria*	*Sulfobacillus*
	O_2	*Euryarchaeota*	*Ferroplasma acidarmanus*
Chemoorganotrophy	O_2	*Alphaproteobacteria*	*Pedomicrobium ferrugineum*
	O_2	*Betaproteobacteria*	*Leptothrix*
Photolithotrophy	CO_2	*Alphaproteobacteria*	*Rhodovulum*
	CO_2	*Chloroflexi*	*Chlorobium ferrooxidans*

Acidophilic iron-oxidizing microbes, such as *Leptospirillum* (*Nitrospirae* phylum) and *Ferroplasma* (*Euryarchaeota*), play a key role in generating acidity in runoff from abandoned mines, called "acid mine drainage." These microbes belong to simple communities consisting of just a few taxa, which make them ideal targets for metagenomic and other omic studies (Aliaga Goltsman et al., 2015).

The problem of acid mine drainage starts with the oxidation of iron sulfur minerals, such as pyrite (FeS_2):

$$FeS_2 + 14Fe^{3+} + 8H_2O \rightarrow 15Fe^{2+} + 2SO_4^{2-} + 16H^+ \quad (13.4)$$

producing sulfuric acid (H_2SO_4). Acid production is minimal in the absence of iron-oxidizing microbes because concentrations of ferric iron (Fe^{3+}) are low relative to pyrite. Production of ferric iron is slow because ferrous iron (Fe^{2+}) is stable when the pH is low and iron-oxidizing microbes are absent. Ferric iron is needed because it is a stronger catalyst than the other possibility, oxygen. However, acidophilic iron-oxidizing bacteria oxidize Fe^{2+} back to Fe^{3+}, leading to more pyrite oxidation and faster acid production. The end result is the huge environmental damage caused by acid mine drainage.

Other iron-oxidizing bacteria are anoxygenic photoautotrophs. These bacteria use Fe^{2+} as a source of electrons for reducing CO_2, in place of H_2O or H_2S used by oxygenic and anoxygenic sulfur-oxidizing photoautotrophs, respectively. Experimental studies have indicated that four ferrous atoms are needed to reduce one molecule of CO_2 following the equation (Ehrenreich and Widdel, 1994):

$$4Fe^{2+} + CO_2 + 4H^+ + light \rightarrow CH_2O + 4Fe^{3+} + H_2O \quad (13.5).$$

The organisms carrying out phototrophic iron oxidation, known as photoferrotrophs, are closely related to purple sulfur, purple non-sulfur, and green bacteria, which carry out sulfur oxidation and anoxygenic photosynthesis (Chapter 6 and 11). Their distribution is limited today to the few environments with high Fe^{2+} but low sulfide concentrations and adequate light.

Even so, phototrophic iron oxidizers are of interest to geomicrobiologists exploring the evolution of early life on Earth. Iron-oxidizing anoxygenic phototrophs may have been one of the first photosynthetic organisms in the biosphere, predating cyanobacteria. If they were abundant and active enough, the oxidized iron produced by phototrophic iron oxidization may explain "banded iron formations" in the Precambrian (Figure 13.5), when other data indicate oxygen concentrations were too low for aerobic iron-oxidizing microbes to contribute much (Posth et al., 2013; Crowe et al., 2008). The alternative explanation for banded iron formations, oxygen-dependent chemolithotrophic iron oxidation, requires not only oxygen production by cyanobacteria, but also mechanisms to prevent oxygen from building up to measurable levels in the atmosphere. Iron- and sulfide-fueled anoxygenic photosynthesis may have been critical in setting oxygen levels during the Proterozoic (Johnston et al., 2009).

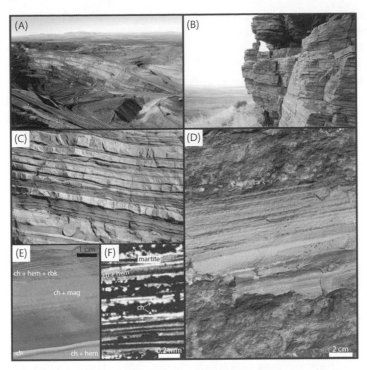

Figure 13.5 Photographs of banded iron formations in Western Australia. Aerial view of formations along the southern ridge of Mt. Tom Price (A). Formation at Kuruman Kop (B). Banding in the Joffre Member (C). Close-up of the Kuruman formation, showing iron-rich and iron-poor millimeter-thick bands (D). Alternating bands of chert (ch) and magnetite (mag), and chert–hematite–riebeckite (hem = hematite, an iron-rich mineral, and rbk = riebeckite, a sodium and iron-rich mineral) (E). A high magnification of the bands. Martite is a variety of hematite (F). Banded iron formations came about in the Precambrian about 3.3 to 1.8 billion years ago. These striking geological formations have yielded important clues about the birth of an oxygen-rich atmosphere during the Precambrian. Photographs from Konhauser et al. (2017) and used with permission of the publisher. Color version is available at www.oup.co.uk/companion/kirchman.

Magnetite and magnetotactic bacteria

The previous section described the formation of iron minerals by Fe^{2+} oxidation. Another important iron mineral, magnetite (Fe_3O_4), can be produced as a by-product of Fe(III) reduction coupled to organic carbon oxidation (Chapter 11). Geologists are very interested in magnetite because it records changes in the direction and intensity of the Earth's magnetic field over geological time. These changes are very useful for understanding plate tectonics and other geological processes. Magnetite is the most magnetic of all minerals on the planet, but only magnetite crystals of the right size are single-domain magnets with magnetic properties. The presence of magnetite in a Martian meteor was initially taken as a sign of life on Mars (McKay et al., 1996), but subsequent work demonstrated that all signs of

Martian life on the meteor could be explained by abiotic mechanisms.

The magnetite-producing bacteria can be divided into those that do not purposely use magnetite (it is just a by-product) and those that use magnetite in their ecophysiology. The two best-studied iron-reducing bacteria which produce magnetite as a by-product are *Geobacter metallireducens* and *Shewanella putrefaciens*. The Fe(II) resulting from Fe(III) reduction by these bacteria abiotically forms mostly small magnetite crystals outside the cell, unaligned in chains or other formations (Pósfai et al., 2013). Most of these crystals are too small to have magnetic properties, and these Fe(III)-reducing bacteria do not respond to magnetic fields.

In contrast to being just a by-product, magnetite is an essential and unique feature of the ecophysiology of magnetotactic bacteria. These bacteria are from several taxa, the best known of which are in the *Proteobacteria*, although a genomic survey found evidence for representatives in other phyla, including *Nitrospirae* and *Planctomycetes* (Lin et al., 2017). Magnetotactic bacteria have intracellular membrane-lined structures, called magnetosomes, which carry magnetite or greigite or both. As implied by the name, magnetotactic bacteria swim towards or away from the Earth's magnetic poles due to the arrangement of the magnetosomes in relationship with the Earth's magnetic field lines. Even dead magnetotactic bacterial cells become aligned with the magnetic field. The magnetism of magnetite in magnetotactic bacteria is due to its chemical purity and having the right size, typically 35–120 nm. Because of size and chemical characteristics, magnetite crystals in magnetotactic bacteria are permanent, single-domain magnets at ambient temperatures.

The advantage of magnetotactic behavior for microbes is not fully known. It is especially unclear for the few eukaryotes with magnetite. For bacteria, one hypothesis is that magnetotaxis helps the bacterium determine which way is up and which is down (Figure 13.6). By coupling magnetotaxis with mechanisms for sensing oxygen (aerotaxis), these bacteria potentially find microhabitats with optimal oxygen (many are microaerophilic) or hydrogen sulfide concentrations necessary for chemolithotrophy. One sign of a problem with this hypothesis is that magnetotactic bacteria are found at the equator, where magnetotaxis cannot be used to distinguish up from down because the magnetic field lines are horizontal to the plane of the Earth's surface. In any case, there must be some advantage to having magnetosomes because cells devote a large amount of iron and energy to their synthesis.

Manganese nodules and manganese-oxidizing bacteria

Mn(III) oxyhydroxides and Mn(IV) oxides can co-occur with Fe(III) oxides, and commonly precipitate together to form ferromanganese deposits, known as "manganese nodules," on the sea floor, in soils, and in lakes. The bottom of the Pacific Ocean is thought to be covered with 10^{12} tons of nodules, and Oneida Lake, New York alone has 10^6 tons. Along with their commercial value, manganese minerals provide highly reactive surfaces that mediate abiotic transformations of other metals and compounds, both natural and contaminants (Borch et al., 2010). Because the abiotic oxidation of Mn^{2+} is slow, most of the manganese in nodules and other mineral formations is thought to be the by-product of manganese-oxidizing bacteria with some help from fungi (Hansel and Learman, 2015). Manganese-oxidizing bacteria become coated in manganese oxides or have appendages, sheaths, or spore coats where manganese oxides precipitate.

What is odd about manganese-oxidizing bacteria and fungi is that it is not entirely clear why they do it (Tebo et al., 2010). There is little evidence that microbes gain any energy from manganese oxidation, even though the theoretical gain is about 50 kJ mol^{-1}, depending on environmental conditions and assumptions, about the same or even more than gained by iron oxidation, from which we know bacteria harvest energy. Spores of a *Bacillus* strain can oxidize manganese, evidence that active cells are not even necessary. However, in other cases, manganese oxidation appears to be a specific process mediated by specific enzymes and bacteria (Figure 13.7). Some fungi and bacteria use a different mechanism and oxidize manganese using superoxide that is produced extracellularly (Hansel et al., 2012).

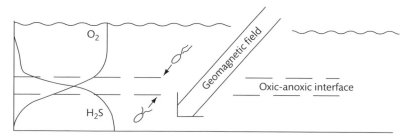

Figure 13.6 One explanation for why microbes have magnetotaxis. By moving along the geomagnetic field, magnetotactic microbes are hypothesized to be more adept at finding the oxic–anoxic interface, the micro-habitat optimal for their growth. Based on Bazylinski and Frankel (2004).

If bacteria and fungi do not gain any energy directly from the reaction, why do they bother? Several answers have been proposed. The manganese precipitates that coat manganese-oxidizing microbes may protect them from UV light and from reactive oxygen species, such as superoxide and other strong oxidants formed by UV-driven photochemistry. Of course, this explanation would not account for microbes with manganese coats in dark habitats. Another explanation that would be applicable to all habitats is that the manganese coating may ward off predators and viruses. Another possible gain is related to energy production. The Mn(IV) oxides and hydroxides from manganese oxidation may help to oxidize recalcitrant organic material, making it more labile for use by microbes. Some fungi appear to use Mn(IV) oxides to degrade lignin (Chapter 7).

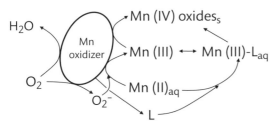

Figure 13.7 Mechanism for Mn(II) oxidation via two one-electron steps. Mn(II) is first oxidized to Mn(III) and then to Mn(IV) oxides. The subscripts "s" and "aq" indicate solid and aqueous (dissolved), respectively. When iron concentrations are low, bacteria produce organic ligands (L) that can promote the oxidation of Mn(II) to Mn(III)-L. Another mechanism involves the production of superoxide (O_2^-) that oxidizes Mn(II) to Mn(III). Based on Geszvain et al. (2012).

Box 13.1 Microbes are not omnipotent

One theme running through this book is the huge metabolic diversity of bacteria, archaea, and protists. Microbes seem to be able to do everything, even gaining energy from reactions that at first seem thermodynamically impossible. But there are some rare cases of microbes apparently failing to take advantage of an energy-generating reaction. One such case is manganese oxidation. Reduced manganese is also not used as an electron source (reductant) for anoxygenic photosynthesis, nor is ammonium, in contrast with reduced iron. However, it is conceivable that microbial ecologists have not looked in the right place or have not done the proper experiment. Decades after work with sulfur anoxygenic photosynthesizing bacteria, Griffin et al. (2007) demonstrated that some photosynthesizing bacteria use nitrite as a reductant.

Carbonate minerals

While iron and manganese mineral deposits are large and interesting, carbonate minerals are even more abundant and important in the carbon cycle. The formation of carbon-rich minerals traps carbon in huge, long-lived pools, in contrast with carbon dioxide fixation into organic carbon by primary producers (Figure 13.8). Carbonate minerals make up the largest pool of carbon in the biosphere, orders of magnitude larger than the carbon in living organisms, detrital organic material including soil organic carbon, and atmospheric carbon dioxide.

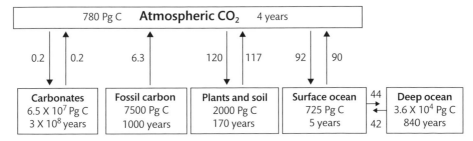

Figure 13.8 Carbon pools and residence times of carbon in carbonates, fossil fuels ("fossil carbon"), terrestrial, and oceanic pools. The numbers by the arrows are fluxes (Pg C per year). Residence times were calculated by dividing the pool size by the flux. Data from Sundquist and Visser (2004) and Houghton (2007).

Consequently, the residence time of a carbon atom in carbonate pools is extremely long, on the order of 300 million years, much longer than the 500 to 10,000-year timescale of organic pools (Sundquist and Visser, 2004). Artificially promoting carbonate mineral formation is a way of capturing carbon dioxide before it is released to the atmosphere. This and other "carbon sequestration" actions are being examined (and to a limited extent, being done) to slow the increase in atmospheric carbon dioxide and to minimize its impact on our climate.

Several microbes and microbial processes are directly or indirectly involved in the formation of the most abundant carbonate mineral, calcium carbonate ($CaCO_3$). These processes affect CO_2 or HCO_3^-, both of which are involved in calcium carbonate formation as described by the equation

$$Ca^{2+} + 2HCO_3^- \rightarrow CaCO_3 + CO_2 \qquad (13.6).$$

(Note the release of CO_2 by calcium carbonate precipitation. A similar equation explains why concrete manufacturing contributes to the build-up of atmospheric CO_2 and global warming.) A critical parameter in thinking about carbonate mineral formation is the solubility product (K_{sp}), defined as

$$K_{sp} = \left[Ca^{2+}\right]_{eq}\left[CO_3^{2-}\right]_{eq} \qquad (13.7)$$

where $[Ca^{2+}]_{eq}$ and $[CO_3^{2-}]_{eq}$ indicate the concentrations of Ca^{2+} and CO_3^{2-} at equilibrium with a solid calcium carbonate mineral. A lower value of K_{sp} indicates a greater tendency for the mineral to form. Solubility constants for common carbonate minerals vary by several orders of magnitude, and even minerals with the same chemical make-up can have different constants (Table 13.3). Another, less formal way of expressing solubility is the maximum concentration at which a mineral can be dissolved in water. A sodium salt of carbonate (natron) will stay dissolved even if several kilograms are added to a liter of water, while a liter can hold less than 2 g of calcium carbonate before it precipitates.

Two expressions are used to evaluate how close a particular solution is to being saturated with respect to calcium carbonate. Above saturation, precipitation and mineral formation is possible. Geomicrobiologists use the saturation index (SI), which is

$$SI = \log\left(\left[Ca^{2+}\right]\left[CO_3^{2-}\right]\right)/K_{sp} \qquad (13.8).$$

Experiments have demonstrated that SI has to be greater than about 1, equivalent to $[Ca^{2+}][CO_3^{2-}] = 10$, in order for calcium carbonate to form (Visscher and Stolz, 2005). Chemical oceanographers use a different expression for the degree of saturation (Ω),

$$\begin{aligned}\Omega &= \left[Ca^{2+}\right]\left[CO_3^{2-}\right]/\left[Ca^{2+}\right]_{sat}\left[CO_3^{2-}\right]_{sat} \\ &= \left[Ca^{2+}\right]\left[CO_3^{2-}\right]/K_{sp}\end{aligned} \qquad (13.9)$$

where $[Ca^{2+}]_{sat}$ and $[CO_3^{2-}]_{sat}$ indicate concentrations at which Ca^{2+} and CO_3^{2-} are in equilibrium with $CaCO_3$. The surface ocean with a pH of about 8.2 is now supersaturated for calcium carbonate (carbonate stays as a solid rather than dissolving), and Ω values for the main forms of calcium carbonate, calcite, and aragonite, are now about 5.6 and 3.7, respectively (Doney et al., 2009).

Table 13.3 Some carbonate-containing minerals potentially formed by biomineralization. The solubility constants K_{sp} were calculated, except the measured value for natron (Morse and Mackenzie, 1990). The maximum soluble concentration data are from Weast (1987). Above the maximum concentration at 20 °C, the mineral precipitates. See the main text and Kendall et al. (2012) for information about the microbes potentially involved in forming these minerals.

Mineral	Formula	$-\log(K_{sp})$	Maximum soluble conc (g L^{-1})	Microbes
Aragonite	$CaCO_3$	8.1	1.5	Cyanobacteria
Calcite	$CaCO_3$	8.3	1.4	Cyanobacteria
Siderite	$FeCO_3$	10.5	6.7	Iron reducers
Dolomite	$MgCa(CO_3)_2$	17.1	<1	Sulfate reducers
Magnesite	$MgCO_3$	8.2	10.6	Actinomycetes
Rhodochrosite	$MnCO_3$	10.5	6.5	Manganese reducers
Natron	$Na_2CO_3 \cdot 10H_2O$	0.8	7100	Sulfate reducers
Strontianite	$SrCO_3$	8.8	1.1	Various

Figure 13.9 Modern stromatolites in the Hamelin Pool Marine Nature Reserve, Shark Bay in Western Australia. Each is about 0.5 m in diameter. See also Suosaari et al. (2016). Photograph by Jennifer Glass and used with permission. Color version is available at www.oup.co.uk/companion/kirchman.

Box 13.2 Fingerprints from early life on Earth

An argument for studying modern stromatolites and microbial mats is to understand ancient ones and to explore questions about the beginning of life during the Precambrian, nearly four billion years ago (Ga). Some of the earliest signs of life have been found in fossil stromatolites such as the 3.5 Ga Warrawoona Group in Australia (Knoll et al., 2016; Dupraz et al., 2009). Ancient stromatolites have yielded several clues about early life, starting with their laminated structure, reminiscent of modern-day microbial mats. Perhaps most telling, ancient stromatolites have yielded microbe-like fossils ("microfossils") with morphologies similar to those of modern filamentous cyanobacteria, or perhaps anoxygenic photosynthetic bacteria such as *Chloroflexus*. Other evidence of microbial life at this time comes from carbon and sulfur isotopes and organic biomarkers indicative of cyanobacteria. Because of their size and areal coverage, microbial mats were crucial in the evolution of life and the development of the atmosphere during Earth's early days.

However, these Ω values have been decreasing over the years due to ocean acidification caused by increasing atmospheric carbon dioxide, which imperils organisms with carbonate shells and walls, particularly those composed of the more soluble form of calcium carbonate, aragonite, the form found in corals (Chapter 14). Other examples of carbonate-containing organisms include coccolithophorids (Chapter 6) and other marine protists (foraminifera). These organisms are involved in biologically controlled biomineralization.

Several other organisms are involved in biologically influenced biomineralization of calcium carbonate. In particular, the role of cyanobacteria in promoting carbonate formation has been extensively examined. Some species can form mats made up of layer upon layer of cyanobacteria, other microbes, and calcium carbonate and other inorganic material (Chapter 6). Some laminated mats grow to a size large enough to be called "stromatolites"; "microbialites" is a more general term to describe mats varying in organization and size. Carbonate formation in these mats is usually mostly mediated by cyanobacteria and diatoms, but a metagenomic study found evidence of other photosynthetic eukaryotes, anoxygenic photosynthetic bacteria, and sulfate reducers contributing to mat formation in an alkaline, brackish lake in Mexico (Saghaï et al., 2015). While a few are still living today (Figure 13.9), the heyday of stromatolites was at the beginning of the biosphere when life may have been dominated by these microbial mats (Box 13.2).

Phosphorus minerals

While carbonate mineral formation has a direct impact on the carbon cycle, phosphorus minerals indirectly affect it by influencing primary production and other microbial processes requiring phosphorus. The formation and deposition of calcium phosphate minerals, in particular apatite, is thought to be the largest route by which phosphorus is captured over geological timescales and stored in sediments and soils (Ingall, 2010). Apatite refers to a group of minerals that includes hydroxyapatite, fluorapatite, and chlorapatite, with their respective formulas being $Ca_{10}(PO_4)_6(OH)_2$, $Ca_{10}(PO_4)_6F_2$, and $Ca_{10}(PO_4)_6Cl_2$. Apatite and other calcium phosphate minerals do not form as readily as may be expected from the high concentrations of calcium and phosphate found in sediments. Release of phosphate from decomposing organic material may promote apatite formation, but another mechanism seems necessary.

A study provided strong evidence of the role of biologically induced biomineralization of apatite in the Benguela upwelling system off the coast of Namibia (Goldhammer et al., 2010). These investigators showed that phosphate, traced with its radioactive form ($^{33}PO_4^{2-}$), was incorporated into apatite when sulfide-oxidizing bacteria were present, but not when these bacteria were absent. The reaction was quick, with ^{33}P appearing in apatite within 48 hours. The mechanism involves formation of polyphosphate that either chelates Ca^{2+}, eventually forming apatite, or is hydrolyzed to PO_4^{3-} leading to the precipitation of apatite (Figure 13.10). Regardless of

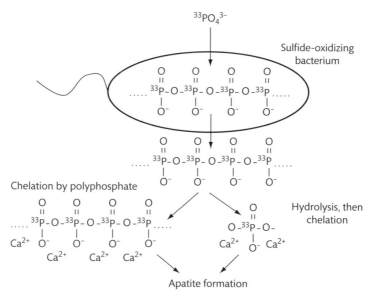

Figure 13.10 Formation of apatite following the uptake of phosphate (traced by adding $^{33}PO_4^{3-}$) by sulfide-oxidizing bacteria. Phosphate taken up in phosphorus-rich environments is stored as polyphosphate, which is hydrolyzed back to phosphate intracellularly or extracellularly. The released phosphate then chelates Ca^{2+} (Goldhammer et al., 2010). Alternatively, the polyphosphate may be released and directly chelates Ca^{2+} without prior hydrolysis (Ingall, 2010).

the precise mechanism, sequestration of phosphate in apatite in anoxic sediments is surprising, because usually phosphate is released from sediments under these conditions, at least in non-marine systems. The key may be the presence of large sulfide-oxidizing bacteria that promote apatite formation in organic-rich marine sediments but are absent in non-marine anoxic sediments and soils. Polyphosphate from diatoms has also been shown to be important in apatite formation in sediments (Diaz et al., 2008).

Weathering and mineral dissolution by microbes

While microbes and abiotic reactions are taking ions out of solution to form minerals, other microbes and abiotic reactions are breaking them up and putting ions back into solution. The breaking up or the "dissolution" of primary minerals and the formation of secondary minerals is called "weathering" by geologists and geomicrobiologists (Uroz et al., 2009). In terms of the carbon cycle, the most important weathering reaction is the dissolution of rocks by carbon dioxide, with the following reaction being one example:

$$CO_2 + 2H_2O + CaAl_2Si_2O_8 \rightarrow Al_2Si_2O_5(OH)_4 \\ + CaCO_3 \quad (13.10).$$

Hidden in Equation 13.10 is the formation of the weak acid, carbonic acid (H_2CO_3), from CO_2 combining with water. It is carbonic acid and its protons that do the weathering of the mineral anorthite ($CaAl_2Si_2O_8$) in Equation 13.10, forming the secondary clay mineral kaolinite ($Al_2Si_2O_5(OH)_4$).

As illustrated before (Figure 13.8), the end result is the removal of carbon dioxide from the atmosphere into a geological reservoir, analogous to fixation of atmospheric carbon dioxide by autotrophic organisms into a biological reservoir, the organic material of cells. Both weathering and microbial carbon dioxide fixation affect carbon dioxide concentrations in the atmosphere, but the huge difference between the two is the timescale. While the biotic part of the carbon cycle operates on the day to year timescale, corresponding to the lifespan of microbes and higher plants, the removal of atmospheric carbon dioxide by geological processes occurs over thousands to millions of years. These geological processes, however, are affected by microbes.

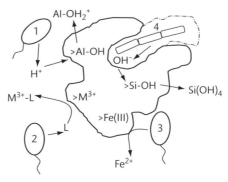

Figure 13.11 Mechanisms of mineral dissolution by microbes. 1: acidity (protons); 2: ligands (L), including both low molecular weight compounds and extracellular polymers; 3: iron reduction; 4: hydroxides. Similar to what is depicted for the hydroxide mechanism (4), microbes can bore into minerals and rocks and excrete extracellular polymers (the dashed line in mechanism 4) which traps dissolution-promoting compounds. The ">" for the metals, such as >Al–OH, indicates a link with other elements in the solid mineral. For the second mechanism, microbes may release weathering agents to access phosphate instead of a metal (M^{3+}).

Equation 13.10 is just one of many weathering reactions. Microbes have a role in most if not all of these reactions. Figure 13.11 summarizes some of the mechanisms by which microbes affect weathering and mineral dissolution, more colorfully called "eating rocks." Analogous to the distinction between biologically induced and biologically influenced biomineralization, microbes have both direct and indirect roles in mineral dissolution and weathering. We have already seen that many microbes produce or consume carbon dioxide, thus indirectly affecting mineral dissolution as indicated in Equation 13.10. At the other extreme, microbes dissolve minerals in order to access needed nutrients, such as phosphate bound up in apatite. Also, anaerobic heterotrophic bacteria contribute to the dissolution of iron-bearing minerals by reducing Fe(III). Whatever the mechanism, microbes are important in many weathering reactions.

Dissolution by acid and base production

Microbial weathering starts when a fresh rock surface is colonized by bacteria, fungi, cyanobacteria, and eukaryotic algae. The types of microbes colonizing rock surfaces vary depending on the mineral composition and other environmental properties. DNA fingerprint methods have shown differences in community structure of the bacteria colonizing muscovite, plagioclase, potassium-feldspar, and quartz (Gleeson et al., 2006). These bacteria contribute to dissolution through the production of acids, organic chelators, and in some cases hydrogen cyanide (HCN). In addition to epilithic microbes on the rock surface, other, endolithic microbes proliferate in cracks and crevices of rocks to escape the harsh environment of exposed surfaces; life on rock surfaces is tough because of low water availability, exposure to full sunlight, and limited availability of nutrients. Fungal hyphae are particularly adept at exploiting narrow channels within rocks and between mineral boundaries (Robson, 2017). Because of acid production and the release of chelators, these microbes can drastically change rock surfaces, ranging from simple etching and pitting to more extensive disruption of the rock. Once altered by microbial action, physical forces can more easily erode the rocks and expose new reactive surfaces.

Microscopic analyses and cultivation-independent approaches have found diverse communities of eukaryotic algae and cyanobacteria on and inside rocks. Studies using clone libraries of rRNA genes have shown that this microbial photoautotrophic community varies with the type of rock and with exposure to different amounts of sunlight (Hallmann et al., 2013). Endolithic photoautotrophs contribute to the dissolution of sandstones and other silicate-bearing rocks by a different mechanism than seen for heterotrophic microbes (Büdel et al.,

---

Box 13.3 Microbes and monuments

Microbes colonize minerals regardless of whether the mineral is part of a rock in nature or of a structure built by humans. When the latter, microbes contribute to the deterioration of stone monuments, statues, and buildings, threatening archaeological and historic sites (Gadd, 2017). Less visible but perhaps even more important is the effect of microbes on concrete underground. Concrete sewer pipes are attacked by acids and hydrogen sulfide produced by sulfate-reducing bacteria. Oxidation of hydrogen sulfide by other microbes leads to sulfuric acid production and further corrosion of underground concrete.

2004). Photosynthesis by these microbes raises the pH to over 10 in the local micro-environment surrounding the endolithic photoautotroph cell. The resulting OH^- ions cause deprotonation of SiOH bonds and loss of soluble ions from the solid rock. Because of their need for light, endolithic algae are active only within millimeters of the rock surface and can grow only on exposed rocks not covered by soil or vegetation.

Rock-dwelling photoautotrophic microbes may be especially important in deserts where few higher plants can grow. These microbes, along with lichens, mosses, and liverworts, make up "biological soil crusts," which

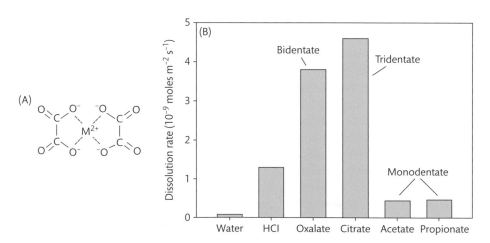

Figure 13.12 Dissolution by organic acids. Oxalate binding to a metal (M^{2+}) (A). Dissolution by inorganic and organic acids of bytownite (B). Mono-, di-, and tridentate refer to organic acids with one, two, or three acid groups. Data from Welch and Ullman (1993).

are complex communities living on soil surfaces of arid habitats. One of the most extreme arid habitats is the Atacama Desert in Chile, where rain does not fall for years, sometimes decades. The Atacama soil crust and rock-dwelling microbes may be a model for life on Mars (Bull et al., 2016).

Dissolution by low and high molecular weight chelators

Acid-producing microbes affect weathering reactions by releasing organic anions as well as protons. These anions can chelate metals, thus increasing the solubility of minerals in the surrounding solution and promoting dissolution from the solid phase. Bidentate (two acid groups) and tridentate (three acid groups) organic acids are more effective than those with only one acidic moiety (Figure 13.12). A good example is oxalic acid, which can occur in concentrations high enough to mediate mineral dissolution, but even gluconate with its single acidic functional group enhances dissolution of silicate minerals. Organic acids can be released by fermenting bacteria (Chapter 11) or by other bacteria, partially oxidizing organic material, such as the incomplete oxidation of glucose to gluconate. Chelation as well as acidity of organic acids contribute to weathering reactions.

Organic ligands released by microbes to retrieve phosphate or essential trace metals can also contribute to mineral dissolution. One class of ligands, "siderophores," have a high affinity for iron and are used by microbes to access this metal when concentrations are low. As with the organic acids, these organic ligands promote solubilization of the metal from the solid phase. Siderophores have been extensively examined because of their ecological relevance, but also because pathogenic bacteria use siderophores to access iron in humans and other mammals.

In addition to low molecular weight chelators, microbes can affect mineral dissolution and weathering reactions by the production of extracellular polysaccharides and other polymers. Experiments with isolated polymers have demonstrated direct effects on mineral dissolution (Welch et al., 1999). Polymers with acidic groups promote more mineral dissolution than do neutral polymers. Other polymers inhibit dissolution by coating and blocking reactive sites on the mineral. Even if these polymers do not have a direct role, they indirectly affect rates by trapping protons and hydroxides close to the mineral surface, leading to higher local concentrations than would otherwise occur. In dry environments, the water retained by polymers promotes fracturing, hydrolysis, and other chemical reactions that eventually break up rocks.

Dissolution by lichens

Lichens contribute substantially to mineral dissolution, and are a fascinating example of a microbial symbiosis. Some lichens superficially resemble mosses (a small flowerless plant), reflected in some of their common names, such as reindeer moss. But the individual members of the lichen symbiosis are microbes, which collectively are large enough to be seen with the naked eye. Lichens were thought to be a symbiosis between a fungus, usually an ascomycete, and usually one, sometimes two photosynthetic microbes (Figure 13.13). The photosynthetic microbe is most often a green alga (Chlorophyta), but it can be a cyanobacterium. A metatranscriptomic study found evidence of another fungus partner, a yeast in the Basidiomycota (Spribille et al., 2016). Lichens grow on just about any surface on land:

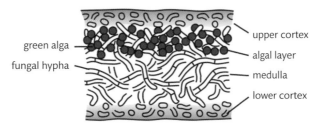

green alga
fungal hypha

upper cortex
algal layer
medulla
lower cortex

Figure 13.13 One model for the lichen symbiosis. Fungal hypha make up the medulla as well as the upper cortex exposed to the external environment, and the lower cortex anchoring the lichen to the surface. Some lichens attach via the medulla to the surface. Diagram by M. Piepenbring and used with permission.

rocks, soil, bark, leaves, hanging from tree branches, and even on other lichens.

The fungus partner in the symbiosis appears to be mainly responsible for the dissolution of minerals in rocks and soils (Asplund and Wardle, 2017). The fungi can secrete polyphenolic acids to solubilize minerals and gain access to phosphate in order to support growth and carbon fixation by the photoautotrophic partner, which in turn provides organic carbon for the fungi. Both partners can produce H^+ during respiration, adding to the weathering process. Using data on net carbon fixation rates to estimate lichen metabolic rates, Porada et al. (2014) suggested that weathering by lichens was similar to total weathering by both biotic and abiotic mechanisms. Lichens are important in weathering human-built structures as well as natural ones. The US National Park service has had to clean lichens from Mt. Rushmore to minimize deterioration of the massive presidential monument.

The deep biosphere and the geomicrobiology of fossil fuels

The minerals and rocks inhabited by microbes discussed so far are easily reached, more or less, except perhaps for those on mountains and giant monuments. Microbes and microbe–mineral interactions are also very important in another environment, the deep biosphere, that can be sampled only from boreholes, deep mines and caves, and oil drilling rigs. This environment has been defined as any habitat greater than one meter below soil or ocean sediments, but the term is more commonly applied to habitats tens to thousands of meters below the Earth's surface (Edwards et al., 2012). Microbial life is usually detected in deep subsurface environments if the temperature is hospitable (Colwell and D'Hondt, 2013). Microbes have been found living in micron-sized fractures and pores in rocks and sandstone up to about 85 °C (Jørgensen and Marshall, 2016). The initial estimates were too high (Whitman et al., 1998), but even more conservative numbers indicate that as much as 20% of the Earth's biomass may be in the deep biosphere (McMahon and Parnell, 2014). The relative abundance of bacteria is either greater than, equal to, or less than archaeal abundance, depending on the study (Colwell and D'Hondt, 2013). Fungi and viruses are much less abundant but are detectable, and metazoans have been found in deep gold mines of South Africa (Borgonie et al., 2015).

As mentioned in Chapter 8, a large fraction of the bacteria and archaea are active in deep subsurface environments, albeit metabolizing at very low rates. What fuels the deep biosphere, a topic of much research, varies with the microbe and location (Figure 13.14).

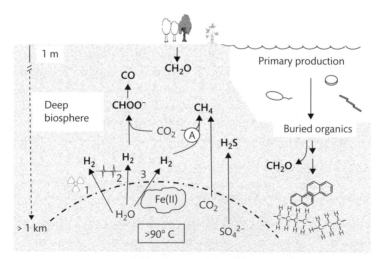

Figure 13.14 Energy sources (in bold) supporting microbial activity in the deep biosphere. Hydrogen gas (H_2) is produced by three abiotic mechanisms: (1) radiolytic splitting of water exposed to radioactive minerals; (2) splitting of water by physical shearing of silicate minerals, and (3) reduction of water by Fe(II)-rich minerals. The other energy-producing mechanisms are also abiotic, except for methane production by methanogenic archaea ("A") and the organic material (CH_2O) from surface primary production. Based on Colman et al. (2017) and Colwell and D'Hondt (2013).

A small portion of organic material from surface, light-driven primary production can be buried and survive degradation long enough to support activity of heterotrophic microbes in subsurface environments. Some of that organic material may be substantially modified over geological time to become natural gas and petroleum, which can also support deep microbial life.

Most intriguing are compounds abiotically produced by geological forces. Hydrogen gas (H_2) can be formed by abiotic mechanisms in the deep, especially hot biosphere and can directly support microbial activity (Colman et al., 2017). The genes encoding for H_2 utilization (hydrogenases) are ten-fold more common in deep surface genomes than in their surface counterparts. The H_2 can also react with CO_2 via the Fischer–Tropsch process to form formate ($HCOO^-$) and eventually carbon monoxide (CO), both potential energy sources for microbes. In subsurface environments without organic carbon from surface primary production, the organic carbon must come from chemoautotrophic fixation. These organisms are called "lithoautotrophs" in the subsurface literature.

rRNA and omic studies have shown that autotrophic sulfate-reducing bacteria and archaea are abundant in deep subsurface environments (Colman et al., 2017). Unlike the heterotrophic sulfate reducers discussed in Chapter 11, these sulfate reducers use H_2, CO, or $HCOO^-$ as electron donors and CO_2 as their carbon source, making them chemolithoautotrophs. In terrestrial subsurface environments far from seawater, sulfate can come from abiotic pyrite oxidation. Methanogens have also been found, as well as the archaea in the ANME clades known to be involved in anaerobic methane oxidation (Chapter 11). Some of these archaea have been given exotic names, such as *Bathyarchaeota* and *Hadesarchaea*, reflecting their subterranean origin. Many deep biosphere bacteria are in *Firmicutes*, *Deltaproteobacteria*, *Chloroflexi*, and other now familiar phyla, although their physiologies likely differ from those seen in the environments discussed so far in this book. Other bacteria are from poorly known phyla, such as *Aminicenantes*, also referred to as OP8.

The deep biosphere remains largely unexplored because it is technically challenging and expensive to sample. The technical challenges include sampling hard-to-access habitats without contamination from the surface; contamination is a big concern because microbial abundance and biomass are so low in subsurface environments. The sampling can require expensive equipment and tools more familiar to oil drilling roughnecks than to scientists. In spite of the challenges, work in the deep biosphere has yielded many surprises about a profoundly important and novel ecosystem.

Microbes in petroleum reservoirs

In addition to basic science reasons, geomicrobiologists have been motivated and financially supported to work on the deep biosphere because of the many practical problems caused by or ameliorated by deep subsurface microbes. One problem is the impact of microbes on the quality and quantity of fossil fuels in subsurface reservoirs. While microbial processes contributed substantially to the formation of these hydrocarbons millions of years ago, they also degrade petroleum hydrocarbons and create problems for the petroleum industry.

Microbes are found in many but not all petroleum reservoirs. An essential ingredient for making a petroleum reservoir hospitable for life is water (Head, 2015). The reservoir hydrocarbons exclude water, pushing it into isolated microhabitats where microbes may flourish.

Box 13.4 An early proponent of a deep but also hot biosphere

Thomas Gold (1920–2004) was an astrophysicist who also thought about processes potentially occurring deep below the Earth's surface (Colman et al., 2017). In the 1950s, he proposed that abiotic mechanisms produced fossil fuels and a nearly endless supply of natural gas and petroleum. Long before microbes had been found in deep subsurface environments, Gold also hypothesized that energy from geothermal processes, not organic material from sunlight-driven primary production, supports large microbial communities in what he called the "deep, hot biosphere," a term he used in the title of a high profile publication and a book where he also speculated about life on other planets (Gold, 1999). Although mostly wrong about the abiotic origin of fossil fuels, Gold was prescient about microbes in deep subsurface environments, a hot topic today in geomicrobiology.

The oil–water interface is where bacterial abundance can be highest (Figure 13.15). Water itself is essential for sustaining microbial life, but it also brings with it phosphate and other inorganic nutrients. These nutrients may limit especially aerobic degradation of hydrocarbons. The water can also have enough sulfate to support the activity of sulfate-reducing bacteria. Microbes are not found in high temperature reservoirs or those that have experienced "paleopasteurization" in their geological past. Sulfate reduction and methanogenesis have not been detected in reservoirs hotter than 80–90 °C. Hypersaline conditions can also inhibit microbial activity in reservoirs.

Sulfate reduction in oil reservoirs causes several problems for the oil industry. An end product of sulfate reduction, the gas hydrogen sulfide, is a health hazard to oil extraction workers and is part of microbially influenced corrosion; all types of corrosion, including the big contribution from microbes, cost the oil and gas industry about USD $1.4 billion in 2013 (Head, 2015). Sulfide-contaminated or "sour" oil is more expensive to refine. One strategy for controlling souring is to add nitrate to stimulate the growth of nitrate-reducing bacteria. Chemolithotrophic nitrate-reducers would oxidize the sulfides back to sulfate, whereas the heterotrophic ones

would outcompete the sulfate-reducing bacteria for organic compounds and limit sulfide production. Also, sulfate-reducers could be inhibited by any nitrite produced during nitrate reduction.

Sulfate-reducing bacteria and also methanogenic archaea are key in the anaerobic degradation of oil in petroleum reservoirs (Head, 2015). Once thought to be impossible unless the hydrocarbon carried a halogen, anaerobic processes are now known to be more important than aerobic ones in petroleum reservoirs. Anaerobic degradation was first shown to be possible with a sulfate-reducer, *Desulfococcus oleovorans* HxD3 in the *Deltaproteobacteria*, the class of *Proteobacteria* with many other sulfate-reducers. Cultivation-independent studies have found, however, that the most abundant bacteria in petroleum reservoirs are in the *Firmicutes* or the *Epsilonproteobacteria* (Head et al., 2014). *Firmicutes* and abundant, non-sulfate-reducing bacteria may be the primary hydrocarbon degraders, the latter producing acetate and other compounds used by sulfate-reducing bacteria. When sulfate concentrations are low, methanogenic archaea take over, reminiscent of the anaerobic food chain discussed in Chapter 11. As with anaerobic degradation in other low-sulfate environments, oil is degraded anaerobically by a syntrophic consortium of

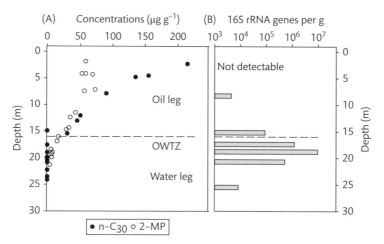

Figure 13.15 Hydrocarbon concentrations and microbial abundance in a heavy oil reservoir. Concentrations of a saturated hydrocarbon (N–C$_{30}$ alkane) and an aromatic compound (2-methylphenanthrene, 2-MP) are given per gram of oil (A). Prokaryotic 16S rRNA gene abundance per gram of sediment was estimated by quantitative PCR (B). The dashed line is the top of the oil–water transition zone (OWTZ), marking the boundary between where oil ("oil leg") or water ("water leg") dominates the column. The depth is for the reservoir and does not include the top shale cap rock which extends another 600 m to the surface. Data from Bennett et al. (2013).

microbes, consisting of organisms that break hydrocarbons down eventually to acetate and H_2, followed by the consumption of those two compounds by methanogens. Removal of the two end products makes the thermodynamics for hydrocarbon degradation more favorable.

Anaerobic degradation of hydrocarbons is much slower than aerobic degradation. Modeling studies have found that rate constants for anaerobic degradation in petroleum reservoirs are very low, 10^{-6}–10^{-7} y^{-1} (Larter et al., 2003). (An amino acid may have a rate constant of 10^4 y^{-1}.) Yet these incredibly slow rates, acting over one to two million years, are fast enough to account for the conversion of light oil to heavy oil and bitumen, two forms of petroleum that make up the bulk of oil on the planet (Head, 2015). The oil sands of western Canada alone are estimated to contain 900 billion barrels, more than double the amount in the conventional oil fields of Saudi Arabia and Kuwait. The microbe-driven conversion of light to heavy oil is one of many examples in geomicrobiology of how microbes metabolizing even slowly at micron scales over long time periods have large impacts on Earth material and geology.

Microbes and oil spills: Deepwater Horizon

Once hydrocarbon gases and oil reach surface environments, either via natural seeps or during an oil spill, different microbes and processes take over. These processes are mainly aerobic and thus complement the previous discussion about the anaerobic processes occurring in subsurface petroleum reservoirs. Here we focus on one oil spill already mentioned in this book: the Deepwater Horizon oil spill in the northern Gulf of Mexico. That focus is justified if only because the spill is the largest so far and was a major environmental disaster. Also unique for a spill, the oil streamed out of a wellhead 1500 m below the ocean surface, deeper than any other spill to date. Even more important for this book, the Deepwater Horizon oil spill occurred when many powerful cultivation-independent approaches were sufficiently developed to apply to the problem. The microbial ecology of the spill touches on several themes discussed already in this book, most notably relationships between community structure, diversity, and biogeochemical processes.

Cultivation-independent approaches have yielded many insights into the response of microbial communities to the spill and how degradation by specific bacterial taxa decreased concentrations and changed the chemical make-up of the hydrocarbons that spewed out from the wellhead (Figure 13.16). A single taxon in the gammaproteobacterial family *Oceanospirillaceae*, according to 16S rRNA gene data, dominated bacterial communities in the oil plume emanating from the wellhead soon after the blowout (Hazen et al., 2010). These bacteria were most closely related to known psychrophilic oil-degrading bacteria. It is curious that oil was apparently degraded in the deep cold water of the Gulf more so than in subarctic surface waters following another large spill, that caused by an oil tanker, the *Exxon Valdez*, which ran aground in Prince William Sound, Alaska, in 1989 (King et al., 2015). Metatranscriptomic and single-cell genomic work confirmed the important role of *Oceanospirillaceae* in degrading alkanes in the oil plume streaming from the wellhead.

As time went on and the spill was being brought under control, the bacterial community changed as hydrocarbon concentrations decreased and the composition of the spilled oil shifted from alkanes to more aromatic compounds such as benzene and toluene (Figure 13.16). This shift is consistent with other work showing that hydrocarbons are degraded in a particular order, ranging from easily degraded linear *n*-alkanes to very recalcitrant polynuclear aromatics and asphaltenes (Van Hamme et al., 2003). In response to this shift in hydrocarbons, taxa in the gammaproteobacterial genus *Colwellia* became abundant, perhaps because of their capacity to use benzene, as shown by stable isotope probing work. Methanotrophic bacteria also became abundant at this time, and oxidation of methane and other light hydrocarbons may explain the low oxygen associated with the oil plume, although there are alternative explanations (Dubinsky et al., 2013). Finally, after the well was "shut in" and the spill temporarily stopped on July 15, 2010, still another gammaproteobacterial taxon, the family *Alteromonadaceae*, took over, followed later by *Flavobacteriaceae* (in the *Bacteroidetes*). These taxa are thought to have responded to the higher molecular weight organic material from organisms stimulated by the oil spill hydrocarbons. The organic carbon from the spilled oil entered the planktonic food chain of the Gulf, evident from the very low ^{14}C levels in Gulf organisms (Chanton et al., 2012); in contrast with modern

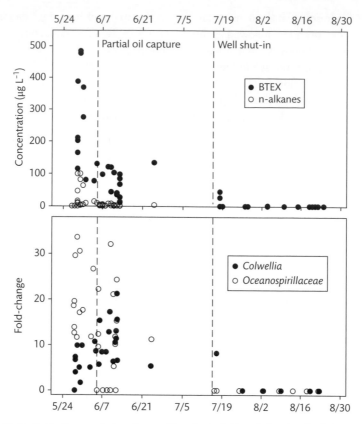

Figure 13.16 Hydrocarbon concentrations and abundance of key bacterial taxa following the Deepwater Horizon oil spill which started April 20, 2010. Oil capture efforts reduced the flow of oil by 50% or more beginning June 5. All flow was stopped with the well "shut-in" on July 15. BTEX refers to four aromatic hydrocarbons: benzene, toluene, ethyl-benzene, and xylene. Data from Dubinsky et al. (2013).

organic carbon, fossil fuels have very low levels of ^{14}C after millions of years of radioactive decay since the carbon was fixed by photosynthesis.

The work on the Deepwater Horizon and other spills has focused on bacteria, but some studies have examined archaea and eukaryotes. *Thaumarchaeota* do make up a large fraction of the microbial community in the deep waters of the Gulf of Mexico, as in other deep oceans (Chapter 4), but archaea did not appear to have responded to the spilled oil (King et al., 2015). Some fungi are capable of hydrocarbon degradation and apparently fared better than metazoans in sediments affected by the spill. Protist grazers may have been adversely affected by the dispersants, such as Corext, extensively used during the Deepwater Horizon spill to help solubilize the spilled oil. Whether the dispersants actually enhance overall oil

degradation or not is controversial (Tremblay et al., 2017; Kleindienst et al., 2015).

The good news is that hydrocarbon concentrations are back to normal in the waters of the Gulf of Mexico, thanks in part to degradation by microbes and natural "attenuation" (King et al., 2015). This is another example of bioremediation. Only a small fraction of the spilled oil was collected by skimming (about 3%) or removed by burning (5%) in spite of the millions of dollars spent on those efforts. The bad news is that spilled oil has been buried in sediments and will persist for decades to come, as has been the case for previous spills, due to the slow rates of anaerobic hydrocarbon degradation. Microbial ecologists will continue to be needed to understand the impact of oil spills on food web dynamics and biogeochemical processes in the Gulf of Mexico and elsewhere.

Summary

1. Geomicrobiology provides many examples of how seemingly small-scale processes mediated by microbes have large impacts on Earth's geology when carried out for thousands and even millions of years.

2. Microbes mediate the precipitation of minerals from dissolved constituents, a process called "biomineralization." Biomineralization affects geological formations and has left a fingerprint of early life on the planet.

3. In "biologically induced biomineralization," the microbial role may be indirect, unlike "biologically controlled biomineralization" in which microbes actively regulate every aspect of the mineralization process.

4. One example of biomineralization is the oxidation of ferrous iron carried out by chemolithotrophic and phototrophic microbes. Acidophilic bacteria are mainly responsible for iron oxidation in low pH environments, whereas microaerophilic bacteria and abiotic oxidation each account for about half of the total rate in near-neutral environments.

5. Microbes are involved in the weathering and dissolution of minerals and rocks, a complex suite of reactions that are important in removing carbon dioxide from the atmosphere over geological timescales.

6. Bacteria can degrade oil in petroleum reservoirs or after it has been released into the environment by spills. Anaerobic degradation is very slow, but still fast enough to account for the production of heavy oil and bitumen in reservoirs. Aerobic degradation in surface environments is faster, but it is not fast enough to prevent environmental damage by oil spills.

CHAPTER 14

Symbioses and microbes

One view of the microbial world is that it is filled with organisms carrying out crucial biogeochemical processes alone, sometimes in competition with one another, always in danger of being eaten, lysed by a virus, or killed off by UV light or desiccation. Life is a four billion year war, it has been said. All true, but previous chapters did mention examples of cooperation among microbes and between microbes and larger organisms. We have seen the reliance of acetogens on methanogens (syntrophy) and the partnership of sulfate reducers with methanogens in the anaerobic degradation of methane. Chapter 12 pointed out that the nitrogen-fixing bacteria inside eukaryotic algae and plants contribute substantially to global N_2 fixation. All of these are examples of symbiotic relationships, the focus of this chapter. Here we will discuss more examples of symbiotic relationships between microbes and large eukaryotes.

The examples just mentioned were brought up in previous chapters because the symbiosis is important to a particular biogeochemical process, the main reason for this book to explore symbiotic relationships in greater detail. However, there are other reasons. Symbiosis is a big topic in microbial ecology and is becoming more so in the study of large organisms, including humans. Because symbioses are so common and crucial, the study of large organisms would be incomplete without considering symbiotic microbes. Finally, symbioses are fascinating, with many examples of the wonderful, the weird, and the exotic.

There are at least two definitions of symbiosis. One is that symbiosis is the association between different species in persistent and close contact in which all members receive some benefit (Douglas, 2010). This definition,

which is the same as "mutualism," fits most of the examples discussed in this chapter. Another definition, the original one proposed by Anton de Bary in 1879, is that "symbiosis" covers the entire spectrum of interactions between organisms, ranging from those in which the organisms are indifferent to each other to outright antagonism, as in parasitic relationships (Figure 14.1). One advantage of de Bary's definition is that "symbiosis" could be used to describe a relationship between two organisms even if the nature of that relationship is unknown. But also according to de Bary's definition, the relationships between humans and malaria-causing protozoa and between potatoes and late blight (this potato pathogen caused the Irish famine in the nineteenth century) would be examples of symbioses. The inclusion of parasitism and pathogenicity does not fit with the everyday use of "symbiosis" and may be more confusing than illuminating.

Pathogenicity is relevant to the discussion of mutualistic relationships because pathogens as well as parasites

	Impact on:	
	Host	Microbe
Commensalism	0	+
Pathogenesis	–	+
Parasitism	–	+
Mutualism	+	+

Figure 14.1 Terms related to symbiosis. The plus and minus indicate that the relationship is beneficial to or has a negative effect on the symbiotic partner, while zero (0) indicates no impact. One definition of "symbiosis" would cover all of these interactions, whereas another definition is equivalent to mutualism.

Processes in Microbial Ecology. Second Edition. David L. Kirchman. Oxford University Press (2018). © David L. Kirchman 2018.
DOI 10.1093/oso/9780198789406.001.0001

can evolve into symbiotic microbes that provide benefits to their hosts. For example, some members of the fungal family Clavicipitaceae are parasites of grasses, while others have evolved into symbionts of grasses (Suh et al., 2001). In exchange for organic compounds, these fungal symbionts produce alkaloids that help grasses fend off herbivores. Another example also comes from the Clavicipitaceae. Most species of the genus *Cordiceps* are pathogens of insects, but others are symbionts that provide nitrogen and steroids to their insect hosts. In these cases, phylogenetic analyses of genomic and other data indicate an evolution from pathogenicity to symbiosis (Gibson and Hunter, 2010; Fan et al., 2015), perhaps through a stage in which negative impacts lessen as the host builds up resistance to the invading microbe. To fend off complete eviction, the microbe evolves to provide useful services to the host, reasons for its retention by the host. The end result is a mutualistic relationship. Not all pathogens turn into friends, nor do all mutualistic microbes start off as pathogens. In some cases, however, pathogens and mutualistic microbes have evolved mechanisms for the intimate interactions encountered while inhabiting a larger host.

Ubiquity of symbiosis and the exceptions

Table 14.1 lists the several symbioses discussed in this chapter, but the table could be much longer because symbiotic relationships are so common in biology. One important set of symbioses missing from the table are those between humans and microbes. Chapter 1 mentioned that the human body has more microbial cells than human cells, mainly because of the huge number of microbes in our gastrointestinal tract. Our gut microbiome has been implicated in health issues ranging from obesity to mental illness, as vividly described by Ed Yong in *I Contain Multitudes*, a popular science book packed with references to publications in the scientific literature (Yong, 2016). Our skin is also home to many commensal bacteria and yeasts that live on keratin and other proteins, secreted oils, and lipids. As mentioned in Chapter 1, the human microbiome is now being intensively explored because of the role of microbes in our health and well-being. Even if the human body were a natural environment like a grassland or lake, its symbioses deserve more space than is possible in this book.

Given that symbiotic relationships are so common, it is noteworthy when organisms are not involved in symbiosis. The exception among microbes is archaea. There is only one close symbiosis between an archaeon and eukaryote (a protist, to be discussed further) that has been discovered to date. Also, although archaea along with bacteria are prominent residents in sponges (Taylor et al., 2007) and in the gastrointestinal tracts of animals, these archaea–eukaryote interactions are not as close as seen between bacteria and eukaryotes. Likewise, there are

Table 14.1 The symbiotic relationships between large eukaryotes and microbes discussed in this chapter. "All" includes archaea, bacteria, and protozoa. "Organic C" is organic carbon. The host also provides other nutrients in these cases.

Eukaryote	Symbiont	Microbial taxa	Benefit to	
			Host	Symbiont
Ruminants	All	Various	Organic C	Organic C
Termites	All	Various	Organic C	Organic C
Leaf-cutting ants	Fungi and bacteria	Various	Organic C	Organic C
Aphids	Bacteria	*Buchnera*	Organic C	Organic C
Mealybugs	Bacteria	*Tremblaya, Moranella*	Organic C	Organic C
Corals	Algae	*Symbiodinium*	Organic C	Nutrients
Tubeworms	Bacteria	Various	Organic C	Sulfide
Mussels	Bacteria	Various	Organic C	Methane
Legumes	Bacteria	*Rhizobium*	Ammonium	Organic C
Alder	Bacteria	*Frankia*	Ammonium	Organic C
Plants	Fungi	Mycorrhizal fungi	Nutrients	Organic C
Bobtail squid	Bacteria	*Aliivibrio*	Bioluminescence	Organic C

few examples of pathogenic archaea (Bang and Schmitz, 2015). It is unclear why archaea do not form close relationships, negative or positive, with eukaryotes.

Another interesting exception is the paucity of intracellular symbioses between microbes and vertebrates (Douglas, 2010). Nearly all mutualistic relationships between these organisms are cases of "ectosymbioses," meaning the symbiont is on the outside of the host, in contrast with "endosymbiosis" in which the microbe is inside the host. The prominent gut microbiomes of vertebrates are arguably not inside their hosts, at least not in the same way as microbes are inside their insect or plant hosts. One clear exception is the symbiosis between a green alga and the spotted salamander; the algal cells live within salamander tissue and cells, according to 18S rRNA gene work and microscopic detection of chlorophyll *a* autofluorescence (Kerney et al., 2011). The intracellular and other types of endosymbioses to be discussed are between microbes and invertebrates and between microbes and plants. Vertebrates may have few intracellular symbionts because of their highly developed immune system, in contrast with the primitive immune systems of invertebrates and plants. The vertebrate immune system is designed to keep out and destroy microbial intruders, thus erecting a barrier perhaps too high for evolution to surmount. Pathogenic microbes have developed the necessary biochemical machinery to enter vertebrate cells, but the eventual end result is death, either of the host cell or of the invading microbe.

Symbionts convert inedible organic material into food for their hosts

One common theme of many symbiotic relationships is that microbes enable the host to thrive on diets that otherwise would be incomplete or even completely inedible. In some cases, the larger organism could potentially carry out a metabolic function, hydrolyzing cellulose for example, but has outsourced that function to a microbial symbiont or two. In other cases, no eukaryote has the capacity to carry out a particular function, such as dissimilatory sulfide oxidation or N_2 fixation, and the host has no choice but to rely on microbes. In nearly all of the examples discussed, the host organism attains high abundance and widespread distribution thanks in large part to their microbial symbionts.

The rumen microbiome

The human gut microbiome is complex and important in ensuring our good health, but we could in theory exist without it. Not so for ruminants, the dominant herbivores in terrestrial ecosystems. These animals have especially complex microbiomes that are essential in digesting grass, hay, and other plant material. Some examples of ruminants include deer, bison, and giraffes, as well as several domesticated animals, such as cattle and sheep. There are some aquatic analogs to ruminants. The digestive tracts of herbivorous fish are home to many microbes that degrade plant material by metabolic pathways seen in ruminants (Clements et al., 2014). Minke whales have a multi-chambered stomach system akin to ruminants for digesting herring.

The rumen is a large stomach-like pouch consisting of several muscular sacs, containing about 10^{10}–10^{11} bacteria and 10^6 protozoa per gram of rumen fluid, as well as fungi and archaea (Russell and Rychlik, 2001). In a metagenomic study of the bovine rumen, 90–95% of all gene sequences were from bacteria, 2–4% from archaea, and 1–2% from eukaryotes (Brulc et al., 2009), whereas a metatranscriptomic study found 77%, 22%, and about 1% for bacteria, eukaryotes, and archaea, respectively (Comtet-Marre et al., 2017). The rumen is the main site where ingested plant material is digested and converted to compounds assimilated by the ruminant. The degradation of plant material in the rumen (Figure 14.2) is reminiscent of the anaerobic food chain (Chapter 11).

Plant polymers, such as cellulose, are hydrolyzed to glucose and other monomers, which are then fermented to organic acids, most notably acetate, propionate, and butyrate. The organic acids are transported across the rumen wall and assimilated by the ruminant. The microbial cells are also broken down, providing protein and other nutrients for the animal. The symbiotic microbes are essential for converting cellulose, other plant biopolymers, and complex organic material, which the animal alone could not digest, into organic compounds that it can use.

In addition to ruminants, cellulose-degrading microbes are found in many vertebrates and invertebrates eating plant material in both terrestrial and aquatic environments. These microbes were probably essential in the evolution of herbivores from carnivore predecessors (Ley et al., 2008; Colston and Jackson, 2016). Herbivory

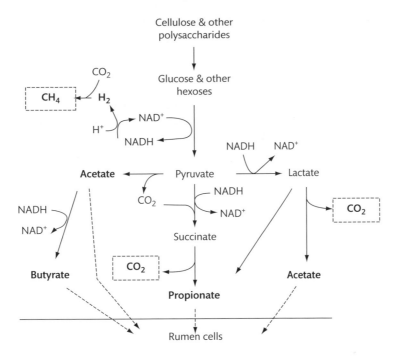

Figure 14.2 Polysaccharide degradation by prokaryotes in the rumen. Methane and carbon dioxide are removed by eructations and flatuses from the ruminant. Based on Russell and Rychlik (2001).

is key to the diversification of mammals during evolution and to their success in the biosphere, made possible by the mammalian gut microbiome. Gut microbes are also essential to the success of many detritivores in soils, lakes, and the oceans. These animals could not do their main job in the carbon cycle and in other ecosystem processes were it not for microbes.

Termite–microbe symbioses

Like ruminants, termites and many other insects are dependent on microbial symbionts to live on plant material and to thrive in spite of imperfect diets. Symbiotic relationships with microbes help to explain the great abundance and high diversity of insects in terrestrial habitats. The examples of insect–microbe symbioses to be discussed in this chapter focus on the benefits that the symbiont provides to the host. However, some interactions with microbes can be complicated if not detrimental to the insect host. For example, the alphaproteobacterium *Wolbachia*, which infects up to 66% of the 1–10 million insect species, has several complex, often negative effects

on insect reproduction (Jiggins, 2016). Manipulation of host insect reproduction by microbes is common. Here we focus on insect–microbe interactions involved in host nutrition.

Termites can live in wood thanks to symbiotic bacteria, archaea, and protists that degrade and transform the ligno-cellulose of wood to organic compounds usable by the termite. The role of the microbe varies with the type of termite. The higher termites rely on bacteria, whereas flagellates stand out as being the intriguing microbial symbionts of lower termites. These protists are very abundant in the hindguts of lower termites, making up as much as half of the termite's total mass (Brune and Stingl, 2006). Studies using cultivation-independent methods have found more than one type of flagellate in these termites, depending on the termite species (Brune, 2014). Some have only a few (three flagellate species in *Coptotermes formosanus*), while others have several (19 species in *Hodotermopsis japonica*), each having different roles in digestion. Termites also contribute to their digestion of cellulose by synthesizing endoglucanases that hydrolyze polysaccharides in the midgut, before the material is passed on to the hindgut. While the amount of digestion carried out by the midgut versus the hindgut varies with the termite species, microbes are important in all regions.

Early experiments in the 1920s found that lower termites could not live if their protists were killed. Robert Hungate then suggested in the 1940s that protists hydrolyze cellulose and produce acetate and hydrogen gas in the anoxic micro-environment of the termite hindgut. Later work confirmed this hypothesis, adding that lactate and other organic compounds may also be produced (Brune, 2014). It is thought that cellulose and other polysaccharides from wood particles are hydrolyzed in the protist's food vacuoles, yielding simple sugars that are converted to pyruvate by glycolysis (Figure 14.3). The pyruvate is then metabolized to acetate and hydrogen gas in the hydrogenosome, a specialized protist organelle that evolved from mitochondria (Chapter 11), where ATP is synthesized. The hydrogen gas consumption by methanogens lowers concentrations and makes the reaction more thermodynamically favorable in the hydrogenosome.

The bacteria in lower termites are involved in the latter stages of wood degradation but not in hydrolyzing wood polymers; particles of wood ingested by the termites are engulfed by the protists before they reach any bacteria (Brune, 2014). Rather, the role of the bacteria is to provide nitrogen by fixing N_2 or recycling ammonium produced as waste by the termites, both important processes in the nitrogen-poor wood habitat. From either N_2 fixation or recycling, the ammonium is used by bacteria to synthesize amino acids and vitamins that eventually are released to the host. The most numerous hindgut bacteria, spirochaetes, in the *Spirochaetes* phylum, carry out acetogenesis using H_2 and CO_2, and probably contribute to fermentation as well. What is striking about spirochaetes is their long (up to 100 μm) corkscrew-shaped cell, which twists as its swims using flagella running along its length. The spirochaetes are mainly free-living in the hindgut fluid where their large size and motility help them to maintain their position in the hindgut habitat.

What is remarkable about the other bacteria in the hindgut is their location: they are mainly attached to the outside of the protist or inside its cytoplasm. Note the latter are endosymbionts within the protists which in turn are symbionts within the termites. The ectosymbiotic

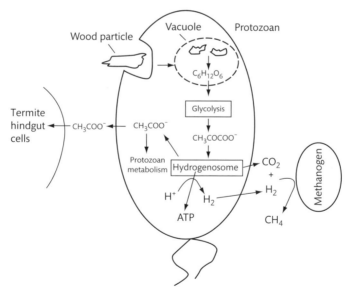

Figure 14.3 Model of cellulose degradation by protozoa in the hindguts of lower termites. Cellulase and other enzymes needed to hydrolyze wood to glucose ($C_6H_{12}O_6$) and other sugars are thought to be from the protist. Glucose is then partially oxidized to pyruvate (CH_3COCOO^-) which is then oxidized to acetate (CH_3COO^-), coupled to hydrogen gas production in the hydrogenosome. Acetate taken up by hindgut cells is the main fuel for termite metabolism. Based on Brune and Stingl (2006). See also Brune (2014).

bacteria can form a dense cover on the protist, lining up in regular rows of tightly packed cells (Figure 14.4). Some are spirochaetes, but most belong predominately to two other phyla, *Elusimicrobia*, which used to be called "Termite Group 1," and *Bacteroidetes*. These symbiotic bacteria seem to have important roles in the N cycle. The genes and transcripts for N_2 fixation (*nif*) have been found in ectosymbiotic *Bacteroidetes* attached to flagellates picked out from termites (Desai and Brune, 2012).

Termites acquired symbiotic protists about 150 million years ago, and then the higher termites lost their protists about 60 million years ago and evolved other strategies for degrading plant biopolymers (Brune, 2014). Because of these other strategies, the higher termites can use many types of plant litter and even highly degraded organic detritus, more than just the wood fed on by the lower termites. Consequently, the higher termites are more abundant and diverse than the lower termites, accounting for 80% of all termite species. The different strategies depend on different termite gut anatomies and hindgut microbiomes whose composition varies with the termite diet and species. For example, the hindgut microbiome of termites living on a cellulose-rich diet consists of spirochaetes and members of the *Fibrobacteres* and the related candidate phylum TG3, whereas termites feeding on dung or humus have large numbers of *Firmicutes*. Regardless, all of the endosymbiotic

protist duties in the lower termites have been taken on by bacteria in the higher termites.

Another noteworthy process carried out by symbiotic gut microbes is the production of two potent greenhouse gases, methane and nitrous oxide. Methanogens are members of the microbiome in termites and other insects, and in many vertebrate gastrointestinal tracts, including those of humans. Methane production by symbiotic methanogens in termites, ruminants, and other organisms contribute over 20% of all methane production, more than the methane released from leaky natural gas pipes (Conrad, 2009; Kirschke et al., 2013). The production of another greenhouse gas by termites, nitrous oxide (Brune, 2014), is not as well appreciated but is probably as important as methane production by these insects.

Ant-tended fungal gardens

As is the case for ruminant and termite symbioses, fungi help some ants to live on otherwise indigestible plant material. The role of the fungi in this symbiosis, however, differs from what we have seen so far for the microbes in ruminants and termites. More broadly, fungus–insect symbioses differ from bacterial–insect symbioses in several respects (Gibson and Hunter, 2010). For starters, compared with bacteria, few fungal species are symbionts in insects; many fungi have been found in insect guts but

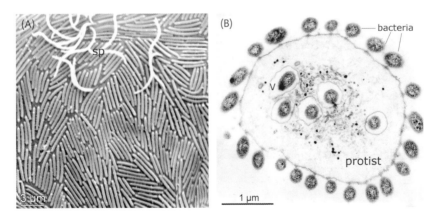

Figure 14.4 The protist *Staurojoenina sp.* with its ectosymbiotic bacteria from the termite *Neotermes cubanus*. A scanning electron micrograph illustrates the dense coverage by the many rod-shaped bacteria with a few spirochaetes ("sp"), the white filaments at the top of the photomicrograph (A). A transmission electron micrograph shows these bacteria and the protist in cross section (B). In addition to the ectosymbiotic bacteria, inside the protist are bacteria in five separate food vacuoles ("v"). Photomicrographs provided by Renate Radek and used with permission. See also Maaß and Radek (2006).

they may not be essential residents there. The few fungi that are truly symbionts are not usually intracellular symbionts, but rather they occur between cells in the insect or are maintained or "farmed" in the external environment (Figure 14.5). The few inter- and intracellular symbiotic fungi are true yeasts in the Saccharomycotina, often more cautiously referred to as "yeast-like symbionts" (YLS). Cell size does not necessarily prevent fungi from forming endosymbiotic relationships with more insects, as many symbiotic fungi are not that much bigger than symbiotic bacteria.

A more likely explanation is that extracellular enzymes produced by fungi are too destructive to be maintained intracellularly (Gibson and Hunter, 2010). The symbiotic yeasts (or YLS) do not produce copious amounts of extracellular enzymes. For the few well-studied cases, the genomes of endosymbiotic fungi do not become smaller over evolutionary time, unlike those of endosymbiotic bacteria. For both symbiotic bacteria and fungi, however, the metabolic capacities of the microbe enable the insect to exploit an imperfect diet. One example is the reliance of several ant species on fungal farms and associated bacteria.

One group carrying out fungus farming are leaf-cutter ants in the genera *Atta* and *Acromyrmex* (Hölldobler and Wilson, 1990). These ants strip bushes or grasses of leaves and carry them back to the ant nest on long marches along the forest floor, each ant bearing a leaf fragment much larger and heavier than its body. Once back in the nest, the ants cut the leaves into small pieces 1–2 mm in

diameter, and then they chew these small fragments until they are wet and spongy, sometimes topping them off with a small drop of ant anal fluid containing hydrolytic enzymes and nitrogen-rich fertilizer. After lining the leaf fragments into a garden, the ants inoculate the new leaf fragments with fungal mycelia from an older part of the garden. In ways not completely understood, N$_2$-fixing bacteria come into the relationship (Pinto-Tomas et al., 2009), providing much needed nitrogen to complement the carbon-rich plant material. The transplanted fungi quickly grow, covering the leaf fragments within a day. The fungi are then eaten by the ants.

The ant colony is dependent on the fungus as it is the sole food source for most colony members; only some of the workers can supplement their fungal diet with plant sap (Zientz et al., 2005). The ants apparently have lost some enzymatic capabilities, such as the biosynthesis of some amino acids, leaving those to the fungi to carry out (Suen et al., 2011). The fungi and leaf-cutting ants appear to have evolved together, resulting in tight relationships between the two (Mehdiabadi et al., 2012).

The fungi farmed by the ants are mainly in the tribe Leucocoprineae, which consists of two genera, *Leucoagaricus* and *Leucocoprinus*, in the family Lepiotaceae (Mehdiabadi and Schultz, 2010). Some of these can live outside the garden, making them facultative symbionts, while others cannot (Mehdiabadi et al., 2012). Analysis of rRNA genes have indicated that each garden consists of a single fungal strain (Zientz et al., 2005). This lack of diversity helps the fungus to avoid the cost of competition,

(A)

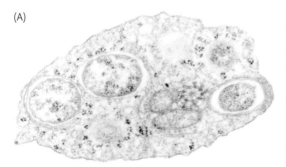

(B)

Fungi

Fungi

Figure 14.5 Examples of intracellular (endosymbiotic) and extracellular symbiotic relationships between microbes and their insect hosts. Cells of the bacterium *Wolbachia* within a sperm cell of the moth *Cadra cautella*. Courtesy of Scott L. O'Neill, Monash University, Creative Commons Attribution 4.0 license (doi:10.1371/journal.pbio.0020076) (A). The beetle *Beaverium insulindicus* with eggs deposited in its fungal garden. Courtesy of Jiri Hulcr and used with permission (B). Color version of panel B is available at www.oup.co.uk/companion/kirchman.

leading to higher fungal yield and more food for the ant. Both the fungus and ant help to maintain the monoculture of the garden. The fungus does its share by outcompeting fungal strains from other gardens. For their part, the ants use their antennae to distinguish their fungus from alien fungi, and they prevent sexual reproduction by the farmed fungi. Asexual reproduction results in higher fungal yields and also maintains low diversity within the fungal garden, contributing again to higher fungal yields.

A negative consequence of low diversity is that fungal gardens are prone to invasion by the parasitic ascomycete fungus in the genus *Escovopsis*. This fungus has not been found outside of ant fungal gardens and is dependent on the farmed fungi in the garden (de Man et al., 2016). If left unchecked, the parasitic fungus would penetrate and kill the farmed fungi. To defend against it, the fungus-gardening ants are covered by actinomycetes, bacteria in the *Actinobacteria* phylum that produce antibiotics to inhibit the parasitic fungi while minimizing harm to the farmed fungi (Cafaro et al., 2011). These antibiotics help both the fungus and ant to ward off parasites and pathogens (Figure 14.6).

Thanks to the metabolic capacities of the microbes, leaf-cutter ants are the dominant herbivores in tropical savannahs and rainforest and have huge roles in structuring these ecosystems. The New World leaf-cutter ants are replaced in the Old World tropics by some species of termites (Macrotermitinae) that also cultivate fungal gardens (Brune, 2014). These termites are serious pests in the tropics, but the fruiting bodies (mushrooms) of the fungi protruding from termite mounds are a highly prized delicacy in Asia and Africa. Fungus-gardening termites do not occur in the Americas.

Symbionts provide dietary supplements: the aphid–*Buchnera* symbiosis

Aphids and other Hemiptera face a different problem from the one termites and leaf-cutting ants have solved with their microbial symbionts. Unlike termites and their diet of wood or plant detritus, aphids can digest the compounds found in their main food source, the sap or phloem of vascular plants. But this diet is quite unbalanced; it is rich in sugars but poor in essential amino acids. The missing vital compounds are provided by symbiotic bacteria, as shown by tracing the movement of radioactive compounds from symbionts to the host. Aphids "cured" of their symbiotic bacteria by antibiotics cannot survive without the missing compounds added back to their diet (Bennett and Moran, 2015). The radiation of Hemiptera ("true bugs") into various habitats probably depended on early acquisition of bacterial symbionts in order to subsist on vascular plant phloem. Today, almost all hemipterans harbor symbionts, often endosymbionts found in specialized cells, the bacteriocytes. These microbes are often called the primary symbionts because the partnership is obligatory for both microbe and insect.

The main primary endosymbiotic bacteria in aphids belong to the gammaproteobacterial genus, *Buchnera*. The aphid–*Buchnera* symbiosis is thought to be at least 80 million years old, based on analysis of material preserved in amber, while extrapolation from 16S rRNA gene sequences suggest ages of 150–250 million years (Moran et al., 2008). The estimated age for a particular aphid species and its bacterial symbiont in the *Buchnera* species complex varies (Figure 14.7). The coevolution of the aphid and bacterium has been aided by the vertical transmission of the endosymbiotic bacteria to aphid eggs (Box 14.1). Aphids and other insects may have other, secondary symbiotic bacteria that provide essential compounds not supplied in sufficient amounts by the primary symbionts. Secondary symbionts can also help ward off parasites. But some of these secondary symbionts resemble pathogens more than mutualistic primary symbionts, or at least are not essential to the insect host.

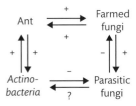

Figure 14.6 Symbiotic relationships among ants, fungi, and bacteria. The relationship between the ant and the farmed fungi is probably mutualistic (two positive arrows) as is the one between the ant and the antibiotic-producing *Actinobacteria*. The parasitic fungi gain from feeding on the farmed fungi (positive arrow) while the latter loses out (negative arrow) in their parasitic or pathogenic relationship.

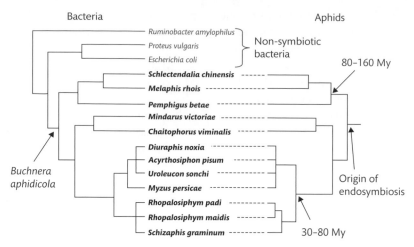

Bacteria

Aphids

Figure 14.7 Phylogenetic trees of the bacterial symbionts (*Buchnera*) and their aphid hosts. Based on Moran and Baumann (1994) and used with permission of the publisher. See also Nováková et al. (2013).

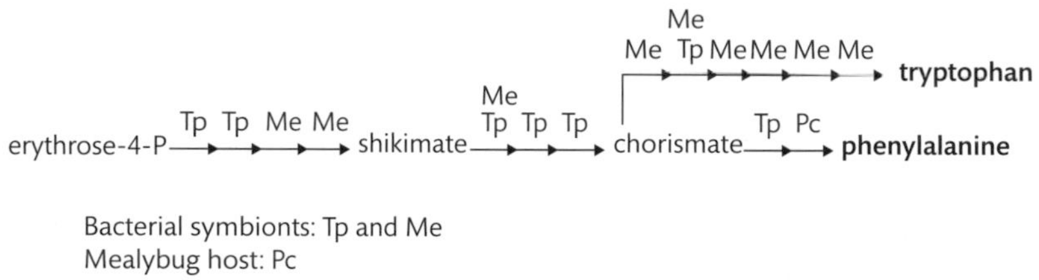

erythrose-4-P →Tp→Tp→Me→Me→ shikimate →Me→Tp→Tp→Tp→ chorismate →Tp→Pc→ **phenylalanine**

→Me→Me→Tp→Me→Me→Me→Me→ **tryptophan**

Bacterial symbionts: Tp and Me
Mealybug host: Pc

Figure 14.8 Synthesis of phenylalanine and tryptophan by the citrus mealybug *Planococcus citri* (Pc) and its two bacterial symbionts, *Tremblaya princeps* (Tp) and *Moranella endobia* (Me). The letters indicate the organism providing the enzyme catalyzing each step (the arrows). Based on McCutcheon and von Dohlen (2011).

Box 14.1 Transmission of symbionts

An endosymbiont like *Buchnera* is vertically transmitted to new hosts in which the symbiont is transferred from mother to offspring without being released into the environment. When transmission is horizontal, the symbionts are acquired from the environment. Newly hatched termites, for example, acquire their symbionts by ingesting the symbiont-rich excreta of adult termites (Brune and Stingl, 2006).

The aphid–*Buchnera* symbiosis has evolved to the point that the bacterium is close to being an organelle completely integrated into the host's biology. This integration is evident in the symbiont's genome. The genome of *Buchnera* exemplifies three interrelated features of genomes in many bacterial endosymbionts: (1) small genome size; (2) high AT content (the AT content of *Buchnera* is 74–80% (Moya et al., 2009), much higher than the 50% in the genomes of free-living bacteria); and (3) rapid evolution. The genomes of *Buchnera* and *Wigglesworthia* found in tsetse flies are only 0.45–0.66 Mb and 0.7 Mb, respectively, and the *Carsonella* genome in psyllids (jumping plant lice) is even smaller at 0.16 Mb. The smallest genome of a free-living bacterium is at least twice as big (Chapter 5). Endosymbiotic bacteria can have small genomes because some of the genes essential for an independent lifestyle are no longer necessary in the host environment or when the host has taken over a particular function. Any genes not essential for maintaining the symbiosis can tolerate more mutations, potentially leading to the complete loss of the gene. High AT content results from the bias of mutations to AT from CG

(Hershberg and Petrov, 2010) and may be favored because of the lower energetic cost of synthesizing ATP and TTP compared with GTP and CTP.

The loss of genes from an endosymbiont does not necessarily mean the insect host takes on all of the function without any help from the bacterium. Analyses of gene expression suggest that the aphid and *Buchnera* both contribute to the synthesis of essential amino acids (Hansen and Moran, 2011). An even more fascinating example is the citrus mealybug and its symbionts (McCutcheon and von Dohlen, 2011). This insect has two bacterial symbionts: the betaproteobacterium *Tremblaya princeps*, which has inside a gammaproteobacterium, *Moranella endobia*. (Insects often have more than one symbiont, although not necessarily nested inside of them.) Both mealybug symbionts have very small genomes, 0.139 and 0.538 Mb, respectively. Like the aphid–*Buchnera* symbiosis, the mealybug and its two symbionts share duties in critical metabolic pathways. The synthesis of the essential amino acids phenylalanine and tryptophan, for example, is split between the three organisms (Figure 14.8). This model is supported by reverse-transcriptase PCR analyses of gene expression (McCutcheon and von Dohlen, 2011).

Symbiotic relationships supported by light or reduced sulfur

In all of the examples discussed so far, the energy and carbon source sustaining the symbiotic relationship is the organic material produced by a higher plant. Termites and aphids depend on their symbiotic microbes to sub-

sist on the wood or sap directly produced by a plant, while ants eat the farmed fungus, which in turn subsists on plant leaves or other plant material. Other than providing the organic material as wood, leaves, or sap, the plant is not directly involved in the symbiosis.

In contrast, a few symbiotic microbes are themselves primary producers and directly synthesize the organic material to support the growth of the larger organism. Like the examples mentioned so far, the first symbiosis discussed next also involves light-dependent primary production, but this time by a symbiotic microbe. The second is even more unusual because the energy source is not light but reduced sulfur compounds.

Coral reefs: massive structures built with microbial help
One of the best known and spectacular symbioses is that between the dinoflagellate *Symbiodinium*, also called zooxanthellae, and corals, members of the invertebrate phylum Cnidaria (Figure 14.9). It is in large part due to this symbiosis that coral reefs can form one of the largest, most beautiful and diverse biological structures in the biosphere. It is another example of an invertebrate flourishing in an environment, in this case oligotrophic waters, because of organic material provided by the microbial symbiont. Corals depend on the symbiotic dinoflagellate for organic carbon, although most corals can feed on zooplankton as well. The algae also recycle waste by-products, most importantly ammonium, from coral metabolism, thus retaining nitrogen and other elements that otherwise limit growth in the coral's oligotrophic habitat. Many studies have focused on the

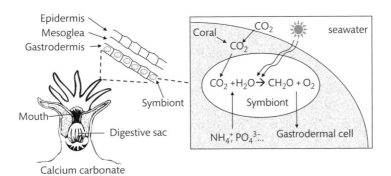

Figure 14.9 Anatomy of a coral polyp with a close-up view of the coral surface tissue and an even closer view of a gastrodermal cell with a symbiotic dinoflagellate. Each coral cell can have several symbiotic cells. Drawn with input from Mark Warner.

dinoflagellate symbiont, but corals in fact have complex microbiomes with many other microbes (Bourne et al., 2016).

Coral reefs are now threatened by many human-caused or human-exacerbated problems that disrupt the dinoflagellate–coral symbiosis. One problem is the warming of the oceans due to increases in carbon dioxide and other greenhouse gases in the atmosphere (Chapter 1). Even a degree or two increase in water temperature could have drastic effects on coral reefs and their symbionts, because these organisms live near their maximum temperature in tropical waters. When temperatures rise above that maximum, the corals become "bleached" due to the expulsion (or escape) of the algal symbiont, leaving behind a colorless cnidarian. The coral can recover if temperatures return to normal, but they might die instead and decimate the reef.

Increasing atmospheric carbon dioxide is causing another problem for corals and other carbonate-synthesizing organisms: ocean acidification. As mentioned in Chapter 1, more carbon dioxide in the atmosphere means more carbon dioxide dissolved in seawater and lower pH. The pH of the North Pacific Ocean has already dropped by about 30% and is predicted to decrease by 0.3–0.4 pH units by 2100 (Doney et al., 2009). Acidification causes problems because it inhibits precipitation of the calcium carbonate needed by many corals and other organisms (Chapter 13). Rising water temperatures, dropping pH, and many other problems are resulting in large declines in coral reefs worldwide. The survival of corals and coral reefs in the face of these environmental problems depends on a healthy symbiosis with the dinoflagellate symbiont and other microbes (Bourne et al., 2016).

Chemosynthetic symbioses in marine invertebrates
In 1977, geologists in the submersible Alvin were hunting in waters near the Galápagos Islands for hydrothermal vents, cracks in the ocean bottom, hypothesized to be the source of anomalously warm waters and heavy metals. What they found was much more astonishing: luxurious communities of shrimp, crabs, clams, and plant-like echinoderms, with meter-high tube-like creatures towering over all, waving in the hot water spewing from the vents. This riot of life was in stark contrast with the desert that is most of the ocean floor under deep waters,

thousands of meters from the surface. The source of energy and carbon supporting the rich vent community was at first a mystery. Little organic material reaches the bottom from primary production at the surface, certainly not enough to fuel the dense and diverse communities at vents.

Later work revealed that the entire vent community is based on sulfide oxidation by nonsymbiotic and symbiotic chemoautotrophic bacteria. In contrast with anoxic systems where microbial sulfate reduction supplies the sulfide (Chapter 11), the sulfide at hydrothermal vents comes primarily from the reduction of sulfate by geochemical mechanisms (Figure 14.10). Seawater seeps into the ocean floor where it is superheated to 350 °C or higher, reducing sulfate to sulfide. The reduced sulfur and metals as well as other compounds in the hot vent fluid then gush out at vents, where they are oxidized by abiotic mechanisms and chemolithotrophy. Sulfide oxidation supports the synthesis of organic carbon by chemolithoautotrophy, often called just "chemosynthesis." Chemolithoautotrophic bacteria form the base of the food chain at vents and support the rich biological community that was seen from Alvin's portholes.

Initial work focused on the large tube-like creatures, eventually called *Riftia pachyptila*, members of the Annelida phylum. These tubeworms puzzled zoologists because they lack an obvious digestive tract and are too large to live on the low concentrations of dissolved organic compounds found at vents. Suspecting that *Riftia* relied on chemoautotrophy, zoologists found CO_2 fixation and sulfide oxidation activity in tubeworm tissue. This enzymatic activity was initially thought to be carried out by tubeworm cells, leading zoologists to conclude that they had found the first "chemoautotrophic animal." About the same time, however, microbial ecologists showed that bacteria, not tubeworm cells, were carrying out the measured CO_2 fixation and sulfide oxidation (Box 14.2). Stable carbon isotope data made clear that tubeworms depend on endosymbiotic bacteria for nutrition.

Tubeworms have a unique structure, the trophosome, designed to support bacterial symbionts (Stewart and Cavanaugh, 2006). This organ consists of blood vessels, coelomic fluid, and bacteriocytes housing the endosymbionts. One gram of trophosome tissue has about 10^9 bacterial cells, taking up 15–35% of the trophosome volume. The morphology of the endosymbiotic bacteria

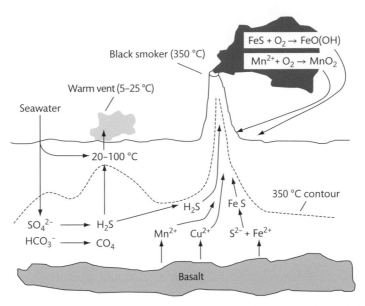

$$FeS + O_2 \rightarrow FeO(OH)$$
$$Mn^{2+} + O_2 \rightarrow MnO_2$$

Black smoker (350 °C)

Warm vent (5–25 °C)

Seawater

350 °C contour

20–100 °C

H_2S Fe S

$SO_4^{2-} \longrightarrow H_2S$
$HCO_3^- \longrightarrow CO_4$

Mn^{2+} Cu^{2+} $S^{2-} + Fe^{2+}$

Basalt

Figure 14.10 Structure of hydrothermal vents. Vents occur at spreading zones between continental plates where the ocean floor is splitting apart, allowing seawater to permeate far enough into the subsurface to be heated geothermally. Compounds in the seawater are reduced abiotically; one example is sulfate (SO_4^{2-})→ hydrogen sulfide (H_2S). The hydrogen sulfide and reduced metals, such as Fe^{2+} and Mn^{2+}, from basalt are carried back to the surface where they are oxidized either abiotically or by chemolithotrophic reactions, both using oxygen as the electron acceptor. The resulting metal oxides precipitate onto the vent chimney or the ocean floor.

Box 14.2 Bacteria on the brain

Although established zoologists had the first look at tubeworms, it took a graduate student with a background in microbial ecology to come up with the endosymbiotic chemolithoautotroph hypothesis. Colleen Cavanaugh got her first clue while listening to a zoologist, Meredith Jones, lecturing about mouthless and gutless tubeworms and their strange anatomy, in a lecture at Harvard University. During his presentation, Jones showed a photograph of a dissected tubeworm, noting that he had found numerous sulfur granules within their trophosome tissue. He mentioned that the function of the trophosome was not known. While sitting in the lecture hall, Cavanaugh thought "symbiosis." She reckoned that the worms fed on symbiotic chemosynthetic bacteria, much like corals living off their symbiotic photosynthetic algae. Cells in the trophosome looked like bacteria she had seen in other scanning electron micrographs. But it took hard work and more definitive evidence to prove this hypothesis. Transmission electron microscopy (TEM) revealed the presence of the two double-layer membranes typical of Gram-negative bacteria, which was confirmed by detection of lipopolysaccharide (Cavanaugh et al., 1981). This evidence along with the enzymatic data strongly supported the endosymbiotic chemolithoautotroph hypothesis, one of the most fascinating chapters in microbial ecology and in all biology.

---

varies with location within the trophosome, due to chemical gradients or growth stages of the bacteria. Tubeworm blood carries sulfide, oxygen, and nitrate to the bacteriocytes where chemolithoautotrophy occurs (Figure 14.11). Nitrate, which is reduced to ammonium by the symbiont, supplies the nitrogen needed by both the host and symbiont, and also serves as an electron acceptor when oxygen concentrations are low. As payback

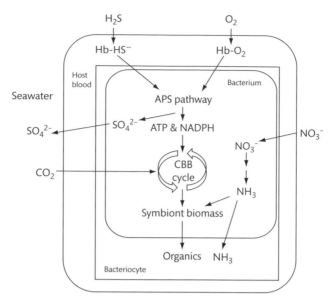

Figure 14.11 Relationships between the endosymbiotic bacteria and a tubeworm. Both H_2S and oxygen are carried to the bacterium via hemoglobin (Hb). Sulfide is oxidized via the adenosine 5'-phosphosulfate (APS) pathway, with oxygen as the electron acceptor, yielding ATP and NADPH. These are used by the bacteria to fix CO_2 and synthesize organic compounds via the Calvin–Benson–Bassham (CBB) cycle. Host metabolism is supported by organic compounds leaked from the symbiont or by symbiont biomass directly. Based on Stewart et al. (2005).

for servicing the bacterial symbiont, the tubeworm host gets organic compounds leaked from the symbiotic bacteria, supplemented from time to time by digesting some of the bacteria.

After the discovery of endosymbionts in tubeworms, chemolithoautotrophic symbiotic bacteria were found in many other invertebrates living in sulfide-rich environments. This endosymbiotic relationship is now known to occur in over six metazoan phyla and in ciliates (Stewart et al., 2005). One example is the mudflat bivalve *Solemya velum*, examined soon after the initial tubeworm studies, although its tiny gut had long puzzled zoologists; a close relative, *S. reidi*, has no gut at all. Other gutless invertebrates with sulfide-oxidizing endosymbionts include oligochaete worms. Some of these worms appear to migrate between sediment zones, collecting sulfide in the anoxic, sulfide-rich zone, then swimming to the sulfide-poor but oxygen-rich zone where the sulfide could be oxidized (Dubilier et al., 2006). Other gutless oligochaetes have sulfide-oxidizing symbionts in spite of living in environments without high sulfide concentrations. The sulfide-oxidizing symbionts may depend on other symbiotic bacteria that

carry out sulfate reduction and produce sulfide to be used by chemolithoautotrophs.

For all of these symbiotic relationships, chemolithoautotrophic bacteria use only reduced inorganic sulfur, not other typical chemolithotrophic substrates, with some exceptions. Substrate availability and energetic yield may explain why substrates like ammonium and ferrous iron are not used to support symbiotic relationships; ammonium concentrations and the energy yield of ferrous iron may be too low. One exception is symbionts using H_2 in tubeworms, mussels, and shrimp living at hydrothermal vents (Petersen et al., 2011). Other bacterial symbionts oxidize H_2 but also carbon monoxide (CO) to support carbon dioxide fixation in a gutless marine worm living in seagrass sediments (Kleiner et al., 2015). We know the most about another type of symbiotic chemoautotrophic bacteria: aerobic methanotrophs that oxidize methane.

These methanotrophic bacteria live in mussels at cold seeps with high methane fluxes, such as in the Gulf of Mexico (Cavanaugh et al., 1987). Like hydrothermal vents, cold seeps are natural springs from which methane and other low molecular weight hydrocarbons leak out into the oceans from subterranean oil and gas reservoirs.

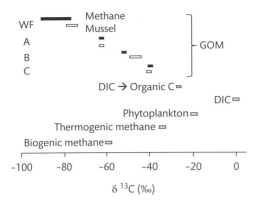

Figure 14.12 ^{13}C content of methane (solid bar) and mussels (open bar) at four sites (WF, A, B, and C) in the Gulf of Mexico (GOM). The mussel ^{13}C is much more negative than the value predicted for chemoautotrophic microbes fixing CO_2 from dissolved inorganic carbon (DIC) ("DIC→Organic C"). Biogenic methane can have even more negative $\delta^{13}C$ values (down to $-100°/_{oo}$) than given here for the Gulf. Data from Petersen and Dubilier (2009) and Schaefer et al. (2016).

Some of the same type of data were collected for cold seep mussels as had been for the hydrothermal vent tubeworms. Although these data indicated the presence of symbiotic bacteria in the seep mussels, other data indicated that the bacteria could not be sulfur oxidizers. The ^{13}C content was very low and could be explained only if biogenic methane was the source (Figure 14.12). (The lighter ^{12}C isotope is favored over the heavier ^{13}C during biotic methanogenesis and abiotic methane production by geochemical reactions, producing methane with highly depleted ^{13}C and very negative $\delta^{13}C$ values.) Enzymatic and gene assays for methane oxidation as well as probing for methanotroph 16S rRNA by fluorescence in situ hybridization confirmed that the symbionts were methanotrophs.

Symbiotic microbes provide inorganic nutrients to plants

The types of symbiotic relationships established between microbes and animals are also seen in plants with their symbiotic microbes. As with animals, these relationships range from commensal, ectosymbioses to mutualisms with close endosymbiotic relationships. Analogous to the human skin, exposed surfaces of terrestrial plant leaves and stems are microhabitats for many types of

microbes, quite important for plant growth (Vacher et al., 2016; Laforest-Lapointe et al., 2017). These plant surface microhabitats are collectively referred to as the "phyllosphere," while the zone around roots is the rhizosphere. Cultivation-dependent approaches of the phyllosphere have focused on bacteria such as *Pseudomonas syringae*, famous for its role in facilitating the formation of ice crystals in near-freezing weather; less ice forms on leaves coated with "ice-minus" mutants of *P. syringae*. As usual, cultivation-independent methods turn up much more diverse communities.

Probably even more important than microbes in the phyllosphere, however, are the interactions between microbes and plant roots. Symbiotic relationships between microbes and plants via roots are very common. As much as 90% of all plant species have microbial symbionts (Parniske, 2008; Martin et al., 2016). The symbionts found in or around roots are fungi and diazotrophic bacteria.

Diazotrophic bacteria–plant symbiosis

This symbiosis is important in agriculture and for many plants and photoautotrophic protists in environments limited by the supply of nitrogen. Several plants and protists harbor symbiotic heterotrophic bacteria or cyanobacteria capable of N_2 fixation. One example is *Frankia*, a heterotrophic bacterium, which forms symbioses with 194 species in eight dicot families, including woody shrubs and trees growing on nitrogen-limited land (Benson and Silvester, 1993; Barka et al., 2016). Symbiotic *Frankia* are housed in ball-like structures or "root nodules," as big as 10 cm in diameter sticking out from the root.

Another major group of diazotrophic bacteria form symbiotic relationships with plants in the Fabaceae family, commonly known as legumes, the third-largest family of flowering plants. Among crop plants, legumes include clover, soybeans, and peas, while examples of wild legumes include some flowering plants (clover, lupines, and wild indigo) and trees such as black locus (*Robinia pseudoacacia*) and redbud (*Cercis canadensis)*. The best-known legume bacteria are in the alphaproteobacterial order *Rhizobiales* (Figure 14.13), including the genus *Rhizobium*, but N_2-fixing bacteria in other genera and even other proteobacterial divisions can also form symbioses with legumes. Collectively all of these bacteria are called rhizobia.

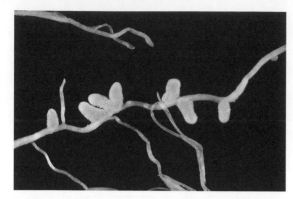

Figure 14.13 A root nodule made by *Sinorhizobium meliloti* in the *Rhizobiales* on a legume root (*Medicago italic*). Image by Ninjatacoshell, used under the Creative Commons Attribution-ShareAlike 3.0 Unported.

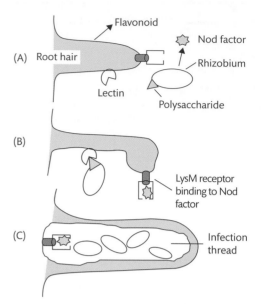

Figure 14.14 Establishment of the legume–rhizobium symbiosis. The interaction begins with the release of a flavonoid specific for the targeted rhizobium, which responds by releasing a Nod factor (A). A secondary factor in establishing the root hair–rhizobium partnership is the binding of a lectin to a polysaccharide on the bacterial cell surface. All signaling events lead to the curling of the root hair and other responses (B). The end result is the formation of the infection thread and proliferation of rhizobial bacteria (C). Based on Downie (2010).

The root nodule is the end result of several biochemical exchanges between the plant root and rhizobia (Zipfel and Oldroyd, 2017). Among several possible rhizobial species, only a few strains can form successful symbiotic relationships with any particular legume species. The courtship starts with the release of flavonoids by the plant specific for its compatible rhizobium (Figure 14.14); flavonoids are multi-ring compounds with a 2-phenyl-1,4-benzopyrone backbone. The compatible rhizobium responds by releasing the Nod factor, an acylated oligosaccharide. The Nod factor binds to a membrane-associated receptor on the root hair, which in turn triggers other biochemical events in the root. Another component of the legume–rhizobium courtship is the binding of lectins on the root hair surface to specific carbohydrate moieties on the cell surface of compatible rhizobia. These various signaling processes eventually lead to morphological changes both in the root hair and the bacterium. The transformed bacteria are called "bacteroids."

The plant creates conditions in root nodules to facilitate N_2 fixation by symbiotic rhizobia. It produces leghemoglobin that binds to oxygen, thus controlling levels of this nitrogenase-poisoning gas (Chapter 12) and coloring root nodules rust red. The outer cortex of the root nodule is oxic, but oxygen concentrations are much lower (<25 nM) in the center (Udvardi and Poole, 2013). In spite of low free oxygen, rhizobial bacteria can continue to respire and generate the ATP needed for N_2 fixation and the rest of rhizobial metabolism using the oxygen

delivered by leghemoglobin. The plant also releases malate or succinate or both to fuel rhizobial metabolism, and in return, the symbiotic rhizobial bacteria release mostly ammonium and in some cases amino acids to the plant root (Dunn, 2015). The high energetic cost to the plant in supplying the organic carbon is more than offset by the gain of the fixed nitrogen.

Fungus–plant symbioses

Symbiotic diazotrophic bacteria supply only nitrogen to their plant hosts and cannot offer any help in securing other necessary nutrients or water, potentially limiting plant growth in soils. Plants partially solve this problem by forming symbiotic relationships with fungi. The near ubiquity of fungus–plant symbioses explains the previously mentioned high fraction (>90%) of plants with microbial symbionts. In addition to supplying nutrients, fungal symbionts may also help plants fend off pathogenic fungi and bacteria and survive drought (Mommer et al., 2016), but it is possible that the fungi provide no services at all for the plant, at least for some time periods or habitats. Perhaps the plant gains something from the fungi when environmental conditions change over time or the plant is introduced to a new habitat. But it simply may not be worth the cost to expel the fungus if it is doing the plant no harm (commensalism). In any event, fungi are generally thought to be essential for the success of plants in terrestrial ecosystems. Both fossil and DNA evidence indicate that symbiotic fungi–plant relationships formed 400–460 million years ago when plants began to colonize the land (Humphreys et al., 2010). The successful invasion of terrestrial habitats by plants is thought to be due in part to mycorrhizal fungi. "Mycorrhizal" is derived from Greek for fungus and root.

The various types of mycorrhizal fungus–plant symbioses differ in how the fungus interacts with the plant root, among other features (Table 14.2). Ectomycorrhizal fungi were discovered and examined first because they form large fruiting bodies, such as mushrooms, puffballs, and truffles. These symbiotic fungi are outside of the epidermal root cells (they do not interact directly with plant cytoplasm), hence the "ecto" prefix, but they can weave their way between the root epidermal and cortical cell walls, forming a structure called the Hartig net. Another defining characteristic is that these symbiotic fungi form a sheath or mantle around the root tip.

A more common type of plant symbiotic fungi, arbuscular mycorrhizae, are endosymbionts that do penetrate the root cell wall, associating with the plant plasma membrane. Arbuscular mycorrhizae earn their name from the formation of an arbuscule, a fungal structure inside root cells (Figure 14.15). The term was first coined by Isobel Gallaud in 1905, who thought the structure had vine-like features of an arbor. Still another type, ectendomycorrhizae, can penetrate the root cells like arbuscular mycorrhizae, while also forming a Hartig net like ectomycorrhizal fungi. Overall, however, arbuscular mycorrhizae are the most common, accounting for over 85% of all known fungi–plant symbioses among angiosperms (Brundrett, 2009).

Initiation of arbuscular mycorrhizal symbiosis is similar to how the legume–rhizobium symbiosis starts off. Instead of flavonoids in the case of legume–rhizobium symbioses, plant roots initiate the mycorrhizal symbiosis by excreting strigolactones, which in turn trigger release of a Myc factor by the fungi, analogous to the Nod factor of rhizobia. This signaling between plant and fungi initiates other changes in both organisms, as seen in the legume–rhizobium symbiosis.

Unlike rhizobia, arbuscular mycorrhizal fungi can infect all plants receptive to these fungi (Smith and Read, 2008). This promiscuity has been demonstrated in "pot" experiments in which individual plants grown in pots were successfully inoculated with a variety of arbuscular mycorrhizal fungi. Observations in the field suggest a more specific relationship between plants and fungi than seen in the controlled pot experiments (Martínez-García et al., 2015). In one study, fungal communities associated with grasses from the same species were more similar to each other than to the fungi associated with different grass species (Vandenkoornhuyse et al., 2003). Other field studies, however, found weak relationships between

Table 14.2 Three types of mycorrhizal fungi. The taxa abbreviations are Glomero (Glomeromycota), Basidio (Basidiomycota), Asco (Ascomycota), Bryo (Bryophyta), Pterido (Pteridophyta), Gymno (Gymnospermae), and Angio (Angiospermae). Based on Smith and Read (2008), which discusses four other types of mycorrhizal fungi.

Characteristic	Arbuscular mycorrhiza	Ectomycorrhiza	Ectendomycorrhiza
Intracellular symbiosis	Yes	No	Yes
Mantle around root	No	Yes	Yes or No
Hartig net*	Absent	Present	Present
Above ground fruiting bodies	No	Yes	No
Fungi with septate**	No	Yes	Yes
Fungal taxa	Glomero	Basidio, Asco	Basidio, Asco
Plant taxa	Bryo, Pterido, Gymno, Angio	Gymno, Angio	Gymno, Angio

* A Hartig net, named after the nineteenth-century German plant pathologist, Robert Hartig, is a network of fungal hyphae in and around plant roots.

** Septate is the septum dividing fungal cells.

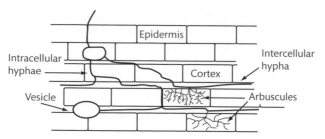

Figure 14.15 Arbuscular mycorrhizal fungus infecting a plant root.

plant and arbuscular mycorrhizal fungal diversity (Horn et al., 2017).

Several types of experiments and field observations support the hypothesis that arbuscular mycorrhizal fungi help the host plant to acquire nutrients. The fungi in effect become an extension of the plant roots, going where roots cannot because of size; the diameter of a typical fungal hypha is 2–10 μm versus >300 μm for root hairs. Consequently, symbiotic fungi can access a ten- to a thousand-fold greater area with potential nutrients over what can be achieved by roots alone. Symbiotic fungi may also use organic forms of nutrients not otherwise available to plants. Field observations suggest that symbiotic fungi are most useful to the plant when soil nutrient concentrations are low, such as during the late stages of plant community succession. Plants can survive without mycorrhizal fungi if the supply of nutrients is adequate.

But arbuscular mycorrhizal fungi cannot survive for long away from their host plant and the organic compounds it supplies (Smith and Read, 2008). The transfer of organic material from the plant host to symbiotic fungi was demonstrated with ^{14}C tracer experiments. The plant was exposed to $^{14}CO_2$ and the synthesized ^{14}C-organic material was followed into the fungi. These experiments and others with ^{13}C showed that the plant supplies glucose and possibly other hexoses to the fungi. As much as 20% of total plant photosynthate may be transferred to arbuscular mycorrhizal fungi. Because of this high organic input from the plant, mycorrhizal fungal biomass is large, 3–20% of root biomass. Labeling experiments with ^{13}C have also revealed the transfer of organic material along a "common mycelia network" linking plants that share a mycorrhizal fungus (Klein et al., 2016). The network may link plants from the same or different species, depending on the plants and the fungi. These networks

may even "warn" neighboring plants of an herbivore attack (Babikova et al., 2013).

Non-nutritional roles: Bioluminescent squid and a quorum-sensing bacterium

In nearly all of the symbiotic relationships discussed so far, the microbial symbiont is instrumental in the nutrition of the host. The microbe turns inedible organic material into compounds that the host can consume (ruminants, termites, and fungal gardening ants), provides limiting nutrients (plants), helps with the synthesis of essential compounds like amino acids missing from the host's diet (aphids), or directly synthesizes organic material for the host using light or chemical energy (corals, tubeworms, and mussels).

But microbes also have other types of symbiotic relationships with larger organisms, including humans. Commensal microbes on our skin and in our digestive tract help to prevent pathogenic microbes from gaining a foothold. Microbes may have similar roles in insects. Symbiotic bacteria in termites help fend off pathogenic bacteria and fungi that otherwise would be a serious problem for these social insects living in highly dense colonies (Peterson and Scharf, 2016). Bacteria may also confer some resistance in aphids to parasites and promote tolerance of high temperature (Tsuchida et al., 2010). Endosymbiont bacteria in the genus *Rickettsiella* affect body color of one aphid species, perhaps lowering predation by ladybird beetles.

Predation also figures into the next example of symbiosis to be discussed here. The example also illustrates an important mechanism, "quorum sensing," by which microbes communicate with each other.

Many large organisms both on land and in water rely on symbiotic bacteria to generate light or "bioluminescence."

In the case of many marine invertebrates and fishes, the symbionts provide benefits for the host organism not directly connected to nutrition (Haddock et al., 2010). The anglerfish does rely on symbiotic bacteria for feeding by dangling in front of its mouth a tentacle filled with symbiotic bioluminescent bacteria designed to lure in unsuspecting prey. More commonly, symbiotic bioluminescent bacteria are important components of defenses against predators or for attracting mates. Many of the 43 families of known bioluminescent fish are thought to gain their bioluminescence from symbiotic bacteria. The biochemical machinery for bioluminescence may have evolved 40–50 times independently among metazoans and microbes.

Squids in the families Sepiolidae and Loliginidae use bioluminescence provided by symbiotic bacteria, although most species in the 70 genera of bioluminescent squids make their own light (Haddock et al., 2010). One of the best-known species with bioluminescent symbionts is *Euprymna scolopes*, commonly known as the Hawaiian bobtail squid. This cephalopod burrows in the sand of shallow reef flats during the day and then emerges in the early evening to feed (Figure 14.16). If it were not for bioluminescence, the squid would appear as a dark

object backlit by moonlight, an easy target for a predator waiting below in deeper waters. The squid breaks up its dark silhouette by projecting bioluminescent light downward from its light organ. Thanks to a shutter made from a black ink sack and a yellow filter over the light organ, incredibly, the squid can control the intensity and color of the bioluminescence in order to match the background light intensity and color visible where the squid swims.

The bioluminescence comes from the bacterium *Aliivibrio fischeri* colonizing the light organ. (*A. fischeri* was once called *Vibrio fischeri*.) This bacterium or closely related strains account for the bioluminescence of several squid species and monocentrid fishes, while *Photobacterium leiognathi* and relatives are primarily symbionts for leiognathid, apogonid, and morid fishes (Haddock et al., 2010). *Aliivibrio*, *Vibrio*, and, *Photobacterium* are three closely related genera in the gammaproteobacterial family *Vibrionaceae*. In exchange for providing marine fish and invertebrates with bioluminescence, the bacterial symbionts gain a safe, nutrient-rich home.

The symbiosis starts when a newly hatched squid picks up its symbiotic vibrios from the surrounding seawater (McFall-Ngai, 2014). In spite of its initial low abundance, *A. fischeri* establishes its dominance in its squid host, by

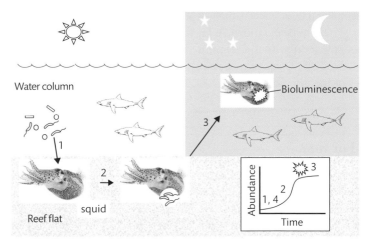

Figure 14.16 A day in the life of the Hawaiian bobtail squid. The squid burrows in the sand flat during the day and comes out to feed in the water column at night. A newly hatched squid first acquires its symbiotic bioluminescence bacteria from the surrounding water (Step 1). Older squid have symbiotic bacteria left over after having expelled most at the end of the night (Step 4, not shown). During the day (Step 2), the bacteria grow and increase in abundance (see inserted graph) housed in the squid's light organ. But bioluminescence does not begin until the bacteria reach a critical population level (Step 3), at which time the squid emerges from the sand and ventures out into the open water. Before dawn, the squid expels most of the symbiotic bacteria and burrows back in the sand to start another day. Picture of the squid, which is only a few centimeters long, courtesy of M. McFall-Ngai and used with permission.

mechanisms that are not completely understood, during the first few hours of the squid's life. The bacterium induces changes in squid cells that terminate the symbiont inoculation phase. Bacteria in older squid are the leftovers from the previous day; the squid releases about 90% of their vibrios each day at dawn, and it expels by unknown mechanisms vibrio strains that are not producing enough bioluminescence (Haddock et al., 2010). Whether freshly inoculated or the leftovers, A. fischeri starts to increase in abundance in the light organ, feasting on the amino acid-rich broth provided by the squid. As the day progresses, vibrio abundance increases until levels are sufficient for bioluminescence to begin. It is then safe (or safer) for the bobtail squid to emerge from the sand and venture out into open water to feed. Because of the bioluminescence, the squid blends in with the ambient light and is less obvious to its many fish predators.

The bacteria sense it is time to turn on bioluminescence based on their cell density, not sunlight or other cues. This detection of population levels, called "quorum sensing," is used by many bacteria in other situations, such as *Pseudomonas aeruginosa* in biofilm formation and *Rhizobium leguminosarum* in root nodulation (Fuqua et al., 1996). The details of the genetic and biochemical machinery vary among different bacteria, but many of the main features are exemplified by A. fischeri, the first bacterial quorum sensing system to be described. A. fischeri uses two complementary quorum-sensing systems (Dunlap, 2014; Lupp and Ruby, 2005). One, the *lux* system, is involved in later stages of the symbiosis and triggers

bioluminescence (Figure 14.17). A. fischeri uses the gene *luxI* to produce a signaling compound, N-3-oxo-hexanoyl homoserine, which is an N-acyl-homoserine lactone (AHL). Concentrations of this signaling compound remain

Low cell density

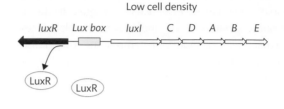

High cell density

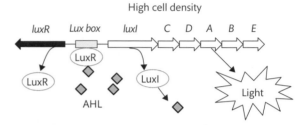

Figure 14.17 Regulation of bioluminescence in A. fischeri by quorum sensing. Each box is a gene in the Lux operon with the arrow indicating the direction of transcription and the width indicating the relative strength of transcription. When cell abundance is low, the autoinducer (AHL) does not bind to the sensing protein (LuxR), transcription of the genes for bioluminescence (*luxCDABE*) is low, and there is no bioluminescence. When abundance is high, AHL concentrations are high enough that it binds to LuxR, allowing for high transcription and the production of bioluminescence. Quorum sensing is used by several other bacteria to regulate many other facets of metabolism.

Box 14.3 Friend or foe?

While A. fischeri is beneficial to the bobtail squid, other vibrios, such as *V. parahaemolyticus* and *V. vulnificus*, are pathogenic to larger organisms, including humans. Vibrios have several biochemical features for interacting with eukaryotes both positively and negatively, many of which were revealed by comparing the genome of A. fischeri and the cholera-causative pathogen, *V. cholerae* (Ruby et al., 2005). The squid symbiotic vibrio has several genes for making Type IV pili similar to those found in *V. cholerae*. These cell surface-associated structures are used by bacteria to colonize surfaces,

with some being essential for *V. cholerae*'s pathogenicity, while others are needed by A. fischeri for normal colonization of the light organ. A. fischeri also has several genes similar to toxin-producing genes of *V. cholerae*. The comparison of the two vibrio genomes suggests that the shared genes are behind a mutualistic relationship with squids in the case of A. fischeri and a pathogenic relationship with humans in the case of *V. cholerae*. The similarities between the two vibrios illustrate the continuum in microbe–eukaryote interactions, from pathogenicity to mutualism.

low because of diffusion when vibrio cell numbers are low. When population levels reach a threshold of about 10^{10} cells per ml, seen only in the light organ, AHL concentrations can build up to levels high enough for AHL to bind to a sensing protein encoded by *luxR*. Formation of the AHL–LuxR complex stimulates more AHL synthesis. For that reason, AHL is called an "autoinducer"; it induces higher production of itself and thus amplifies the quorum-sensing signal. Most importantly, the AHL–LuxR complex also turns on the *lux* operon, leading to the production of bioluminescence.

Concluding remarks

The examples of symbioses discussed here illustrate how microbes are essential for the success of larger organisms in the biosphere and for facilitating the contribution by larger organisms to biogeochemical processes. Even when a large organism seems to be the main character and star in the story, microbes are still around, sometimes only behind the scenes, always indispensable in making many things possible. All of life in our world depends on the processes carried out by microbes.

Summary

1. Microbes form close physical relationships with many large organisms, including invertebrates, vertebrates, and higher plants. These relationships range from commensalism to mutualistic symbioses.

2. Tight symbiotic relationships often cause morphological and genetic changes in both the eukaryotic host and the microbial symbiont. Bacterial symbionts lose unneeded genes when a function is provided by the host, resulting in much smaller genomes than seen in free-living bacteria.

3. Humans, other vertebrates, and invertebrates benefit from hosting symbiotic microbes in their gastrointestinal tract, where microbes aid in digesting otherwise unavailable food. Termites, for example, rely on bacteria, archaea, and protists to use cellulose and other complex polysaccharides in wood, which the insect cannot digest alone. The bacterium *Buchnera* provides essential amino acids for its aphid host feeding on vascular plant phloem.

4. Some symbiotic microbes are the direct source of organic carbon for their metazoan hosts. These microbes include photoautotrophic algae in the case of corals or chemolithoautotrophic bacteria in the case of hydrothermal vent tube worms, some mudflat bivalves, and mussels living at methane seeps. The diverse, rich community at hydrothermal vents is based on primary production by chemolithoautotrophs using sulfide produced by geochemical reactions.

5. Nearly all higher terrestrial plants host symbiotic microbes. Legumes depend on the nitrogen from symbiotic diazotrophic bacteria, while many other plants use mycorrhizal fungi to acquire phosphate and other plant nutrients.

6. Symbiotic microbes provide many other services for their hosts besides securing limiting nutrients, digesting otherwise inedible food, or supplementing incomplete diets. One example is the use of symbiotic bacteria by marine fish and invertebrates to generate bioluminescence as camouflage against predators or as an attractant for mates.

References

Aliaga Goltsman, D. S., Comolli, L. R., Thomas, B. C., and Banfield, J. F. (2015). Community transcriptomics reveals unexpected high microbial diversity in acidophilic biofilm communities. *The ISME Journal*, 9, 1014–23.

Alster, C. J., Baas, P., Wallenstein, M. D., Johnson, N. G., and Von Fischer, J. C. (2016). Temperature sensitivity as a microbial trait using parameters from macromolecular rate theory. *Frontiers in Microbiology*, 7, 1821.

Amin, S. A., Hmelo, L. R., Van Tol, H. M., Durham, B. P., Carlson, L. T., Heal, K. R., Morales, R. L., Berthiaume, C. T., Parker, M. S., Djunaedi, B., Ingalls, A. E., Parsek, M. R., Moran, M. A., and Armbrust, E. V. (2015). Interaction and signalling between a cosmopolitan phytoplankton and associated bacteria. *Nature*, 522, 98–101.

Anderson, M. J., Crist, T. O., Chase, J. M., Vellend, M., Inouye, B. D., Freestone, A. L., Sanders, N. J., Cornell, H. V., Comita, L. S., Davies, K. F., Harrison, S. P., Kraft, N. J. B., Stegen, J. C., and Swenson, N. G. (2011). Navigating the multiple meanings of β diversity: a roadmap for the practicing ecologist. *Ecology Letters*, 14, 19–28.

Anderson, O. R., Lee, J. M., and Mcguire, K. (2016). Experimental evidence that fungi are dominant microbes in carbon content and growth response to added soluble organic carbon in moss-rich tundra soil. *Journal of Eukaryotic Microbiology*, 63, 363–6.

Arndt, N. T. and Nisbet, E. G. (2012). Processes on the young earth and the habitats of early life. *Annual Review of Earth and Planetary Sciences*, 40, 521–49.

Asplund, J. and Wardle, D. A. (2017). How lichens impact on terrestrial community and ecosystem properties. *Biological Reviews*, 92, 1720–38.

Atkinson, D., Ciotti, B. J., and Montagnes, D. J. S. (2003). Protists decrease in size linearly with temperature: *ca.* 2.5% $°C^{-1}$. *Proceedings of the Royal Society of London. Series B: Biological Sciences*, 270, 2605–11.

Babbin, A. R., Keil, R. G., Devol, A. H., and Ward, B. B. (2014). Organic matter stoichiometry, flux, and oxygen control nitrogen loss in the ocean. *Science*, 344, 406–8.

Babikova, Z., Gilbert, L., Bruce, T. J. A., Birkett, M., Caulfield, J. C., Woodcock, C., Pickett, J. A., and Johnson, D. (2013). Underground signals carried through common mycelial networks warn neighbouring plants of aphid attack. *Ecology Letters*, 16, 835–43.

Bagby, S. C., Reddy, C. M., Aeppli, C., Fisher, G. B., and Valentine, D. L. (2017). Persistence and biodegradation of oil at the ocean floor following Deepwater Horizon. *Proceedings of the National Academy of Sciences of the United States of America*, 114, E9–18.

Bailly, J., Fraissinet-Tachet, L., Verner, M.-C., Debaud, J.-C., Lemaire, M., Wesolowski-Louvel, M., and Marmeisse, R. (2007). Soil eukaryotic functional diversity, a metatranscriptomic approach. *The ISME Journal*, 1, 632–42.

Baldauf, S. L. (2003). The deep roots of eukaryotes. *Science*, 300, 1703–6.

Baltar, F., Palovaara, J., Unrein, F., Catala, P., Hornak, K., Simek, K., Vaque, D., Massana, R., Gasol, J. M., and Pinhassi, J. (2016). Marine bacterial community structure resilience to changes in protist predation under phytoplankton bloom conditions. *The ISME Journal*, 10, 568–81.

Banerjee, S., Kirkby, C. A., Schmutter, D., Bissett, A., Kirkegaard, J. A., and Richardson, A. E. (2016). Network analysis reveals functional redundancy and keystone taxa amongst bacterial and fungal communities during organic matter decomposition in an arable soil. *Soil Biology and Biochemistry*, 97, 188–98.

Bang, C. and Schmitz, R. A. (2015). Archaea associated with human surfaces: not to be underestimated. *FEMS Microbiology Reviews*, 39, 631–48.

Barberán, A. and Casamayor, E. O. (2010). Global phylogenetic community structure and beta-diversity patterns in surface bacterioplankton metacommunities. *Aquatic Microbial Ecology*, 59, 1–10.

Barberán, A., Casamayor, E. O., and Fierer, N. (2014). The microbial contribution to macroecology. *Frontiers in Microbiology*, 5, 203.

Barberán, A., Ramirez, K. S., Leff, J. W., Bradford, M. A., Wall, D. H., and Fierer, N. (2014). Why are some microbes more ubiquitous than others? Predicting the habitat breadth of soil bacteria. *Ecology Letters*, 17, 794–802.

Bárcenas-Moreno, G., Gómez-Brandón, M., Rousk, J., and Bååth, E. (2009). Adaptation of soil microbial communities to temperature: comparison of fungi and bacteria in a laboratory experiment. *Global Change Biology*, 15, 2950–7.

Barka, E. A., Vatsa, P., Sanchez, L., Gaveau-Vaillant, N., Jacquard, C., Klenk, H.-P., Clément, C., Ouhdouch, Y., and Van Wezel, G. P. (2016). Taxonomy, physiology, and natural products of *Actinobacteria*. *Microbiology and Molecular Biology Reviews*, 80, 1–43.

Barrangou, R. and Horvath, P. (2017). A decade of discovery: CRISPR functions and applications, *Nature Microbiology* 2, 17092.

Barron, A. R., Wurzburger, N., Bellenger, J. P., Wright, S. J., Kraepiel, A. M. L., and Hedin, L. O. (2009). Molybdenum limitation of asymbiotic nitrogen fixation in tropical forest soils. *Nature Geoscience*, 2, 42–5.

Bates, S. T., Berg-Lyons, D., Caporaso, J. G., Walters, W. A., Knight, R., and Fierer, N. (2011). Examining the global distribution of dominant archaeal populations in soil. *The ISME Journal*, 5, 908–17.

Bates, S. T., Clemente, J. C., Flores, G. E., Walters, W. A., Parfrey, L. W., Knight, R., and Fierer, N. (2013). Global biogeography of highly diverse protistan communities in soil. *The ISME Journal*, 7, 652–9.

Bazylinski, D. A. and Frankel, R. B. (2004). Magnetosome formation in prokaryotes. *Nature Reviews Microbiology*, 2, 217–30.

Behrenfeld, M. J. (2010). Abandoning Sverdrup's critical depth hypothesis on phytoplankton blooms. *Ecology*, 91, 977–89.

Béjà, O., Aravind, L., Koonin, E. V., Suzuki, M. T., Hadd, A., Nguyen, L. P., Jovanovich, S., Gates, C. M., Feldman, R. A., Spudich, J. L., Spudich, E. N., and DeLong, E. F. (2000). Bacterial rhodopsin: Evidence for a new type of phototrophy in the sea. *Science*, 289, 1902–6.

Beman, J. M., Chow, C.-E., King, A. L., Feng, Y., Fuhrman, J. A., Andersson, A., Bates, N. R., Popp, B. N., and Hutchins, D. A. (2011). Global declines in oceanic nitrification rates as a consequence of ocean acidification. *Proceedings of the National Academy of Sciences of the United States of America*, 108, 208–13.

Benner, R., Moran, M. A., and Hodson, R. E. (1986). Biogeochemical cycling of lignocellulosic carbon in marine and freshwater ecosystems: relative contributions of procaryotes and eucaryotes. *Limnology and Oceanography*, 31, 89–100.

Bennett, B., Adams, J. J., Gray, N. D., Sherry, A., Oldenburg, T. B. P., Huang, H., Larter, S. R., and Head, I. M. (2013). The controls on the composition of biodegraded oils in the deep subsurface— Part 3. The impact of microorganism distribution on petroleum geochemical gradients in biodegraded petroleum reservoirs. *Organic Geochemistry*, 56, 94–105.

Bennett, G. M. and Moran, N. A. (2015). Heritable symbiosis: The advantages and perils of an evolutionary rabbit hole. *Proceedings of the National Academy of Sciences of the United States of America*, 112, 10169–76.

Benson, D. R. and Silvester, W. B. (1993). Biology of *Frankia* strains, actinomycete symbionts of actinorhizal plants. *Microbiology and Molecular Biology Reviews*, 57, 293–319.

Berg, B. and Laskowski, R. (2006). *Litter Decomposition: A Guide to Carbon and Nutrient Turnover*, 428 pp. Elsevier, Boston, MA.

Bergh, O., Borsheim, K. Y., Bratbak, G., and Heldal, M. (1989). High abundance of viruses found in aquatic environments. *Nature*, 340, 467–8.

Bernard, L., Chapuis-Lardy, L., Razafimbelo, T., Razafindrakoto, M., Pablo, A.-L., Legname, E., Poulain, J., Bruls, T., O'donohue, M., Brauman, A., Chotte, J.-L., and Blanchart, E. (2012). Endogeic earthworms shape bacterial functional communities and affect organic matter mineralization in a tropical soil. *The ISME Journal*, 6, 213–22.

Berner, R. A. (1999). Atmospheric oxygen over Phanerozoic time. *Proceedings of the National Academy of Sciences of the United States of America*, 96, 10955–7.

Berner, R. A. and Kothavala, Z. (2001). GEOCARB III: A revised model of atmospheric CO_2 over phanerozoic time. *American Journal of Science*, 301, 182–204.

Berney, C., Romac, S., Mahe, F., Santini, S., Siano, R., and Bass, D. (2013). Vampires in the oceans: predatory cercozoan amoebae in marine habitats. *The ISME Journal*, 7, 2387–99.

Berruti, A., Desirò, A., Visentin, S., Zecca, O., and Bonfante, P. (2017). ITS fungal barcoding primers versus 18S AMF-specific primers reveal similar AMF-based diversity patterns in roots and soils of three mountain vineyards. *Environmental Microbiology Reports*, 9, 658–67.

Bertilsson, S., Berglund, O., Karl, D. M., and Chisholm, S. W. (2003). Elemental composition of marine *Prochlorococcus* and *Synechococcus*: Implications for the ecological stoichiometry of the sea. *Limnology and Oceanography*, 48, 1721–31.

Berube, P. M., Coe, A., Roggensack, S. E., and Chisholm, S. W. (2016). Temporal dynamics of *Prochlorococcus* cells with the potential for nitrate assimilation in the subtropical Atlantic and Pacific oceans. *Limnology and Oceanography*, 61, 482–95.

Bianchi, T. S. and Canuel, E. A. (2011). *Chemical Biomarkers in Aquatic Ecosystems*, pp. 392. Princeton University Press, Princeton, NJ.

Bianchi, T. S., Thornton, D. C. O., Yvon-Lewis, S. A., King, G. M., Eglinton, T. I., Shields, M. R., Ward, N. D., and Curtis, J. (2015). Positive priming of terrestrially derived dissolved organic matter in a freshwater microcosm system. *Geophysical Research Letters*, 42, 5460–7.

Bidle, K. D. (2015). The molecular ecophysiology of programmed cell death in marine phytoplankton. *Annual Review of Marine Science*, 7, 341–75.

Biersmith, A. and Benner, R. (1998). Carbohydrates in phytoplankton and freshly produced dissolved organic matter. *Marine Chemistry*, 63, 131–44.

Biller, S. J., Berube, P. M., Lindell, D., and Chisholm, S. W. (2015). *Prochlorococcus*: the structure and function of collective diversity. *Nature Reviews Microbiology*, 13, 13–27.

Blagodatskaya, E. and Kuzyakov, Y. (2013). Active microorganisms in soil: Critical review of estimation criteria and approaches. *Soil Biology and Biochemistry*, 67, 192–211.

Blattner, F. R., Plunkett, G., III, Bloch, C. A., Perna, N. T., Burland, V., Riley, M., Collado-Vides, J., Glasner, J. D., Rode, C. K., Mayhew, G. F., Gregor, J., Davis, N. W., Kirkpatrick, H. A.,

Goeden, M. A., Rose, D. J., Mau, B., and Shao, Y. (1997). The complete genome sequence of *Escherichia coli* K-12. *Science*, 277, 1453–62.

Blazewicz, S. J., Barnard, R. L., Daly, R. A., and Firestone, M. K. (2013). Evaluating rRNA as an indicator of microbial activity in environmental communities: Limitations and uses. *The ISME Journal*, 7, 2061–8.

Blazewicz, S. J., Schwartz, E., and Firestone, M. K. (2014). Growth and death of bacteria and fungi underlie rainfall-induced carbon dioxide pulses from seasonally dried soil. *Ecology*, 95, 1162–72.

Bochdansky, A. B., Clouse, M. A., and Herndl, G. J. (2017). Eukaryotic microbes, principally fungi and labyrinthulomycetes, dominate biomass on bathypelagic marine snow. *The ISME Journal*, 11, 362–73.

Boetius, A., Ravenschlag, K., Schubert, C. J., Rickert, D., Widdel, F., Gieseke, A., Amann, R., Jorgensen, B. B., Witte, U., and Pfannkuche, O. (2000). A marine microbial consortium apparently mediating anaerobic oxidation of methane. *Nature*, 407, 623–6.

Bohannan, B. J. M. and Lenski, R. E. (1999). Effect of prey heterogeneity on the response of a model food chain to resource enrichment. *The American Naturalist*, 153, 73–82.

Bongers, L. (1970). Energy generation and utilization in hydrogen bacteria. *Journal of Bacteriology*, 104, 145–51.

Bonkowski, M. (2004). Protozoa and plant growth: the microbial loop in soil revisited. *New Phytologist*, 162, 617–31.

Bonnain, C., Breitbart, M., and Buck, K. N. (2016). The Ferrojan Horse Hypothesis: iron–virus interactions in the ocean. *Frontiers in Marine Science*, 3, 82.

Boras, J. A., Sala, M. M., Baltar, F., Aristegui, J., Duarte, C. M., and Vaque, D. (2010). Effect of viruses and protists on bacteria in eddies of the Canary Current region (subtropical northeast Atlantic). *Limnology and Oceanography*, 55, 885–98.

Boras, J. A., Sala, M. M., Vázquez-Domínguez, E., Weinbauer, M. G., and Vaqué, D. (2009). Annual changes of bacterial mortality due to viruses and protists in an oligotrophic coastal environment (NW Mediterranean). *Environmental Microbiology*, 11, 1181–93.

Borch, T., Kretzschmar, R., Kappler, A., Cappellen, P. V., Ginder-Vogel, M., Voegelin, A., and Campbell, K. (2010). Biogeochemical redox processes and their impact on contaminant dynamics. *Environmental Science & Technology*, 44, 15–23.

Borgonie, G., Linage-Alvarez, B., Ojo, A. O., Mundle, S. O. C., Freese, L. B., Van Rooyen, C., Kuloyo, O., Albertyn, J., Pohl, C., Cason, E. D., Vermeulen, J., Pienaar, C., Litthauer, D., Van Niekerk, H., Van Eeden, J., Lollar, B. S., Onstott, T. C., and Van Heerden, E. (2015). Eukaryotic opportunists dominate the deep-subsurface biosphere in South Africa. *Nature Communications*, 6, 8952.

Boschker, H. T. S. and Middelburg, J. J. (2002). Stable isotopes and biomarkers in microbial ecology. *FEMS Microbiology Ecology*, 40, 85–95.

Bourke, M. F., Marriott, P. J., Glud, R. N., Hasler-Sheetal, H., Kamalanathan, M., Beardall, J., Greening, C., and Cook, P. L.
M. (2017). Metabolism in anoxic permeable sediments is dominated by eukaryotic dark fermentation. *Nature Geoscience*, 10, 30–5.

Bourne, D. G., Morrow, K. M., and Webster, N. S. (2016). Insights into the coral microbiome: underpinning the health and resilience of reef ecosystems. *Annual Review of Microbiology*, 70, 317–40.

Boyd, P. W., Jickells, T., Law, C. S., Blain, S., Boyle, E. A., Buesseler, K. O., Coale, K. H., Cullen, J. J., De Baar, H. J. W., Follows, M., Harvey, M., Lancelot, C., Levasseur, M., Owens, N. P. J., Pollard, R., Rivkin, R. B., Sarmiento, J., Schoemann, V., Smetacek, V., Takeda, S., Tsuda, A., Turner, S., and Watson, A. J. (2007). Mesoscale iron enrichment experiments 1993–2005: Synthesis and future directions. *Science*, 315, 612–17.

Bradford, M. A., Wieder, W. R., Bonan, G. B., Fierer, N., Raymond, P. A., and Crowther, T. W. (2016). Managing uncertainty in soil carbon feedbacks to climate change. *Nature Climate Change*, 6, 751–8.

Breitbart, M. (2012). Marine viruses: truth or dare. *Annual Review of Marine Science*, 4, 425–48.

Breuillin-Sessoms, F., Venterea, R. T., Sadowsky, M. J., Coulter, J. A., Clough, T. J., and Wang, P. (2017). Nitrification gene ratio and free ammonia explain nitrite and nitrous oxide production in urea-amended soils. *Soil Biology and Biochemistry*, 111, 143–53.

Brewer, T. E., Handley, K. M., Carini, P., Gilbert, J. A., and Fierer, N. (2016). Genome reduction in an abundant and ubiquitous soil bacterium "*Candidatus* Udaeobacter copiosus." *Nature Microbiology*, 2, 16198.

Bristow, L. A., Dalsgaard, T., Tiano, L., Mills, D. B., Bertagnolli, A. D., Wright, J. J., Hallam, S. J., Ulloa, O., Canfield, D. E., Revsbech, N. P., and Thamdrup, B. (2016). Ammonium and nitrite oxidation at nanomolar oxygen concentrations in oxygen minimum zone waters. *Proceedings of the National Academy of Sciences of the United States of America*, 113, 10601–6.

Brock, T. D. (1966). *Principles of Microbial Ecology*, 306 pp. Prentice-Hall, Inc, Englewood Cliffs, NJ.

Brulc, J. M., Antonopoulos, D. A., Berg Miller, M. E., Wilson, M. K., Yannarell, A. C., Dinsdale, E. A., Edwards, R. E., Frank, E. D., Emerson, J. B., Wacklin, P., Coutinho, P. M., Henrissat, B., Nelson, K. E., and White, B. A. (2009). Gene-centric metagenomics of the fiber-adherent bovine rumen microbiome reveals forage specific glycoside hydrolases. *Proceedings of the National Academy of Sciences of the United States of America*, 106, 1948–53.

Brum, J. R., Ignacio-Espinoza, J. C., Roux, S., Doulcier, G., Acinas, S. G., Alberti, A., Chaffron, S., Cruaud, C., De Vargas, C., Gasol, J. M., Gorsky, G., Gregory, A. C., Guidi, L., Hingamp, P., Iudicone, D., Not, F., Ogata, H., Pesant, S., Poulos, B. T., Schwenck, S. M., Speich, S., Dimier, C., Kandels-Lewis, S., Picheral, M., Searson, S., Coordinators, T. O., Bork, P., Bowler, C., Sunagawa, S., Wincker, P., Karsenti, E., and Sullivan, M. B. (2015). Patterns and ecological drivers of ocean viral communities. *Science*, 348, 1261498.

Brum, J. R. and Sullivan, M. B. (2015). Rising to the challenge: accelerated pace of discovery transforms marine virology. *Nature Reviews Microbiology*, 13, 147–59.

Brundrett, M. (2009). Mycorrhizal associations and other means of nutrition of vascular plants: understanding the global diversity of host plants by resolving conflicting information and developing reliable means of diagnosis. *Plant and Soil*, 320, 37–77.

Brune, A. (2014). Symbiotic digestion of lignocellulose in termite guts. *Nature Reviews Microbiology*, 12, 168–80.

Brune, A. and Stingl, U. (2006). Prokaryotic symbionts of termite gut flagellates: Phylogenetic and metabolic implications of a tripartite symbiosis. In Overmann, J., ed. *Molecular Basis of Symbiosis*, pp. 39–60. Springer, Berlin; Heidelberg.

Buckel, W. (1999). Anaerobic energy metabolism. In Lengeler, J. W., Drews, G., and Schlegel, H. G., eds. *Biology of the Prokaryotes*, pp. 278–326. Backwell Science, Thieme.

Büdel, B., Weber, B., Kühl, M., Pfanz, H., Sültemeyer, D., and Wessels, D. (2004). Reshaping of sandstone surfaces by cryptoendolithic cyanobacteria: bioalkalization causes chemical weathering in arid landscapes. *Geobiology*, 2, 261–8.

Bugg, T. D. H., Ahmad, M., Hardiman, E. M., and Rahmanpour, R. (2011). Pathways for degradation of lignin in bacteria and fungi. *Natural Product Reports*, 28, 1883–96.

Bull, A. T., Asenjo, J. A., Goodfellow, M., and Gómez-Silva, B. (2016). The Atacama Desert: technical resources and the growing importance of novel microbial diversity. *Annual Review of Microbiology*, 70, 215–34.

Burge, C. A., Eakin, C. M., Friedman, C. S., Froelich, B., Hershberger, P. K., Hofmann, E. E., Petes, L. E., Prager, K. C., Weil, E., Willis, B. L., Ford, S. E., and Harvell, C. D. (2014). Climate change influences on marine infectious diseases: implications for management and society. *Annual Review of Marine Science*, 6, 249–77.

Burki, F., Kaplan, M., Tikhonenkov, D. V., Zlatogursky, V., Minh, B. Q., Radaykina, L. V., Smirnov, A., Mylnikov, A. P., and Keeling, P. J. (2016). Untangling the early diversification of eukaryotes: a phylogenomic study of the evolutionary origins of Centrohelida, Haptophyta and Cryptista. *Proceedings of the Royal Society B: Biological Sciences*, 283, 20152802.

Burow, L. C., Woebken, D., Marshall, I. P. G., Lindquist, E. A., Bebout, B. M., Prufert-Bebout, L., Hoehler, T. M., Tringe, S. G., Pett-Ridge, J., Weber, P. K., Spormann, A. M., and Singer, S. W. (2013). Anoxic carbon flux in photosynthetic microbial mats as revealed by metatranscriptomics. *The ISME Journal*, 7, 817–29.

Cafaro, M. J., Poulsen, M., Little, A. E. F., Price, S. L., Gerardo, N. M., Wong, B., Stuart, A. E., Larget, B., Abbot, P., and Currie, C. R. (2011). Specificity in the symbiotic association between fungus-growing ants and protective *Pseudonocardia* bacteria. *Proceedings of the Royal Society B: Biological Sciences*, 278, 1814–22.

Campbell, B. J., Yu, L., Heidelberg, J. F., and Kirchman, D. L. (2011). Activity of abundant and rare bacteria in a coastal ocean. *Proceedings of the National Academy of Sciences of the United States of America*, 108, 12776–81.

Candy, R. M., Blight, K. R., and Ralph, D. E. (2009). Specific iron oxidation and cell growth rates of bacteria in batch culture. *Hydrometallurgy*, 98, 148–55.

Canfield, D. E., Thamdrup, B., and Kristensen, E. (2005). *Aquatic Geomicrobiology*, 640 pp. Elsevier Academic Press, San Diego.

Capone, D. G. (2000). The marine microbial nitrogen cycle. In Kirchman, D. L., ed. *Microbial Ecology of the Oceans*, pp. 455–93. Wiley-Liss, New York.

Carere, C. R., Hards, K., Houghton, K. M., Power, J. F., Mcdonald, B., Collet, C., Gapes, D. J., Sparling, R., Boyd, E. S., Cook, G. M., Greening, C., and Stott, M. B. (2017). Mixotrophy drives niche expansion of verrucomicrobial methanotrophs. *The ISME Journal*, 11, 2599–610.

Carey, C. J., Dove, N. C., Beman, J. M., Hart, S. C., and Aronson, E. L. (2016a). Meta-analysis reveals ammonia-oxidizing bacteria respond more strongly to nitrogen addition than ammonia-oxidizing archaea. *Soil Biology and Biochemistry*, 99, 158–66.

Carey, J. C., Tang, J., Templer, P. H., Kroeger, K. D., Crowther, T. W., Burton, A. J., Dukes, J. S., Emmett, B., Frey, S. D., Heskel, M. A., Jiang, L., Machmuller, M. B., Mohan, J., Panetta, A. M., Reich, P. B., Reinsch, S., Wang, X., Allison, S. D., Bamminger, C., Bridgham, S., Collins, S. L., De Dato, G., Eddy, W. C., Enquist, B. J., Estiarte, M., Harte, J., Henderson, A., Johnson, B. R., Larsen, K. S., Luo, Y., Marhan, S., Melillo, J. M., Peñuelas, J., Pfeifer-Meister, L., Poll, C., Rastetter, E., Reinmann, A. B., Reynolds, L. L., Schmidt, I. K., Shaver, G. R., Strong, A. L., Suseela, V., and Tietema, A. (2016b). Temperature response of soil respiration largely unaltered with experimental warming. *Proceedings of the National Academy of Sciences of the United States of America*, 113, 13797–802.

Carini, P., Van Mooy, B. a. S., Thrash, J. C., White, A., Zhao, Y., Campbell, E. O., Fredricks, H. F., and Giovannoni, S. J. (2015). SAR11 lipid renovation in response to phosphate starvation. *Proceedings of the National Academy of Sciences of the United States of America*, 112, 7767–72.

Caron, D. A., Alexander, H., Allen, A. E., Archibald, J. M., Armbrust, E. V., Bachy, C., Bell, C. J., Bharti, A., Dyhrman, S. T., Guida, S. M., Heidelberg, K. B., Kaye, J. Z., Metzner, J., Smith, S. R., and Worden, A. Z. (2017). Probing the evolution, ecology and physiology of marine protists using transcriptomics. *Nature Reviews Microbiology*, 15, 6–20.

Caron, D. A., Countway, P. D., Jones, A. C., Kim, D. Y., and Schnetzer, A. (2012). Marine protistan diversity. *Annual Review of Marine Science*, 4, 467–93.

Carotenuto, Y., Dattolo, E., Lauritano, C., Pisano, F., Sanges, R., Miralto, A., Procaccini, G., and Ianora, A. (2014). Insights into the transcriptome of the marine copepod *Calanus helgolandicus* feeding on the oxylipin-producing diatom *Skeletonema marinoi*. *Harmful Algae*, 31, 153–62.

Carson, J. K., Gonzalez-Quinones, V., Murphy, D. V., Hinz, C., Shaw, J. A., and Gleeson, D. B. (2010). Low pore connectivity increases bacterial diversity in soil. *Applied and Environmental Microbiology*, 76, 3936–42.

Castellano, M. J., Mueller, K. E., Olk, D. C., Sawyer, J. E., and Six, J. (2015). Integrating plant litter quality, soil organic matter stabilization, and the carbon saturation concept. *Global Change Biology*, 21, 3200–9.

Catalán, N., Kellerman, A. M., Peter, H., Carmona, F., and Tranvik, L. J. (2015). Absence of a priming effect on dissolved organic carbon degradation in lake water. *Limnology and Oceanography*, 60, 159–68.

Cavanaugh, C. M., Gardiner, S. L., Jones, M. L., Jannasch, H. W., and Waterbury, J. B. (1981). Prokaryotic cells in the hydrothermal vent tube worm *Riftia pachyptila* Jones—Possible chemoautotrophic symbionts. *Science*, 213, 340–2.

Cavanaugh, C. M., Levering, P. R., Maki, J. S., Mitchell, R., and Lidstrom, M. E. (1987). Symbiosis of methylotrophic bacteria and deep-sea mussels. *Nature*, 325, 346–8.

Cebrian, J. (1999). Patterns in the fate of production in plant communities. *The American Naturalist*, 154, 449–68.

Chan, C. S., Mcallister, S. M., Leavitt, A. H., Glazer, B. T., Krepski, S. T., and Emerson, D. (2016). The architecture of iron microbial mats reflects the adaptation of chemolithotrophic iron oxidation in freshwater and marine environments. *Frontiers in Microbiology*, 7, 796.

Chanton, J. P., Cherrier, J., Wilson, R. M., Sarkodee-Adoo, J., Bosman, S., Mickle, A., and Graham, W. M. (2012). Radiocarbon evidence that carbon from the Deepwater Horizon spill entered the planktonic food web of the Gulf of Mexico. *Environmental Research Letters*, 7, 045303.

Chapin, F. S., Matson, P. A., and Mooney, H. A. (2002). *Principles of Terrestrial Ecosystem Ecology*, 436 pp. Springer, New York.

Chen, B., Liu, H., and Lau, M. (2010). Grazing and growth responses of a marine oligotrichous ciliate fed with two nanoplankton: does food quality matter for micrograzers? *Aquatic Ecology*, 44, 113–19.

Chen, J., Luo, Y., Xia, J., Jiang, L., Zhou, X., Lu, M., Liang, J., Shi, Z., Shelton, S., and Cao, J. (2015a). Stronger warming effects on microbial abundances in colder regions. *Scientific Reports*, 5, 18032.

Chen, L.-X., Hu, M., Huang, L.-N., Hua, Z.-S., Kuang, J.-L., Li, S.-J., and Shu, W.-S. (2015b). Comparative metagenomic and metatranscriptomic analyses of microbial communities in acid mine drainage. *The ISME Journal*, 9, 1579–92.

Cheng, L., Zhang, N., Yuan, M., Xiao, J., Qin, Y., Deng, Y., Tu, Q., Xue, K., Van Nostrand, J. D., Wu, L., He, Z., Zhou, X., Leigh, M. B., Konstantinidis, K. T., Schuur, E. a. G., Luo, Y., Tiedje, J. M., and Zhou, J. (2017). Warming enhances old organic carbon decomposition through altering functional microbial communities. *The ISME Journal*, 11, 1825–35.

Chisholm, S. W. (1992). Phytoplankton size. In Falkowski, P. G. and Woodhead, A. D., eds. *Primary Productivity and Biogeochemical Cycles in the Sea*, pp. 214–37. Plenum Press, New York.

Chisholm, S. W., Olson, R. J., Zettler, E. R., Goericke, R., Waterbury, J. B., and Welschmeyer, N. A. (1988). A novel free-living prochlorophyte abundant in the oceanic euphotic zone. *Nature*, 334, 340–3.

Chistoserdova, L., Vorholt, J. A., and Lidstrom, M. E. (2005). A genomic view of methane oxidation by aerobic bacteria and anaerobic archaea. *Genome Biology*, 6, 208.

Choma, M., Bárta, J., Šantrůčková, H., and Urich, T. (2016). Low abundance of Archaeorhizomycetes among fungi in soil metatranscriptomes. *Scientific Reports*, 6, 38455.

Chow, C.-E. T. and Suttle, C. A. (2015). Biogeography of viruses in the sea. *Annual Review of Virology*, 2, 41–66.

Christias, C., Couvaraki, C., Georgopoulos, S. G., Macris, B., and Vomvoyanni, V. (1975). Protein content and amino acid composition of certain fungi evaluated for microbial protein production. *Applied Microbiology*, 29, 250–4.

Chrzanowski, T. H. and Foster, B. L. L. (2014). Prey element stoichiometry controls ecological fitness of the flagellate *Ochromonas danica*. *Aquatic Microbial Ecology*, 71, 257–69.

Church, M., J., Wai, B., Karl, D. M., and DeLong, E. F. (2010). Abundances of crenarchaeal *amoA* genes and transcripts in the Pacific Ocean. *Environmental Microbiology*, 12, 679–88.

Church, M. J., DeLong, E. F., Ducklow, H. W., Karner, M. B., Preston, C. M., and Karl, D. M. (2003). Abundance and distribution of planktonic Archaea and Bacteria in the waters west of the Antarctic Peninsula. *Limnology and Oceanography*, 48, 1893–902.

Clements, K. D., Angert, E. R., Montgomery, W. L., and Choat, J. H. (2014). Intestinal microbiota in fishes: what's known and what's not. *Molecular Ecology*, 23, 1891–8.

Cleveland, C. C. and Liptzin, D. (2007). C:N:P stoichiometry in soil: is there a "Redfield ratio" for the microbial biomass? *Biogeochemistry*, 85, 235–52.

Colatriano, D., Ramachandran, A., Yergeau, E., Maranger, R., Gélinas, Y., and Walsh, D. A. (2015). Metaproteomics of aquatic microbial communities in a deep and stratified estuary. *Proteomics*, 15, 3566–79.

Cole, J. J., Pace, M. L., Carpenter, S. R., and Kitchell, J. F. (2000). Persistence of net heterotrophy in lakes during nutrient addition and food web manipulations. *Limnology and Oceanography*, 45, 1718–30.

Coleman, D. C. (2008). From peds to paradoxes: Linkages between soil biota and their influences on ecological processes. *Soil Biology and Biochemistry*, 40, 271–89.

Coleman, D. C. and Wall, D. H. (2015). Soil fauna: Occurrence, biodiversity, and roles in ecosystem function. In Paul, E. A., ed. *Soil Microbiology and Biochemistry*, Fourth ed, pp. 112–49. Academic Press, London, UK; Waltham, MA, USA.

Collins, S. M., Surette, M., and Bercik, P. (2012). The interplay between the intestinal microbiota and the brain. *Nature Reviews Microbiology*, 10, 735–42.

Colman, D. R., Poudel, S., Stamps, B. W., Boyd, E. S., and Spear, J. R. (2017). The deep, hot biosphere: Twenty-five years of retrospection. *Proceedings of the National Academy of Sciences of the United States of America*, 114, 6895–903.

Colson, P., La Scola, B., Levasseur, A., Caetano-Anolles, G., and Raoult, D. (2017). Mimivirus: leading the way in the discovery of giant viruses of amoebae. *Nature Reviews Microbiology*, 15, 243–54.

Colston, T. J. and Jackson, C. R. (2016). Microbiome evolution along divergent branches of the vertebrate tree of life: what is known and unknown. *Molecular Ecology*, 25, 3776–800.

Colwell, F. S. and D'Hondt, S. (2013). Nature and extent of the Deep Biosphere. *Reviews in Mineralogy and Geochemistry*, 75, 547–74.

Comolli, L. R., Baker, B. J., Downing, K. H., Siegerist, C. E., and Banfield, J. F. (2008). Three-dimensional analysis of the structure and ecology of a novel, ultra-small archaeon. *The ISME Journal*, 3, 159–67.

Comtet-Marre, S., Parisot, N., Lepercq, P., Chaucheyras-Durand, F., Mosoni, P., Peyretaillade, E., Bayat, A. R., Shingfield, K. J., Peyret, P., and Forano, E. (2017). Metatranscriptomics reveals the active bacterial and eukaryotic fibrolytic communities in the rumen of dairy cow fed a mixed diet. *Frontiers in Microbiology*, 8, 67.

Conrad, R. (2009). The global methane cycle: recent advances in understanding the microbial processes involved. *Environmental Microbiology Reports*, 1, 285–92.

Cook, P. F., Reichmuth, C., Rouse, A. A., Libby, L. A., Dennison, S. E., Carmichael, O. T., Kruse-Elliott, K. T., Bloom, J., Singh, B., Fravel, V. A., Barbosa, L., Stuppino, J. J., Van Bonn, W. G., Gulland, F. M. D., and Ranganath, C. (2015). Algal toxin impairs sea lion memory and hippocampal connectivity, with implications for strandings. *Science*, 350, 1545–7.

Costa, E., Pérez, J., and Kreft, J.-U. (2006). Why is metabolic labour divided in nitrification? *Trends in Microbiology*, 14, 213–19.

Cotner, J. B., Ammerman, J. W., Peele, E. R., and Bentzen, E. (1997). Phosphorus-limited bacterioplankton growth in the Sargasso Sea. *Aquatic Microbial Ecology*, 13, 141–9.

Cottrell, M. T. and Kirchman, D. L. (2000). Natural assemblages of marine proteobacteria and members of the *Cytophaga–Flavobacter* cluster consuming low- and high-molecular-weight dissolved organic matter. *Applied and Environmental Microbiology*, 66, 1692–7.

Cottrell, M. T. and Kirchman, D. L. (2016). Transcriptional control in marine copiotrophic and oligotrophic bacteria with streamlined genomes. *Applied and Environmental Microbiology*, 82, 6010–18.

Cottrell, M. T., Malmstrom, R. R., Hill, V., Parker, A. E., and Kirchman, D. L. (2006). The metabolic balance between autotrophy and heterotrophy in the western Arctic Ocean. *Deep Sea Research*, 53, 1831–44.

Cottrell, M. T., Wood, D. N., Yu, L. Y., and Kirchman, D. L. (2000). Selected chitinase genes in cultured and uncultured marine bacteria in the alpha- and gamma-subclasses of the proteo-bacteria. *Applied and Environmental Microbiology*, 66, 1195–201.

Couradeau, E., Karaoz, U., Lim, H. C., Nunes Da Rocha, U., Northen, T., Brodie, E., and Garcia-Pichel, F. (2016). Bacteria increase arid-land soil surface temperature through the production of sunscreens. *Nature Communications*, 7, 10373.

Crotty, F. V., Adl, S. M., Blackshaw, R. P., and Murray, P. J. (2013). Measuring soil protist respiration and ingestion rates using stable isotopes. *Soil Biology and Biochemistry*, 57, 919–21.

Crowe, S. A., Jones, C., Katsev, S., Magen, C., O'neill, A. H., Sturm, A., Canfield, D. E., Haffner, G. D., Mucci, A., Sundby, B., and Fowle, D. A. (2008). Photoferrotrophs thrive in an Archean Ocean analogue. *Proceedings of the National Academy of Sciences of the United States of America*, 105, 15938–43.

Culley, A. I., Lang, A. S., and Suttle, C. A. (2006). Metagenomic analysis of coastal RNA virus communities. *Science*, 312, 1795–8.

Dacks, J. B. and Doolittle, W. F. (2001). Reconstructing/deconstructing the earliest eukaryotes: how comparative genomics can help. *Cell*, 107, 419–25.

Dadon-Pilosof, A., Conley, K. R., Jacobi, Y., Haber, M., Lombard, F., Sutherland, K. R., Steindler, L., Tikochinski, Y., Richter, M., Glöckner, F. O., Suzuki, M. T., West, N. J., Genin, A., and Yahel, G. (2017). Surface properties of SAR11 bacteria facilitate grazing avoidance. *Nature Microbiology*, 2, 1608–15.

Daims, H., Lebedeva, E. V., Pjevac, P., Han, P., Herbold, C., Albertsen, M., Jehmlich, N., Palatinszky, M., Vierheilig, J., Bulaev, A., Kirkegaard, R. H., Von Bergen, M., Rattei, T., Bendinger, B., Nielsen, P. H., and Wagner, M. (2015). Complete nitrification by *Nitrospira* bacteria. *Nature*, 528, 504–9.

Daims, H., Lücker, S., and Wagner, M. (2016). A new perspective on microbes formerly known as nitrite-oxidizing bacteria. *Trends in Microbiology*, 24, 699–712.

Dalsgaard, T., Stewart, F. J., Thamdrup, B., De Brabandere, L., Revsbech, N. P., Ulloa, O., Canfield, D. E., and DeLong, E. F. (2014). Oxygen at nanomolar levels reversibly suppresses process rates and gene expression in anammox and denitrification in the oxygen minimum zone off northern Chile. *mBio*, 5, e01966–14.

Damon, C., Lehembre, F., Oger-Desfeux, C., Luis, P., Ranger, J., Fraissinet-Tachet, L., and Marmeisse, R. (2012). Metatranscriptomics reveals the diversity of genes expressed by eukaryotes in forest soils. *PLoS One*, 7, e28967.

Darjany, L. E., Whitcraft, C. R., and Dillon, J. G. (2014). Lignocellulose-responsive bacteria in a southern California salt marsh identified by stable isotope probing. *Frontiers in Microbiology*, 5, 263.

Daubin, V., Moran, N. A., and Ochman, H. (2003). Phylogenetics and the cohesion of bacterial genomes. *Science*, 301, 829–32.

Davidson, E. A. and Janssens, I. A. (2006). Temperature sensitivity of soil carbon decomposition and feedbacks to climate change. *Nature*, 440, 165–73.

Davidson, E. A., Janssens, I. A., and Luo, Y. Q. (2006). On the variability of respiration in terrestrial ecosystems: moving beyond Q_{10}. *Global Change Biology*, 12, 154–64.

Davidson, E. A., Samanta, S., Caramori, S. S., and Savage, K. (2012). The dual Arrhenius and Michaelis–Menten kinetics model for decomposition of soil organic matter at hourly to seasonal time scales. *Global Change Biology*, 18, 371–84.

de Man, T. J. B., Stajich, J. E., Kubicek, C. P., Teiling, C., Chenthamara, K., Atanasova, L., Druzhinina, I. S., Levenkova, N., Birnbaum, S. S. L., Barribeau, S. M., Bozick, B. A., Suen, G., Currie, C. R., and Gerardo, N. M. (2016). Small genome of the fungus E*scovopsis weberi*, a specialized disease agent of ant agriculture. *Proceedings of the National Academy of Sciences of the United States of America*, 113, 3567–72.

de Vargas, C., Audic, S., Henry, N., Decelle, J., Mahé, F., Logares, R., Lara, E., Berney, C., Le Bescot, N., Probert, I., Carmichael, M., Poulain, J., Romac, S., Colin, S., Aury, J.-M., Bittner, L., Chaffron, S., Dunthorn, M., Engelen, S., Flegontova, O., Guidi, L., Horák, A., Jaillon, O., Lima-Mendez, G., Lukeš, J., Malviya, S., Morard, R., Mulot, M., Scalco, E., Siano, R., Vincent, F., Zingone, A., Dimier, C., Picheral, M., Searson, S., Kandels-Lewis, S., Coordinators, T. O., Acinas, S. G., Bork, P., Bowler, C., Gorsky, G., Grimsley, N., Hingamp, P., Iudicone, D., Not, F., Ogata, H., Pesant, S., Raes, J., Sieracki, M. E., Speich, S., Stemmann, L., Sunagawa, S., Weissenbach, J., Wincker, P., and Karsenti, E. (2015). Eukaryotic plankton diversity in the sunlit ocean. *Science*, 348, 1261605.

DeAngelis, K. M., Chivian, D., Fortney, J. L., Arkin, A. P., Simmons, B., Hazen, T. C., and Silver, W. L. (2013). Changes in microbial dynamics during long-term decomposition in tropical forests. *Soil Biology and Biochemistry*, 66, 60–8.

DeAngelis, K. M., Pold, G., Topcuoglu, B. D., Van Diepen, L. T. A., Varney, R. M., Blanchard, J., Melillo, J., and Frey, S. (2015). Long-term forest soil warming alters microbial communities in temperate forest soils. *Frontiers in Microbiology*, 6, 104.

del Giorgio, P. A. and Gasol, J. M. (2008). Physiological structure and single-cell activity in marine bacterioplankton. In Kirchman, D. L., ed. *Microbial Ecology of the Ocean*, Second ed, pp. 243–98. John Wiley & Sons, New York.

Delgado-Baquerizo, M., Reich, P. B., Khachane, A. N., Campbell, C. D., Thomas, N., Freitag, T. E., Abu Al-Soud, W., Sørensen, S., Bardgett, R. D., and Singh, B. K. (2017). It is elemental: soil nutrient stoichiometry drives bacterial diversity. *Environmental Microbiology*, 19, 1176–88.

Delwiche, C. F. (1999). Tracing the thread of plastid diversity through the tapestry of life. *The American Naturalist*, 154, S164–77.

Derelle, E., Ferraz, C., Rombauts, S., Rouze, P., Worden, A. Z., Robbens, S., Partensky, F., Degroeve, S., Echeynie, S., Cooke, R., Saeys, Y., Wuyts, J., Jabbari, K., Bowler, C., Panaud, O.,

Piegu, B., Ball, S. G., Ral, J. P., Bouget, F. Y., Piganeau, G., De Baets, B., Picard, A., Delseny, M., Demaille, J., Van De Peer, Y. and Moreau, H. (2006). Genome analysis of the smallest free-living eukaryote *Ostreococcus tauri* unveils many unique features. *Proceedings of the National Academy of Sciences of the United States of America*, 103, 11647–52.

Des Marais, D. J., D'amilio, E., Farmer, J. D., Jørgensen, B. B., Palmisano, A. C., and Pierson, B. K. (1992). Case study of a modern microbial mat-building community: the submerged cyanobacterial mats of Guerrero Negro, Baja California Sur, Mexico. In Schopf, J. W. and Klein, C., eds. *The Proterozoic Biosphere: A Multidisciplinary Study*, pp. 325–33. Cambridge University Press, Cambridge; New York.

Desai, M. S. and Brune, A. (2012). Bacteroidales ectosymbionts of gut flagellates shape the nitrogen-fixing community in dry-wood termites. *The ISME Journal*, 6, 1302–13.

Devol, A. H. (2015). Denitrification, anammox, and N_2 production in marine sediments. *Annual Review of Marine Science*, 7, 403–23.

Diaz, J., Ingall, E., Benitez-Nelson, C., Paterson, D., De Jonge, M. D., Mcnulty, I., and Brandes, J. A. (2008). Marine polyphosphate: A key player in geologic phosphorus sequestration. *Science*, 320, 652–5.

Dohnalkova, A. C., Marshall, M. J., Arey, B. W., Williams, K. H., Buck, E. C., and Fredrickson, J. K. (2011). Imaging hydrated microbial extracellular polymers: comparative analysis by electron microscopy. *Applied and Environmental Microbiology*, 77, 1254–62.

Dolan, J. R., Montagnes, D. J. S., Agatha, S., Coats, D. W., and Stoecker, D. K. (eds.) (2013). *The Biology and Ecology of Tintinnid Ciliates: Models for Marine Plankton*, 296 pp. Wiley-Blackwell. Chichester, West Sussex; Hoboken, NJ.

Dolan, J. R., Yang, E. J., Kang, S.-H., and Rhee, T. S. (2016). Declines in both redundant and trace species characterize the latitudinal diversity gradient in tintinnid ciliates. *The ISME Journal*, 10, 2174–83.

Doney, S. C., Fabry, V. J., Feely, R. A., and Kleypas, J. A. (2009). Ocean acidification: The other CO_2 problem. *Annual Review of Marine Science*, 1, 169–92.

Dong, H.-P., Hong, Y.-G., Lu, S., and Xie, L.-Y. (2014). Metaproteomics reveals the major microbial players and their biogeochemical functions in a productive coastal system in the northern South China Sea. *Environmental Microbiology Reports*, 6, 683–95.

Dong, L. F., Sobey, M. N., Smith, C. J., Rusmana, I., Phillips, W., Stott, A., Osborn, A. M., and Nedwell, D. B. (2011). Dissimilatory reduction of nitrate to ammonium, not denitrification or anammox, dominates benthic nitrate reduction in tropical estuaries. *Limnology and Oceanography*, 56, 279–91.

Donoghue, P. C. J. and Antcliffe, J. B. (2010). Early life: Origins of multicellularity. *Nature*, 466, 41–2.

Douglas, A. E. (2010). *The Symbiotic Habit*, pp. ix, 202. Princeton University Press, Princeton, NJ.

Downie, J. A. (2010). The roles of extracellular proteins, polysaccharides and signals in the interactions of rhizobia with legume roots. *FEMS Microbiology Reviews*, 34, 150–70.

Drake, H. L., Gößner, A. S., and Daniel, S. L. (2008). Old acetogens, new light. *Annals of the New York Academy of Sciences*, 1125, 100–28.

Dubilier, N., Blazejak, A., and Rühland, C. (2006). Symbioses between bacteria and gutless marine oligochaetes. In Overmann, J., ed. *Molecular Basis of Symbiosis*, pp. 251–75. Springer, Berlin; Heidelberg.

Dubinsky, E. A., Conrad, M. E., Chakraborty, R., Bill, M., Borglin, S. E., Hollibaugh, J. T., Mason, O. U., M. Piceno, Y., Reid, F. C., Stringfellow, W. T., Tom, L. M., Hazen, T. C., and Andersen, G. L. (2013). Succession of hydrocarbon-degrading bacteria in the aftermath of the Deepwater Horizon oil spill in the Gulf of Mexico. *Environmental Science & Technology*, 47, 10860–7.

Ducklow, H. W., Purdie, D. A., Williams, P. J. L., and Davies, J. M. (1986). Bacterioplankton: a sink for carbon in a coastal marine plankton community. *Science*, 232, 865–7.

Dunlap, P. (2014). Biochemistry and genetics of bacterial bioluminescence. In Thouand, G. and Marks, R., eds. *Bioluminescence: Fundamentals and Applications in Biotechnology—Volume 1*, pp. 37–64. Springer, Berlin; Heidelberg.

Dunn, M. F. (2015). Key roles of microsymbiont amino acid metabolism in rhizobia–legume interactions. *Critical Reviews in Microbiology*, 41, 411–51.

Dupont, A. Ö. C., Griffiths, R. I., Bell, T., and Bass, D. (2016). Differences in soil micro-eukaryotic communities over soil pH gradients are strongly driven by parasites and saprotrophs. *Environmental Microbiology*, 18, 2010–24.

Dupraz, C., Reid, R. P., Braissant, O., Decho, A. W., Norman, R. S., and Visscher, P. T. (2009). Processes of carbonate precipitation in modern microbial mats. *Earth-Science Reviews*, 96, 141–62.

Dürichen, H., Siegmund, L., Burmester, A., Fischer, M. S., and Wöstemeyer, J. (2016). Ingestion and digestion studies in *Tetrahymena pyriformis* based on chemically modified microparticles. *European Journal of Protistology*, 52, 45–57.

Dusenbery, D. B. (1997). Minimum size limit for useful locomotion by free-swimming microbes. *Proceedings of the National Academy of Science of the United States of America*, 94, 10949–54.

Edwards, I. P., Zak, D. R., Kellner, H., Eisenlord, S. D., and Pregitzer, K. S. (2011). Simulated atmospheric N deposition alters fungal community composition and suppresses ligninolytic gene expression in a northern hardwood forest. *PLoS One*, 6, e20421.

Edwards, K. J., Becker, K., and Colwell, F. (2012). The deep, dark energy biosphere: intraterrestrial life on earth. *Annual Review of Earth and Planetary Sciences*, 40, 551–68.

Ehrenreich, A. and Widdel, F. (1994). Anaerobic oxidation of ferrous iron by purple bacteria, a new type of phototrophic metabolism. *Applied and Environmental Microbiology*, 60, 4517–26.

Elliott, T. A. and Gregory, T. R. (2015). What's in a genome? The C-value enigma and the evolution of eukaryotic genome content. *Philosophical Transactions of the Royal Society B: Biological Sciences*, 370, 20140331.

Elser, J. J., Acharya, K., Kyle, M., Cotner, J., Makino, W., Markow, T., Watts, T., Hobbie, S., Fagan, W., Schade, J., Hood, J., and Sterner, R. W. (2003). Growth rate–stoichiometry couplings in diverse biota. *Ecology Letters*, 6, 936–43.

Elser, J. J., Fagan, W. F., Denno, R. F., Dobberfuhl, D. R., Folarin, A., Huberty, A., Interlandi, S., Kilham, S. S., Mccauley, E., Schulz, K. L., Siemann, E. H., and Sterner, R. W. (2000). Nutritional constraints in terrestrial and freshwater food webs. *Nature*, 408, 578–80.

Emerson, D., Field, E., Chertkov, O., Davenport, K., Goodwin, L., Munk, C., Nolan, M., and Woyke, T. (2013). Comparative genomics of freshwater Fe-oxidizing bacteria: implications for physiology, ecology, and systematics. *Frontiers in Microbiology*, 4, 254.

Emerson, D., Fleming, E. J., and Mcbeth, J. M. (2010). Iron-oxidizing bacteria: An environmental and genomic perspective. *Annual Review of Microbiology*, 64, 561–83.

Endo, H., Sugie, K., Yoshimura, T., and Suzuki, K. (2016). Response of spring diatoms to CO_2 availability in the western North Pacific as determined by next-generation sequencing. *PLoS One*, 11, e0154291.

Engelhardt, T., Orsi, W. D., and Jørgensen, B. B. (2015). Viral activities and life cycles in deep subseafloor sediments. *Environmental Microbiology Reports*, 7, 868–73.

Etheridge, D. M., Steele, L. P., Langenfelds, R. L., Francey, R. J., Barnola, J. M., and Morgan, V. I. (1996). Natural and anthropogenic changes in atmospheric CO_2 over the last 1000 years from air in Antarctic ice and firn. *Journal of Geophysical Research*, 101, 4115–28.

Ettwig, K. F., Butler, M. K., Le Paslier, D., Pelletier, E., Mangenot, S., Kuypers, M. M. M., Schreiber, F., Dutilh, B. E., Zedelius, J., De Beer, D., Gloerich, J., Wessels, H. J. C. T., Van Alen, T., Luesken, F., Wu, M. L., Van De Pas-Schoonen, K. T., Op Den Camp, H. J. M., Janssen-Megens, E. M., Francoijs, K.-J., Stunnenberg, H., Weissenbach, J., Jetten, M. S. M., and Strous, M. (2010). Nitrite-driven anaerobic methane oxidation by oxygenic bacteria. *Nature*, 464, 543–8.

Fabian, J., Zlatanovic, S., Mutz, M., and Premke, K. (2017). Fungal–bacterial dynamics and their contribution to terrigenous carbon turnover in relation to organic matter quality. *The ISME Journal*, 11, 415–25.

Falkowski, P. G. and Oliver, M. J. (2007). Mix and match: how climate selects phytoplankton. *Nature Reviews Microbiology*, 5, 813–19.

Fan, H.-W., Noda, H., Xie, H.-Q., Suetsugu, Y., Zhu, Q.-H., and Zhang, C.-X. (2015). Genomic analysis of an ascomycete fungus from the rice planthopper reveals how it adapts to an endosymbiotic lifestyle. *Genome Biology and Evolution*, 7, 2623–34.

Fang, J., Zhang, L., and Bazylinski, D. A. (2010). Deep-sea piezosphere and piezophiles: geomicrobiology and biogeochemistry. *Trends in Microbiology*, 18, 413–22.

Farmer, I. S. and Jones, C. W. (1976). The effect of temperature on the molar growth yield and maintenance requirement of *Escherichia coli* W during aerobic growth in continuous culture. *FEBS Letters*, 67, 359–63.

Faruque, S. M., Biswas, K., Udden, S. M. N., Ahmad, Q. S., Sack, D. A., Nair, G. B., and Mekalanos, J. J. (2006). Transmissibility of cholera: In vivo-formed biofilms and their relationship to infectivity and persistence in the environment. *Proceedings of the National Academy of Sciences of the United States of America*, 103, 6350–5.

Fenchel, T. (1987). *Ecology of Protozoa*, 197 pp. Science Tech Publishers, Madison, Wisconsin.

Fenchel, T. and Blackburn, T. H. (1979). *Bacteria and Mineral Cycling*, 225 pp. Academic Press, London.

Fennel, K. (2017). Biogeochemistry: Ocean hotspots of nitrogen loss. *Nature*, 551, 305–6.

Fernandez-Gomez, B., Richter, M., Schuler, M., Pinhassi, J., Acinas, S. G., Gonzalez, J. M., and Pedrós-Alió, C. (2013). Ecology of marine Bacteroidetes: a comparative genomics approach. *The ISME Journal*, 7, 1026–37.

Fierer, N. (2017). Embracing the unknown: disentangling the complexities of the soil microbiome. *Nature Reviews Microbiology*, 15, 579–90.

Fierer, N. and Jackson, R. B. (2006). The diversity and biogeography of soil bacterial communities. *Proceedings of the National Academy of Sciences of the United States of America*, 103, 626–31.

Fierer, N., Lauber, C. L., Ramirez, K. S., Zaneveld, J., Bradford, M. A., and Knight, R. (2012). Comparative metagenomic, phylogenetic and physiological analyses of soil microbial communities across nitrogen gradients. *The ISME Journal*, 6, 1007–17.

Finke, N., Vandieken, V., and Jørgensen, B. B. (2007). Acetate, lactate, propionate, and isobutyrate as electron donors for iron and sulfate reduction in Arctic marine sediments, Svalbard. *FEMS Microbiology Ecology*, 59, 10–22.

Finkel, Z. V., Follows, M. J., Liefer, J. D., Brown, C. M., Benner, I., and Irwin, A. J. (2016). Phylogenetic diversity in the macromolecular composition of microalgae. *PLoS One*, 11, e0155977.

Flemming, H.-C. and Wingender, J. (2010). The biofilm matrix. *Nature Reviews Microbiology*, 8, 623–33.

Fodelianakis, S., Pitta, P., Thingstad, T. F., Kasapidis, P., Karakassis, I., and Ladoukakis, E. D. (2014). Phosphate addition has minimal short-term effects on bacterioplankton community structure of the P-starved Eastern Mediterranean. *Aquatic Microbial Ecology*, 72, 98–104.

Fouilland, E. and Mostajir, B. (2010). Revisited phytoplanktonic carbon dependency of heterotrophic bacteria in freshwaters, transitional, coastal and oceanic waters. *FEMS Microbiology Ecology*, 73, 419–29.

Fowler, D., Steadman, C. E., Stevenson, D., Coyle, M., Rees, R. M., Skiba, U. M., Sutton, M. A., Cape, J. N., Dore, A. J., Vieno, M., Simpson, D., Zaehle, S., Stocker, B. D., Rinaldi, M., Facchini, M. C., Flechard, C. R., Nemitz, E., Twigg, M., Erisman, J. W., Butterbach-Bahl, K., and Galloway, J. N. (2015). Effects of global change during the 21st century on the nitrogen cycle. *Atmospheric Chemistry and Physics*, 15, 13849–93.

Freing, A., Wallace, D. W. R., and Bange, H. W. (2012). Global oceanic production of nitrous oxide. *Philosophical Transactions of the Royal Society B: Biological Sciences*, 367, 1245–55.

Frias-Lopez, J., Thompson, A., Waldbauer, J., and Chisholm, S. W. (2009). Use of stable isotope-labelled cells to identify active grazers of picocyanobacteria in ocean surface waters. *Environmental Microbiology*, 11, 512–25.

Froelich, P. N., Klinkhammer, G. P., Bender, M. L., Luedtke, N. A., Heath, G. R., Cullen, D., Dauphin, P., Hammond, D., Hartman, B., and Maynard, V. (1979). Early oxidation of organic matter in pelagic sediments of the Eastern Equatorial Atlantic–suboxic diagenesis. *Geochimica et Cosmochimica Acta*, 43, 1075–90.

Fuchs, G. (2011). Alternative pathways of carbon dioxide fixation: Insights into the early evolution of life? *Annual Review of Microbiology*, 65, 631–58.

Fuqua, C., Winans, S. C., and Greenberg, E. P. (1996). Census and consensus in bacterial ecosystems: The LuxR-LuxI family of quorum-sensing transcriptional regulators. *Annual Review of Microbiology*, 50, 727–51.

Fussel, J., Lam, P., Lavik, G., Jensen, M. M., Holtappels, M., Gunter, M., and Kuypers, M. M. M. (2012). Nitrite oxidation in the Namibian oxygen minimum zone. *The ISME Journal*, 6, 1200–9.

Gadd, G. M. (2017). Geomicrobiology of the built environment. *Nature Microbiology*, 2, 16275.

Gao, H., Schreiber, F., Collins, G., Jensen, M. M., Kostka, J. E., Lavik, G., De Beer, D., Zhou, H.-Y., and Kuypers, M. M. M. (2010). Aerobic denitrification in permeable Wadden Sea sediments. *The ISME Journal*, 4, 417–26.

Garcia, J.-L., Patel, B. K. C., and Ollivier, B. (2000). Taxonomic, phylogenetic, and ecological diversity of methanogenic archaea. *Anaerobe*, 6, 205–26.

Garcia, N. S., Bonachela, J. A., and Martiny, A. C. (2016). Interactions between growth-dependent changes in cell size, nutrient supply and cellular elemental stoichiometry of marine *Synechococcus*. *The ISME Journal*, 10, 2715–24.

Garcia, S. L., Mcmahon, K. D., Martinez-Garcia, M., Srivastava, A., Sczyrba, A., Stepanauskas, R., Grossart, H.-P., Woyke, T., and Warnecke, F. (2013). Metabolic potential of a single cell belonging to one of the most abundant lineages in freshwater bacterioplankton. *The ISME Journal*, 7, 137–47.

Gasol, J. M. and Morán, X. A. G. (2016). Flow cytometric determination of microbial abundances and its use to obtain indices of community structure and relative activity. In McGenity, T. J., Timmis, K. N., and Nogales, B., eds. *Hydrocarbon and Lipid Microbiology Protocols: Single-cell and Single-molecule Methods*, pp. 159–87. Springer-Verlag, Berlin, Heidelberg.

Gause, G. F. (1964). *The Struggle for Existence*, pp. ix, 163. Hafner, New York.

Gauthier, M., Bradley, R. L., and Šimek, M. (2015). More evidence that anaerobic oxidation of methane is prevalent in soils: Is it time to upgrade our biogeochemical models? *Soil Biology and Biochemistry*, 80, 167–74.

Gebremikael, M. T., Steel, H., Bert, W., Maenhout, P., Sleutel, S., and De Neve, S. (2015). Quantifying the contribution of entire free-living nematode communities to carbon mineralization under contrasting C and N availability. *PLoS One*, 10, e0136244.

Geisen, S., Cornelia, B., Jörg, R., and Michael, B. (2014). Soil water availability strongly alters the community composition of soil protists. *Pedobiologia*, 57, 205–13.

Geisen, S., Koller, R., Hünninghaus, M., Dumack, K., Urich, T., and Bonkowski, M. (2016). The soil food web revisited: Diverse and widespread mycophagous soil protists. *Soil Biology and Biochemistry*, 94, 10–18.

Geisen, S., Mitchell, E. a. D., Wilkinson, D. M., Adl, S., Bonkowski, M., Brown, M. W., Fiore-Donno, A. M., Heger, T. J., Jassey, V. E. J., Krashevska, V., Lahr, D. J. G., Marcisz, K., Mulot, M., Payne, R., Singer, D., Anderson, O. R., Charman, D. J., Ekelund, F., Griffiths, B. S., Rønn, R., Smirnov, A., Bass, D., Belbahri, L., Berney, C., Blandenier, Q., Chatzinotas, A., Clarholm, M., Dunthorn, M., Feest, A., Fernández, L. D., Foissner, W., Fournier, B., Gentekaki, E., Hájek, M., Helder, J., Jousset, A., Koller, R., Kumar, S., La Terza, A., Lamentowicz, M., Mazei, Y., Santos, S. S., Seppey, C. V. W., Spiegel, F. W., Walochnik, J., Winding, A., and Lara, E. (2017). Soil protistology rebooted: 30 fundamental questions to start with. *Soil Biology and Biochemistry*, 111, 94–103.

Geisen, S., Tveit, A. T., Clark, I. M., Richter, A., Svenning, M. M., Bonkowski, M., and Urich, T. (2015). Metatranscriptomic census of active protists in soils. *The ISME Journal*, 9, 2178–90.

Geszvain, K., Butterfield, C., Davis, Richard E., Madison, Andrew S., Lee, S.-W., Parker, Dorothy L., Soldatova, A., Spiro, Thomas G., Luther, George W. III, and Tebo, Bradley M. (2012). The molecular biogeochemistry of manganese(II) oxidation. *Biochemical Society Transactions*, 40, 1244–8.

Ghabrial, S. A., Castón, J. R., Jiang, D., Nibert, M. L., and Suzuki, N. (2015). 50-plus years of fungal viruses. *Virology*, 479–80, 356–68.

Ghosh, D., Roy, K., Williamson, K. E., White, D. C., Wommack, K. E., Sublette, K. L., and Radosevich, M. (2008). Prevalence of lysogeny among soil bacteria and presence of 16S rRNA and *trzN* genes in viral-community DNA. *Applied and Environmental Microbiology*, 74, 495–502.

Gibbons, S. M., Caporaso, J. G., Pirrung, M., Field, D., Knight, R., and Gilbert, J. A. (2013). Evidence for a persistent microbial seed bank throughout the global ocean. *Proceedings of the National Academy of Sciences of the United States of America*, 110, 4651–5.

Gibson, C. M. and Hunter, M. S. (2010). Extraordinarily widespread and fantastically complex: comparative biology of endosymbiotic bacterial and fungal mutualists of insects. *Ecology Letters*, 13, 223–34.

Gies, E. A., Konwar, K. M., Beatty, J. T., and Hallam, S. J. (2014). Illuminating microbial dark matter in meromictic Sakinaw Lake. *Applied and Environmental Microbiology*, 80, 6807–18.

Gifford, S. M., Sharma, S., Booth, M., and Moran, M. A. (2013). Expression patterns reveal niche diversification in a marine microbial assemblage. *The ISME Journal*, 7, 281–98.

Gifford, S. M., Sharma, S., and Moran, M. A. (2014). Linking activity and function to ecosystem dynamics in a coastal bacterioplankton community. *Frontiers in Microbiology*, 5, 185.

Gilbert, J. A., Field, D., Swift, P., Newbold, L., Oliver, A., Smyth, T., Somerfield, P. J., Huse, S., and Joint, I. (2009). The seasonal structure of microbial communities in the Western English Channel. *Environmental Microbiology*, 11, 3132–9.

Gillooly, J. F., Brown, J. H., West, G. B., Savage, V. M., and Charnov, E. L. (2001). Effects of size and temperature on metabolic rate. *Science*, 293, 2248–51.

Gimmler, A., Korn, R., De Vargas, C., Audic, S., and Stoeck, T. (2016). The Tara Oceans voyage reveals global diversity and distribution patterns of marine planktonic ciliates. *Scientific Reports*, 6, 33555.

Giner, C. R., Forn, I., Romac, S., Logares, R., De Vargas, C., and Massana, R. (2016). Environmental sequencing provides reasonable estimates of the relative abundance of specific picoeukaryotes. *Applied and Environmental Microbiology*, 82, 4757–66.

Giovannoni, S. J. (2017). SAR11 bacteria: The most abundant plankton in the oceans. *Annual Review of Marine Science*, 9, 231–55.

Giovannoni, S. J., Cameron Thrash, J., and Temperton, B. (2014). Implications of streamlining theory for microbial ecology. *The ISME Journal*, 8, 1553–65.

Giovannoni, S. J., Tripp, H. J., Givan, S., Podar, M., Vergin, K. L., Baptista, D., Bibbs, L., Eads, J., Richardson, T. H., Noordewier, M., Rappe, M. S., Short, J. M., Carrington, J. C., and Mathur, E. J. (2005). Genome streamlining in a cosmopolitan oceanic bacterium. *Science*, 309, 1242–5.

Glass, J. B., Wolfe-Simon, F., Elser, J. J., and Anbar, A. D. (2010). Molybdenum-nitrogen co-limitation in freshwater and coastal heterocystous cyanobacteria. *Limnology and Oceanography*, 55, 667–76.

Gleeson, D., Kennedy, N., Clipson, N., Melville, K., Gadd, G., and Mcdermott, F. (2006). Characterization of bacterial community structure on a weathered pegmatitic granite. *Microbial Ecology*, 51, 526–34.

Godwin, C. M. and Cotner, J. B. (2015). Stoichiometric flexibility in diverse aquatic heterotrophic bacteria is coupled to differences in cellular phosphorus quotas. *Frontiers in Microbiology*, 6, 159.

Gold, T. (1999). *The Deep Hot Biosphere*, pp. xiv, 235. Copernicus, New York.

Goldberg, I., Rock, J. S., Ben-Bassat, A., and Mateles, R. I. (1976). Bacterial yields on methanol, methylamine, formaldehyde, and formate. *Biotechnology and Bioengineering*, 18, 1657–68.

Goldhammer, T., Bruchert, V., Ferdelman, T. G., and Zabel, M. (2010). Microbial sequestration of phosphorus in anoxic upwelling sediments. *Nature Geoscience*, 3, 557–61.

Goldman, J. C., Caron, D. A., and Dennett, M. R. (1987a). Regulation of gross growth efficiency and ammonium regeneration in bacteria by substrate C:N ratio. *Limnology and Oceanography*, 32, 1239–52.

Goldman, J. G., Caron, D. A., and Dennett, M. R. (1987b). Nutrient cycling in a microflagellate food chain: IV. Phytoplankton-microflagellate interactions. *Marine Ecology Progress Series*, 38, 75–87.

Gómez-Consarnau, L., Akram, N., Lindell, K., Pedersen, A., Neutze, R., Milton, D. L., González, J., and Pinhassi, J. (2010). Proteorhodopsin phototrophy promotes survival of marine bacteria during starvation. *PLoS Biology*, 8, e1000358.

Gong, J., Dong, J., Liu, X., and Massana, R. (2013). Extremely high copy numbers and polymorphisms of the rDNA operon estimated from single cell analysis of oligotrich and peritrich ciliates. *Protist*, 164, 369–79.

Gottlieb, D. and Van Etten, J. L. (1966). Changes in fungi with age I. Chemical composition of *Rhizoctonia solani* and *Sclerotium bataticola*. *Journal of Bacteriology*, 91, 161–8.

Grabowski, R. C., Droppo, I. G., and Wharton, G. (2011). Erodibility of cohesive sediment: The importance of sediment properties. *Earth-Science Reviews*, 105, 101–20.

Graf, D. R. H., Jones, C. M., and Hallin, S. (2014). Intergenomic comparisons highlight modularity of the denitrification pathway and underpin the importance of community structure for N$_2$O emissions. *PLoS One*, 9, e114118.

Griffin, B. M., Schott, J., and Schink, B. (2007). Nitrite, an electron donor for anoxygenic photosynthesis. *Science*, 316, 1870.

Grimmett, I. J., Shipp, K. N., Macneil, A., and Bärlocher, F. (2013). Does the growth rate hypothesis apply to aquatic hyphomycetes? *Fungal Ecology*, 6, 493–500.

Gross, L. (2006). Bacterial fimbriae designed to stay with the flow. *PLoS Biology*, 4, e314.

Grossmann, L., Jensen, M., Heider, D., Jost, S., Glucksman, E., Hartikainen, H., Mahamdallie, S. S., Gardner, M., Hoffmann, D., Bass, D., and Boenigk, J. (2016). Protistan community analysis: Key findings of a large-scale molecular sampling. *The ISME Journal*, 10, 2269–79.

Gruber, N. and Galloway, J. N. (2008). An Earth-system perspective of the global nitrogen cycle. *Nature*, 451, 293–6.

Gweon, H. S., Bailey, M. J., and Read, D. S. (2017). Assessment of the bimodality in the distribution of bacterial genome sizes. The ISME Journal., 11, 821–4.

Haddock, S. H. D., Moline, M. A., and Case, J. F. (2010). Bioluminescence in the sea. *Annual Review of Marine Science*, 2, 443–93.

Hager, T. (2008). *The Alchemy of Air: A Jewish Genius, a Doomed Tycoon, and the Scientific Discovery that Fed the World but Fueled the Rise of Hitler*, pp. xvii, 316. Harmony Books, New York.

Hagström, Å., Pommier, T., Rohwer, F., Simu, K., Stolte, W., Svensson, D., and Zweifel, U. L. (2002). Use of 16S ribosomal DNA for delineation of marine bacterioplankton species. *Applied and Environmental Microbiology*, 68, 3628–33.

Hallin, S., Philippot, L., Löffler, F. E., Sanford, R. A., and Jones, C. M. (2018). Genomics and ecology of novel N$_2$O-reducing microorganisms. *Trends in Microbiology*, 26, 43–55.

Hallmann, C., Stannek, L., Fritzlar, D., Hause-Reitner, D., Friedl, T., and Hoppert, M. (2013). Molecular diversity of phototrophic biofilms on building stone. *FEMS Microbiology Ecology*, 84, 355–72.

Hamberger, A., Horn, M. A., Dumont, M. G., Murrell, J. C., and Drake, H. L. (2008). Anaerobic consumers of monosaccharides in a moderately acidic fen. *Applied and Environmental Microbiology*, 74, 3112–20.

Hansel, C. M. and Learman, D. R. (2015). Geomicrobiology of manganese. In Ehrlich, H. L., Newman, D. K., and Kappler, A., eds. *Ehrlich's Geomicrobiology*, sixth ed, pp. 401–52. CRC Press, Boca Raton.

Hansel, C. M., Zeiner, C. A., Santelli, C. M., and Webb, S. M. (2012). Mn(II) oxidation by an ascomycete fungus is linked to superoxide production during asexual reproduction. *Proceedings of the National Academy of Sciences of the United States of America*, 109, 12621–5.

Hansell, D. A. (2013). Recalcitrant dissolved organic carbon fractions. *Annual Review of Marine Science*, 5, 421–45.

Hansen, A. K. and Moran, N. A. (2011). Aphid genome expression reveals host-symbiont cooperation in the production of amino acids. *Proceedings of the National Academy of Sciences of the United States of America*, 108, 2849–54.

Hansen, J., Ruedy, R., Sato, M., and Lo, K. (2010). Global surface temperature change. *Reviews of Geophysics*, 48, RG4004.

Hansen, P. J. (1991). Quantitative importance and trophic role of heterotrophic dinoflagellates in a coastal pelagial food web. *Marine Ecology Progress Series*, 73, 253–61.

Hansman, R. L., Griffin, S., Watson, J. T., Druffel, E. R. M., Ingalls, A. E., Pearson, A., and Aluwihare, L. I. (2009). The radiocarbon signature of microorganisms in the mesopelagic ocean. *Proceedings of the National Academy of Sciences of the United States of America*, 106, 6513–18.

Hanson, C. A., Fuhrman, J. A., Horner-Devine, M. C., and Martiny, J. B. H. (2012a). Beyond biogeographic patterns:

processes shaping the microbial landscape. *Nature Reviews Microbiology*, 10, 497–506.

Hanson, P. J., Edwards, N. T., Garten, C. T., and Andrews, J. A. (2000). Separating root and soil microbial contributions to soil respiration: A review of methods and observations. *Biogeochemistry*, 48, 115–46.

Hanson, T. E., Alber, B. E., and Tabita, F. R. (2012b). Phototrophic CO_2 fixation: recent insights into ancient metabolisms. In Burnap, R. L. and Vermaas, W., eds. *Functional Genomics and Evolution of Photosynthetic Systems*, pp. 225–51. Springer, Dordrecht, The Netherlands.

Haro-Moreno, J. M., Rodriguez-Valera, F., López-García, P., Moreira, D., and Martin-Cuadrado, A.-B. (2017). New insights into marine group III Euryarchaeota, from dark to light. *The ISME Journal*, 11, 1102–17.

Haroon, M. F., Hu, S., Shi, Y., Imelfort, M., Keller, J., Hugenholtz, P., Yuan, Z., and Tyson, G. W. (2013). Anaerobic oxidation of methane coupled to nitrate reduction in a novel archaeal lineage. *Nature*, 500, 567–70.

Harrison, J. P., Gheeraert, N., Tsigelnitskiy, D., and Cockell, C. S. (2013). The limits for life under multiple extremes. *Trends in Microbiology*, 21, 204–12.

Hartman, W. H. and Richardson, C. J. (2013). Differential nutrient limitation of soil microbial biomass and metabolic quotients (qCO_2): Is there a biological stoichiometry of soil microbes? *PLoS One*, 8, e57127.

Hassell, J. M., Begon, M., Ward, M. J., and Fèvre, E. M. (2017). Urbanization and disease emergence: dynamics at the wildlife–livestock–human interface. *Trends in Ecology & Evolution*, 32, 55–67.

Hayden, C. J. and Beman, J. M. (2016). Microbial diversity and community structure along a lake elevation gradient in Yosemite National Park, California, USA. *Environmental Microbiology*, 18, 1782–91.

Hayer, M., Schwartz, E., Marks, J. C., Koch, B. J., Morrissey, E. M., Schuettenberg, A. A., and Hungate, B. A. (2016). Identification of growing bacteria during litter decomposition in freshwater through $H_2{}^{18}O$ quantitative stable isotope probing. *Environmental Microbiology Reports*, 8, 975–82.

Hazen, T. C., Dubinsky, E. A., Desantis, T. Z., Andersen, G. L., Piceno, Y. M., Singh, N., Jansson, J. K., Probst, A., Borglin, S. E., Fortney, J. L., Stringfellow, W. T., Bill, M., Conrad, M. E., Tom, L. M., Chavarria, K. L., Alusi, T. R., Lamendella, R., Joyner, D. C., Spier, C., Baelum, J., Auer, M., Zemla, M. L., Chakraborty, R., Sonnenthal, E. L., D'haeseleer, P., Holman, H.-Y. N., Osman, S., Lu, Z., Van Nostrand, J. D., Deng, Y., Zhou, J., and Mason, O. U. (2010). Deep-sea oil plume enriches indigenous oil-degrading bacteria. *Science*, 330, 204–8.

He, Y., Li, M., Perumal, V., Feng, X., Fang, J., Xie, J., Sievert, S. M., and Wang, F. (2016). Genomic and enzymatic evidence for acetogenesis among multiple lineages of the archaeal phylum Bathyarchaeota widespread in marine sediments. *Nature Microbiology*, 1, 16035.

Head, I. M. (2015). Geomicrobiology of fossil fuels. In Ehrlich, H. L., Newman, D. K., and Kappler, A., eds. *Ehrlich's Geomicrobiology*, sixth ed, pp. 565–622. CRC Press, Boca Raton.

Head, I. M., Gray, N. D., and Larter, S. R. (2014). Life in the slow lane; biogeochemistry of biodegraded petroleum containing reservoirs and implications for energy recovery and carbon management. *Frontiers in Microbiology*, 5, 566.

Hedrich, S., Schlömann, M., and Johnson, D. B. (2011). The iron-oxidizing proteobacteria. *Microbiology*, 157, 1551–64.

Hengl, T., De Jesus, J. M., Macmillan, R. A., Batjes, N. H., Heuvelink, G. B. M., Ribeiro, E., Samuel-Rosa, A., Kempen, B., Leenaars, J. G. B., Walsh, M. G., and Gonzalez, M. R. (2014). SoilGrids1km—Global soil information based on automated mapping. *PLoS One*, 9, e105992.

Hershberg, R. and Petrov, D. A. (2010). Evidence that mutation is universally biased towards AT in bacteria. *PLoS Genetics*, 6, e1001115.

Hesse, C. N., Mueller, R. C., Vuyisich, M., Gallegos-Graves, L. V., Gleasner, C. D., Zak, D. R., and Kuske, C. R. (2015). Forest floor community metatranscriptomes identify fungal and bacterial responses to N deposition in two maple forests. *Frontiers in Microbiology*, 6, 337.

Heywood, J. L., Sieracki, M. E., Bellows, W., Poulton, N. J., and Stepanauskas, R. (2011). Capturing diversity of marine heterotrophic protists: one cell at a time. *The ISME Journal*, 5, 674–84.

Hinrichs, K. U., Hayes, J. M., Sylva, S. P., Brewer, P. G., and DeLong, E. F. (1999). Methane-consuming archaebacteria in marine sediments. *Nature*, 398, 802–5.

Ho, T.-Y., Quigg, A., Finkel, Z. V., Milligan, A. J., Wyman, K., Falkowski, P. G., and Morel, F. M. M. (2003). The elemental composition of some marine phytoplankton. *Journal of Phycology*, 39, 1145–59.

Hobbie, J. E., Daley, R. J., and Jasper, S. (1977). Use of Nuclepore filters for counting bacteria by fluorescence microscopy. *Applied and Environmental Microbiology*, 33, 1225–8.

Hoehler, T. M. and Jørgensen, B. B. (2013). Microbial life under extreme energy limitation. *Nature Reviews Microbiology*, 11, 83–94.

Hoffmaster, A. R., Ravel, J., Rasko, D. A., Chapman, G. D., Chute, M. D., Marston, C. K., De, B. K., Sacchi, C. T., Fitzgerald, C., Mayer, L. W., Maiden, M. C. J., Priest, F. G., Barker, M., Jiang, L., Cer, R. Z., Rilstone, J., Peterson, S. N., Weyant, R. S., Galloway, D. R., Read, T. D., Popovic, T., and Fraser, C. M. (2004). Identification of anthrax toxin genes in a *Bacillus cereus* associated with an illness resembling inhalation anthrax. *Proceedings of the National Academy of Sciences of the United States of America*, 101, 8449–54.

Högberg, P. and Read, D. J. (2006). Towards a more plant physiological perspective on soil ecology. *Trends in Ecology & Evolution*, 21, 548–54.

Hoiczyk, E. and Hansel, A. (2000). Cyanobacterial cell walls: News from an unusual prokaryotic envelope. *Journal of Bacteriology*, 182, 1191–9.

Hölldobler, B. and Wilson, E. O. (1990). *The Ants*, pp. xii, 732. Harvard University Press, Cambridge, MA.

Horn, S., Hempel, S., Verbruggen, E., Rillig, M. C., and Caruso, T. (2017). Linking the community structure of arbuscular mycorrhizal fungi and plants: a story of interdependence? *The ISME Journal*, 11, 1400–11.

Houghton, R. A. (2007). Balancing the global carbon budget. *Annual Review of Earth and Planetary Sciences*, 35, 313–47.

Howard-Varona, C., Hargreaves, K. R., Abedon, S. T., and Sullivan, M. B. (2017). Lysogeny in nature: mechanisms, impact and ecology of temperate phages. *The ISME Journal*, 11, 1511–20.

Howe, A. T., Bass, D., Vickerman, K., Chao, E. E., and Cavalier-Smith, T. (2009). Phylogeny, taxonomy, and astounding genetic diversity of Glissomonadida ord. nov., the dominant gliding zooflagellates in soil (Protozoa: Cercozoa). *Protist*, 160, 159–89.

Hu, H.-W., Chen, D., and He, J.-Z. (2015). Microbial regulation of terrestrial nitrous oxide formation: understanding the biological pathways for prediction of emission rates. *FEMS Microbiology Reviews*, 39, 729–49.

Hug, L. A., Baker, B. J., Anantharaman, K., Brown, C. T., Probst, A. J., Castelle, C. J., Butterfield, C. N., Hernsdorf, A. W., Amano, Y., Ise, K., Suzuki, Y., Dudek, N., Relman, D. A., Finstad, K. M., Amundson, R., Thomas, B. C., and Banfield, J. F. (2016). A new view of the tree of life. *Nature Microbiology*, 1, 16048.

Hughes, J. A. and Gooday, A. J. (2004). Associations between living benthic foraminifera and dead tests of *Syringammina fragilissima* (*Xenophyophorea*) in the Darwin Mounds region (NE Atlantic). *Deep Sea Research Part I: Oceanographic Research Papers*, 51, 1741–58.

Humbert, S., Zopfi, J., and Tarnawski, S.-E. (2012). Abundance of anammox bacteria in different wetland soils. *Environmental Microbiology Reports*, 4, 484–90.

Humphreys, C. P., Franks, P. J., Rees, M., Bidartondo, M. I., Leake, J. R., and Beerling, D. J. (2010). Mutualistic mycorrhiza-like symbiosis in the most ancient group of land plants. *Nature Communications*, 1, 103.

Hunt, H. W., Coleman, D. C., Ingham, E. R., Ingham, R. E., Elliott, E. T., Moore, J. C., Rose, S. L., Reid, C. P. P., and Morley, C. R. (1987). The detrital food web in a shortgrass prairie. *Biology and Fertility of Soils*, 3, 57–68.

Hutchins, D. A. and Boyd, P. W. (2016). Marine phytoplankton and the changing ocean iron cycle. *Nature Climate Change*, 6, 1072–9.

Hutchinson, G. E. (1961). The paradox of the plankton. *The American Naturalist*, 95, 137–45.

Ingall, E. D. (2010). Biogeochemistry: Phosphorus burial. *Nature Geoscience*, 3, 521–2.

Ingraham, J. L. (2010). *March of the Microbes: Sighting the Unseen*, pp. x, 326. Belknap Press of Harvard University Press, Cambridge, MA.

Ingraham, J. L., Maaloe, O., and Neidhardt, F. C. (1983). *Growth of the Bacterial Cell*, pp. xi, 435. Sinauer Assoc. Inc., Sunderland, MA.

Inselsbacher, E., Hinko-Najera Umana, N., Stange, F. C., Gorfer, M., Schuller, E., Ripka, K., Zechmeister-Boltenstern, S., Hood-Novotny, R., Strauss, J., and Wanek, W. (2010). Short-term competition between crop plants and soil microbes for inorganic N fertilizer. *Soil Biology and Biochemistry*, 42, 360–72.

Iovieno, P. and Bååth, E. (2008). Effect of drying and rewetting on bacterial growth rates in soil. *FEMS Microbiology Ecology*, 65, 400–7.

Jaakkola, S. T., Ravantti, J. J., Oksanen, H. M., and Bamford, D. H. (2016). Buried alive: microbes from ancient halite. *Trends in Microbiology*, 24, 148–60.

Jannasch, H. W., Eimhjell. K, Wirsen, C. O., and Farmanfa. A (1971). Microbial degradation of organic matter in deep sea. *Science*, 171, 672–5.

Jannasch, H. W. and Jones, G. E. (1959). Bacterial populations in sea water as determined by different methods of enumeration. *Limnology and Oceanography*, 4, 128–39.

Janssen, P. H. (2006). Identifying the dominant soil bacterial taxa in libraries of 16S rRNA and 16S rRNA genes. *Applied and Environmental Microbiology*, 72, 1719–28.

Jeong, H. J., Ha, J. H., Yoo, Y. D., Park, J. Y., Kim, J. H., Kang, N. S., Kim, T. H., Kim, H. S., and Yih, W. H. (2007). Feeding by the *Pfiesteria*-like heterotrophic dinoflagellate *Luciella masanensis*. *Journal of Eukaryotic Microbiology*, 54, 231–41.

Jeong, H. J., Kim, J. S., Park, J. Y., Kim, J. H., Kim, S., Lee, I., Lee, S. H., Ha, J. H., and Yih, W. H. (2005). *Stoeckeria algicida* n. gen., n. sp (Dinophyceae) from the coastal waters off Southern Korea: Morphology and small subunit ribosomal DNA gene sequence. *Journal of Eukaryotic Microbiology*, 52, 382–90.

Jetten, M. S. M. (2001). New pathways for ammonia conversion in soil and aquatic systems. *Plant and Soil*, 230, 9–19.

Jewell, T. N. M., Karaoz, U., Brodie, E. L., Williams, K. H., and Beller, H. R. (2016). Metatranscriptomic evidence of pervasive and diverse chemolithoautotrophy relevant to C, S, N and Fe cycling in a shallow alluvial aquifer. *The ISME Journal*, 10, 2106–17.

Jezbera, J., Jezberová, J., Koll, U., Horňák, K., Šimek, K., and Hahn, M. W. (2012). Contrasting trends in distribution of four major planktonic betaproteobacterial groups along a pH gradient of epilimnia of 72 freshwater habitats. *FEMS Microbiology Ecology*, 81, 467–79.

Ji, Q. and Ward, B. B. (2017). Nitrous oxide production in surface waters of the mid-latitude North Atlantic Ocean. *Journal of Geophysical Research: Oceans*, 122, 2612–21.

Jia, Z. J. and Conrad, R. (2009). Bacteria rather than Archaea dominate microbial ammonia oxidation in an agricultural soil. *Environmental Microbiology*, 11, 1658–71.

Jiggins, F. M. (2016). Open questions: how does *Wolbachia* do what it does? *BMC Biology*, 14, 92.

Joergensen, R. G. and Wichern, F. (2008). Quantitative assessment of the fungal contribution to microbial tissue in soil. *Soil Biology and Biochemistry*, 40, 2977–91.

Johnson, P. T. J., Stanton, D. E., Preu, E. R., Forshay, K. J., and Carpenter, S. R. (2006). Dining on disease: how interactions between infection and environment affect predation risk. *Ecology*, 87, 1973–80.

Johnston, D. T., Wolfe-Simon, F., Pearson, A., and Knoll, A. H. (2009). Anoxygenic photosynthesis modulated Proterozoic oxygen and sustained Earth's middle age. *Proceedings of the National Academy of Sciences of the United States of America*, 106, 16925–9.

Jones, R. T., Robeson, M. S., Lauber, C. L., Hamady, M., Knight, R., and Fierer, N. (2009). A comprehensive survey of soil acidobacterial diversity using pyrosequencing and clone library analyses. *The ISME Journal*, 3, 442–53.

Jonsson, P. R. (1986). Particle size selection, feeding rates and growth dynamics of marine planktonic oligotrichous ciliates (Ciliophora: Oligotrichina). *Marine Ecology Progress Series*, 33, 265–77.

Jørgensen, B. B. (2006). Bacteria and marine biogeochemistry. In Schulz, H. D. and Zabel, M., eds. *Marine Geochemistry*, pp. 173–203. Springer-Verlag, Berlin.

Jørgensen, B. B. and Des Marais, D. J. (1986). Competition for sulfide among colorless and purple sulfur bacteria in cyanobacterial mats. *FEMS Microbiology Ecology*, 2, 179–86.

Jørgensen, B. B. and Marshall, I. P. G. (2016). Slow microbial life in the seabed. *Annual Review of Marine Science*, 8, 311–32.

Kallenbach, C. M., Frey, S. D., and Grandy, A. S. (2016). Direct evidence for microbial-derived soil organic matter formation and its ecophysiological controls. *Nature Communications*, 7, 13630.

Kamble, P. N. and Bååth, E. (2016). Comparison of fungal and bacterial growth after alleviating induced N-limitation in soil. *Soil Biology and Biochemistry*, 103, 97–105.

Kamp, A., Høgslund, S., Risgaard-Petersen, N., and Stief, P. (2015). Nitrate storage and dissimilatory nitrate reduction by eukaryotic microbes. *Frontiers in Microbiology*, 6, 1492.

Kamp, A., Stief, P., Bristow, L. A., Thamdrup, B., and Glud, R. N. (2016). Intracellular nitrate of marine diatoms as a driver of anaerobic nitrogen cycling in sinking aggregates. *Frontiers in Microbiology*, 7, 1669.

Kamp, A., Stief, P., and Schulz-Vogt, H. N. (2006). Anaerobic sulfide oxidation with nitrate by a freshwater *Beggiatoa* enrichment culture. *Applied and Environmental Microbiology*, 72, 4755–60.

Kappler, A., Emerson, D., Gralnick, J. A., Roden, E. E., and Muehe, E. M. (2015). Geomicrobiology of iron. In Ehrlich, H. L., Newman, D. K., and Kappler, A., eds. *Ehrlich's Geomicrobiology*, sixth ed, pp. 344–99. CRC Press, Boca Raton.

Karner, M. B., DeLong, E. F., and Karl, D. M. (2001). Archaeal dominance in the mesopelagic zone of the Pacific Ocean. *Nature*, 409, 507–10.

Kato, S., Ohkuma, M., Powell, D. H., Krepski, S. T., Oshima, K., Hattori, M., Shapiro, N., Woyke, T., and Chan, C. S. (2015). Comparative genomic insights into ecophysiology of neutrophilic, microaerophilic iron oxidizing bacteria. *Frontiers in Microbiology*, 6, 1265.

Keller, J. K. and Bridgham, S. D. (2007). Pathways of anaerobic carbon cycling across an ombrotrophic–minerotrophic peatland gradient. *Limnology and Oceanography*, 52, 96–107.

Kellerman, A. M., Kothawala, D. N., Dittmar, T., and Tranvik, L. J. (2015). Persistence of dissolved organic matter in lakes related to its molecular characteristics. *Nature Geoscience*, 8, 454–7.

Kelly, D. P. (1999). Thermodynamic aspects of energy conservation by chemolithotrophic sulfur bacteria in relation to the sulfur oxidation pathways. *Archives of Microbiology*, 171, 219–29.

Kendall, B., Anbar, A. D., Kappler, A., and Konhauser, K. O. (2012). The global iron cycle. In Knoll, A. H., Canfield, D. E., and Konhauser, K., eds. *Fundamentals of Geobiology*, pp. 65–92. Wiley-Blackwell, Chichester, UK; Hoboken, NJ.

Kerney, R., Kim, E., Hangarter, R. P., Heiss, A. A., Bishop, C. D., and Hall, B. K. (2011). Intracellular invasion of green algae in a salamander host. *Proceedings of the National Academy of Sciences of the United States of America*, 108, 6497–502.

Kerr, B., West, J., and Bohannanm, B. J. M. (2008). Bacteriophages: models for exploring basic principles of ecology. In Abedon, S. T., ed. *Bacteriophage Ecology: Population Growth, Evolution, and Impact of Bacterial Viruses*, pp. 31–63. Cambridge University Press, Cambridge.

Khatiwala, S., Tanhua, T., Mikaloff Fletcher, S., Gerber, M., Doney, S. C., Graven, H. D., Gruber, N., Mckinley, G. A., Murata, A., Ríos, A. F., and Sabine, C. L. (2013). Global ocean storage of anthropogenic carbon. *Biogeosciences*, 10, 2169–91.

Kielak, A. M., Barreto, C. C., Kowalchuk, G. A., Van Veen, J. A., and Kuramae, E. E. (2016). The ecology of *Acidobacteria*: moving beyond genes and genomes. *Frontiers in Microbiology*, 7, 744.

Kiene, R. P. and Linn, L. J. (2000). Distribution and turnover of dissolved DMSP and its relationship with bacterial production and dimethylsulfide in the Gulf of Mexico. *Limnology and Oceanography*, 45, 849–61.

Kimura, M., Jia, Z.-J., Nakayama, N., and Asakawa, S. (2008). Ecology of viruses in soils: Past, present and future perspectives. *Soil Science & Plant Nutrition*, 54, 1–32.

King, G. M., Kostka, J. E., Hazen, T. C., and Sobecky, P. A. (2015). Microbial responses to the Deepwater Horizon oil spill: from coastal wetlands to the deep sea. *Annual Review of Marine Science*, 7, 377–401.

King, J. D. and White, D. C. (1977). Muramic acid as a measure of microbial biomass in estuarine and marine samples. *Applied and Environmental Microbiology*, 33, 777–83.

Kirchman, D. L. (2016). Growth rates of microbes in the oceans. *Annual Review of Marine Science*, 8, 285–309.

Kirchman, D. L. and Hanson, T. E. (2013). Bioenergetics of photoheterotrophic bacteria in the oceans. *Environmental Microbiology Reports*, 5, 188–99.

Kirschbaum, M. U. F. (2013). Seasonal variations in the availability of labile substrate confound the temperature dependence of organic matter decomposition. *Soil Biology and Biochemistry*, 57, 568–76.

Kirschke, S., Bousquet, P., Ciais, P., Saunois, M., Canadell, J. G., Dlugokencky, E. J., Bergamaschi, P., Bergmann, D., Blake, D. R., Bruhwiler, L., Cameron-Smith, P., Castaldi, S., Chevallier, F., Feng, L., Fraser, A., Heimann, M., Hodson, E. L., Houweling, S., Josse, B., Fraser, P. J., Krummel, P. B., Lamarque, J.-F., Langenfelds, R. L., Le Quere, C., Naik, V., O'doherty, S., Palmer, P. I., Pison, I., Plummer, D., Poulter, B., Prinn, R. G., Rigby, M., Ringeval, B., Santini, M., Schmidt, M., Shindell, D. T., Simpson, I. J., Spahni, R., Steele, L. P., Strode, S. A., Sudo, K., Szopa, S., Van Der Werf, G. R., Voulgarakis, A., Van Weele, M., Weiss, R. F., Williams, J. E., and Zeng, G. (2013). Three decades of global methane sources and sinks. *Nature Geoscience*, 6, 813–23.

Kivelson, D. and Tarjus, G. (2001). H_2O below 277 K: A novel picture. *Journal of Physical Chemistry B*, 105, 6620–7.

Klein, T., Siegwolf, R. T. W., and Körner, C. (2016). Belowground carbon trade among tall trees in a temperate forest. *Science*, 352, 342–4.

Kleindienst, S., Grim, S., Sogin, M., Bracco, A., Crespo-Medina, M., and Joye, S. B. (2016). Diverse, rare microbial taxa responded to the Deepwater Horizon deep-sea hydrocarbon plume. *The ISME Journal*, 10, 400–15.

Kleindienst, S., Seidel, M., Ziervogel, K., Grim, S., Loftis, K., Harrison, S., Malkin, S. Y., Perkins, M. J., Field, J., Sogin, M. L., Dittmar, T., Passow, U., Medeiros, P. M., and Joye, S. B. (2015). Chemical dispersants can suppress the activity of natural oil-degrading microorganisms. *Proceedings of the National Academy of Sciences of the United States of America*, 112, 14900–5.

Kleiner, M., Wentrup, C., Holler, T., Lavik, G., Harder, J., Lott, C., Littmann, S., Kuypers, M. M. M., and Dubilier, N. (2015). Use of carbon monoxide and hydrogen by a bacteria–animal symbiosis from seagrass sediments. *Environmental Microbiology*, 17, 5023–35.

Klironomos, J. N. and Hart, M. M. (2001). Food-web dynamics: Animal nitrogen swap for plant carbon. *Nature*, 410, 651–2.

Knoll, A. H., Bergmann, K. D., and Strauss, J. V. (2016). Life: the first two billion years. *Philosophical Transactions of the Royal Society B: Biological Sciences*, 371, 20150493.

Knowles, B., Bailey, B., Boling, L., Breitbart, M., Cobián-Güemes, A., Del Campo, J., Edwards, R., Felts, B., Grasis, J., Haas, A. F., Katira, P., Kelly, L. W., Luque, A., Nulton, J., Paul, L., Peters, G., Robinett, N., Sandin, S., Segall, A., Silveira, C., Youle, M., and Rohwer, F. (2017). Variability and host density independence in inductions-based estimates of environmental lysogeny. *Nature Microbiology*, 2, 17064.

Knowles, B., Silveira, C. B., Bailey, B. A., Barott, K., Cantu, V. A., Cobián-Güemes, A. G., Coutinho, F. H., Dinsdale, E. A., Felts, B., Furby, K. A., George, E. E., Green, K. T., Gregoracci, G. B., Haas, A. F., Haggerty, J. M., Hester, E. R., Hisakawa, N., Kelly, L. W., Lim, Y. W., Little, M., Luque, A., Mcdole-Somera, T., Mcnair, K., De Oliveira, L. S., Quistad, S. D., Robinett, N. L., Sala, E., Salamon, P., Sanchez, S. E., Sandin, S., Silva, G. G. Z., Smith, J., Sullivan, C., Thompson, C., Vermeij, M. J. A., Youle,

M., Young, C., Zgliczynski, B., Brainard, R., Edwards, R. A., Nulton, J., Thompson, F., and Rohwer, F. (2016). Lytic to temperate switching of viral communities. *Nature*, 531, 466–70.

Kolber, Z. S., Van Dover, C. L., Niederman, R. A., and Falkowski, P. G. (2000). Bacterial photosynthesis in surface waters of the open ocean. *Nature*, 407, 177–9.

Konhauser, K. O. (2007). *Introduction to Geomicrobiology*, 425 pp. Blackwell Publishing, Malden, MA.

Konhauser, K. O., Planavsky, N. J., Hardisty, D. S., Robbins, L. J., Warchola, T. J., Haugaard, R., Lalonde, S. V., Partin, C. A., Oonk, P. B. H., Tsikos, H., Lyons, T. W., Bekker, A., and Johnson, C. M. (2017). Iron formations: A global record of Neoarchaean to Palaeoproterozoic environmental history. *Earth-Science Reviews*, 172, 140–77.

Koonin, E. V., Dolja, V. V., and Krupovic, M. (2015). Origins and evolution of viruses of eukaryotes: The ultimate modularity. *Virology*, 479–80, 2–25.

Koonin, E. V. and Wolf, Y. I. (2008). Genomics of bacteria and archaea: the emerging dynamic view of the prokaryotic world. *Nucleic Acids Research*, 36, 6688–719.

Kozlowski, J. A., Stieglmeier, M., Schleper, C., Klotz, M. G., and Stein, L. Y. (2016). Pathways and key intermediates required for obligate aerobic ammonia-dependent chemolithotrophy in bacteria and Thaumarchaeota. *The ISME Journal*, 10, 1836–45.

Kremer, C. T., Thomas, M. K., and Litchman, E. (2017). Temperature- and size-scaling of phytoplankton population growth rates: Reconciling the Eppley curve and the metabolic theory of ecology. *Limnology and Oceanography*, 62, 1658–70.

Kristensen, D. M., Mushegian, A. R., Dolja, V. V., and Koonin, E. V. (2010). New dimensions of the virus world discovered through metagenomics. *Trends in Microbiology*, 18, 11–19.

Krupovic, M., Prangishvili, D., Hendrix, R. W., and Bamford, D. H. (2011). Genomics of bacterial and archaeal viruses: dynamics within the prokaryotic virosphere. *Microbiology and Molecular Biology Reviews*, 75, 610–35.

Kubitschek, H. E. (1969). Growth during the bacterial cell cycle: Analysis of cell size distribution. *Biophysical Journal*, 9, 792–809.

Kump, L. R. (2008). The rise of atmospheric oxygen. *Nature*, 451, 277–8.

Kunoh, T., Matsumoto, S., Nagaoka, N., Kanashima, S., Hino, K., Uchida, T., Tamura, K., Kunoh, H., and Takada, J. (2017). Amino group in *Leptothrix* sheath skeleton is responsible for direct deposition of Fe(III) minerals onto the sheaths. *Scientific Reports*, 7, 6498.

Kuzyakov, Y., Friedel, J. K., and Stahr, K. (2000). Review of mechanisms and quantification of priming effects. *Soil Biology and Biochemistry*, 32, 1485–98.

Labonte, J. M., Swan, B. K., Poulos, B., Luo, H., Koren, S., Hallam, S. J., Sullivan, M. B., Woyke, T., Eric Wommack, K., and Stepanauskas, R. (2015). Single-cell genomics-based analysis of virus–host interactions in marine surface bacterioplankton. *The ISME Journal*, 9, 2386–99.

Laforest-Lapointe, I., Paquette, A., Messier, C., and Kembel, S. W. (2017). Leaf bacterial diversity mediates plant diversity and ecosystem function relationships. *Nature*, 546, 145–7.

Lami, R., Cottrell, M. T., Ras, J., Ulloa, O., Obernosterer, I., Claustre, H., Kirchman, D. L., and Lebaron, P. (2007). High abundances of aerobic anoxygenic photosynthetic bacteria in the South Pacific Ocean. *Applied and Environmental Microbiology*, 73, 4198–205.

Landry, M. R. and Hassett, R. P. (1982). Estimating the grazing impact of marine micro-zooplankton. *Marine Biology*, 67, 283–8.

Lankiewicz, T. S., Cottrell, M. T., and Kirchman, D. L. (2016). Growth rates and rRNA content of four marine bacteria in pure cultures and in the Delaware estuary. *The ISME Journal*, 10, 823–32.

Larter, S., Wilhelms, A., Head, I., Koopmans, M., Aplin, A., Di Primio, R., Zwach, C., Erdmann, M., and Telnaes, N. (2003). The controls on the composition of biodegraded oils in the deep subsurface—part 1: biodegradation rates in petroleum reservoirs. *Organic Geochemistry*, 34, 601–13.

Lauber, C. L., Hamady, M., Knight, R., and Fierer, N. (2009). Pyrosequencing-based assessment of soil pH as a predictor of soil bacterial community structure at the continental scale. *Applied and Environmental Microbiology*, 75, 5111–20.

Lauro, F. M., Chastain, R. A., Blankenship, L. E., Yayanos, A. A., and Bartlett, D. H. (2007). The unique 16S rRNA genes of piezophiles reflect both phylogeny and adaptation. *Applied and Environmental Microbiology*, 73, 838–45.

Lauro, F. M., McDougald, D., Thomas, T., Williams, T. J., Egan, S., Rice, S., Demaere, M. Z., Ting, L., Ertan, H., Johnson, J., Ferriera, S., Lapidus, A., Anderson, I., Kyrpides, N., Munk, A. C., Detter, C., Han, C. S., Brown, M. V., Robb, F. T., Kjelleberg, S., and Cavicchioli, R. (2009). The genomic basis of trophic strategy in marine bacteria. *Proceedings of the National Academy of Sciences of the United States of America*, 106, 15527–33.

Ledin, M. (2000). Accumulation of metals by microorganisms—processes and importance for soil systems. *Earth-Science Reviews*, 51, 1–31.

Lee, C. C., Hu, Y., and Ribbe, M. W. (2010). Vanadium nitrogenase reduces CO. *Science*, 329, 642.

Lee, Z. M. and Schmidt, T. M. (2014). Bacterial growth efficiency varies in soils under different land management practices. *Soil Biology and Biochemistry*, 69, 282–90.

Leff, J. W., Nemergut, D. R., Grandy, A. S., O'Neill, S. P., Wickings, K., Townsend, A. R., and Cleveland, C. C. (2012). The effects of soil bacterial community structure on decomposition in a tropical rain forest. *Ecosystems*, 15, 284–98.

Lehmann, J. and Kleber, M. (2015). The contentious nature of soil organic matter. *Nature*, 528, 60–8.

Leininger, S., Urich, T., Schloter, M., Schwark, L., Qi, J., Nicol, G. W., Prosser, J. I., Schuster, S. C., and Schleper, C. (2006). Archaea predominate among ammonia-oxidizing prokaryotes in soils. *Nature*, 442, 806–9.

Lenhart, K., Klintzsch, T., Langer, G., Nehrke, G., Bunge, M., Schnell, S., and Keppler, F. (2016). Evidence for methane production by the marine algae *Emiliania huxleyi*. *Biogeosciences*, 13, 3163–74.

Lennon, J. T. and Jones, S. E. (2011). Microbial seed banks: the ecological and evolutionary implications of dormancy. *Nature Reviews Microbiology*, 9, 119–30.

Lenski, R. E., Wiser, M. J., Ribeck, N., Blount, Z. D., Nahum, J. R., Morris, J. J., Zaman, L., Turner, C. B., Wade, B. D., Maddamsetti, R., Burmeister, A. R., Baird, E. J., Bundy, J., Grant, N. A., Card, K. J., Rowles, M., Weatherspoon, K., Papoulis, S. E., Sullivan, R., Clark, C., Mulka, J. S., and Hajela, N. (2015). Sustained fitness gains and variability in fitness trajectories in the long-term evolution experiment with *Escherichia coli*. *Proceedings of the Royal Society of London B: Biological Sciences*, 282. DOI: 10.1098/rspb.2015.2292

Lesen, A., Juhl, A., and Anderson, R. (2010). Heterotrophic microplankton in the lower Hudson River Estuary: potential importance of naked, planktonic amebas for bacterivory and carbon flux. *Aquatic Microbial Ecology*, 61, 45–56.

Levy-Booth, D. J., Prescott, C. E., and Grayston, S. J. (2014). Microbial functional genes involved in nitrogen fixation, nitrification and denitrification in forest ecosystems. *Soil Biology and Biochemistry*, 75, 11–25.

Ley, R. E., Hamady, M., Lozupone, C., Turnbaugh, P. J., Ramey, R. R., Bircher, J. S., Schlegel, M. L., Tucker, T. A., Schrenzel, M. D., Knight, R., and Gordon, J. I. (2008). Evolution of mammals and their gut microbes. *Science*, 320, 1647–51.

Li, M., Baker, B. J., Anantharaman, K., Jain, S., Breier, J. A., and Dick, G. J. (2015). Genomic and transcriptomic evidence for scavenging of diverse organic compounds by widespread deep-sea archaea. *Nature Communications*, 6, 8933.

Lin, S., Zhang, H., Zhuang, Y., Tran, B., and Gill, J. (2010). Spliced leader-based metatranscriptomic analyses lead to recognition of hidden genomic features in dinoflagellates. *Proceedings of the National Academy of Sciences of the United States of America*, 107, 20033–8.

Lin, W., Pan, Y., and Bazylinski, D. A. (2017). Diversity and ecology of and biomineralization by magnetotactic bacteria. *Environmental Microbiology Reports*, 9, 345–56.

Lindell, D., Jaffe, J. D., Johnson, Z. I., Church, G. M., and Chisholm, S. W. (2005). Photosynthesis genes in marine viruses yield proteins during host infection. *Nature*, 438, 86–9.

Lipson, D. A. (2015). The complex relationship between microbial growth rate and yield and its implications for ecosystem processes. *Frontiers in Microbiology*, 6, 615.

Liu, J., McBride, M. J., and Subramaniam, S. (2007). Cell surface filaments of the gliding bacterium *Flavobacterium johnsoniae* revealed by cryo-electron tomography. *Journal of Bacteriology*, 189, 7503–6.

Liu, K. K. and Kaplan, I. R. (1984). Denitrification rates and availability of organic-matter in marine environments. *Earth and Planetary Science Letters*, 68, 88–100.

Liu, Z., Hu, S. K., Campbell, V., Tatters, A. O., Heidelberg, K. B., and Caron, D. A. (2017). Single-cell transcriptomics of small microbial eukaryotes: limitations and potential. *The ISME Journal*, 11, 1282–5.

Lloyd-Smith, J. O. (2017). Infectious diseases: Predictions of virus spillover across species. *Nature*, 546, 603–4.

Locey, K. J. and Lennon, J. T. (2016). Scaling laws predict global microbial diversity. *Proceedings of the National Academy of Sciences of the United States of America*, 113, 5970–5.

Logan, B. E. (1999). *Environmental Transport Processes*, 654 pp. Wiley-Interscience, New York.

Loman, N. J. and Pallen, M. J. (2015). Twenty years of bacterial genome sequencing. *Nature Reviews Microbiology*, 13, 787–94.

Lønborg, C., Middelboe, M., and Brussaard, C. D. (2013). Viral lysis of *Micromonas pusilla*: impacts on dissolved organic matter production and composition. *Biogeochemistry*, 116, 231–40.

Long, R. A. and Azam, F. (2001). Microscale patchiness of bacterioplankton assemblage richness in seawater. *Aquatic Microbial Ecology*, 26, 103–13.

López-García, P., Rodríguez-Valera, F., Pedrós-Alió, C., and Moreira, D. (2001). Unexpected diversity of small eukaryotes in deep-sea Antarctic plankton. *Nature*, 409, 603–7.

Lopez-Sangil, L., Rousk, J., Wallander, H., and Casals, P. (2011). Microbial growth rate measurements reveal that land-use abandonment promotes a fungal dominance of SOM decomposition in grazed Mediterranean ecosystems. *Biology and Fertility of Soils*, 47, 129–38.

Louis, B. P., Maron, P.-A., Viaud, V., Leterme, P., and Menasseri-Aubry, S. (2016). Soil C and N models that integrate microbial diversity. *Environmental Chemistry Letters*, 14, 331–44.

Lovley, D. R. (2017). Happy together: microbial communities that hook up to swap electrons. *The ISME Journal*, 11, 327–36.

Lozupone, C. A. and Knight, R. (2007). Global patterns in bacterial diversity. *Proceedings of the National Academy of Sciences of the United States of America*, 104, 11436–40.

Luef, B., Frischkorn, K. R., Wrighton, K. C., Holman, H.-Y. N., Birarda, G., Thomas, B. C., Singh, A., Williams, K. H., Siegerist, C. E., Tringe, S. G., Downing, K. H., Comolli, L. R., and Banfield, J. F. (2015). Diverse uncultivated ultra-small bacterial cells in groundwater. *Nature Communications*, 6, 6372.

Lundkvist, M., Grue, M., Friend, P. L., and Flindt, M. R. (2007). The relative contributions of physical and microbiological factors to cohesive sediment stability. *Continental Shelf Research*, 27, 1143–52.

Lupp, C. and Ruby, E. G. (2005). *Vibrio fischeri* uses two quorum-sensing systems for the regulation of early and late colonization factors. *Journal of Bacteriology*, 187, 3620–9.

Maaß, A. and Radek, R. (2006). The gut flagellate community of the termite *Neotermes cubanus* with special reference to *Staurojoenina* and *Trichocovina hrdyi* nov. gen. nov. sp. *European Journal of Protistology*, 42, 125–41.

Madigan, M. T., Martinko, J. M., Stahl, D. A., and Clark, D. P. (2012). *Brock Biology of Microorganisms*, 1043 pp. Benjamin Cummings, Boston, MA.

Madsen, E. L. (2016). *Environmental Microbiology: From Genomes to Biogeochemistry*, pp. ix, 577. Wiley-Blackwell, Hoboken, N.J.

Maeda, K., Spor, A., Edel-Hermann, V., Heraud, C., Breuil, M.-C., Bizouard, F., Toyoda, S., Yoshida, N., Steinberg, C., and Philippot, L. (2015). N_2O production, a widespread trait in fungi. *Scientific Reports*, 5, 9697.

Malarkey, J., Baas, J. H., Hope, J. A., Aspden, R. J., Parsons, D. R., Peakall, J., Paterson, D. M., Schindler, R. J., Ye, L., Lichtman, I. D., Bass, S. J., Davies, A. G., Manning, A. J., and Thorne, P. D. (2015). The pervasive role of biological cohesion in bedform development. *Nature Communications*, 6, 6257.

Malmstrom, R. R., Cottrell, M. T., Elifantz, H., and Kirchman, D. L. (2005). Biomass production and assimilation of dissolved organic matter by SAR11 bacteria in the Northwest Atlantic Ocean. *Applied Environmental Microbiology*, 71, 2979–86.

Mandelstam, J., McQuillen, K., and Dawes, I. (1982). *Biochemistry of Bacterial Growth*, 449 pp. John Wiley & Sons, New York.

Manzoni, S., Čapek, P., Mooshammer, M., Lindahl, B. D., Richter, A., and Šantrůčková, H. (2017). Optimal metabolic regulation along resource stoichiometry gradients. *Ecology Letters*, 20, 1182–91.

Martens-Habbena, W., Berube, P. M., Urakawa, H., De La Torre, J. R., and Stahl, D. A. (2009). Ammonia oxidation kinetics determine niche separation of nitrifying Archaea and Bacteria. *Nature*, 461, 976–9.

Martin-Creuzburg, D. and Von Elert, E. (2009). Ecological significance of sterols in aquatic food webs. In Kainz, M., Brett, M. T., and Arts, M. T., eds. *Lipids in Aquatic Ecosystems*, pp. 43–64. Springer, New York.

Martin, F., Kohler, A., Murat, C., Veneault-Fourrey, C., and Hibbett, D. S. (2016). Unearthing the roots of ectomycorrhizal symbioses. *Nature Reviews Microbiology*, 14, 760–73.

Martin, W. F. and Sousa, F. L. (2016). Early microbial evolution: The age of anaerobes. *Cold Spring Harbor Perspectives in Biology*, 8, a018127.

Martínez-García, L. B., Richardson, S. J., Tylianakis, J. M., Peltzer, D. A., and Dickie, I. A. (2015). Host identity is a dominant driver of mycorrhizal fungal community composition during ecosystem development. *New Phytologist*, 205, 1565–76.

Marzano, S.-Y. L. and Domier, L. L. (2016). Novel mycoviruses discovered from metatranscriptomics survey of soybean phyllosphere phytobiomes. *Virus Research*, 213, 332–42.

Massana, R. (2015). Getting specific: making taxonomic and ecological sense of large sequencing data sets. *Molecular Ecology*, 24, 2904–6.

Massana, R., Gobet, A., Audic, S., Bass, D., Bittner, L., Boutte, C., Chambouvet, A., Christen, R., Claverie, J.-M., Decelle, J., Dolan, J. R., Dunthorn, M., Edvardsen, B., Forn, I., Forster, D., Guillou, L., Jaillon, O., Kooistra, W. H. C. F., Logares, R., Mahé, F., Not, F., Ogata, H., Pawlowski, J., Pernice, M. C., Probert, I., Romac, S., Richards, T., Santini, S., Shalchian-Tabrizi, K., Siano, R., Simon, N., Stoeck, T., Vaulot, D., Zingone, A., and de Vargas, C. (2015). Marine protist diversity in European coastal waters and sediments as revealed by high-throughput sequencing. *Environmental Microbiology*, 17, 4035–49.

Massana, R. and Logares, R. (2013). Eukaryotic versus prokaryotic marine picoplankton ecology. *Environmental Microbiology*, 15, 1254–61

Mayer, F. (1999). Cellular and subcellular organization of prokaryotes. In Lengeler, J. W., Drews, G., and Schlegel, H. G., eds. *Biology of the Prokaryotes*, pp. 20–46. Blackwell Science, New York.

Mayumi, D., Mochimaru, H., Tamaki, H., Yamamoto, K., Yoshioka, H., Suzuki, Y., Kamagata, Y., and Sakata, S. (2016). Methane production from coal by a single methanogen. *Science*, 354, 222–5.

McAnulty, M. J., G. Poosarla, V., Kim, K.-Y., Jasso-Chávez, R., Logan, B. E., and Wood, T. K. (2017). Electricity from methane by reversing methanogenesis. *Nature Communications*, 8, 15419.

McCutcheon, John P. and Von Dohlen, Carol D. (2011). An interdependent metabolic patchwork in the nested symbiosis of mealybugs. *Current Biology*, 21, 1366–72.

McFall-Ngai, M. (2014). Divining the essence of symbiosis: insights from the squid-vibrio model. *PLoS Biology*, 12, e1001783.

McGlynn, S. E., Chadwick, G. L., Kempes, C. P., and Orphan, V. J. (2015). Single cell activity reveals direct electron transfer in methanotrophic consortia. *Nature*, 526, 531–5.

McKay, D. S., Gilson, E. K. J., Thomas-Keprta, K. L., Vali, H., Romanek, C. S., Clemett, S. J., Chillier, X. D. F., Maechling, C. R., and Zare, R. N. (1996). Search for past life on Mars: Possible relic biogenic activity in Martian meteorite ALH84001. *Science*, 273, 924–30.

McMahon, S. and Parnell, J. (2014). Weighing the deep continental biosphere. *FEMS Microbiology Ecology*, 87, 113–20.

Medina, L. E., Taylor, C. D., Pachiadaki, M. G., Henríquez-Castillo, C., Ulloa, O., and Edgcomb, V. P. (2017). A review of protist grazing below the photic zone emphasizing studies of oxygen-depleted water columns and recent applications of in situ approaches. *Frontiers in Marine Science*, 4, 105.

Medini, D., Donati, C., Tettelin, H., Masignani, V., and Rappuoli, R. (2005). The microbial pan-genome. *Current Opinion in Genetics & Development*, 15, 589–94.

Mehdiabadi, N. J., Mueller, U. G., Brady, S. G., Himler, A. G., and Schultz, T. R. (2012). Symbiont fidelity and the origin of species in fungus-growing ants. *Nature Communications*, 3, 840.

Mehdiabadi, N. J. and Schultz, T. R. (2010). Natural history and phylogeny of the fungus-farming ants (Hymenoptera: Formicidae: Myrmicinae: Attini). *Myrmecological News*, 13, 37–55.

Meisner, A., Rousk, J., and Bååth, E. (2015). Prolonged drought changes the bacterial growth response to rewetting. *Soil Biology and Biochemistry*, 88, 314–22.

Méndez-García, C., Peláez, A. I., Mesa, V., Sánchez, J., Golyshina, O. V., and Ferrer, M. (2015). Microbial diversity and metabolic networks in acid mine drainage habitats. *Frontiers in Microbiology*, 6, 475.

Miltner, A., Bombach, P., Schmidt-Brücken, B., and Kästner, M. (2012). SOM genesis: microbial biomass as a significant source. *Biogeochemistry*, 111, 41–55.

Mitchell, J. G. and Kogure, K. (2006). Bacterial motility: links to the environment and a driving force for microbial physics. *FEMS Microbiology Ecology*, 55, 3–16.

Mitra, A., Flynn, K. J., Tillmann, U., Raven, J. A., Caron, D., Stoecker, D. K., Not, F., Hansen, P. J., Hallegraeff, G., Sanders, R., Wilken, S., McManus, G., Johnson, M., Pitta, P., Våge, S., Berge, T., Calbet, A., Thingstad, F., Jeong, H. J., Burkholder, J., Glibert, P. M., Granéli, E., and Lundgren, V. (2016). Defining planktonic protist functional groups on mechanisms for energy and nutrient acquisition; incorporation of diverse mixotrophic strategies. *Protist*, 167, 106–20.

Mojica, K. D. A., Huisman, J., Wilhelm, S. W., and Brussaard, C. P. D. (2016). Latitudinal variation in virus-induced mortality of phytoplankton across the North Atlantic Ocean. *The ISME Journal*, 10, 500–13.

Mommer, L., Hinsinger, P., Prigent-Combaret, C., and Visser, E. J. W. (2016). Advances in the rhizosphere: stretching the interface of life. *Plant and Soil*, 407, 1–8.

Montagnes, D. J. S., Barbosa, A. B., Boenigk, J., Davidson, K., Jürgens, K., Macek, M., Parry, J. D., Roberts, E. C., and Simek, K. (2008). Selective feeding behaviour of key free-living protists: Avenues for continued study. *Aquatic Microbial Ecology*, 53, 83–98.

Mopper, K., Kieber, D. J., and Stubbins, A. (2015). Marine photochemistry of organic matter: processes and impacts. In Hansell, D. A. and Carlson, C. A., eds. *Biogeochemistry of Marine Dissolved Organic Matter*, Second ed, pp. 389–450. Academic Press, Boston.

Moran, M. A. (2015). The global ocean microbiome. *Science*, 350, aac8455.

Moran, M. A., Reisch, C. R., Kiene, R. P., and Whitman, W. B. (2012). Genomic insights into bacterial DMSP transformations. *Annual Review of Marine Science*, 4, 523–42.

Moran, M. A., Satinsky, B., Gifford, S. M., Luo, H., Rivers, A., Chan, L.-K., Meng, J., Durham, B. P., Shen, C., Varaljay, V. A., Smith, C. B., Yager, P. L., and Hopkinson, B. M. (2013). Sizing up metatranscriptomics. *The ISME Journal*, 7, 237–43.

Moran, N. and Baumann, P. (1994). Phylogenetics of cytoplasmically inherited microorganisms of arthropods. *Trends in Ecology & Evolution*, 9, 15–20.

Moran, N. A., McCutcheon, J. P., and Nakabachi, A. (2008). Genomics and evolution of heritable bacterial symbionts. *Annual Review of Genetics*, 42, 165–90.

Moriarty, D. J. W. (1977). Improved method using muramic acid to estimate biomass of bacteria in sediments. *Oecologia*, 26, 317–23.

Morling, K., Raeke, J., Kamjunke, N., Reemtsma, T., and Tittel, J. (2017). Tracing aquatic priming effect during microbial decomposition of terrestrial dissolved organic carbon in chemostat experiments. *Microbial Ecology*, 74, 534–49.

Morono, Y., Terada, T., Nishizawa, M., Ito, M., Hillion, F., Takahata, N., Sano, Y., and Inagaki, F. (2011). Carbon and nitrogen assimilation in deep subseafloor microbial cells. *Proceedings of the National Academy of Sciences of the United States of America*, 108, 18295–300.

Morris, B. E. L., Henneberger, R., Huber, H., and Moissl-Eichinger, C. (2013). Microbial syntrophy: interaction for the common good. *FEMS Microbiology Reviews*, 37, 384–406.

Morrissey, E. M., Mau, R. L., Schwartz, E., Caporaso, J. G., Dijkstra, P., Van Gestel, N., Koch, B. J., Liu, C. M., Hayer, M., McHugh, T. A., Marks, J. C., Price, L. B., and Hungate, B. A. (2016). Phylogenetic organization of bacterial activity. *The ISME Journal*, 10, 2336–40.

Morrissey, E. M., Mau, R. L., Schwartz, E., McHugh, T. A., Dijkstra, P., Koch, B. J., Marks, J. C., and Hungate, B. A. (2017). Bacterial carbon use plasticity, phylogenetic diversity and the priming of soil organic matter. *The ISME Journal*, 11, 1890–9.

Morse, J. W. and Mackenzie, F. T. (1990). *Geochemistry of Sedimentary Carbonates*, pp. xvi, 707. Elsevier, Amsterdam; New York.

Mosbaek, F., Kjeldal, H., Mulat, D. G., Albertsen, M., Ward, A. J., Feilberg, A., and Nielsen, J. L. (2016). Identification of syntrophic acetate-oxidizing bacteria in anaerobic digesters by combined protein-based stable isotope probing and metagenomics. *The ISME Journal*, 10, 2405–18.

Mouginot, C., Kawamura, R., Matulich, K. L., Berlemont, R., Allison, S. D., Amend, A. S., and Martiny, A. C. (2014). Elemental stoichiometry of fungi and bacteria strains from grassland leaf litter. *Soil Biology and Biochemistry*, 76, 278–85.

Moya, A., Gil, R., Latorre, A., Peretó, J., Pilar Garcillán-Barcia, M., and de la Cruz, F. (2009). Toward minimal bacterial cells: evolution vs. design. *FEMS Microbiology Reviews*, 33, 225–35.

Müller, A. L., De Rezende, J. R., Hubert, C. R. J., Kjeldsen, K. U., Lagkouvardos, I., Berry, D., Jørgensen, B. B., and Loy, A. (2013). Endospores of thermophilic bacteria as tracers of microbial dispersal by ocean currents. *The ISME Journal*, 8, 1153.

Müller, C., Laughlin, R. J., Spott, O., and Rütting, T. (2014). Quantification of N_2O emission pathways via a ^{15}N tracing model. *Soil Biology and Biochemistry*, 72, 44–54.

Müller, M., Mentel, M., Van Hellemond, J. J., Henze, K., Woehle, C., Gould, S. B., Yu, R.-Y., Van Der Giezen, M., Tielens, A. G. M., and Martin, W. F. (2012). Biochemistry and evolution of anaerobic energy metabolism in eukaryotes. *Microbiology and Molecular Biology Reviews*, 76, 444–95.

Murdoch, W. W., Briggs, C. J. and Nisbet, R. M. (2003). *Consumer-resource Dynamics*, pp. xiii, 462. Princeton University Press, Princeton.

Muyzer, G. and Stams, A. J. M. (2008). The ecology and biotechnology of sulphate-reducing bacteria. *Nature Reviews Microbiology*, 6, 441–54.

Myhre, G., Shindell, D., Bréon, F.-M., Collins, W., Fuglestvedt, J., Huang, J., Koch, D., Lamarque, J.-F., Lee, D., Mendoza, B., Nakajima, T., Robock, A., Stephens, G., Takemura, T., and Zhan, H. (2014). Anthropogenic and natural radiative forcing. In Stocker, T., Qin, D., Plattner, G.-K., Tignor, M., Allen, S. K., Boschung, J., Nauels, A., Xia, Y., Bex, V., and Midgle, P. M., eds. *Climate Change 2013: The Physical Science Basis: Working Group I Contribution to the Fifth Assessment Report of the Intergovernmental Panel on Climate Change*, pp. 659–740. Cambridge University Press, Cambridge, UK and New York.

Nealson, K. H. and Rowe, A. R. (2016). Electromicrobiology: realities, grand challenges, goals and predictions. *Microbial Biotechnology*, 9, 595–600.

Needham, D. M. and Fuhrman, J. A. (2016). Pronounced daily succession of phytoplankton, archaea and bacteria following a spring bloom. *Nature Microbiology*, 1, 16005.

Neidhardt, F. C., Ingraham, J. L., and Schaechter, M. (1990). *Physiology of the Bacterial Cell: A Molecular Approach*, p. xii, 506 pp. Sinauer Associates, Sunderland, MA.

Nekola, J. C. and White, P. S. (1999). The distance decay of similarity in biogeography and ecology. *Journal of Biogeography*, 26, 867–78.

Nemergut, D. R., Schmidt, S. K., Fukami, T., O'Neill, S. P., Bilinski, I. M., Stanish, L. F., Knelman, J. E., Darcy, J. L., Lynch, R. C., Wickey, P., and Ferrenberg, S. (2013). Patterns and processes of microbial community assembly. *Microbiology and Molecular Biology Reviews*, 77, 342–56.

Neubauer, S. C., Emerson, D., and Megonigal, J. P. (2002). Life at the energetic edge: Kinetics of circumneutral iron oxidation by lithotrophic iron-oxidizing bacteria isolated from the wetland-plant rhizosphere. *Applied and Environmental Microbiology*, 68, 3988–95.

Newton, R. J., Jones, S. E., Eiler, A., McMahon, K. D., and Bertilsson, S. (2011). A guide to the natural history of freshwater lake bacteria. *Microbiology and Molecular Biology Reviews*, 75, 14–49.

Nisbet, E. G., Dlugokencky, E. J., and Bousquet, P. (2014). Methane on the rise—again. *Science*, 343, 493–5.

Nisbet, R. E. R., Fisher, R., Nimmo, R. H., Bendall, D. S., Crill, P. M., Gallego-Sala, A. V., Hornibrook, E. R. C., López-Juez, E., Lowry, D., Nisbet, P. B. R., Shuckburgh, E. F., Sriskantharajah, S., Howe, C. J., and Nisbet, E. G. (2009). Emission of methane from plants. *Proceedings of the Royal Society B: Biological Sciences*, 276, 1347–54.

Nováková, E., Hypša, V., Klein, J., Foottit, R. G., Von Dohlen, C. D., and Moran, N. A. (2013). Reconstructing the phylogeny of aphids (Hemiptera: Aphididae) using DNA of the obligate symbiont *Buchnera aphidicola*. *Molecular Phylogenetics and Evolution*, 68, 42–54.

Offre, P., Spang, A., and Schleper, C. (2013). Archaea in biogeochemical cycles. *Annual Review of Microbiology*, 67, 437–57.

Ogawa, H., Amagai, Y., Koike, I., Kaiser, K., and Benner, R. (2001). Production of refractory dissolved organic matter by bacteria. *Science*, 292, 917–20.

Oikonomou, A., Pachiadaki, M., and Stoeck, T. (2014). Protistan grazing in a meromictic freshwater lake with anoxic bottom water. *FEMS Microbiology Ecology*, 87, 691–703.

Okamoto, N., Chantangsi, C., Horák, A., Leander, B. S., and Keeling, P. J. (2009). Molecular phylogeny and description of the novel katablepharid *Roombia truncata* gen. et sp. nov., and establishment of the Hacrobia taxon nov. *PLoS One*, 4, e7080.

Olsen, G. J. and Woese, C. R. (1993). Ribosomal RNA: a key to phylogeny. *The FASEB Journal*, 7, 113–23.

Omori, Y., Tanimoto, H., Inomata, S., Wada, S., Thume, K., and Pohnert, G. (2015). Enhancement of dimethylsulfide production by anoxic stress in natural seawater. *Geophysical Research Letters*, 42, 4047–53.

Onstott, T. C. (2017). *Deep Life: The Hunt for the Hidden Biology of Earth, Mars, and Beyond*, pp. xvii, 486. Princeton University Press, Princeton, NJ.

Orchard, V. A. and Cook, F. J. (1983). Relationship between soil respiration and soil moisture. *Soil Biology and Biochemistry*, 15, 447–53.

Oren, A. (1999). Bioenergetic aspects of halophilism. *Microbiology and Molecular Biology Reviews*, 63, 334–48.

Orphan, V. J., House, C. H., Hinrichs, K. U., McKeegan, K. D., and DeLong, E. F. (2001). Methane-consuming archaea revealed by directly coupled isotopic and phylogenetic analysis. *Science*, 293, 484–7.

Oshiki, M., Satoh, H., and Okabe, S. (2016). Ecology and physiology of anaerobic ammonium oxidizing bacteria. *Environmental Microbiology*, 18, 2784–96.

Ostfeld, R. S., Keesing, F., and Eviner, V. T. (eds.) (2008). *Infectious Disease Ecology: Effects of Ecosystems on Disease and of Disease on Ecosystems*, 504 pp. Princeton University Press, Princeton.

Ottesen, E. A., Young, C. R., Eppley, J. M., Ryan, J. P., Chavez, F. P., Scholin, C. A., and DeLong, E. F. (2013). Pattern and synchrony of gene expression among sympatric marine microbial populations. *Proceedings of the National Academy of Sciences of the United States of America*, 110, E488–497.

Ovchinnikov, S., Park, H., Varghese, N., Huang, P.-S., Pavlopoulos, G. A., Kim, D. E., Kamisetty, H., Kyrpides, N. C., and Baker, D. (2017). Protein structure determination using metagenome sequence data. *Science*, 355, 294–8.

Overmann, J., Abt, B., and Sikorski, J. (2017). Present and future of culturing bacteria. *Annual Review of Microbiology*, 71, 711–30.

Overmann, J. and Garcia-Pichel, F. (2006). The phototrophic way of life. In Dworkin, M., Falkow, S., Rosenberg, E., Schleifer, K.-H.,

and Stackebrandt, E., eds. *The Prokaryotes: Volume 2: Ecophysiology and Biochemistry*, pp. 32–85. Springer, New York.

Pacton, M., Wacey, D., Corinaldesi, C., Tangherlini, M., Kilburn, M. R., Gorin, G. E., Danovaro, R., and Vasconcelos, C. (2014). Viruses as new agents of organomineralization in the geological record. *Nature Communications*, 5, 4298.

Paerl, H. W. and Otten, T. G. (2013). Harmful cyanobacterial blooms: causes, consequences, and controls. *Microbial Ecology*, 65, 995–1010.

Paez-Espino, D., Eloe-Fadrosh, E. A., Pavlopoulos, G. A., Thomas, A. D., Huntemann, M., Mikhailova, N., Rubin, E., Ivanova, N. N., and Kyrpides, N. C. (2016). Uncovering Earth's virome. *Nature*, 536, 425–30.

Pagaling, E., Wang, H., Venables, M., Wallace, A., Grant, W. D., Cowan, D. A., Jones, B. E., Ma, Y., Ventosa, A., and Heaphy, S. (2009). Microbial biogeography of six salt lakes in Inner Mongolia, China, and a salt lake in Argentina. *Applied and Environmental Microbiology*, 75, 5750–60.

Pande, S. and Kost, C. (2017). Bacterial unculturability and the formation of intercellular metabolic networks. *Trends in Microbiology*, 25, 349–61.

Parikka, K. J., Le Romancer, M., Wauters, N., and Jacquet, S. (2017). Deciphering the virus-to-prokaryote ratio (VPR): insights into virus-host relationships in a variety of ecosystems. *Biological Reviews*, 92, 1081–100.

Parkes, R. J., Gibson, G. R., Muellerharvey, I., Buckingham, W. J., and Herbert, R. A. (1989). Determination of the substrates for sulfate-reducing bacteria within marine and estuarine sediments with different rates of sulfate reduction. *Journal of General Microbiology*, 135, 175–87.

Parniske, M. (2008). Arbuscular mycorrhiza: the mother of plant root endosymbioses. *Nature Reviews Microbiology*, 6, 763–75.

Pascal, P.-Y., Dupuy, C., Richard, P., Mallet, C., Châtelet, E. A. du, and Niquil, N. (2009). Seasonal variation in consumption of benthic bacteria by meio- and macrofauna in an intertidal mudflat. *Limnology and Oceanography*, 54, 1048–59.

Paul, E. A. (2016). The nature and dynamics of soil organic matter: Plant inputs, microbial transformations, and organic matter stabilization. *Soil Biology and Biochemistry*, 98, 109–26.

Pawlowski, J. A. N. and Burki, F. (2009). Untangling the phylogeny of amoeboid protists. *Journal of Eukaryotic Microbiology*, 56, 16–25.

Payne, J. L., Boyer, A. G., Brown, J. H., Finnegan, S., Kowalewski, M., Krause, R. A., Lyons, S. K., Mcclain, C. R., Mcshea, D. W., Novack-Gottshall, P. M., Smith, F. A., Stempien, J. A., and Wang, S. C. (2009). Two-phase increase in the maximum size of life over 3.5 billion years reflects biological innovation and environmental opportunity. *Proceedings of the National Academy of Sciences of the United States of America*, 106, 24–7.

Peay, K. G., Kennedy, P. G., and Talbot, J. M. (2016). Dimensions of biodiversity in the Earth mycobiome. *Nature Reviews Microbiology*, 14, 434–47.

Pérez, M. a. P., Moreira-Turcq, P., Gallard, H., Allard, T., and Benedetti, M. F. (2011). Dissolved organic matter dynamic in the Amazon basin: Sorption by mineral surfaces. *Chemical Geology*, 286, 158–68.

Persson, J., Fink, P., Goto, A., Hood, J. M., Jonas, J., and Kato, S. (2010). To be or not to be what you eat: regulation of stoichiometric homeostasis among autotrophs and heterotrophs. *Oikos*, 119, 741–51.

Petersen, J. M. and Dubilier, N. (2009). Methanotrophic symbioses in marine invertebrates. *Environmental Microbiology Reports*, 1, 319–35.

Petersen, J. M., Zielinski, F. U., Pape, T., Seifert, R., Moraru, C., Amann, R., Hourdez, S., Girguis, P. R., Wankel, S. D., Barbe, V., Pelletier, E., Fink, D., Borowski, C., Bach, W., and Dubilier, N. (2011). Hydrogen is an energy source for hydrothermal vent symbioses. *Nature*, 476, 176–80.

Peterson, B. F. and Scharf, M. E. (2016). Lower termite associations with microbes: synergy, protection, and interplay. *Frontiers in Microbiology*, 7, 422.

Pfeffer, C., Larsen, S., Song, J., Dong, M., Besenbacher, F., Meyer, R. L., Kjeldsen, K. U., Schreiber, L., Gorby, Y. A., El-Naggar, M. Y., Leung, K. M., Schramm, A., Risgaard-Petersen, N. and Nielsen, L. P. (2012). Filamentous bacteria transport electrons over centimetre distances. *Nature*, 491, 218–21.

Philippot, L., Andersson, S. G. E., Battin, T. J., Prosser, J. I., Schimel, J. P., Whitman, W. B., and Hallin, S. (2010). The ecological coherence of high bacterial taxonomic ranks. *Nature Reviews Microbiology*, 8, 523–9.

Pietikainen, J., Pettersson, M., and Bååth, E. (2005). Comparison of temperature effects on soil respiration and bacterial and fungal growth rates. *FEMS Microbiology Ecology*, 52, 49–58.

Pinto-Tomas, A. A., Anderson, M. A., Suen, G., Stevenson, D. M., Chu, F. S. T., Cleland, W. W., Weimer, P. J., and Currie, C. R. (2009). Symbiotic nitrogen fixation in the fungus gardens of leaf-cutter ants. *Science*, 326, 1120–3.

Pollierer, M. M., Dyckmans, J., Scheu, S., and Haubert, D. (2012). Carbon flux through fungi and bacteria into the forest soil animal food web as indicated by compound-specific ^{13}C fatty acid analysis. *Functional Ecology*, 26, 978–90.

Pomeroy, L. R. (1974). The ocean food web—a changing paradigm. *Bioscience*, 24, 499–504.

Porada, P., Weber, B., Elbert, W., Pöschl, U., and Kleidon, A. (2014). Estimating impacts of lichens and bryophytes on global biogeochemical cycles. *Global Biogeochemical Cycles*, 28, 71–85.

Pósfai, M., Lefèvre, C., Trubitsyn, D., Bazylinski, D., and Frankel, R. (2013). Phylogenetic significance of composition and crystal morphology of magnetosome minerals. *Frontiers in Microbiology*, 4, 344.

Posth, N. R., Konhauser, K. O., and Kappler, A. (2013). Microbiological processes in banded iron formation deposition. *Sedimentology*, 60, 1733–54.

Prangishvili, D. (2013). The wonderful world of archaeal viruses. *Annual Review of Microbiology*, 67, 565–85.

Prather, M. J. and Holmes, C. D. (2017). Overexplaining or under-explaining methane's role in climate change. *Proceedings of the National Academy of Sciences of the United States of America*, 114, 5324–6.

Proctor, L. M. and Fuhrman, J. A. (1990). Viral mortality of marine bacteria and cyanobacteria. *Nature*, 343, 60–2.

Prosser, J. I. and Nicol, G. W. (2012). Archaeal and bacterial ammonia-oxidisers in soil: the quest for niche specialisation and differentiation. *Trends in Microbiology*, 20, 523–31.

Purcell, E. M. (1977). Life at low Reynolds number. *American Journal of Physics*, 45, 3–11.

Qin, W., Carlson, L. T., Armbrust, E. V., Devol, A. H., Moffett, J. W., Stahl, D. A., and Ingalls, A. E. (2015). Confounding effects of oxygen and temperature on the TEX86 signature of marine Thaumarchaeota. *Proceedings of the National Academy of Sciences of the United States of America*, 112, 10979–84.

Qin, W., Meinhardt, K. A., Moffett, J. W., Devol, A. H., Virginia Armbrust, E., Ingalls, A. E., and Stahl, D. A. (2017). Influence of oxygen availability on the activities of ammonia-oxidizing archaea. *Environmental Microbiology Reports*, 9, 250–6.

Quammen, D. (2012). *Spillover: Animal Infections and the Next Human Pandemic*, 587 pp. W.W. Norton & Co., New York.

Quax, Tessa E. F., Claassens, Nico j., Söll, D., and van der Oost, J. (2015). Codon bias as a means to fine-tune gene expression. *Molecular Cell*, 59, 149–61.

Rabus, R., Venceslau, S. S., Wöhlbrand, L., Voordouw, G., Wall, J. D., and Pereira, I. A. C. (2015). A post-genomic view of the ecophysiology, catabolism and biotechnological relevance of sulphate-reducing prokaryotes. *Advances in Microbial Physiology*, 66, 55–321.

Raffel, T. R., Halstead, N. T., Mcmahon, T. A., Davis, A. K., and Rohr, J. R. (2015). Temperature variability and moisture synergistically interact to exacerbate an epizootic disease. *Proceedings of the Royal Society B: Biological Sciences*, 282, 20142039.

Ramsay, A. J. (1984). Extraction of bacteria from soil: Efficiency of shaking or ultrasonication as indicated by direct counts and autoradiography. *Soil Biology and Biochemistry*, 16, 475–81.

Randlett, D. L., Zak, D. R., Pregitzer, K. S., and Curtis, P. S. (1996). Elevated atmospheric carbon dioxide and leaf litter chemistry: Influences on microbial respiration and net nitrogen mineralization. *Soil Science Society of America Journal*, 60, 1571–7.

Ravishankara, A. R., Daniel, J. S., and Portmann, R. W. (2009). Nitrous oxide (N_2O): The dominant ozone-depleting substance emitted in the 21st century. *Science*, 326, 123–5.

Redfield, A. C. (1958). The biological control of chemical factors in the environment. *American Scientist*, 46, 205–21.

Reeburgh, W. S. (1980). Anaerobic methane oxidation: Rate depth distributions in Skan Bay sediments. *Earth and Planetary Science Letters*, 47, 345–52.

Reef, R., Ball, M. C., Feller, I. C., and Lovelock, C. E. (2010). Relationships among RNA:DNA ratio, growth and elemental stoichiometry in mangrove trees. *Functional Ecology*, 24, 1064–72.

Regaudie-De-Gioux, A. and Duarte, C. M. (2012). Temperature dependence of planktonic metabolism in the ocean. *Global Biogeochem. Cycles*, 26, GB1015.

Reischke, S., Rousk, J., and Bååth, E. (2014). The effects of glucose loading rates on bacterial and fungal growth in soil. *Soil Biology and Biochemistry*, 70, 88–95.

Repeta, D. J., Ferron, S., Sosa, O. A., Johnson, C. G., Repeta, L. D., Acker, M., DeLong, E. F., and Karl, D. M. (2016). Marine methane paradox explained by bacterial degradation of dissolved organic matter. *Nature Geoscience*, 9, 884–7.

Reyes-Prieto, A., Weber, A. P. M., and Bhattacharya, D. (2007). The origin and establishment of the plastid in algae and plants. *Annual Review of Genetics*, 41, 147–68.

Rii, Y. M., Karl, D. M., and Church, M. J. (2016). Temporal and vertical variability in picophytoplankton primary productivity in the North Pacific Subtropical Gyre. *Marine Ecology Progress Series*, 562, 1–18.

Rinke, C., Schmitz-Esser, S., Loy, A., Horn, M., Wagner, M., and Bright, M. (2009). High genetic similarity between two geographically distinct strains of the sulfur-oxidizing symbiont 'Candidatus Thiobios zoothamnicoli'. *FEMS Microbiology Ecology*, 67, 229–41.

Rinnan, R. and Bååth, E. (2009). Differential utilization of carbon substrates by bacteria and fungi in tundra soil. *Applied and Environmental Microbiology*, 75, 3611–20.

Risgaard-Petersen, N., Langezaal, A. M., Ingvardsen, S., Schmid, M. C., Jetten, M. S. M., Op Den Camp, H. J. M., Derksen, J. W. M., Pina-Ochoa, E., Eriksson, S. P., Nielsen, L. P., Revsbech, N. P., Cedhagen, T., and Van Der Zwaan, G. J. (2006). Evidence for complete denitrification in a benthic foraminifer. *Nature*, 443, 93–6.

Robson, G. D. (2017). Fungi: geoactive agents of metal and mineral transformations. *Environmental Microbiology*, 19, 2533–6.

Rocap, G., Larimer, F. W., Lamerdin, J., Malfatti, S., Chain, P., Ahlgren, N. A., Arellano, A., Coleman, M., Hauser, L., Hess, W. R., Johnson, Z. I., Land, M., Lindell, D., Post, A. F., Regala, W., Shah, M., Shaw, S. L., Steglich, C., Sullivan, M. B., Ting, C. S., Tolonen, A., Webb, E. A., Zinser, E. R., and Chisholm, S. W. (2003). Genome divergence in two *Prochlorococcus* ecotypes reflects oceanic niche differentiation. *Nature*, 424, 1042–7.

Roden, E. E. and Wetzel, R. G. (1996). Organic carbon oxidation and suppression of methane production by microbial Fe(III) oxide reduction in vegetated and unvegetated freshwater wetland sediments. *Limnology and Oceanography*, 41, 1733–48.

Rohrlack, T., Christoffersen, K., Dittmann, E., Nogueira, I., Vasconcelos, V., and Borner, T. (2005). Ingestion of microcystins by *Daphnia*: Intestinal uptake and toxic effects. *Limnology and Oceanography*, 50, 440–8.

Roller, B. R. K., Stoddard, S. F., and Schmidt, T. M. (2016). Exploiting rRNA operon copy number to investigate bacterial reproductive strategies. *Nature Microbiology*, 1, 16160.

Romero-Kutzner, V., Packard, T. T., Berdalet, E., Roy, S. O., Gagné, J. P., and Gómez, M. (2015). Respiration quotient variability: bacterial evidence. *Marine Ecology Progress Series*, 519, 47–59.

Roossinck, M. J. (2011). The good viruses: viral mutualistic symbioses. *Nature Reviews Microbiology*, 9, 99–108.

Rosenthal, L. M., Larsson, K.-H., Branco, S., Chung, J. A., Glassman, S. I., Liao, H.-L., Peay, K. G., Smith, D. P., Talbot, J. M., Taylor, J. W., Vellinga, E. C., Vilgalys, R., and Bruns, T. D. (2017). Survey of corticioid fungi in North American pinaceous forests reveals hyperdiversity, underpopulated sequence databases, and species that are potentially ectomycorrhizal. *Mycologia*, 109, 115–27.

Rousk, J. and Bååth, E. (2011). Growth of saprotrophic fungi and bacteria in soil. *FEMS Microbiology Ecology*, 78, 17–30.

Rousk, J., Brookes, P. C., and Bååth, E. (2009). Contrasting soil pH effects on fungal and bacterial growth suggest functional redundancy in carbon mineralization. *Applied and Environmental Microbiology*, 75, 1589–96.

Rousk, J., Demoling, L. A., Bahr, A., and Bååth, E. (2008). Examining the fungal and bacterial niche overlap using selective inhibitors in soil. *FEMS Microbiology Ecology*, 63, 350–8.

Rousk, J. and Frey, S. D. (2015). Revisiting the hypothesis that fungal-to-bacterial dominance characterizes turnover of soil organic matter and nutrients. *Ecological Monographs*, 85, 457–72.

Rousk, J., Smith, A. R., and Jones, D. L. (2013a). Investigating the long-term legacy of drought and warming on the soil microbial community across five European shrubland ecosystems. *Global Change Biology*, 19, 3872–84.

Rousk, K., Jones, D., and Deluca, T. (2013b). Moss–cyanobacteria associations as biogenic sources of nitrogen in boreal forest ecosystems. *Frontiers in Microbiology*, 4, 150.

Roux, S., Brum, J. R., Dutilh, B. E., Sunagawa, S., Duhaime, M. B., Loy, A., Poulos, B. T., Solonenko, N., Lara, E., Poulain, J., Pesant, S., Kandels-Lewis, S., Dimier, C., Picheral, M., Searson, S., Cruaud, C., Alberti, A., Duarte, C. M., Gasol, J. M., Vaqué, D., Tara Oceans, C., Bork, P., Acinas, S. G., Wincker, P., and Sullivan, M. B. (2016). Ecogenomics and potential biogeochemical impacts of globally abundant ocean viruses. *Nature*, 537, 689–93.

Ruby, E. G., Urbanowski, M., Campbell, J., Dunn, A., Faini, M., Gunsalus, R., Lostroh, P., Lupp, C., Mccann, J., Millikan, D., Schaefer, A., Stabb, E., Stevens, A., Visick, K., Whistler, C., and Greenberg, E. P. (2005). Complete genome sequence of *Vibrio fischeri*: A symbiotic bacterium with pathogenic congeners. *Proceedings of the National Academy of Sciences of the United States of America*, 102, 3004–9.

Ruess, L. and Chamberlain, P. M. (2010). The fat that matters: Soil food web analysis using fatty acids and their carbon stable isotope signature. *Soil Biology and Biochemistry*, 42, 1898–910.

Russell, J. B. and Rychlik, J. L. (2001). Factors that alter rumen microbial ecology. *Science*, 292, 1119–22.

Rütting, T., Boeckx, P., Müller, C., and Klemedtsson, L. (2011). Assessment of the importance of dissimilatory nitrate reduction to ammonium for the terrestrial nitrogen cycle. *Biogeosciences*, 8, 1779–91.

Saghaï, A., Zivanovic, Y., Zeyen, N., Moreira, D., Benzerara, K., Deschamps, P., Bertolino, P., Ragon, M., Tavera, R., López-Archilla, A. I., and López-García, P. (2015). Metagenome-based diversity analyses suggest a significant contribution of non-cyanobacterial lineages to carbonate precipitation in modern microbialites. *Frontiers in Microbiology*, 6, 797.

Saito, M. A., Goepfert, T. J., and Ritt, J. T. (2008). Some thoughts on the concept of colimitation: Three definitions and the importance of bioavailability. *Limnology and Oceanography*, 53, 276–90.

Salazar, G., Cornejo-Castillo, F. M., Benitez-Barrios, V., Fraile-Nuez, E., Alvarez-Salgado, X. A., Duarte, C. M., Gasol, J. M., and Acinas, S. G. (2016). Global diversity and biogeography of deep-sea pelagic prokaryotes. *The ISME Journal*, 10, 596–608.

Sánchez, O., Koblížek, M., Gasol, J. M., and Ferrera, I. (2017). Effects of grazing, phosphorus and light on the growth rates of major bacterioplankton taxa in the coastal NW Mediterranean. *Environmental Microbiology Reports*, 9, 300–9.

Sañudo-Wilhelmy, S. A., Gómez-Consarnau, L., Suffridge, C., and Webb, E. A. (2014). The role of B vitamins in marine biogeochemistry. *Annual Review of Marine Science*, 6, 339–67.

Sañudo-Wilhelmy, S. A., Kustka, A. B., Gobler, C. J., Hutchins, D. A., Yang, M., Lwiza, K., Burns, J., Capone, D. G., Raven, J. A., and Carpenter, E. J. (2001). Phosphorus limitation of nitrogen fixation by *Trichodesmium* in the central Atlantic Ocean. *Nature*, 411, 66–9.

Sarmiento, J. L. and Gruber, N. (2006). *Ocean Biogeochemical Dynamics*, p. xii, 503 pp., 8 p. of plates. Princeton University Press, Princeton, NJ.

Sarmiento, J. L. and Toggweiler, J. R. (1984). A new model for the role of oceans in determining atmospheric pCO_2. *Nature*, 308, 621–4.

Sauvadet, M., Chauvat, M., Cluzeau, D., Maron, P.-A., Villenave, C., and Bertrand, I. (2016). The dynamics of soil micro-food web structure and functions vary according to litter quality. *Soil Biology and Biochemistry*, 95, 262–74.

Schaefer, H., Fletcher, S. E. M., Veidt, C., Lassey, K. R., Brailsford, G. W., Bromley, T. M., Dlugokencky, E. J., Michel, S. E., Miller, J. B., Levin, I., Lowe, D. C., Martin, R. J., Vaughn, B. H., and White, J. W. C. (2016). A 21st century shift from fossil-fuel to biogenic methane emissions indicated by $^{13}CH_4$. *Science*, 352, 80–4.

Schlesinger, W. H. and Bernhardt, E. S. (2013). *Biogeochemistry: An Analysis of Global Change*, pp. xi, 672. Elsevier/Academic Press, Amsterdam; Boston.

Schmidtko, S., Stramma, L., and Visbeck, M. (2017). Decline in global oceanic oxygen content during the past five decades. *Nature*, 542, 335–9.

Schneider, T., Keiblinger, K. M., Schmid, E., Sterflinger-Gleixner, K., Ellersdorfer, G., Roschitzki, B., Richter, A., Eberl, L.,

Zechmeister-Boltenstern, S., and Riedel, K. (2012). Who is who in litter decomposition? Metaproteomics reveals major microbial players and their biogeochemical functions. *The ISME Journal*, 6, 1749–62.

Schott, J., Griffin, B. M., and Schink, B. (2010). Anaerobic phototrophic nitrite oxidation by *Thiocapsa* sp strain KS1 and *Rhodopseudomonas* sp strain LQ17. *Microbiology-SGM*, 156, 2428–37.

Schramm, A., De Beer, D., Wagner, M., and Amann, R. (1998). Identification and activities in situ of *Nitrosospira* and *Nitrospira* spp. as dominant populations in a nitrifying fluidized bed reactor. *Applied and Environmental Microbiology*, 64, 3480–5.

Schulz-Bohm, K., Geisen, S., Wubs, E. R. J., Song, C., De Boer, W., and Garbeva, P. (2017). The prey's scent—Volatile organic compound mediated interactions between soil bacteria and their protist predators. *The ISME Journal*, 11, 817–20.

Schulz, H. N. and Jørgensen, B. B. (2001). Big bacteria. *Annual Review of Microbiology*, 55, 105–37.

Schwartz, D. A. and Lindell, D. (2017). Genetic hurdles limit the arms race between *Prochlorococcus* and the T7-like podoviruses infecting them. *The ISME Journal*, 11, 1836–51.

Scott, J. J., Breier, J. A., Luther, G. W. III, and Emerson, D. (2015). Microbial iron mats at the Mid-Atlantic Ridge and evidence that Zetaproteobacteria may be restricted to iron-oxidizing marine systems. *PLoS One*, 10, e0119284.

Sender, R., Fuchs, S., and Milo, R. (2016). Revised estimates for the number of human and bacteria cells in the body. *PLoS Biology*, 14, e1002533.

Serna-Chavez, H. M., Fierer, N., and Van Bodegom, P. M. (2013). Global drivers and patterns of microbial abundance in soil. *Global Ecology and Biogeography*, 22, 1162–72.

Shapiro, B. J., Levade, I., Kovacikova, G., Taylor, R. K., and Almagro-Moreno, S. (2016). Origins of pandemic *Vibrio cholerae* from environmental gene pools. *Nature Microbiology*, 2, 16240.

Sherr, E. B. and Sherr, B. F. (2000). Marine microbes: An overview. In Kirchman, D. L., ed. *Microbial Ecology of the Oceans*, pp. 13–46. Wiley-Liss, New York.

Sherr, E. B. and Sherr, B. F. (2009). Capacity of herbivorous protists to control initiation and development of mass phytoplankton blooms. *Aquatic Microbial Ecology*, 57, 253–62.

Shi, L., Dong, H., Reguera, G., Beyenal, H., Lu, A., Liu, J., Yu, H.-Q., and Fredrickson, J. K. (2016). Extracellular electron transfer mechanisms between microorganisms and minerals. *Nature Reviews Microbiology*, 14, 651–62.

Singer, E., Bushnell, B., Coleman-Derr, D., Bowman, B., Bowers, R. M., Levy, A., Gies, E. A., Cheng, J.-F., Copeland, A., Klenk, H.-P., Hallam, S. J., Hugenholtz, P., Tringe, S. G., and Woyke, T. (2016). High-resolution phylogenetic microbial community profiling. *The ISME Journal*, 10, 2020–32.

Singer, P. C. and Stumm, W. (1970). Acidic mine drainage: the rate-determining step. *Science*, 167, 1121–3.

Skaar, E. P. (2010). The battle for iron between bacterial pathogens and their vertebrate hosts. *PLoS Pathogens*, 6, e1000949.

Skinner, F. A., Jones, P. C. T., and Mollison, J. E. (1952). A comparison of a direct- and a plate-counting technique for the quantitative estimation of soil micro-organisms. *Microbiology*, 6, 261–71.

Slessarev, E. W., Lin, Y., Bingham, N. L., Johnson, J. E., Dai, Y., Schimel, J. P., and Chadwick, O. A. (2016). Water balance creates a threshold in soil pH at the global scale. *Nature*, 540, 567–9.

Smil, V. (2001). *Enriching the Earth: Fritz Haber, Carl Bosch, and the Transformation of World Food Production*, p. xvii. MIT Press, Cambridge, MA.

Smith, D. R. and Keeling, P. J. (2015). Mitochondrial and plastid genome architecture: Reoccurring themes, but significant differences at the extremes. *Proceedings of the National Academy of Sciences of the United States of America*, 112, 10177–84.

Smith, S. E. and Read, D. J. (2008). *Mycorrhizal Symbiosis*, 787 pp. Academic Press, Boston, MA.

Smith, V. H. (2007). Microbial diversity–productivity relationships in aquatic ecosystems. *FEMS Microbiology Ecology*, 62, 181–6.

Sogin, M. L., Morrison, H. G., Huber, J. A., Welch, D. M., Huse, S. M., Neal, P. R., Arrieta, J. M., and Herndl, G. J. (2006). Microbial diversity in the deep sea and the underexplored "rare biosphere." *Proceedings of the National Academy of Sciences of the United States of America*, 103, 12115–20.

Sørensen, J., Jørgensen, B. B., and Revsbech, N. P. (1979). A comparison of oxygen, nitrate, and sulfate respiration in coastal marine sediments. *Microbial Ecology*, 5, 105–15.

Spang, A., Caceres, E. F., and Ettema, T. J. G. (2017). Genomic exploration of the diversity, ecology, and evolution of the archaeal domain of life. *Science*, 357, 605–9.

Spohn, M., Klaus, K., Wanek, W., and Richter, A. (2016a). Microbial carbon use efficiency and biomass turnover times depending on soil depth—Implications for carbon cycling. *Soil Biology and Biochemistry*, 96, 74–81.

Spohn, M., Pötsch, E. M., Eichorst, S. A., Woebken, D., Wanek, W., and Richter, A. (2016b). Soil microbial carbon use efficiency and biomass turnover in a long-term fertilization experiment in a temperate grassland. *Soil Biology and Biochemistry*, 97, 168–75.

Spribille, T., Tuovinen, V., Resl, P., Vanderpool, D., Wolinski, H., Aime, M. C., Schneider, K., Stabentheiner, E., Toome-Heller, M., Thor, G., Mayrhofer, H., Johannesson, H., and McCutcheon, J. P. (2016). Basidiomycete yeasts in the cortex of ascomycete macrolichens. *Science*, 353, 488–92.

Stackebrandt, E. and Goebel, B. M. (1994). A place for DNA-DNA reassociation and 16S ribosomal-RNA sequence-analysis in the present species definition in bacteriology. *International Journal of Systematic Bacteriology*, 44, 846–9.

Stahl, D. A. and De La Torre, J. R. (2012). Physiology and diversity of ammonia-oxidizing archaea. *Annual Review Of Microbiology*, 66, 83–101.

Stal, L. J. (2009). Is the distribution of nitrogen-fixing cyanobacteria in the oceans related to temperature? *Environmental Microbiology*, 11, 1632–45.

Staley, J. T. and Konopka, A. (1985). Measurement of in situ activities of nonphotosynthetic micoorganisms in aquatic and terrestrial habitats. *Annual Review of Microbiology*, 39, 321–46.

Staroscik, A. M. and Smith, D. C. (2004). Seasonal patterns in bacterioplankton abundance and production in Narragansett Bay, Rhode Island, USA. *Aquatic Microbial Ecology*, 35, 275–82.

Stefan, G., Cornelia, B., Jörg, R., and Michael, B. (2014). Soil water availability strongly alters the community composition of soil protists. *Pedobiologia*, 57, 205–13.

Stegen, G., Pasmans, F., Schmidt, B. R., Rouffaer, L. O., Van Praet, S., Schaub, M., Canessa, S., Laudelout, A., Kinet, T., Adriaensen, C., Haesebrouck, F., Bert, W., Bossuyt, F., and Martel, A. (2017). Drivers of salamander extirpation mediated by *Batrachochytrium salamandrivorans*. *Nature*, 544, 353–6.

Sterner, R. W. and Elser, J. J. (2002). *Ecological Stoichiometry: the Biology of Elements from Molecules to the Biosphere*, 439pp. Princeton University Press, Princeton, NJ.

Stevenson, A. K., Kimble, L. K., Woese, C. R., and Madigan, M. T. (1997). Characterization of new phototrophic heliobacteria and their habitats. *Photosynthesis Research*, 53, 1–12.

Stewart, F. J. and Cavanaugh, C. M. (2006). Symbiosis of thioautotrophic bacteria with *Riftia pachyptila*. In Overmann, J., ed. *Molecular Basis of Symbiosis*, pp. 197–225. Springer, Berlin; Heidelberg.

Stewart, F. J., Newton, I. L. G., and Cavanaugh, C. M. (2005). Chemosynthetic endosymbioses: adaptations to oxic–anoxic interfaces. *Trends in Microbiology*, 13, 439–48.

Stewart, P. S. and Franklin, M. J. (2008). Physiological heterogeneity in biofilms. *Nature Reviews Microbiology*, 6, 199–210.

Stoecker, D. K., Hansen, P. J., Caron, D. A., and Mitra, A. (2017). Mixotrophy in the marine plankton. *Annual Review of Marine Science*, 9, 311–35.

Stolz, J. F. (2017). Gaia and her microbiome. *FEMS Microbiology Ecology*, 93, fiw247.

Stomp, M., Huisman, J., Stal, L. J., and Matthijs, H. C. P. (2007). Colorful niches of phototrophic microorganisms shaped by vibrations of the water molecule. *The ISME Journal*, 1, 271–82.

Straza, T. R. A., Cottrell, M. T., Ducklow, H. W., and Kirchman, D. L. (2009). Geographic and phylogenetic variation in bacterial biovolume as revealed by protein and nucleic acid staining. *Applied and Environmental Microbiology*, 75, 4028–34.

Strom, S. L. (2000). Bacterivory: Interactions between bacteria and their grazers. In Kirchman, D. L., ed. *Microbial Ecology of the Oceans*, pp. 351–86. Wiley-Liss, New York.

Strous, M. and Jetten, M. S. M. (2004). Anaerobic oxidation of methane and ammonium. *Annual Review of Microbiology*, 58, 99–117.

Strous, M., Pelletier, E., Mangenot, S., Rattei, T., Lehner, A., Taylor, M. W., Horn, M., Daims, H., Bartol-Mavel, D., Wincker, P., Barbe, V., Fonknechten, N., Vallenet, D., Segurens, B., Schenowitz-Truong, C., Medigue, C., Collingro, A., Snel, B., Dutilh, B. E., Op Den Camp, H. J. M., Van Der Drift, C., Cirpus, I., Van De Pas-Schoonen, K. T., Harhangi, H. R., Van Niftrik, L., Schmid, M., Keltjens, J., Van De Vossenberg, J., Kartal, B., Meier, H., Frishman, D., Huynen, M. A., Mewes, H. W., Weissenbach, J., Jetten, M. S. M., Wagner, M., and Le Paslier, D. (2006). Deciphering the evolution and metabolism of an anammox bacterium from a community genome. *Nature*, 440, 790–4.

Stumm, W. and Morgan, J. J. (1981). *Aquatic Chemistry*, 780 pp. John Wiley & Sons, New York.

Suen, G., Teiling, C., Li, L., Holt, C., Abouheif, E., Bornberg-Bauer, E., Bouffard, P., Caldera, E. J., Cash, E., Cavanaugh, A., Denas, O., Elhaik, E., Fave, M. J., Gadau, J. R., Gibson, J. D., Graur, D., Grubbs, K. J., Hagen, D. E., Harkins, T. T., Helmkampf, M., Hu, H., Johnson, B. R., Kim, J., Marsh, S. E., Moeller, J. A., Munoz-Torres, M. C., Murphy, M. C., Naughton, M. C., Nigam, S., Overson, R., Rajakumar, R., Reese, J. T., Scott, J. J., Smith, C. R., Tao, S., Tsutsui, N. D., Viljakainen, L., Wissler, L., Yandell, M. D., Zimmer, F., Taylor, J., Slater, S. C., Clifton, S. W., Warren, W. C., Elsik, C. G., Smith, C. D., Weinstock, G. M., Gerardo, N. M., and Currie, C. R. (2011). The genome sequence of the leaf-cutter ant *Atta cephalotes* reveals insights into its obligate symbiotic lifestyle. *PLoS Genetics*, 7, e1002007.

Suh, S.-O., Noda, H., and Blackwell, M. (2001). Insect symbiosis: derivation of yeast-like endosymbionts within an entomopathogenic filamentous lineage. *Molecular Biology and Evolution*, 18, 995–1000.

Sunagawa, S., Coelho, L. P., Chaffron, S., Kultima, J. R., Labadie, K., Salazar, G., Djahanschiri, B., Zeller, G., Mende, D. R., Alberti, A., Cornejo-Castillo, F. M., Costea, P. I., Cruaud, C., d'Ovidio, F., Engelen, S., Ferrera, I., Gasol, J. M., Guidi, L., Hildebrand, F., Kokoszka, F., Lepoivre, C., Lima-Mendez, G., Poulain, J., Poulos, B. T., Royo-Llonch, M., Sarmento, H., Vieira-Silva, S., Dimier, C., Picheral, M., Searson, S., Kandels-Lewis, S., Tara Ocean Coordinators Bowler, C., de Vargas, C., Gorsky, G., Grimsley, N., Hingamp, P., Iudicone, D., Jaillon, O., Not, F., Ogata, H., Pesant, S., Speich, S., Stemmann, L., Sullivan, M. B., Weissenbach, J., Wincker, P., Karsenti, E., Raes, J., Acinas, S. G., and Bork, P. (2015). Structure and function of the global ocean microbiome. *Science*, 348, 1261359.

Sunda, W., Kieber, D. J., Kiene, R. P., and Huntsman, S. (2002). An antioxidant function for DMSP and DMS in marine algae. *Nature*, 418, 317–20.

Sundquist, E. T. and Visser, K. (2004). The geological history of the carbon cycle. In Holland, H. D. and Turekian, K. K., eds. *Biogeochemistry*, 1st ed, pp. 425–72. Elsevier Pergamon, Amsterdam; Boston.

Suosaari, E. P., Reid, R. P., Playford, P. E., Foster, J. S., Stolz, J. F., Casaburi, G., Hagan, P. D., Chirayath, V., Macintyre, I. G., Planavsky, N. J., and Eberli, G. P. (2016). New multi-scale perspectives on the stromatolites of Shark Bay, Western Australia. *Scientific Reports*, 6, 20557.

Suseela, V., Conant, R. T., Wallenstein, M. D., and Dukes, J. S. (2012). Effects of soil moisture on the temperature sensitivity of heterotrophic respiration vary seasonally in an old-field climate change experiment. *Global Change Biology*, 18, 336–48.

Suttle, C. A. (2005). Viruses in the sea. *Nature*, 437, 356–61.

Suttle, C. A. and Chen, F. (1992). Mechanisms and rates of decay of marine viruses in seawater. *Applied and Environmental Microbiology*, 58, 3721–9.

Suzuki, H., Macdonald, J., Syed, K., Salamov, A., Hori, C., Aerts, A., Henrissat, B., Wiebenga, A., Vankuyk, P. A., Barry, K., Lindquist, E., Labutti, K., Lapidus, A., Lucas, S., Coutinho, P., Gong, Y., Samejima, M., Mahadevan, R., Abou-Zaid, M., De Vries, R. P., Igarashi, K., Yadav, J. S., Grigoriev, I. V., and Master, E. R. (2012). Comparative genomics of the white-rot fungi, *Phanerochaete carnosa* and *P. chrysosporium*, to elucidate the genetic basis of the distinct wood types they colonize. *BMC Genomics*, 13, 444.

Swan, B. K., Tupper, B., Sczyrba, A., Lauro, F. M., Martinez-Garcia, M., González, J. M., Luo, H., Wright, J. J., Landry, Z. C., Hanson, N. W., Thompson, B. P., Poulton, N. J., Schwientek, P., Acinas, S. G., Giovannoni, S. J., Moran, M. A., Hallam, S. J., Cavicchioli, R., Woyke, T., and Stepanauskas, R. (2013). Prevalent genome streamlining and latitudinal divergence of planktonic bacteria in the surface ocean. *Proceedings of the National Academy of Sciences of the United States of America*, 110, 11463–8.

Tabita, F. R., Hanson, T. E., Li, H. Y., Satagopan, S., Singh, J., and Chan, S. (2007). Function, structure, and evolution of the RubisCO-like proteins and their RubisCO homologs. *Microbiology and Molecular Biology Reviews*, 71, 576–99.

Tagliabue, A., Bowie, A. R., Boyd, P. W., Buck, K. N., Johnson, K. S., and Saito, M. A. (2017). The integral role of iron in ocean biogeochemistry. *Nature*, 543, 51–9.

Taipale, S. J., Brett, M. T., Pulkkinen, K., and Kainz, M. J. (2012). The influence of bacteria-dominated diets on *Daphnia magna* somatic growth, reproduction, and lipid composition. *FEMS Microbiology Ecology*, 82, 50–62.

Tamburini, C., Boutrif, M., Garel, M., Colwell, R. R., and Deming, J. W. (2013). Prokaryotic responses to hydrostatic pressure in the ocean—a review. *Environmental Microbiology*, 15, 1262–74.

Tanay, A. and Regev, A. (2017). Scaling single-cell genomics from phenomenology to mechanism. *Nature*, 541, 331–8.

Tarao, M., Jezbera, J., and Hahn, M. W. (2009). Involvement of cell surface structures in size-independent grazing resistance of freshwater Actinobacteria. *Applied and Environmental Microbiology*, 75, 4720–6.

Tardy, V., Mathieu, O., Lévêque, J., Terrat, S., Chabbi, A., Lemanceau, P., Ranjard, L., and Maron, P.-A. (2014). Stability of soil microbial structure and activity depends on microbial diversity. *Environmental Microbiology Reports*, 6, 173–83.

Taylor, M. W., Radax, R., Steger, D., and Wagner, M. (2007). Sponge-associated microorganisms: Evolution, ecology, and biotechnological potential. *Microbiology and Molecular Biology Reviews*, 71, 295–347.

Tebo, B. M., Geszvain, K., and Lee, S.-W. (2010). The molecular geomicrobiology of bacterial manganese(II) oxidation. In Barton, L. L., Mandl, M., and Loy, A., eds. *Geomicrobiology: Molecular and Environmental Perspective*, pp. 285–308. Springer Netherlands, Dordrecht.

Tecon, R. and Or, D. (2017). Biophysical processes supporting the diversity of microbial life in soil. *FEMS Microbiology Reviews*, 41, 599–623.

Tedersoo, L., Bahram, M., Põlme, S., Kõljalg, U., Yorou, N. S., Wijesundera, R., Ruiz, L. V., Vasco-Palacios, A. M., Thu, P. Q., Suija, A., Smith, M. E., Sharp, C., Saluveer, E., Saitta, A., Rosas, M., Riit, T., Ratkowsky, D., Pritsch, K., Põldmaa, K., Piepenbring, M., Phosri, C., Peterson, M., Parts, K., Pärtel, K., Otsing, E., Nouhra, E., Njouonkou, A. L., Nilsson, R. H., Morgado, L. N., Mayor, J., May, T. W., Majuakim, L., Lodge, D. J., Lee, S. S., Larsson, K.-H., Kohout, P., Hosaka, K., Hiiesalu, I., Henkel, T. W., Harend, H., Guo, L.-D., Greslebin, A., Grelet, G., Geml, J., Gates, G., Dunstan, W., Dunk, C., Drenkhan, R., Dearnaley, J., De Kesel, A., Dang, T., Chen, X., Buegger, F., Brearley, F. Q., Bonito, G., Anslan, S., Abell, S., and Abarenkov, K. (2014). Global diversity and geography of soil fungi. *Science*, 346, 1078.

ter Kuile, B. H. and Bonilla, Y. (1999). Influence of growth conditions on RNA levels in relation to activity of core metabolic enzymes in the parasitic protists *Trypanosoma brucei* and *Trichomonas vaginalis*. *Microbiology*, 145, 755–65.

Thamdrup, B. (2012). New pathways and processes in the global nitrogen cycle. *Annual Review of Ecology, Evolution, and Systematics*, 43, 407–28.

Thauer, R. K., Jungermann, K., and Decker, K. (1977). Energy conservation in chemotrophic anaerobic bacteria. *Bacteriological Reviews*, 41, 100–80.

Thingstad, T. F. (2000). Elements of a theory for the mechanisms controlling abundance, diversity, and biogeochemical role of lytic bacterial viruses in aquatic systems. *Limnology and Oceanography*, 45, 1320–8.

Thingstad, T. F., Våge, S., Storesund, J. E., Sandaa, R.-A., and Giske, J. (2014). A theoretical analysis of how strain-specific viruses can control microbial species diversity. *Proceedings of the National Academy of Sciences of the United States of America*, 111, 7813–18.

Thompson, A. W., Foster, R. A., Krupke, A., Carter, B. J., Musat, N., Vaulot, D., Kuypers, M. M. M., and Zehr, J. P. (2012). Unicellular cyanobacterium symbiotic with a single-celled eukaryotic alga. *Science*, 337, 1546–50.

Thomsen, U., Thamdrup, B., Stahl, D. A., and Canfield, D. E. (2004). Pathways of organic carbon oxidation in a deep lacustrine sediment, Lake Michigan. *Limnology and Oceanography*, 49, 2046–57.

Thyrhaug, R., Larsen, A., Thingstad, T. F., and Bratbak, G. (2003). Stable coexistence in marine algal host–virus systems. *Marine Ecology Progress Series*, 254, 27–35.

Tian, J. H., Pourcher, A. M., and Peu, P. (2016). Isolation of bacterial strains able to metabolize lignin and lignin-related compounds. *Letters in Applied Microbiology*, 63, 30–7.

Timmers, P. H. A., Welte, C. U., Koehorst, J. J., Plugge, C. M., Jetten, M. S. M., and Stams, A. J. M. (2017). Reverse methanogenesis and respiration in methanotrophic archaea. *Archaea*, 2017, 22.

Torrella, F. and Morita, R. Y. (1979). Evidence by electron micrographs for a high incidence of bacteriophage particles in the waters of Yaquina Bay, Oregon: ecological and taxonomical implications. *Applied and Environmental Microbiology*, 37, 774–8.

Tranvik, L. J., Downing, J. A., Cotner, J. B., Loiselle, S. A., Striegl, R. G., Ballatore, T. J., Dillon, P., Finlay, K., Fortino, K., Knoll, L. B., Kortelainen, P. L., Kutser, T., Larsen, S., Laurion, I., Leech, D. M., McCallister, S. L., McKnight, D. M., Melack, J. M., Overholt, E., Porter, J. A., Prairie, Y., Renwick, W. H., Roland, F., Sherman, B. S., Schindler, D. W., Sobek, S., Tremblay, A., Vanni, M. J., Verschoor, A. M., Von Wachenfeldt, E., and Weyhenmeyer, G. A. (2009). Lakes and reservoirs as regulators of carbon cycling and climate. *Limnology and Oceanography*, 54, 2298–314.

Tremblay, J., Singh, K., Fern, A., Kirton, E., He, S., Woyke, T., Lee, J., Chen, F., Dangl, J., and Tringe, S. (2015). Primer and platform effects on 16S rRNA tag sequencing. *Frontiers in Microbiology*, 6, 771.

Tremblay, J., Yergeau, E., Fortin, N., Cobanli, S., Elias, M., King, T. L., Lee, K., and Greer, C. W. (2017). Chemical dispersants enhance the activity of oil- and gas condensate-degrading marine bacteria. *The ISME Journal*, 11, 2793–808.

Trumbore, S. (2009). Radiocarbon and soil carbon dynamics. *Annual Review of Earth and Planetary Sciences*, 37, 47–66.

Tsuchida, T., Koga, R., Horikawa, M., Tsunoda, T., Maoka, T., Matsumoto, S., Simon, J.-C., and Fukatsu, T. (2010). Symbiotic bacterium modifies aphid body color. *Science*, 330, 1102–4.

Tveit, A. T., Urich, T., and Svenning, M. M. (2014). Metatranscriptomic analysis of arctic peat soil microbiota. *Applied and Environmental Microbiology*, 80, 5761–72.

Tyrrell, T. (2013). *On Gaia: A Critical Investigation of the Relationship Between Life and Earth*, 311 pp. Princeton University Press, Princeton, NJ.

Udvardi, M. and Poole, P. S. (2013). Transport and metabolism in legume–rhizobia symbioses. *Annual Review of Plant Biology*, 64, 781–805.

Unrein, F., Gasol, J. M., Not, F., Forn, I., and Massana, R. (2014). Mixotrophic haptophytes are key bacterial grazers in oligotrophic coastal waters. *The ISME Journal*, 8, 164–76.

Urich, T., Lanzén, A., Qi, J., Huson, D. H., Schleper, C., and Schuster, S. C. (2008). Simultaneous assessment of soil microbial community structure and function through analysis of the meta-transcriptome. *PLoS One*, 3, e2527.

Uroz, S., Calvaruso, C., Turpault, M.-P., and Frey-Klett, P. (2009). Mineral weathering by bacteria: ecology, actors and mechanisms. *Trends in Microbiology*, 17, 378–87.

Vacher, C., Hampe, A., Porté, A. J., Sauer, U., Compant, S., and Morris, C. E. (2016). The phyllosphere: microbial jungle at the plant–climate interface. *Annual Review of Ecology, Evolution, and Systematics*, 47, 1–24.

Våge, S., Storesund, J. E., and Thingstad, T. F. (2013). SAR11 viruses and defensive host strains. *Nature*, 499, E3–E4.

Valdemarsen, T. and Kristensen, E. (2010). Degradation of dissolved organic monomers and short-chain fatty acids in sandy marine sediment by fermentation and sulfate reduction. *Geochimica et Cosmochimica Acta*, 74, 1593–605.

Valentine, D. L. (2007). Adaptations to energy stress dictate the ecology and evolution of the Archaea. *Nature Reviews Microbiology*, 5, 316.

Valiela, I. (2015). *Marine Ecological Processes*, 698 pp. Springer, New York.

Vallina, S. M., Follows, M. J., Dutkiewicz, S., Montoya, J. M., Cermeno, P., and Loreau, M. (2014). Global relationship between phytoplankton diversity and productivity in the ocean. *Nature Communications*, 5, 5299.

van der Gast, C. J. (2015). Microbial biogeography: The end of the ubiquitous dispersal hypothesis? *Environmental Microbiology*, 17, 544–6.

van der Wal, A., Geydan, T. D., Kuyper, T. W., and De Boer, W. (2013). A thready affair: linking fungal diversity and community dynamics to terrestrial decomposition processes. *FEMS Microbiology Reviews*, 37, 477–94.

van Groenigen, K.-J., Bloem, J., Bååth, E., Boeckx, P., Rousk, J., Bodé, S., Forristal, D., and Jones, M. B. (2010). Abundance, production and stabilization of microbial biomass under conventional and reduced tillage. *Soil Biology and Biochemistry*, 42, 48–55.

Van Hamme, J. D., Singh, A., and Ward, O. P. (2003). Recent advances in petroleum microbiology. *Microbiology and Molecular Biology Reviews*, 67, 503–49.

van Kessel, M. a. H. J., Speth, D. R., Albertsen, M., Nielsen, P. H., Op Den Camp, H. J. M., Kartal, B., Jetten, M. S. M., and Lücker, S. (2015). Complete nitrification by a single microorganism. *Nature*, 528, 555–9.

Van Mooy, B. A. S., Fredricks, H. F., Pedler, B. E., Dyhrman, S. T., Karl, D. M., Koblížek, M., Lomas, M. W., Mincer, T. J., Moore, L. R., Moutin, T., Rappe, M. S., and Webb, E. A. (2009). Phytoplankton in the ocean use non-phosphorus lipids in response to phosphorus scarcity. *Nature*, 458, 69–72.

Vandenkoornhuyse, P., Ridgway, K. P., Watson, I. J., Fitter, A. H., and Young, J. P. W. (2003). Co-existing grass species have distinctive arbuscular mycorrhizal communities. *Molecular Ecology*, 12, 3085–95.

Vandieken, V. and Thamdrup, B. (2013). Identification of acetate-oxidizing bacteria in a coastal marine surface sediment by RNA-stable isotope probing in anoxic slurries and intact cores. *FEMS Microbiology Ecology*, 84, 373–86.

Vardi, A., Van Mooy, B. A. S., Fredricks, H. F., Popendorf, K. J., Ossolinski, J. E., Haramaty, L., and Bidle, K. D. (2009). Viral glycosphingolipids induce lytic infection and cell death in marine phytoplankton. *Science*, 326, 861–5.

Venter, J. C., Remington, K., Heidelberg, J. F., Halpern, A. L., Rusch, D., Eisen, J. A., Wu, D., Paulsen, I., Nelson, K. E., Nelson, W., Fouts, D. E., Levy, S., Knap, A. H., Lomas, M. W., Nealson, K., White, O., Peterson, J., Hoffman, J., Parsons, R., Baden-Tillson, H., Pfannkoch, C., Rogers, Y.-H., and Smith, H. O. (2004). Environmental genome shotgun sequencing of the Sargasso Sea. *Science*, 304, 66–74.

Vieira-Silva, S. and Rocha, E. P. C. (2010). The systemic imprint of growth and its uses in ecological (meta)genomics. *PLoS Genetics*, 6, e1000808.

Visscher, P. T. and Stolz, J. F. (2005). Microbial mats as bioreactors: populations, processes, and products. *Palaeogeography, Palaeoclimatology, Palaeoecology*, 219, 87–100.

Visser, P. M., Verspagen, J. M. H., Sandrini, G., Stal, L. J., Matthijs, H. C. P., Davis, T. W., Paerl, H. W., and Huisman, J. (2016). How rising CO_2 and global warming may stimulate harmful cyanobacterial blooms. *Harmful Algae*, 54, 145–59.

Viviani, D. A. and Church, M. J. (2017). Decoupling between bacterial production and primary production over multiple time scales in the North Pacific Subtropical Gyre. *Deep Sea Research Part I: Oceanographic Research Papers*, 121, 132–42.

Vollrath, S., Behrends, T., and Van Cappellen, P. (2012). Oxygen dependency of neutrophilic Fe(II) oxidation by *Leptothrix* differs from abiotic reaction. *Geomicrobiology Journal*, 29, 550–60.

von Dassow, P., Petersen, T. W., Chepurnov, V. A., and Armbrust, E. V. (2008). Inter-and intraspecific relationships between nuclear DNA content and cell size in selected members of the centric diatom genus *Thalassiosira* (Bacillariophyceae). *Journal of Phycology*, 44, 335–49.

von Elert, E., Martin-Creuzburg, D., and Le Coz, J. R. (2003). Absence of sterols constrains carbon transfer between cyanobacteria and a freshwater herbivore (*Daphnia galeata*). *Proceedings of the Royal Society: Biological Sciences*, 270, 1209–14.

Voroney, R. P. and Heck, R. J. (2015). The soil habitat. In Paul, E. A., ed. *Soil Microbiology, Ecology, and Biochemistry*, third edn., pp. 15–39. Elsevier, Amsterdam.

Voroney, R. P., Paul, E. A., and Anderson, D. W. (1989). Decomposition of wheat straw and stabilization of microbial products. *Canadian Journal of Soil Science*, 69, 63–77.

Vreeland, R. H., Rosenzweig, W. D., and Powers, D. W. (2000). Isolation of a 250 million-year-old halotolerant bacterium from a primary salt crystal. *Nature*, 407, 897–900.

Waldrop, M. P., Zak, D. R., Blackwood, C. B., Curtis, C. D., and Tilman, D. (2006). Resource availability controls fungal diversity across a plant diversity gradient. *Ecology Letters*, 9, 1127–35.

Wang, G., Jagadamma, S., Mayes, M. A., Schadt, C. W., Megan Steinweg, J., Gu, L., and Post, W. M. (2015). Microbial dormancy improves development and experimental validation of ecosystem model. *The ISME Journal*, 9, 226–37.

Wang, S., Radny, D., Huang, S., Zhuang, L., Zhao, S., Berg, M., Jetten, M. S. M., and Zhu, G. (2017). Nitrogen loss by anaerobic ammonium oxidation in unconfined aquifer soils. *Scientific Reports*, 7, 40173.

Ward, B. B. and Van Oostende, N. (2016). Phytoplankton assemblage during the North Atlantic spring bloom assessed from functional gene analysis. *Journal of Plankton Research*, 38, 1135–50.

Wasmund, K., Mußmann, M., and Loy, A. (2017). The life sulfuric: microbial ecology of sulfur cycling in marine sediments. *Environmental Microbiology Reports*, 9, 323–44.

Watson, S. W., Novitsky, T. J., Quinby, H. L., and Valois, F. W. (1977). Determination of bacterial number and biomass in the marine environment. *Applied and Environmental Microbiology*, 33, 940–6.

Weast, R. C. (ed.) 1987. *CRC Handbook of Chemistry and Physics*. CRC Press. Boca Raton.

Wegner, C.-E. and Liesack, W. (2016). Microbial community dynamics during the early stages of plant polymer breakdown in paddy soil. *Environmental Microbiology*, 18, 2825–42.

Weinbauer, M. G. and Höfle, M. G. (1998). Size-specific mortality of lake bacterioplankton by natural virus communities. *Aquatic Microbial Ecology*, 15, 103–13.

Weisgall, J. M. (1994). *Operation Crossroads: The Atomic Tests at Bikini Atoll*, pp. xvii, 415. Naval Institute Press, Annapolis, MD.

Welch, R. A., Burland, V., Plunkett, G., Redford, P., Roesch, P., Rasko, D., Buckles, E. L., Liou, S. R., Boutin, A., Hackett, J., Stroud, D., Mayhew, G. F., Rose, D. J., Zhou, S., Schwartz, D. C., Perna, N. T., Mobley, H. L. T., Donnenberg, M. S., and Blattner, F. R. (2002). Extensive mosaic structure revealed by the complete genome sequence of uropathogenic *Escherichia coli*. *Proceedings of the National Academy of Sciences of the United States of America*, 99, 17020–4.

Welch, S. A., Barker, W. W., and Banfield, J. F. (1999). Microbial extracellular polysaccharides and plagioclase dissolution. *Geochimica et Cosmochimica Acta*, 63, 1405–19.

Welch, S. A. and Ullman, W. J. (1993). The effect of organic acids on plagioclase dissolution rates and stoichiometry. *Geochimica et Cosmochimica Acta*, 57, 2725–36.

Whitman, W. B., Boone, D. R., Koga, Y., and Keswani, J. (2001). Taxonomy of methanogenic *Archaea*. In Boone, D. R., Castenholz, R. W., and Garrity, G. M., eds. *Bergey's Manual of Systematic Bacteriology*, pp. 211–13. Springer, New York.

Whitman, W. B., Coleman, D. C., and Wiebe, W. J. (1998). Prokaryotes: The unseen majority. *Proceedings of the National Academy of Sciences of the United States of America*, 95, 6578–83.

Whittaker, R. H. (1969). New concepts of kingdoms of organisms. *Science*, 163, 150–60.

Wier, A. M., Sacchi, L., Dolan, M. F., Bandi, C., Macallister, J., and Margulis, L. (2010). Spirochete attachment ultrastructure: Implications for the origin and evolution of cilia. *Biological Bulletin*, 218, 25–35.

Wigington, C. H., Sonderegger, D., Brussaard, C. P. D., Buchan, A., Finke, J. F., Fuhrman, J. A., Lennon, J. T., Middelboe, M., Suttle, C. A., Stock, C., Wilson, W. H., Wommack, K. E., Wilhelm, S. W., and Weitz, J. S. (2016). Re-examination of the relationship between marine virus and microbial cell abundances. *Nature Microbiology*, 1, 15024.

Wilhelm, S. W., Brigden, S. M., and Suttle, C. A. (2002). A dilution technique for the direct measurement of viral production: A comparison in stratified and tidally mixed coastal waters. *Microbial Ecology*, 43, 168–73.

Williams, K. H., Bargar, J. R., Lloyd, J. R., and Lovley, D. R. (2013). Bioremediation of uranium-contaminated groundwater: a systems approach to subsurface biogeochemistry. *Current Opinion in Biotechnology*, 24, 489–97.

Williams, P. J. L. (2000). Heterotrophic bacteria and the dynamics of dissolved organic material. In Kirchman, D. L., ed. *Microbial Ecology of the Oceans*, pp. 153–200. Wiley-Liss, New York.

Williamson, K. E., Radosevich, M., Smith, D. W., and Wommack, K. E. (2007). Incidence of lysogeny within temperate and extreme soil environments. *Environmental Microbiology*, 9, 2563–74.

Williamson, S. J., Houchin, L. A., Mcdaniel, L., and Paul, J. H. (2002). Seasonal variation in lysogeny as depicted by prophage induction in Tampa Bay, Florida. *Applied and Environmental Microbiology*, 68, 4307–14.

Winkelmann, M., Hunger, N., Hüttl, R., and Wolf, G. (2009). Calorimetric investigations on the degradation of water insoluble hydrocarbons by the bacterium *Rhodococcus opacus* 1CP. *Thermochimica Acta*, 482, 12–16.

Wirtz, K. W. (2012). Who is eating whom? Morphology and feeding type determine the size relation between planktonic predators and their ideal prey. *Marine Ecology Progress Series*, 445, 1–12.

Woese, C. R. and Fox, G. E. (1977). Phylogenetic structure of prokaryotic domain—Primary kingdoms. *Proceedings of the National Academy of Sciences of the United States of America*, 74, 5088–90.

Wommack, K. E. and Colwell, R. R. (2000). Virioplankton: Viruses in aquatic ecosystems. *Microbiology and Molecular Biology Reviews*, 64, 69–114.

Wootton, E. C., Zubkov, M. V., Jones, D. H., Jones, R. H., Martel, C. M., Thornton, C. A., and Roberts, E. C. (2007). Biochemical prey recognition by planktonic protozoa. *Environmental Microbiology*, 9, 216–22.

Worden, A. Z. and Not, F. (2008). Ecology and diversity of picoeukaryotes. In Kirchman, D. L., ed. *Microbial Ecology of the Oceans*, Second ed, pp. 159–205. John Wiley & Sons, Hoboken, NJ.

Wright, P. A. (1995). Nitrogen excretion: three end products, many physiological roles. *The Journal of Experimental Biology*, 198, 273–81.

Wu, D., Hugenholtz, P., Mavromatis, K., Pukall, R., Dalin, E., Ivanova, N. N., Kunin, V., Goodwin, L., Wu, M., Tindall, B. J., Hooper, S. D., Pati, A., Lykidis, A., Spring, S., Anderson, I. J., D'haeseleer, P., Zemla, A., Singer, M., Lapidus, A., Nolan, M., Copeland, A., Han, C., Chen, F., Cheng, J.-F., Lucas, S., Kerfeld, C., Lang, E., Gronow, S., Chain, P., Bruce, D., Rubin, E. M., Kyrpides, N. C., Klenk, H.-P., and Eisen, J. A. (2009). A phylogeny-driven genomic encyclopaedia of Bacteria and Archaea. *Nature*, 462, 1056–60.

Wu, W., Logares, R., Huang, B., and Hsieh, C.-H. (2017). Abundant and rare picoeukaryotic sub-communities present contrasting patterns in the epipelagic waters of marginal seas in the northwestern Pacific Ocean. *Environmental Microbiology*, 19, 287–300.

Wymore, A. S., Compson, Z. G., Liu, C. M., Price, L. B., Whitham, T. G., Keim, P., and Marks, J. C. (2013). Contrasting rRNA gene abundance patterns for aquatic fungi and bacteria in response to leaf-litter chemistry. *Freshwater Science*, 32, 663–72.

Xi, D., Bai, R., Zhang, L., and Fang, Y. (2016). Contribution of anammox to nitrogen removal in two temperate forest soils. *Applied and Environmental Microbiology*, 82, 4602–12.

Yang, B., Wang, Y., and Qian, P.-Y. (2016). Sensitivity and correlation of hypervariable regions in 16S rRNA genes in phylogenetic analysis. *BMC Bioinformatics*, 17, 135.

Yang, S. and Gruber, N. (2016). The anthropogenic perturbation of the marine nitrogen cycle by atmospheric deposition: Nitrogen cycle feedbacks and the ^{15}N Haber–Bosch effect. *Global Biogeochemical Cycles*, 30, 1418–40.

Yao, M., Henny, C., and Maresca, J. A. (2016). Freshwater bacteria release methane as a by-product of phosphorus acquisition. *Applied and Environmental Microbiology*, 82, 6994–7003.

Yarza, P., Yilmaz, P., Pruesse, E., Glockner, F. O., Ludwig, W., Schleifer, K.-H., Whitman, W. B., Euzeby, J., Amann, R., and Rossello-Mora, R. (2014). Uniting the classification of cultured and uncultured bacteria and archaea using 16S rRNA gene sequences. *Nature Reviews Microbiology*, 12, 635–45.

Yavitt, J. B. and Lang, G. E. (1990). Methane production in contrasting wetland sites: Response to organic-chemical components of peat and to sulfate reduction. *Geomicrobiology Journal*, 8, 27–46.

Ye, R. W., Averill, B. A., and Tiedje, J. M. (1994). Denitrification-production and consumption of nitric oxide. *Applied and Environmental Microbiology*, 60, 1053–8.

Yilmaz, P., Yarza, P., Rapp, J. Z., and Glöckner, F. O. (2016). Expanding the world of marine bacterial and archaeal clades. *Frontiers in Microbiology*, 6, 1524.

Yong, E. (2016). *I Contain Multitudes: The Microbes Within Us and a Grander View of Life*, 357 pp. HarperCollins, New York.

Yoon, H. S., Price, D. C., Stepanauskas, R., Rajah, V. D., Sieracki, M. E., Wilson, W. H., Yang, E. C., Duffy, S., and Bhattacharya, D. (2011). Single-cell genomics reveals organismal interactions in uncultivated marine protists. *Science*, 332, 714–17.

Yooseph, S., Nealson, K. H., Rusch, D. B., McCrow, J. P., Dupont, C. L., Kim, M., Johnson, J., Montgomery, R., Ferriera, S., Beeson, K., Williamson, S. J., Tovchigrechko, A., Allen, A. E., Zeigler, L. A., Sutton, G., Eisenstadt, E., Rogers, Y.-H., Friedman, R., Frazier, M., and Venter, J. C. (2010). Genomic and functional adaptation in surface ocean planktonic prokaryotes. *Nature*, 468, 60–6.

Yu, G. and Stoltzfus, A. (2012). Population diversity of ORFan genes in *Escherichia coli*. *Genome Biology and Evolution*, 4, 1176–87.

Yuan, H., Ge, T., Chen, C., O'Donnell, A. G., and Wu, J. (2012). Significant role for microbial autotrophy in the sequestration of soil carbon. *Applied and Environmental Microbiology*, 78, 2328–36.

Yvon-Durocher, G., Allen, A. P., Bastviken, D., Conrad, R., Gudasz, C., St-Pierre, A., Thanh-Duc, N., and Del Giorgio, P. A. (2014). Methane fluxes show consistent temperature dependence across microbial to ecosystem scales. *Nature*, 507, 488–91.

Zak, D. R., Tilman, D., Parmenter, R. R., Rice, C. W., Fisher, F. M., Vose, J., Milchunas, D., and Martin, C. W. (1994). Plant production and soil microorganisms in late-successional ecosystems: A continental-scale study. *Ecology*, 75, 2333–47.

Zechmeister-Boltenstern, S., Keiblinger, K. M., Mooshammer, M., Peñuelas, J., Richter, A., Sardans, J., and Wanek, W. (2015). The application of ecological stoichiometry to plant–microbial–soil organic matter transformations. *Ecological Monographs*, 85, 133–55.

Zehnder, A. J. B. and Brock, T. D. (1979). Methane formation and methane oxidation by methanogenic bacteria. *Journal of Bacteriology*, 137, 420–32.

Zehr, J. P., Shilova, I. N., Farnelid, H. M., Muñoz-Maríncarmen, M. D. C., and Turk-Kubo, K. A. (2016). Unusual marine unicellular symbiosis with the nitrogen-fixing cyanobacterium UCYN-A. *Nature Microbiology*, 2, 16214.

Zehr, J. P., Weitz, J. S., and Joint, I. (2017). How microbes survive in the open ocean. *Science*, 357, 646–7.

Zeng, J., Liu, X., Song, L., Lin, X., Zhang, H., Shen, C., and Chu, H. (2016). Nitrogen fertilization directly affects soil bacterial diversity and indirectly affects bacterial community composition. *Soil Biology and Biochemistry*, 92, 41–9.

Zhang, J. and Elser, J. J. (2017). Carbon:nitrogen:phosphorus stoichiometry in fungi: a meta-analysis. *Frontiers in Microbiology*, 8, 1281.

Zhang, J., Müller, C., and Cai, Z. (2015). Heterotrophic nitrification of organic N and its contribution to nitrous oxide emissions in soils. *Soil Biology and Biochemistry*, 84, 199–209.

Zhang, M.-M., Alves, R. J. E., Zhang, D.-D., Han, L.-L., He, J.-Z., and Zhang, L.-M. (2017). Time-dependent shifts in populations and activity of bacterial and archaeal ammonia oxidizers in response to liming in acidic soils. *Soil Biology and Biochemistry*, 112, 77–89.

Zhang, Y. and Gladyshev, V. N. (2010). General trends in trace element utilization revealed by comparative genomic analyses of Co, Cu, Mo, Ni, and Se. *Journal of Biological Chemistry*, 285, 3393–405.

Zhao, M. and Running, S. W. (2010). Drought-induced reduction in global terrestrial net primary production from 2000 through 2009. *Science*, 329, 940–3.

Zhao, Q., Jian, S., Nunan, N., Maestre, F. T., Tedersoo, L., He, J., Wei, H., Tan, X., and Shen, W. (2017). Altered precipitation seasonality impacts the dominant fungal but rare bacterial taxa in subtropical forest soils. *Biology and Fertility of Soils*, 53, 231–45.

Zhao, Y., Temperton, B., Thrash, J. C., Schwalbach, M. S., Vergin, K. L., Landry, Z. C., Ellisman, M., Deerinck, T., Sullivan, M. B., and Giovannoni, S. J. (2013). Abundant SAR11 viruses in the ocean. *Nature*, 494, 357–60.

Zhu, F., Massana, R., Not, F., Marie, D., and Vaulot, D. (2005). Mapping of picoeucaryotes in marine ecosystems with quantitative PCR of the 18S rRNA gene. *FEMS Microbiology Ecology*, 52, 79–92.

Zientz, E., Feldhaar, H., Stoll, S., and Gross, R. (2005). Insights into the microbial world associated with ants. *Archives of Microbiology*, 184, 199–206.

Zimmerman, A. E., Allison, S. D., and Martiny, A. C. (2014). Phylogenetic constraints on elemental stoichiometry and resource allocation in heterotrophic marine bacteria. *Environmental Microbiology*, 16, 1398–410.

Zimmerman, A. E., Martiny, A. C., and Allison, S. D. (2013). Microdiversity of extracellular enzyme genes among sequenced prokaryotic genomes. *The ISME Journal*, 7, 1187–99.

Zimmermann, R. and Meyer-Reil, L.-A. (1974). A new method for fluorescence staining of bacterial populations on membrane filters. *Kieler Meeresforschungen Sonderheft* 30, 24–7.

Zipfel, C. and Oldroyd, G. E. D. (2017). Plant signalling in symbiosis and immunity. *Nature*, 543, 328–36.

Zumft, W. G. (1997). Cell biology and molecular basis of denitrification. *Microbiology and Molecular Biology Reviews*, 61, 533–616.

Index

Page numbers in *italics* refer to figures or tables